Site Characterization and Aggregation of Implanted Atoms in Materials

A series of edited volumes comprising multifaceted studies of contemporary scientific issues by some of the best scientific minds in the world, assembled in cooperation with NATO Scientific Affairs Division.

Series B: Physics

RECENT VOLUMES IN THIS SERIES

This series is published by an international board of publishers in conjunction with NATO Scientific Affairs Division

A Life Sciences	Plenum Publishing Corporation
B Physics	London and New York
C Mathematical and Physical Sciences	D. Reidel Publishing Company Dordrecht and Boston
D Behavioral and Social Sciences	Sijthoff International Publishing Company Leiden
E Applied Sciences	Noordhoff International Publishing Leiden

Site Characterization and Aggregation of Implanted Atoms in Materials

Edited by
A. Perez
Université Claude Bernard Lyon I
Villeurbanne, France

and

R. Coussement
University of Leuven
Leuven, Belgium

PLENUM PRESS • NEW YORK AND LONDON
Published in cooperation with NATO Scientific Affairs Division

Library of Congress Cataloging in Publication Data

Nato Advanced Study Institute, Aleria, France, 1978.
  Site characterization and aggregation of implanted atoms in materials.

  (Nato advanced study institutes series: Series B, Physics; v. 47)
  "Lectures presented at the NATO Advanced Study Institute, held in Aleria, Corsica,
September 10−23, 1978."
  Includes index.
  1. Ion implantation−Congresses. I. Perez A. II. Coussement, Romain. III. Title.
IV. Series.
QC702.7.I55N37 1978   301.31   79-19008
ISBN-13: 978-1-4684-1017-4   e-ISBN-13: 978-1-4684-1015-0
DOI: 10.1007/978-1-4684-1015-0

Lectures presented at the NATO Advanced Study Institute,
held in Aleria, Corsica, September 10−23, 1978

© 1980 Plenum Press, New York
Softcover reprint of the hardcover 1st edition 1980
A Division of Plenum Publishing Corporation
227 West 17th Street, New York, N.Y. 10011

ACKNOWLEDGMENTS

This Advanced Study Institute was sponsored by NATO and co-sponsored by FEBS (Advanced Course No. 63). The Course Directors also acknowledge the financial assistance of the following companies:

Astra Läkemedel AB, Södertalje, Sweden
Bayer Aktiengesellschaft, Leverkusen, West Germany
Beecham Pharmaceuticals, Betchworth, United Kingdom
Boehringer GmbH, Mannheim, West Germany
Calbiochem-Behring Corporation, La Jolla, California, U.S.A.
Grunenthal GmbH, Stolberg, West Germany
International Enzymes, Windsor, United Kingdom
E. Merck, Darmstadt, West Germany
Ortho Pharmaceutical (Canada) Ltd., Ontario, Canada
Roche Products Ltd., Welwyn Garden City, United Kingdom
Sandoz Forschungsinstitut GmbH, Wien, Austria
Schering AG, Berlin, West Germany
Searle & Co., Ltd., High Wycombe, United Kingdom

A.S.I.M.S. ACTIVITIES

The Advanced Study Institute on Materials Science was created in 1973 in the Département de Physique de Matériaux of the University of Lyon. The aim of this Institute was to organize specialized doctoral level courses on materials science in Lyon and International Summer Schools on the same subject at a post-doctoral level in Corsica. Summer Schools published with NATO as the main sponsor are:

Radiation Damage Processes in
  Materials
September 1-15, 1973
Director: C.H.S. Dupuy
Proceedings edited by
  C.H.S. Dupuy
(Noordhoff International
  Publishing Company)

Electrode Processes in Solid State
  Ionics: Theory and Application
  to Energy Conversion and Storage
August 28 – September 9, 1975
Director: M. Kleitz
Proceedings edited by M. Kleitz
  and J. Dupuy
(D. Reidel Publishing Company)

Physics of Nonmetallic Thin Films
August 29 – September 12, 1974
Director: C.H.S. Dupuy
Proceedings edited by
  C.H.S. Dupuy and A. Cachard
(Plenum Publishing Company)

Materials Characterization Using
  Ion Beams
August 29 – September 12, 1976
Director: J.P. Thomas
Proceedings edited by J.P. Thomas
  and A. Cachard
(Plenum Publishing Company)

Microscopic Structure and Dynamics of Liquids
August 28 – September 10, 1977
Director: J. Dupuy
Proceedings edited by J. Dupuy and A.J. Dianoux
(Plenum Publishing Company)

PREFACE

Explosive developments in microelectronics, interest in nuclear
metallurgy, and widespread applications in surface science have all
produced many advances in the field of ion implantation. The research
activity has become so intensive and so broad that the field has
become divided into many specialized subfields.

An Advanced Study Institute, covering the basic and common
phenomena of aggregation, seems opportune for initiating interested
scientists and engineers into these various active subfields since
aggregation usually follows ion implantation. As a consequence,
Drs. Perez, Coussement, Marest, Cachard and I submitted such a pro-
posal to the Scientific Affairs Division of NATO, the approval of
which resulted in the present volume.

For the physicist studying nuclear hyperfine interactions, the
consequences of aggregation of implanted atoms, even at low doses,
need to be taken into account if the results are to be correctly
interpreted. For materials scientists and device engineers, understanding aggregation mechanisms and methods of control is clearly
essential in the tailoring of the end products.

The present Institute, the 7th Advanced Study Institute in
Materials Science and Technology, organized under the direction of
my laboratory in the Department de Physique des Matériaux de Lyon,
was managed in collaboration with a group from Louvain and the
Hyperfine Interaction Group at Lyon. It was a more challenging
one because of the necessity of bringing together a group of people
having little or no previous interactions with each other - physicists, chemists, materials scientists, and engineers. In view of
this, the lecturers and the organizing and scientific committees
necessarily played a more active role in order to insure the success
of this Institute. The task was made easier by the site of Casabianda Village at Corsica because of its excellent facilities and
relative isolation. Living in the same village and having the same
comforts and recreation led to the creation of a common language
and the development of fruitful contacts and discussions. Overall
I believe our efforts achieved the aim of providing the participants

with a broader perspective of the field of ion implantation and a greater awareness of its diverse applications.

In an undertaking such as the present one, the help and support of many groups and people are required, and their contributions are acknowledged with thanks. Special credits go to Messrs. Talberg and Bonaldi, managers of V.I.V.E. Association, and Mr. Caprili, the Director of the Centre Agricole de Casabianda. The support and assistance of Mr. Carlotti, Maire d'Aléria, and Mr. V. Carlotti, Conseiller Général de Moïta Verde deserve special mention. A party given by Mr. Angelini, Maire de Talone, was a timely diversion and we thank him for his hospitality.

The most important contribution to the success of the Institute is of course the financial support of the Scientific Affairs Division of NATO, the Haute Corse Authority and the U.E.R. de Physique of the University Lyon I. It is also fair to say that without the help and consideration of Dr. Kester, manager of the A.S.I. in the NATO Scientific Affairs Division, the 7th ASIMST would not have materialized. He has since taken on a new job and we wish him success in his new undertaking. In particular, we will fully cooperate with and actively participate in his new program of International Transfer in Science and Technology. We also express great confidence in Dr. Di Lullo who is taking over the A.S.I. program.

Finally, the contributions of the lecturers, the scientific committee and the managers, and Drs. Perez, Coussement, Marest, and Cachard, are equally important. To them, the participants and I owe a special gratitude.

C.H.S. Dupuy
Professeur de Physique à l'Université
Claude Bernard Lyon I

CONTENTS

GENERAL INTRODUCTION

A. CACHARD

Departement de Physique des Matériaux

Université Lyon I, 69621 Villeurbanne, France

This book gives the lectures presented at the NATO Summer School held in Corsica in September 1978 on "Site characterization and aggregation of implanted atoms in materials". The aim of such a Summer School was to promote contacts between major international scientists in this field and students at a post doctoral level, and to allow them to deepen their knowledge of ion implantation phenomena. Then the articles of this book are written by specialists but for non specialist people. The purpose of this introduction is to give some general comments about the contents.

When an energetic particle penetrates a solid, the interactions induce series of transient or permanent changes either on the particle itself or on the solid ; and the exhaustive description of all the phenomena involved is a bit complicated.

A first important change occurs when an energetic beam is scattered by a gaseous or a thin solid target : the more or less collisional processes modify the charge states of the ions and give excited states. The study of these excited states is the field of the ion beam spectroscopy which constitutes an important branch of atomic physics. One speaks in terms of beam-gas or beam-foil spectroscopy, depending on the way by which the excited states are obtained. What is interesting in this technique is that in principle any excited state of any atom can be studied.

Other modifications concern the energy change, the spatial distribution and the fraction of the emerging beam which is either forward or back scattered. These informations allow us to go back to the collisional events and then to characterize the target atoms.

1

This constitutes the physical basis of the ion scattering characterization techniques, with their different energy regimes : low
energy (a few keV) which interests the first atomic plane of the
solid ; medium energy (some 100 keV) interesting a few 100 Å under
the surface, and the high energy regime, the so-called Rutherford
backscattering. Evenly, if the particle energy is high enough,
nuclear reactions can occur changing the nature of emerging species.

On the other hand not only the beam particles but also the
atoms of the solid can be excited. The de-excitation processes
give rise to photon emissions (mainly X rays) characteristic of
the solid atoms. All these phenomena (backscattering, nuclear
reactions and ion beam induced X-ray emission) constitute the main
part of the ion beam based techniques for the characterization of
solids and were the subject of a summer school in 1976.

In addition there are other transient phenomena during the
ion beam-solid interaction which are more specific of the solid
state : electronic excitations which imply valence electrons (such
as plasmons or excitons),or vibrationnal excitations involving
molecular vibrations and phonons. But numerous other processes
leave behind them permanent effects which modify the target.

At first, just as they enter the solid and especially if they
are heavy and of relatively low energy, particles induce secondary
emission of neutral or charged species. This sputtering effect
etches more or less the surface and gives rise to changes in surface
structure and composition.

Then along their trajectory, particles create atomic displacements through collision cascades inducing defects whose nature
depends on the energy and the mass of the particle and on the
nature of the solid. These defects interact with the solid and
between each other, leading to important modifications in the materials and in particular breaking the thermodynamical equilibrium.
Finally at the end of their trajectroy, where particles come to
rest, they change the chemical nature of the solid by their presence.

This rapid enumeration of permanent effects shows the great
diversity of this field which constitutes the area of ion implantation.

Historically, the defect creation, that is the destructive
aspect of ion-solid interaction, was first recognized by nuclear
physicists during the development of nuclear reactors and for a
long time they have spoken in terms of radiation damages. Only
fifteen years ago, semi-conductor people initiated the use of ion

implantation in order to improve the performances of devices.
The progress was rapid in this way, and ion implantation studies in
semi-conductors have revealed such an interest and such a quality
that this technique is now currently used in the technology of semiconductors. This subject has been now covered by a number of international conferences and by an abundant literature.

Since a few years the state of quasi-monopoly of semiconductors in ion implantation applications has been broken by
scientists working in other areas : metallurgy, electrochemistry,
corrosion, catalysis, etc... and, as pointed out in the subtitle,
the Summer School was in part devoted to this new field.

These general physical concepts being reminded, I would try
to explain what is behind the title "Site characterization and
aggregation of implanted atoms in materials".

Roughly speaking semi-conductor people are using implantation
for doping : that means low concentrations, thus low doses. But the
interest in other areas is more in higher concentrations and
doses where new effects are revealed which are summarized in the
title as "aggregation". These aggregation effects can take the form
of separate clusters such as metallic colloïds or gas bubbles, or
they can be a mixture in the sense that implanted atoms arrange
themselves within the matrix to form new structures or new chemical
compounds.

But as far as implantation always takes place in a finite
time, it is obvious that before reaching the high dose state,
there is necessarily an intermediate low dose state where the implanted atoms are still dispersed. In this state we are faced with
the problem of atom location that is of the site characterization
of the implanted atoms. Thus as far as the final properties of the
implanted materials are influenced by its evolution during high
dose implantation a complete understanding of general implantation
processes requires a step by step description of the phenomena
involved, even more if you have a tutorial purpose such as in a
summer school.

Then the organization of the lectures and also of the chapters
of this book obeys to this didactic constraint.

At first it appears indispensable to give the basic notions
on ion implantation : physical problems such as stopping powers,
ranges, damages etc ... and technical problems such as beam production and handling.

Then for the characterization of the site of implanted atoms,
the way we followed was to put the accent on the methods with an

emphasis on nuclear methods : hyperfine interaction techniques and channelling.

This second part more or less technical is addressed to non specialist solid state scientists in order to show them to what extent these new techniques are able to give interesting information in their research.

On the contrary, in the third part of the book, the accent is put more on the state of the art in the understanding of the physical phenomena arising at high doses.

The last part shows that some phenomena are sufficiently understood and controlled to allow more or less industrial applications, the interest in which is growing.

Such a position which reflects more or less the specificities of the organizing committee members, corresponds to the willingness to bring in contact scientific people working most often in parallel than in connection on similar subjects. Special attention has been given to ensure comparison between the methods (accuracy and limitations) in the different cases. During the Summer School this has been the occasion of hot discussions.

I would just finish this text by telling that the goal of this Summer School is that hyperfine interaction people and solid state people should extend at least the vocabulary of their scientific language, should deepen the knowledge of their physical field of research and take benefit of their scientific knowledge by personal contacts initiated during the Summer School or induced by the reading of this book.

Part I
Basic Notions on Implantation

BASIC IMPLANTATION PROCESSES

J.A. Davies and L.M. Howe

Solid State Science Branch
Atomic Energy of Canada Limited
Chalk River, Ontario, Canada, K0J 1J0

The main objective of this introductory lecture is to survey the basic processes involved in ion implantation and thus provide a framework for the subsequent lectures and discussions of this summer school. There have been several excellent theoretical surveys of energy loss by Sigmund[1], Chu[2], and most recently by Bonderup[3]. For this reason, and since we are primarily experimentalists, the present lecture notes have a strong experimental bias. The reader is therefore urged to use references 1-3 for a thorough theoretical treatment.

Firstly, let us list some of the unique advantages of ion implantation:

(i)   Precise control of the total number (N) of implanted ions.
(ii)   Independent control of the penetration depth $\bar{R}$.
(iii) Almost all combinations of ion (Z_1, M_1) and target (Z_2, M_2) may be used.
(iv)   A wide concentration range is achievable, with the upper limit generally being set by the sputtering yield rather than by equilibrium solubility.
(v)   It is an ideal "clean room" technique, provided one uses a good cryoshield or UHV chamber. Until recently, this feature had not been properly exploited, since at 10^{-6} torr most accelerator systems have serious (hydro)carbon contamination problems.

In the rest of this lecture, we will discuss the basic collision processes and various other parameters that must be controlled in order to fully exploit these potential advantages. These may be

grouped under four major headings: Number (N), Penetration Depth,
Radiation Damage, and Foreign--Atom Site.

1.   NUMBER OF IMPLANTED IONS

Probably the most crucial single factor in measuring N accurately is the design of the Faraday cup. When a heavy ion is
implanted into a solid, secondary electrons with energies extending
up to several keV are emitted, plus a significant number of photons
and excited neutral (sputtered) atoms. Since the photons and
excited atoms produce additional secondary electrons on impact with
a wall, one cannot rely entirely on electrostatic suppression for
accurate current measurement. The best technique by far is to completely surround the target with a solid shielded Faraday cup,
except of course for the beam entrance aperture. When this is done,
absolute beam current measurement with an accuracy of 1-2% is
readily achievable[4].

Charge exchange along the beam line, uniform sweeping, and beam
heating of the target are also significant factors in determining
accurately the value of N. In each case, adequate control can usually be achieved by proper design of the implantation system, as will
be discussed by Dr. Dearnaley in his lecture on Implantation Proce--
dure.

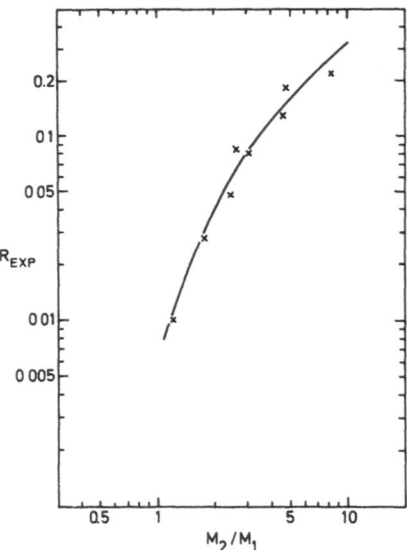

Figure 1 - Experimental reflection coefficients[5] as a function of
target/projectile mass ratio at a fixed energy of 30 keV. The
curve is drawn merely to guide the eye.

However, there exists one basic limitation, which is determined primarily by the physics of the atomic collision process and which is therefore outside the experimentalist's direct control: namely, ion reflection. When the incident ion is heavier than the target atom (i.e. $Z_1 > Z_2$), this backscattered (or reflected) fraction is negligibly small; but if $Z_1 < Z_2$, the reflection losses become significant and can amount to as much as 30% of the implanted current (figure 1) – or even more for non-perpendicular beam inci- dence. Bøttiger[5] has used radiotracer implantations to investigate experimentally the magnitude of this reflection coefficient as a function of mass ratio and of ε, the dimensionless energy parameter which will be defined in section 2. As seen in figure 2, the reflec- tion loss is quite accurately predicted by theory and hence an appropriate correction can be applied in most implantations.

In some cases, additional loss of implanted atoms arises from diffusion (to the surface), especially if the implanted species is volatile. Furthermore, at high doses, sputtering can also cause a major loss to occur, as shown in figure 3. Note that the retention of implanted Xe as a function of dose falls off much sooner in Au

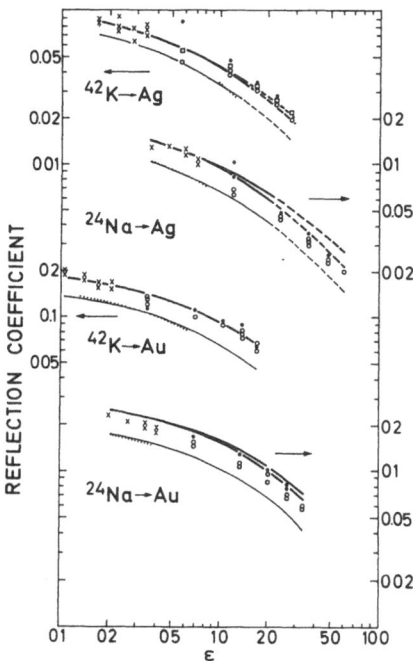

Figure 2 – Energy dependence of the reflection coefficient[5]. Heavy solid lines represent the best theoretical estimate, including surface corrections.

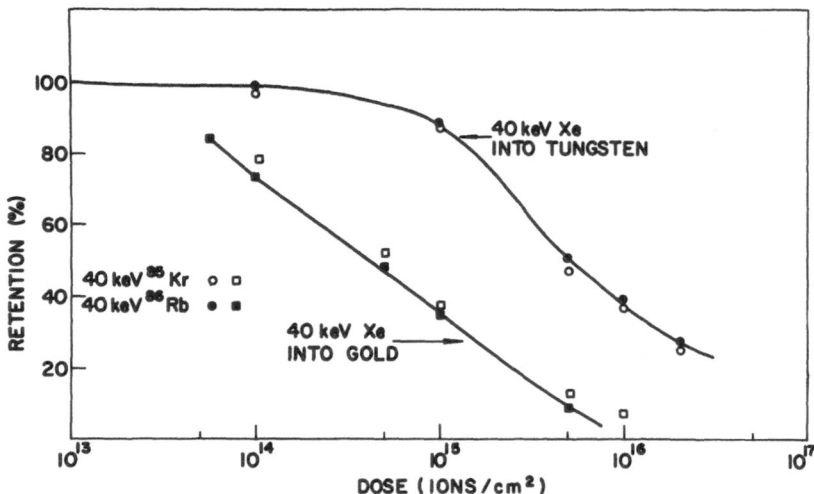

Figure 3 – Dose dependence of the 'sticking factor' for 40-keV Xe in polycrystalline gold and tungsten[6].

(where the sputtering yield is high) than it does in tungsten. Since Kr, Rb and Xe all show almost identical behaviour, the loss in this case is evidently due to sputtering rather than diffusion. Whenever diffusion or sputtering effects are significant, it becomes almost impossible to monitor accurately the implantation dose.

2. PENETRATION DEPTH

A schematic representation of typical range profiles is given in figure 4. In the absence of channeling effects, one obtains an approximately Gaussian profile whose mean value R_p and width can usually be predicted quite accurately (i.e. to ± 20%) as a function of ε, Z_1, and Z_2, provided sputtering and diffusion effects are negligible. Note that a small (shaded) part of the range profile extends back to negative depths; the fractional area of this negative part is indeed the reflection coefficient discussed in section 1. Included in figure 4 are several typical channeled ion profiles in aligned single crystals and also the ideal "channeled" peak which, unfortunately, is rarely observed at implantation energies because of the dominant role of thermal vibrations in scattering (i.e. dechanneling) the incident beam. In the "typical channeled profiles", the maximum penetration depth R_{max} is the only well-defined experimental feature; the height and shape of the inter-vening distribution between R_p and R_{max} depends critically on many parameters that are difficult to predict or control[7]. Hence, in most implantation studies, one tries to minimize or eliminate channeling effects, either by using an amorphous or fine-grained

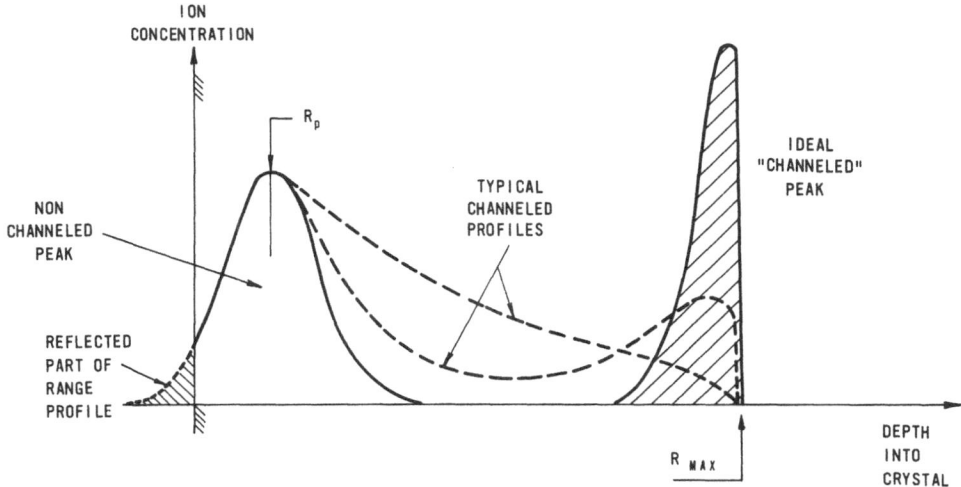

Figure 4 - Typical range profiles (schematic).

polycrystalline target, or by continuously rotating the target
during implantation. In the present section, we will confine our
attention to non-channeled range distributions. A full discussion
of the problem of channeled range profiles is given in reference 7.

Two accurate experimental techniques have been developed over
the years for measuring implantation profiles - namely anodic
oxidation + chemical stripping, and Rutherford backscattering - and
a large amount of experimental data has now been acquired. Generally, these data have an accuracy of 3-5%, but in at least one
case[8] an experimental accuracy of ± 1% has been achieved. There
are also extensive range tabulations[9-11] based on Lindhard's
theoretical treatment[12] of the energy-loss processes involved. As
we shall see below, the agreement between these predicted tabulations and experiment is usually about ± 20%.

2.1 Theoretical Stopping-Power Framework

Figure 5 provides a convenient overview of the main energy loss
processes (nuclear stopping S_N and electronic stopping S_e) over the
entire range of interest in ion bombardment studies. The abscissa
scale ε is the energy expressed in Lindhard's dimensionless Thomas-
Fermi (T.F.) units[12]: i.e., it is the ratio of the T.F. screening
length 'a', defined as $a = 0.8853\ a_0/[Z_1^{2/3} + Z_2^{2/3}]^{1/2}$, to the
distance-of-closest-approach b in an unscreened collision (where
$b = Z_1 Z_2\ e^2 (M_1 + M_2)/M_2\ E$; e is the electronic charge and E is the
initial energy of the incident ion).

The advantages of such an energy scaling are twofold. Firstly, it enables S_N (and associated quantities such as multiple scattering, sputtering, damage production) to be expressed in terms of a universal curve for all combinations of Z_1 and Z_2, as shown in figure 5. Secondly, it permits a clean separation of ion bombardment studies into two widely different regimes: namely,

(i) <u>Ion Implantation Regime</u> - $\varepsilon \lesssim 10$ (and $Z_1 \gtrsim 4$). This consists of relatively slow moving heavy ions, with velocities much smaller than the T.F. velocity $v_0 Z_1^{2/3}$, and is the regime in which S_N and S_e both contribute significantly to the slowing down. Consequently, theoretical treatment here is more complex and less accurate than in regime (ii). It is also the energy region in which sputtering and damage cascade effects are often dominant; hence, sputter-etching techniques such as SIMS and SCANIIR fall within its domain. This is certainly the regime of main interest to the present summer school.

(ii) <u>Nuclear Microanalysis Regime</u> - $\varepsilon \gg 10$ (and low Z_1). In this region, S_N has fallen to an almost negligible and approximately constant fraction ($\sim 10^{-3}$) of S_e. Hence, the trajectory is almost

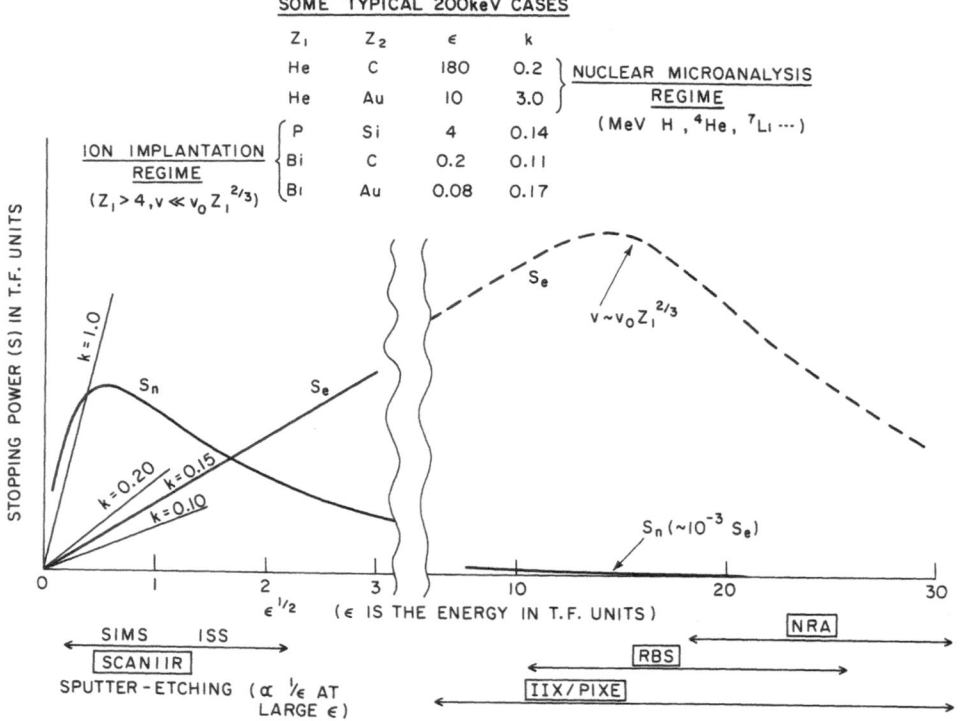

Figure 5 - Stopping power versus energy in LSS units.

linear, sputtering and collision cascade effects are generally
negligible, and a quantitative theoretical framework exists for
accurately predicting energy loss, channeling behaviour, etc.   This
is the energy region widely used for nuclear microanalysis methods
such as Rutherford backscattering (RBS), ion-induced X-rays (IIX
and PIXE), and nuclear reaction analyses (NRA).   Although this
energy region is not of primary interest to the present summer
school, it does form the basis for the channeling technique of
characterizing the location of foreign atoms in a lattice; this will
be discussed in a later chapter by Howe and Davies.

Before leaving figure 5, it should be noted that S_e unfortunately does not exhibit the property of T.F. scaling; hence, unlike
S_N, it cannot be described by a single universal curve.   At low
energies, however, S_e increases linearly with $\varepsilon^{1/2}$ and, as long as
$Z_1 \gtrsim Z_2$, the slope k does fall within the fairly narrow range of
0.14 ± 0.03.   The magnitude of ε and k for a few typical 200-keV
ions has been included in figure 5 to illustrate this point.

In the present chapter, we will restrict ourselves to regime
(i) and consider briefly the basis of Lindhard's theory[12] of nuclear
and electronic stopping.   Following Bohr, he assumes the validity
of classical mechanics in this energy regime and then introduces
the T.F. scaling parameters of length (a), energy (ε = a/b), etc.,
in order to describe the similarity of ion-atom potentials.   With a
few additional approximations, he is finally able to express the
differential scattering cross section dσ in terms of a unique one-
parameter equation: namely,

$$d\sigma = \pi a^2 \; f(t^{1/2}) \cdot dt/2t^{3/2} \tag{1}$$

where $t = \varepsilon^2 \sin^2 \theta/2$. (Θ is the centre-of-mass **scattering angle**).

The function $f(t^{1/2})$ is shown in figure 6 for various
potentials (Bohr, Thomas-Fermi and Lenz-Jensen) together with values
obtained from the experimental cross section data of Loftager[13].   The
sharp peak in the experimental curve occurs at a t value for which
the collision distance is approximately the sum of the inner shell
radii (Ne K-shell + Ar L-shell) of the two colliding atoms; it is
generally associated with a sharp step in the electronic energy
loss, at the onset of shell overlap.   At all other t values, the
agreement between experiment and the Lenz-Jenzen curve is good.
Note that at very large t values, all theoretical curves approach
asymptotically the unscreened (Rutherford) scattering function, i.e.
$0.5 \; t^{1/2}$.

To obtain the universal S_N curve in figure 5, Lindhard[12] carries
out an appropriate integration of equation 1:   i.e. $S_N = \int T \cdot d\sigma$,
where T is the elastic energy loss in the collision.

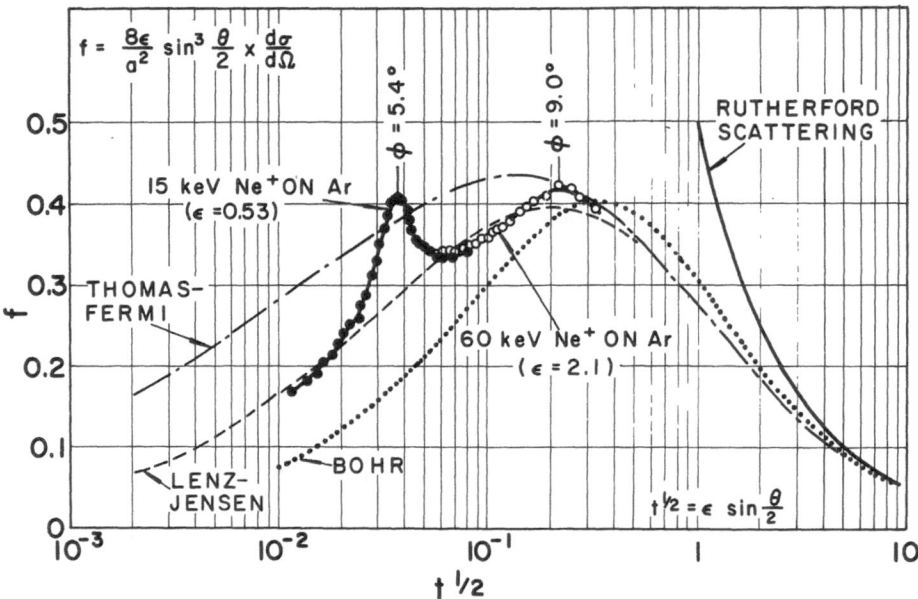

Figure 6 – Comparison of Lindhard's universal scattering function $f(t^{1/2})$ with Loftager's single scattering measurements[13].

The treatment of electronic stopping is much more complex. For a slowly moving ion in an electron gas, one can show quite readily that the force on the ion is proportional to its velocity (and hence that $S_e = k\epsilon^{1/2}$), provided the velocity is much less than the T.F. velocity $v_0 Z_2^{2/3}$ of the target electrons. Lindhard then proceeds via T.F. arguments to obtain as a first order estimate the expression

$$k = \frac{0.0793 \, Z_1^{2/3} \, Z_2^{1/2} \, (M_1 + M_2)^{3/2}}{(Z_1^{2/3} + Z_2^{2/3})^{3/4} \, M_1^{3/2} \, M_2^{1/2}} \tag{2}$$

More recent work, both theoretical[14] and experimental[15], has shown that S_e exhibits oscillations of at least $\pm$ 30% about the smooth Z_1 and Z_2 dependence of equation 2; this oscillatory dependence is believed to be associated with the existence of shell effects in ion–atom radii.

Finally, to estimate theoretically the mean depth, or range, of the implanted ion, one combines the appropriate S_e and S_N curves in figure 5, in order to obtain the total stopping power curve S as a function of ϵ, and then integrates 1/S from the incident energy ϵ_0

to zero. This, of course, gives the total path length ρ, and a
further correction term must be applied to derive ρ_p the projected
range, or penetration depth.

2.2 Comparison With Experiment

Two extensive sets of data over a very wide energy range are
shown in figures 7 and 8. In figure 7, Pringle's high precision
measurements[8] in amorphous Ta_2O_5 are seen to be in excellent agreement with the theoretical curves, based on a T.F. potential, except
for the heavier ions at very low energy (i.e. at ε values $\lesssim 10^{-2}$).
In figure 8, recent data by Combasson, *et al.*[16] over an even wider ε
range show equally good agreement with the T.F. curve down to
$\varepsilon < 10^{-1}$, but at much lower energies the data again lie significantly
higher than the T.F. curve. Evidently, a somewhat weaker potential
(i.e. mid-way between T.F. and Lenz-Jensen) would give the best fit
to the range data.

In summary, one can generally estimate the mean range of an
implanted ion with $\sim$ 20-30% accuracy throughout most of the implan

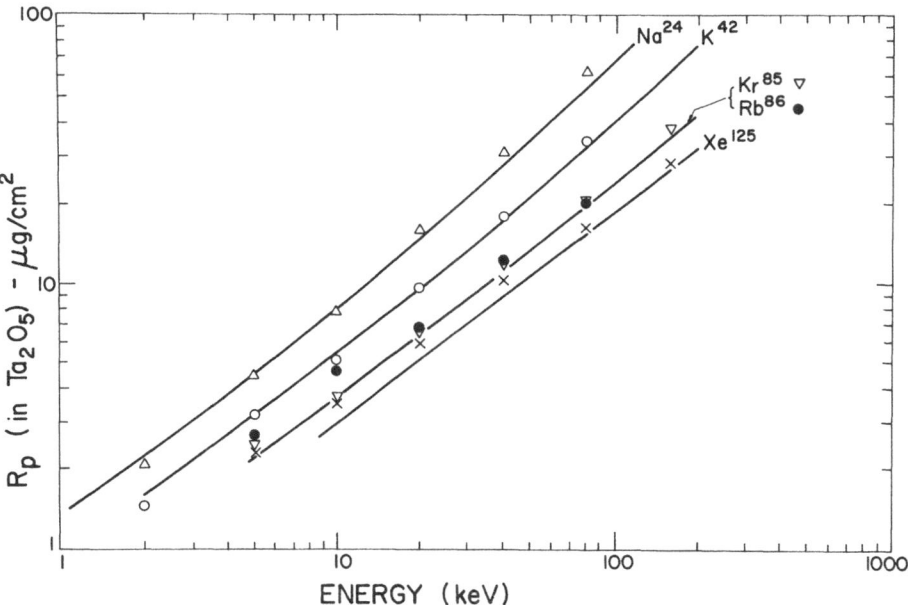

Figure 7 - Projected ranges[8] of various radiotracer ions in amorphous
Ta_2O_5 versus the theoretical curves predicted by the Lindhard,
et al. theory[12].

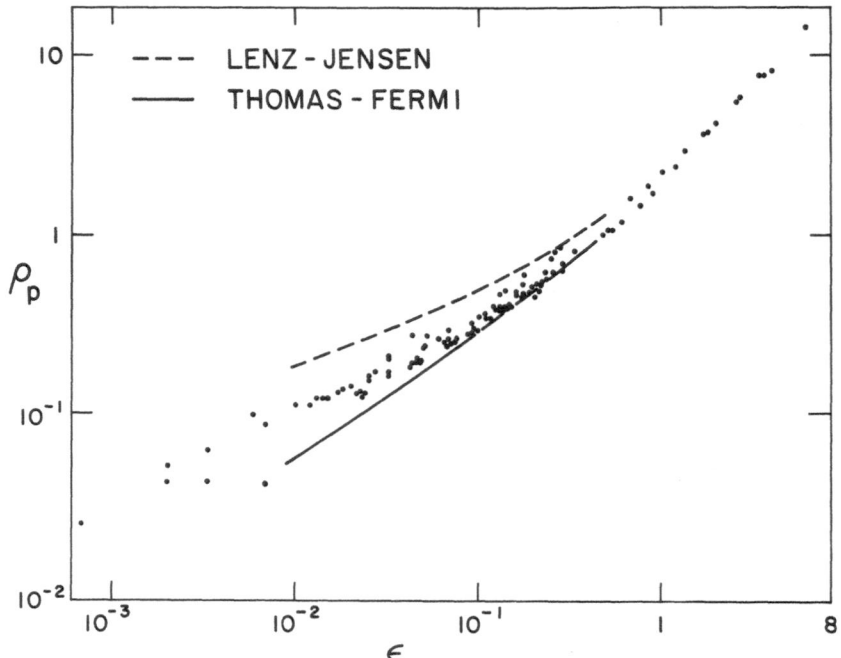

Figure 8 – Collected range data[16] for various heavy ions in Si
versus theoretical curves based on Thomas-Fermi (T.F.) and Lenz-
Jensen (L.J.) potentials.

tation energy regime, provided there are no anomalous diffusion or
crystal lattice (channeling) effects. In a similar manner, but
with slightly poorer accuracy, one can predict the higher order
moments[10] and so derive the entire range profile.

2.3 Penetration Anomalies

We have already discussed briefly the effects of sputtering at
high dose (figure 3) and of channeling in crystalline lattices
(figure 4); obviously, these produce drastic effects and should be
avoided or minimized whenever possible.

Enhanced diffusion is another factor which can seriously
distort the implantation profile, especially in metals of low-
melting point[17] such as Pb, Zn or Cd. A particularly striking
example is illustrated in figure 9, in which the high-dose saturation level for Ni implants in Zn and Cu increases markedly on

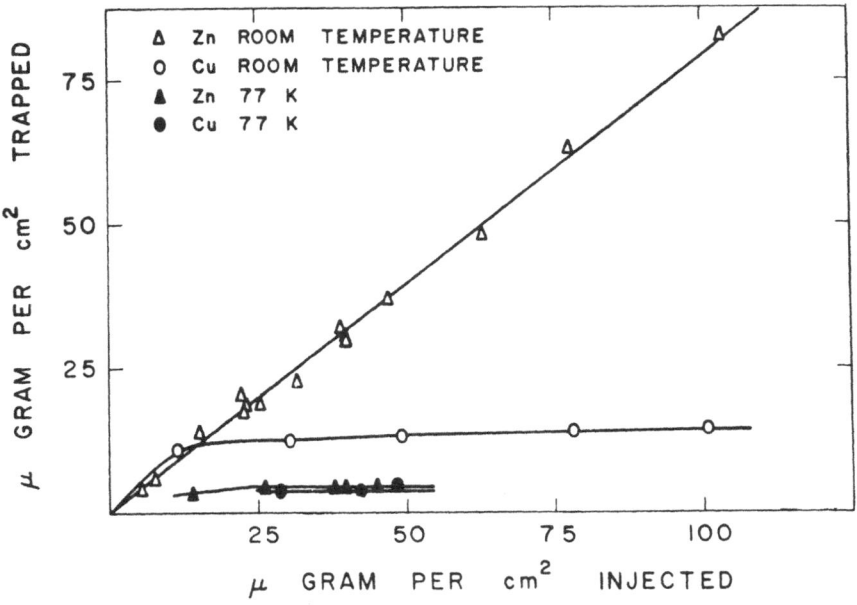

Figure 9 – Amount of Ni retained in Zn and Cu targets at 77 K and at 300 K as a function of implantation dose of 39-keV Ni[+] ions[17].

raising the implantation temperature from 77 K to 300 K. Indeed, in Zn at room temperature there is no evidence of saturation, even at extremely high doses where the total sputtering yield exceeds the theoretical ion range (~ 16 μg cm^{-2}) by at least a factor of ten. Obviously, the implanted Ni atoms are diffusing rapidly to much larger depths during bombardment. Further evidence for this enhanced penetration is clearly seen in the subsequent sputter profiling of these implanted Zn targets (figure 10). Note that the measured profile at 300 K extends to several hundred times the original mean range.

A different type of enhanced diffusion is shown in figure 11, which can occur even in a metal of high melting point, such as tungsten. It involves only a small fraction (typically, a few % or less) of the implanted ions, but again the enhanced penetration can exceed the implant range by factors of 100 or more. Similar implantations at even lower temperature (i.e. 20 K) have shown[19] that the enhanced penetration does not occur unless the crystal is subsequently warmed up above ~ 150 K; hence, it is definitely a diffusion effect. The most reasonable explanation is that a small fraction of the implanted ions come to rest in interstitial sites without creating any neighbouring vacancies, etc., which could have

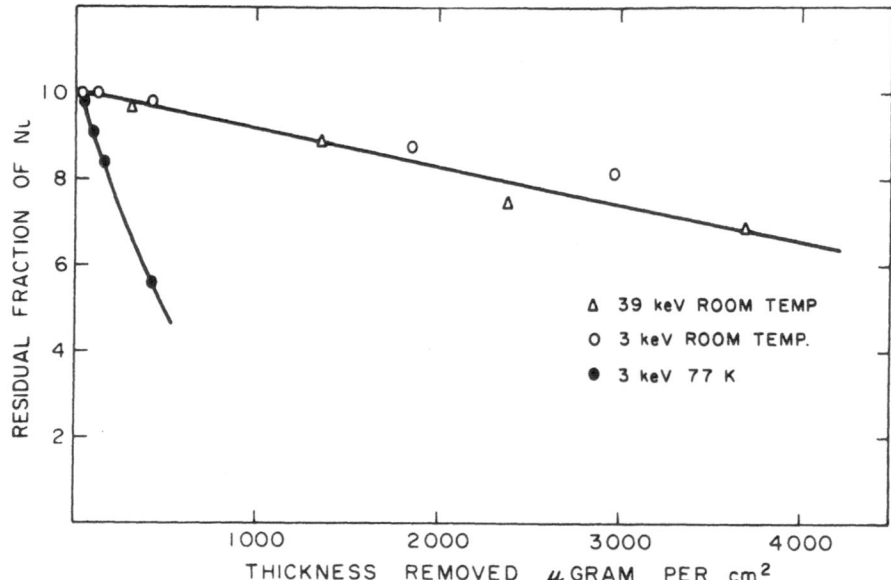

Figure 10 - Residual fraction of Ni in Zn targets as a function of sputter profiling at 77 K and 300 K by 3–keV or 39–keV Ar$^+$ ions[17].

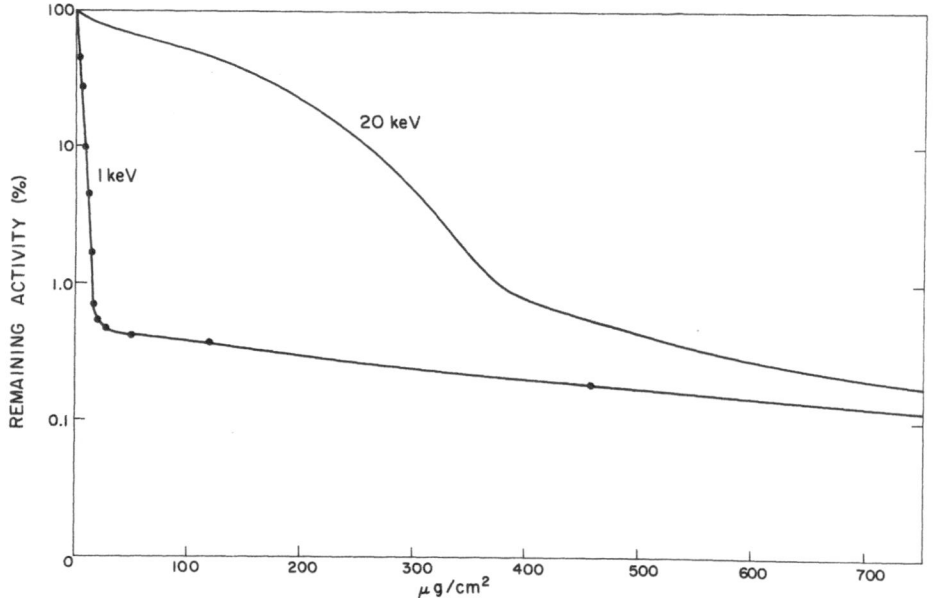

Figure 11 - Integral range profiles[18] for ^{85}Kr injected into <100> W crystals at 1-keV and at 20-keV.

acted as traps, and that these free interstitials are then able to
migrate with very low activation energy.

One other kind of anomaly might perhaps be briefly mentioned:
namely, the trapping of implanted atoms at a phase boundary or
interface. Normally, if the interface is not mobile, this would not
seriously perturb the profile. However, subsequent processing such
as annealing, oxide growth, etc., may cause an interface to move -
and carry the trapped atoms with it. A simple example is shown in
figure 12, where almost 75% of the implanted Kr (during the epit-
axial regrowth of the amorphous Si layer at $\sim$ 600°C) becomes trapped
at the amorphous/crystalline interface and thus is swept out to the
surface, where it escapes. At lower implant doses, where a continuous amorphous layer is not formed, the loss of Kr during this
annealing cycle is completely negligible[20] - i.e. < 1%.

A similar trapping or "snowplow" effect also occurs during
growth of an oxide film on the surface. As the oxide grows and the
metal/oxide interface moves inwards through the implanted ion
profile, ions trapped at this interface will be carried to larger
depths. Mackintosh, et al.[21] have studied this phenomenon during
the anodic oxidation of Al and Si targets, using a wide range of
implanted species, and find that in the majority of cases significant interface trapping does occur. This of course has some
serious consequences, particularly for the anodic oxidation +

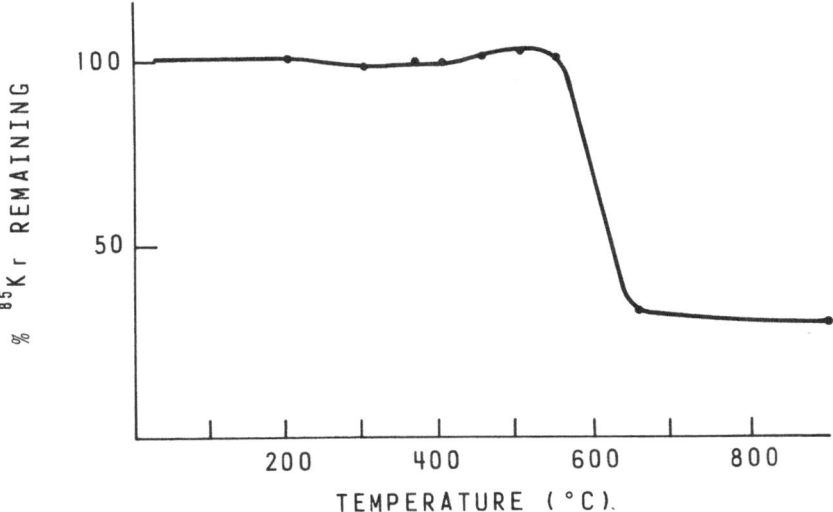

Figure 12 - Residual Kr content versus anneal temperature[20] for an
Si crystal implanted at 50 keV with 7 x 10^{14} Kr ions cm^{-2}.

stripping technique of depth profiling. Fortunately, inert gas ions and alkali metal ions generally do not exhibit significant trapping at the oxide-metal interface and hence these have been widely used for range measurements.

Such interface trapping of implanted ions can also be used to advantage. For example, the effect of yttrium implantation in reducing the corrosion rate of stainless steel persists long after the oxide film thickness has exceeded the implantation range[22]; this is probably due to trapping of yttrium ions at the receeding metal/oxide interface.

In summary, although the initial penetration profile can generally be predicted with reasonable ($\pm$ 20%) accuracy, there are anomalous solid-state effects which in some cases may subsequently distort or even completely dominate the observed distribution.

3. RADIATION DAMAGE

This is the key problem area in almost all implantation studies. It involves a complex interplay between bombardment processes and subsequent anneal processes and cannot adequately be treated in a single introductory lecture. Hence, we consider here only some simple general concepts, and other lectures (such as Dr. Leteutre's) will go into specific aspects in more detail.

The first obvious sub-division of the problem is into bombardment (i.e. atomic collision) and anneal processes:

(a) <u>Bombardment process</u> – this involves the formation of the collision cascade (figure 13) around each individual ion track, and can be treated by the same basic theoretical framework as the ion range (section 2). The main parameters involved here are Z_1, Z_2, E, and channeling, and the total time scale for propagating each cascade is approximately 10^{-13} secs. Hence, implantation provides an almost instantaneous distribution of point defects, etc. within an extremely small, predictable cascade volume.

(b) <u>Anneal processes</u> – the high concentration of defects within a single cascade may cause them to recombine or unite to form larger clusters, loops, voids, etc.; alternatively, they may migrate and interact with the surface or with other cascades or impurity atoms. This is the most complex part of the radiation damage problem, since these defect migration and interaction processes depend strongly on the substrate temperature and on the concentration of other defects. Consequently, the total dose and the dose rate can both play an important role.

Another useful subdivision of the problem is to consider the low-dose and high-dose limits, as illustrated in figure 13.

(a) <u>Low-dose</u>: when the implant dose is sufficiently low ($\lesssim 10^{12}$ ions cm^{-2}), we have individual isolated damage regions around each ion track, with negligible probability of cascade over-lap. Within such regions, the damage level is independent of both the total dose and the dose rate. Hence, in this low-dose limit, the bombardment and anneal processes can be fully separated from one another. Furthermore, one can minimize and even eliminate completely the anneal processes by implanting at very low temperature. When the total dose increases sufficiently that the average separation between cascades becomes comparable to the cascade dimensions, then significant inter-cascade effects start to occur and the problem becomes more complex.

(b) <u>High dose</u>: when the implant dose is sufficiently high ($\gtrsim 10^{14}$ ions cm^{-2}), one gets complete overlap of cascades and an essentially uniform lateral distribution of damage. This damage level will often exhibit a complex dose rate dependence, due to *in situ* annealing effects during the implantation. Again, low temp. can be used to minimize this annealing, but at this high dose level, one will still have complex defect-defect interactions during the implantation, due to overlap of each new cascade with residual defects from the previous ones. At typical implantation currents of 1-10 μA cm^{-2}, the mean time between cascade overlap is approxi-

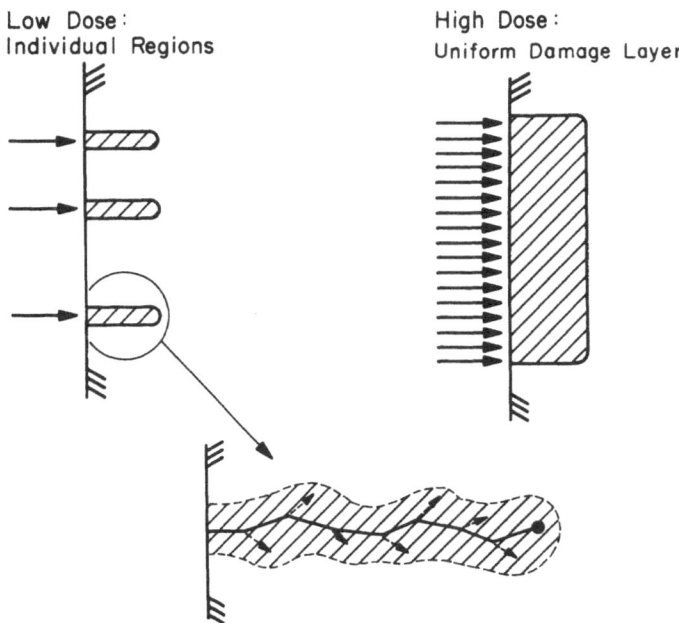

Figure 13 - Lattice disorder schematic.

mately 1 second. Hence, one interesting simplification is to implant
at a sufficiently high temperature to anneal all mobile defects
before the next cascade arrives; this again should eliminate any
dose rate dependence.

For the rest of this lecture, we are going to simplify the
discussion by considering only the isolated collision cascade (i.e.
the low-dose limit). In most cases, we will also restrict ourselves
to low temp. in order to study the basic mechanisms of defect
creation without the further complications of post-implantation
annealing or defect-defect interaction.

3.1 Cascade Theory

The theoretical description of a collision cascade is merely an
extension[23] of Lindhard's Thomas-Fermi description to include the
recoil atoms as well as the primary ion. Again, one separates the
energy loss mechanisms into two types - the nuclear loss $v(E)$ and
the electronic loss $\eta(E)$, with $v(E) + \eta(E) = E_{ion}$. Note that $v(E)$
is usually 20-30% smaller than the total nuclear stopping power of
the primary ion, because the energetic recoils created along the
track of the primary ion lose some of their kinetic energy by
electronic excitation. This small but important distinction between
S_N and $v(E)$ is still not always recognized in the recent
literature.

Winterbon, *et al*.[24] have developed analytical techniques, based
on the transport equation, to provide depth profiles for $v(E)$ and
comprehensive tabulations can be found for example in references 9
and 10. These damage profiles (figure 14) are of course closely
related in shape to the corresponding ion range profiles (figure 4),
including a strong dependence on any channeling effects. They also
exhibit a significant tail extending back to negative depths; this
can be identified as the sputtering efficiency, i.e. the reflected
energy available for removing atoms from the surface.

An alternative procedure for obtaining the depth distribution
of $v(E)$ is to use Monte Carlo techniques to simulate individual
collision cascades and then average the results of several hundred
such cascades. Note the excellent agreement between these two
approaches in figure 14, indicating that one can predict the depth
distribution of $v(E)$ with about the same confidence as the range
distribution. However, to convert this "deposited energy distribution" into a damage distribution is not so easy. The usual procedure
(the so-called Kinchin-Pease method[26]) is to assume that only those
recoils with an energy greater than E_d become displaced, where E_d
is the threshold displacement energy determined from high-energy
electron bombardment. This leads to a fairly simple relationship[27]
between $v(E)$ and the number N_d of displacements produced: namely

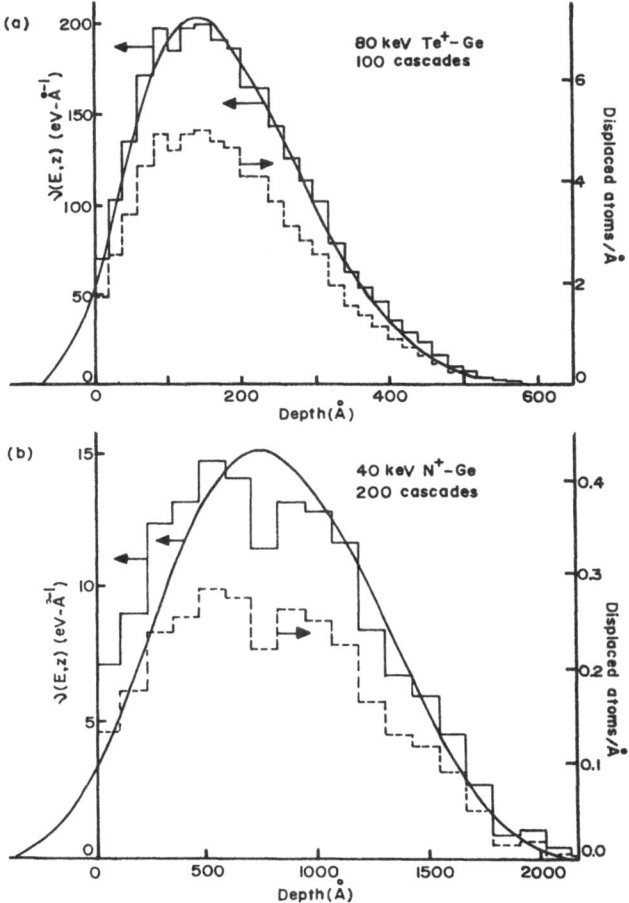

Figure 14 – Comparison of deposited energy distributions[25] ($\nu(E)$) from analytical (solid curve) and Monte Carlo (solid histogram) methods. The displaced atom distribution (dashed histogram) is included.

$$N_d = \frac{0.8\ \nu(E)}{2E_d} \qquad\qquad\qquad (3)$$

The corresponding damage distributions deduced via equation 3 are also included in figure 14.

How well do such predicted damage distributions compare with experiment? Generally, the shape of the distribution is predicted quite accurately, as may be seen in figure 15. However, the absolute magnitude of the damage (i.e. the "conversion factor" from $\nu(E)$ to N_d) is usually not well predicted.

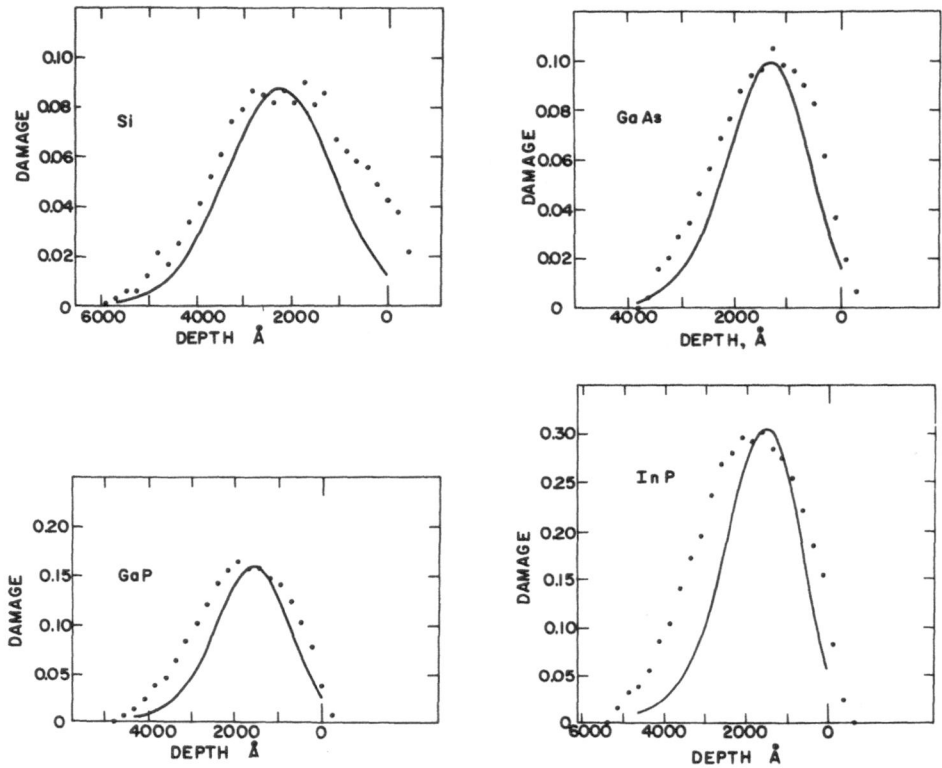

Figure 15 - Comparison of measured (●) and calculated (——) damage
profiles[28] for 300-keV Ar+ implantation in various semiconductors at
30 K. Implantation doses were approximately 5 x 10^{12} ions cm^{-2}.
The height of the calculated curves has been arbitrarily normalized
to coincide with the experimental height.

    In metals, the observed defect levels are up to an order of
magnitude <u>smaller</u> than predicted, suggesting that a large amount of
spontaneous recombination has occurred within the cascade, even at
low temperature.

    In semiconductors on the other hand, the observed damage level
can be almost an order of magnitude <u>greater</u> than predicted,
especially for high-Z ions such as the Tl case in figure 16. Note
that the high-energy portion of each curve bends over to approximately the same slope as the N_{kp} line. This indicates that damage
creation at high energy agrees quite well with equation 3 and that
the discrepancy is occuring at the low-energy end of each ion track.
This is rather strong evidence for the existence of some sort of

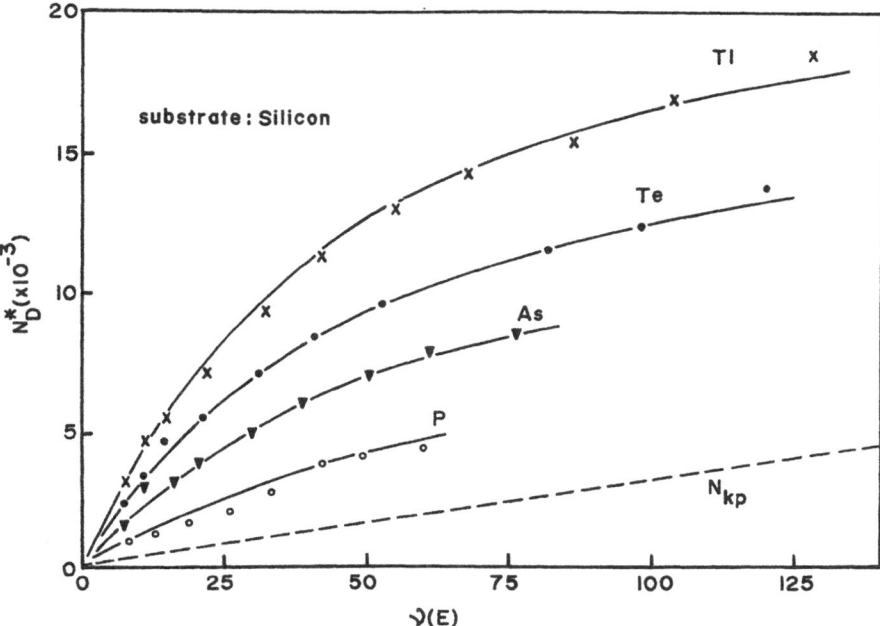

Figure 16 - The total observed number N_d of displaced atoms per
cascade versus $\nu(E)$ in Si at 35 K, as determined by the RBS/
channeling technique[29]. The dashed line N_{kp} is the value predicted
by the Kinchin-Pease relation in equation 3.

non-linear energy density effect such as a "thermal spike". We
shall return to a further discussion of this point in section 3.2.

When applying cascade theory, it is extremely important to
distinguish between the average depth distribution of $\nu(E)$ from a
large number of cascades (figures 14 and 15) and the distribution
within an individual cascade. The latter quantity can only be
obtained by Monte Carlo simulation and several examples are shown in
figure 17, along with the corresponding average distribution for
comparison. When the incident ion is much heavier than the substrate (e.g. Bi → Si case), the primary ion trajectory remains close
to the main central axis of the $\nu(E)$ envelope and hence the damage
distribution within an individual cascade is seen to be distributed
fairly uniformly through most of the volume. In this case, the $\nu(E)$
distribution provides a fairly reasonable approximation for the
individual cascade behaviour. However, at the other extreme of
N^+ → Si, we see that the individual damage cascade occupies only an
extremely small fraction of the corresponding $\nu(E)$ envelope.

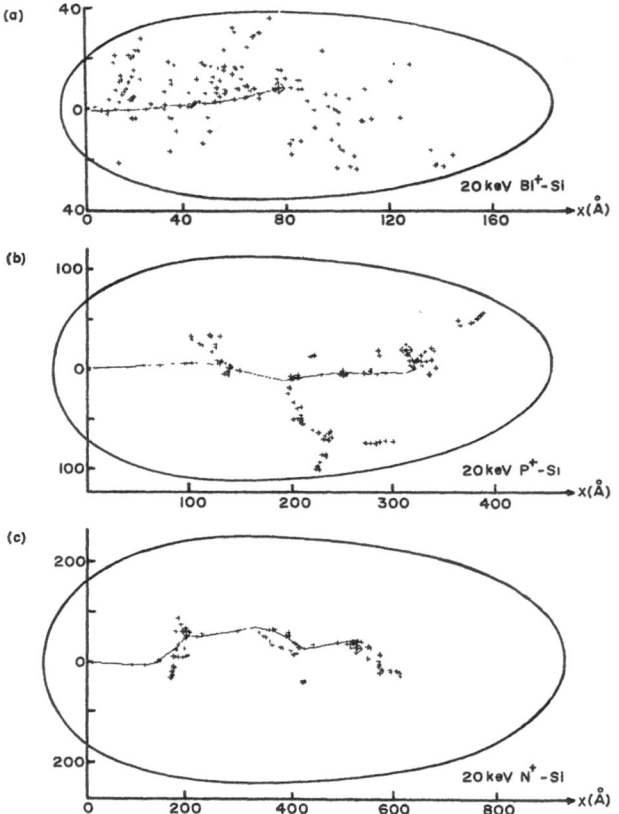

Figure 17 – Relationship between individual (Monte Carlo) cascade
dimensions and the average distribution obtained from the transport
equation[25]. The elliptical curves represent the $\nu(E)$ contour at
10% of $\nu(E)_{max}$. Displaced-atom final locations in the Monte Carlo
cascade are denoted by (+), the ion trajectory by a continuous line
and its final location by (⊕).

        The magnitude of this volume correction factor has been evalu-
ated recently by Walker and Thompson[25] for several different mass
ratios in Si and Ge. Their results are shown in figure 18. For the
Bi in Si case ($M_2/M_1 \sim 0.13$), the volume correction (δ_{corr})^3 is not
much less than unity, but for the N in Si case the corresponding
volume reduction is almost a factor of 100! Obviously, this
correction factor becomes extremely important whenever one wants to
know the instantaneous deposited energy density along an individual
ion track and hence it must be a major concern in many of the
deliberations of this summer school.

3.2 High-Density Cascades

    In applying linear collision cascade theory to obtain the $\nu(E)$
distribution, one is assuming that, in each binary collision between
a target atom and an energetic recoil atom, the target atom is
initially at rest. This is probably a reasonable assumption during
the early stages of cascade propagation, when only a few energetic
recoils are in motion; hence cascade theory is able to predict the
approximate dimensions of the deposited energy distribution. But
simple estimates show that, for most heavy ion implantations, the
final deposited energy density (even if shared equally among all
atoms within the cascade volume) would correspond to a very significant increase in thermal motion. Consider, for example, a 100-keV
ion with a mean range of 400 Å implanted at room temperature. The
total cascade volume contains less than 10^6 atoms and hence the
final deposited-energy density exceeds 0.1 eV/atom. Since the total
cascade propagation time is too short ($\sim 2 \times 10^{-13}$ sec) for normal
quenching mechanisms to remove significant energy, and since 0.1 eV
is much greater than the normal thermal energy of the substrate, it
is obvious that the final stages of the cascade no longer involve
stationary target atoms.

    Recently, Sigmund[30] has suggested that a thermal spike mechanism
might be invoked to handle these high-density cascades and indeed

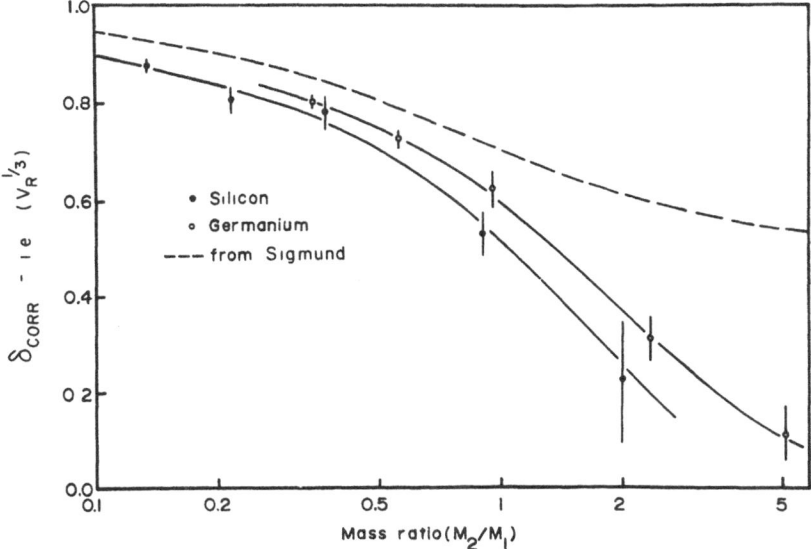

Figure 18 - One-dimensional cascade ratio, δ_{corr}, as a function of
the mass ratio[25]. The dashed curve was obtained analytically by
Sigmund; the points are Monte-Carlo results.

the anomalously large damage levels observed in many semiconductor
implantations (figure 16) are certainly consistent with the formation
(and subsequent rapid quenching) of a pseudo-liquid region around
the end of each heavy ion track.

The topic of "thermal spike" is to be discussed in detail in
one of the subsequent panel discussions. So, in this lecture, we
consider only some very general observations.

The thermal spike concept is an extremely old one, proposed
originally in the 1950's, then discarded, and now resurrected again.
Many people are vehemently opposed to the concept, as evidenced by
the vigorous discussions at this summer school. They point out
(quite correctly) that the quenching times are often too short and
the cascade dimensions too small for Maxwell-Boltzman statistics to
apply; also that the coupling between atomic motion and electronic
exciation is too slow for a true equilibrium to be achieved. Indeed,
a few picoseconds after implantation, one would probably require
two completely different "temperatures" to describe the atomic
vibration level and the electronic excitation (for example, one could
have "hot" atoms and "cold" electrons or vice versa). But this
merely illustrates the complexity of the problem.

Obviously, we cannot directly apply ordinary thermodynamic
arguments, or thermal conductivity equations to such a non-equilibrium situation - but because we cannot properly describe a phenomenon
does not make it disappear! Since the word "temperature" carries
certain specific thermodynamic connotations, perhaps we could
minimize the semantic battle by using some other term such as
"kinetic energy spike" to describe the phenomenon.

The important point to stress here is that, in most ion
implantations, the energy density around the end of each ion track
is far too high for ordinary binary collision theory to apply and
that some sort of collective "hot spot" occurs for at least
10^{-11} - 10^{-12} seconds after the ion comes to rest. Since the main
theme of this summer school is the characterization of the final
site of the implanted ion, an understanding of such spike effects
must be one of our prime concerns.

In concluding this section, it should be noted that in thermal
insulators quenching times become long enough ($\sim 10^{-10}$ secs) that
coupling between atomic and electronic exciation can occur; hence,
even the electronic excitation ($\eta(E)$) may contribute significantly
to lattice "heating". In such cases, an MeV H^+ or He^+ can produce
significant localized heating along each ion track. Recent experiments by Bøttiger, *et al.*[31] on the "sputtering" of frozen Xe (at
20 K) by MeV He^+ provide extremely strong evidence for the existence
of such an evaporation mechanism.

4. ATOM–SITE LOCATION

This is the central theme of our summer school. As in the previous section on Radiation Damage, there are many factors to be considered: namely, the atomic collision process itself, including the role of channeling effects, the subsequent rapid quenching of the "hot" zone around the ion track, and finally any longer term annealing processes that permit mobile defects to interact with the implanted ion.

It is important to distinguish between a localized metastable equilibrium that exists around each implanted-ion site and the true equilibrium distribution corresponding to a fully annealed system. Even the occurrence of a "hot" zone (or so-called "thermal spike") at the end of each ion track would still produce only a localized metastable equilibrium, since the quenching time is so short that defects would migrate no more than a few angstroms.

Consideration of the dynamics of propagation within each collision cascade shows that whenever $Z_1 \sim Z_2$, then direct replacement collisions (in which the incident ion energy falls below the displacement threshold value E_d) can play a significant role. Furthermore, even when Z_1 and Z_2 are vastly different, there is always a vacancy-rich region created along the central primary ion track, with the corresponding lattice interstitials distributed radially some distance away (figure 17). Thus, the implanted ion always comes to rest in a favoured location for becoming trapped by one of these cascade-produced vacancies. At low dose and in the absence of defect migration one should therefore find a large fraction of the implanted ions on substitutional sites. This has been observed in various low-dose room-temperature implantations in semiconductors (figure 19), where 50% or more of the implanted species (B, Bi, Tl, etc.) is found initially on substitutional sites. Subsequent annealing, however, can drastically reduce this substitutional content[33], as mobile defects interact with the implanted species.

In metallic systems, on the other hand, point defects are often mobile well below room temperature and hence will significantly interact with the implanted atoms, unless the implantation temperature is exceedingly low. Kaufmann *et al.*[34] have made an extensive series of room temperature implants in beryllium, and find that the resulting atom site location is determined largely by bulk thermodynamic properties: i.e. that defect migration enables the true equilibrium site distribution to be approached.

One final comment concerns the important role that channeling can sometimes play in determining what site the implanted atom finds. The deeply penetrating "super-tail" in tungsten crystals (figure 11)

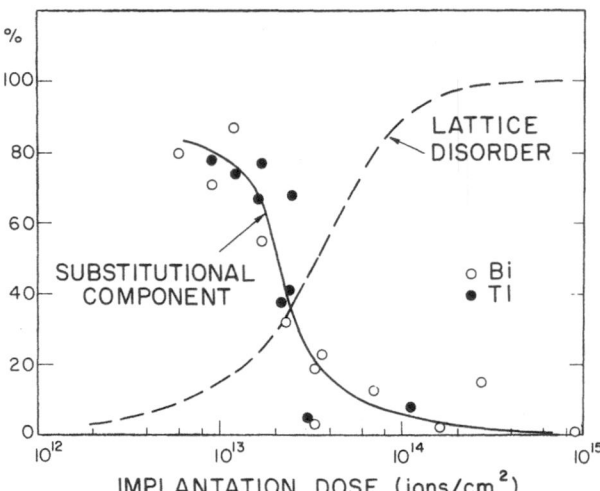

Figure 19 - Dose dependence of the substitutional level of
unannealed Tl and Bi implants in Si at 40 keV and 25°C (ref. 32).

is almost certainly an interstitial diffusion process involving that
small fraction ($\sim$ 1%) of well-channeled ions which slow down without
any violent collisions and hence without creating even a single
vacancy. At least ten widely-different ion species have been
investigated in tungsten[35], and all exhibit this "super-tail"
phenomenon. Furthermore, if the channeled implantation is made at
very low ($\sim$ 20 K) temperature, then an even larger fraction (10-20%)
of the injected ions are retained in interstitial sites.

Since 'atom site location' forms the central theme of this
entire summer school, many of these factors will obviously be considered in much more detail by subsequent lecturers.

REFERENCES

1. P. Sigmund in Radiation Damage Processes in Materials (Corsica Summer School, 1973) pp 2-118.
2. W.K. Chu in Material Characterization Using Ion Beams (Corsica Summer School, 1976) pp 3-34.
3. E. Bonderup, Physics and Applications of Ion Beam Interactions With Solids (Albany Summer School, 1978)
4. J. L'Ecuyer, J.A. Davies and N. Matsunami, Nucl. Instr. Meth. (1979) in press.
5. J. Bøttiger, et al., Rad. Effects 11, 69 and 133 (1971).
6. D.C. Santry, Proc. VIII Int. EMIS Conference (Skövde, Sweden, 1973) p 300.
7. J.A. Davies in Material Characterization Using Ion Beams (Corsica Summer School, 1976) pp 405-428.
8. J.P.S. Pringle, J. Electrochem. Soc. 121, 45 (1974).
9. D. Brice, Ion Implantation Range and Energy Deposition Distributions, Vol. 1 High Energy (Plenum Press, New York, 1975).
10. K.B. Winterbon, Ion Implantation Range and Energy Deposition Distributions, Vol. 2 Low Energy (Plenum Press, New York 1975).
11. W.S. Johnson and J.F. Gibbons, Projected Range Statistics in Semiconductors (Stanford University Press, 1970).
12. J. Lindhard, M. Scharff and H.E. Schiøtt, Kgl. Danske Vid. Selsk. Mat. fys. Medd. 33, No. 14 (1963).
13. P. Loftager, University of Aarhus, private communication.
14. K.B. Winterbon, Can. J. Phys. 46, 2429 (1968).
15. P. Hvelplund and B. Fastrup, Phys. Rev. 165, 408 (1968) and references therein.
16. J.L. Combasson, B.W. Farmery, D. McCulloch, G.W. Nielson and M.W. Thompson, Rad. Effects 36, 149 (1978).
17. H.J. Smith, Rad. Effects 18, 55, 65 and 73 (1973).
18. E.V. Kornelsen, F. Brown, J.A. Davies, B. Domeij and G.R. Piercy, Phys. Rev. 136, A849 (1964).
19. J.A. Davies, L. Eriksson and J.L. Whitton, Can. J. Phys. 46, 573 (1968).
20. J.D. Welch, J.A. Davies, and R.S.C. Cobbold, J. Appl. Phys. 48, 4540 (1977).
21. W.D. Mackintosh, F. Brown and H.H. Plattner, J. Electrochem. Soc. 121, 1282 (1974).
22. J.E. Antill, M.J. Bennett, G. Dearnaley, F.H. Fern, P.D. Goode and J.F. Turner, Proc. III Int. Conf. on Ion Implantation in Semiconductors and Other Materials (Yorktown Heights, N.Y., 1972) p 415.
23. J. Lindhard, V. Nielsen, M. Scharff and P.V. Thomsen, Kgl. Danske Vid. Selsk. Mat. fys. Medd. 33, No. 10 (1963).

24. K.B. Winterbon, P. Sigmund and J.B. Sanders, Kgl. Danske Vid. Selsk. Mat. fys. Medd. 37, No. 14 (1970).
25. R.S. Walker and D.A. Thompson, Rad. Effects 37, 113 (1978).
26. G.H. Kinchin and R.S. Pease, Report Prog. Phys. 18, 1 (1955).
27. P. Sigmund, Appl. Phys. Lett. 14, 114 (1969).
28. J. Bøttiger, J.A. Davies, D.V. Morgan, J.L. Whitton and K.B. Winterbon, Proc. III. Int. Conf. on Ion Implantation in Semiconductors and Other Materials (Yorktown Heights, N.Y., 1972) p 599.
29. D.A. Thompson and R.S. Walker, Rad. Effects 36, 91 (1978).
30. P. Sigmund, Appl. Phys. Lett. 25, 169 (1974).
31. J. Bøttiger, J.A. Davies, J. L'Ecuyer, N. Matsunami and R. Ollerhead, Proc. Int. Conf. on Ion Beam Modification of Materials, Budapest (September 1978) in press.
32. L. Eriksson, G.R. Bellavance and J.A. Davies, Rad. Effects 1, 71 (1969).
33. G. Fladda, K. Bjorkgvist, L. Eriksson and D. Sigurd, Appl. Phys. Lett. 16, 313 (1970).
34. E.N. Kaufmann, R. Vianden, J.R. Chelikowsky and J.C. Phillips, Phys. Rev. Lett. 39, 1671 (1977).
35. L. Eriksson, J.A. Davies and P. Jespersgård, Phys. Rev. 161, 219 (1967).

ION IMPLANTATION PROCEDURE

G. Dearnaley

AERE Harwell

Didcot, England

INTRODUCTION

It is the purpose of these introductory lectures to provide a description of the equipment and experimental methods used in the process of ion implantation. The emphasis will be on the research aspects of the subject, but an account will briefly be given of the somewhat different equipment and objectives involved in several industrial applications.

There can be no doubt that a successful ion implantation facility must incorporate a well-chosen ion source. Obvious though this may seem, nevertheless a large number of systems are being used for ion implantation and in many cases these were originally constructed for some quite different purpose, for example in nuclear research. Without comparing the performance of several different ion sources, it may be difficult to be sure that the one selected is the best. Besides providing an adequate beam intensity, the source must be versatile enough, should be reliable and easy to clean and maintain. The first part of the lecture will therefore review representative examples of the very wide variety of ion sources which have been developed, with comments about their particular features in relation to ion implantation.

It is also important to design the manner in which the ions are extracted from the ion source and focused into the acceleration system. This is only the first part of what should be an integral design of the ion optics throughout the system, and the location of the points of cross-over of the beam need to be considered at the

earliest stage. Only general points of advice can be given, since
individual implantation systems differ in many respects.

Mass analysis, usually by magnetic fields, is almost always
necessary in order to select the appropriate ion species for introduction into the target chamber. Some popular systems will be
described.

The target chamber requires particularly careful design, and
there will be major differences between those intended for research
and the industrial assemblies in which the emphasis is upon
repetitive work and a high throughput. Attention will be drawn to
some of the points which it is necessary to consider.

While the structure of the presentation is related to the
various components of the ion implantation equipment, remarks will
be made concerning the procedure on the basis of over 12 years
experience at Harwell with a variety of ion accelerators both in
research and in commercial applications involving semiconductors
and metals.

The industrial success of ion implantation in the manufacture
of semiconductor devices and the growing number of applications in
other materials, together with the demonstrated value of the
technique for preparing well-characterized specimens for research
all, in my view, ensure that further development and interest in
the field will continue for many years. Novel applications are
stimulating the design of new and more powerful ion sources.
Unexpected effects are being observed under high dose implantation
conditions using high current intensities. This is a field of
research which offers many possibilities for imaginative inter-
disciplinary work.

ION SOURCES

Sidenius, one of the leading workers in this area, recently
described[1] the ion source as "this indispensable but troublesome
part of an accelerator system" and went on to remark that "The
diversity and complexity of both the fundamental ion source
principles and the application requirements have, however, excluded
a complete and real systematic treatment...."

The most common method of creating ions is by subjecting the
vapour of the element to be ionized, or a compound of it, to impact
by energetic electrons. These are normally supplied from a heated
filament and at the pressures used a plasma containing ions and
electrons will be formed. From the boundary of this conducting,
space-charge neutralized zone ions will have to be extracted by
an appropriate electric field.

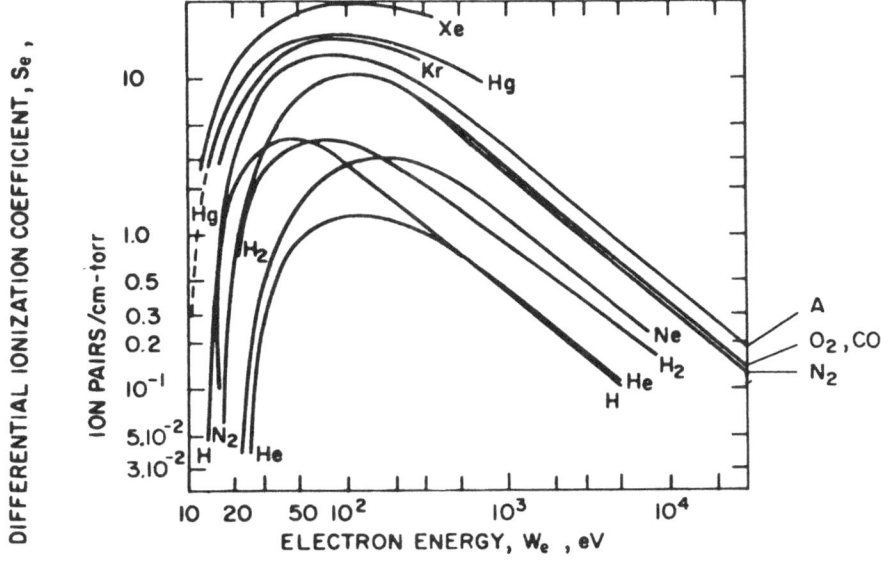

Fig. 1

Figure 1 shows[2] the ionization coefficients, S_e, or the number
of ion pairs produced per unit path length per unit of gas pressure
by electrons of differing energies, W_e. It can be seen that for
the optimum degree of ionization fairly energetic electrons of 50
to several hundred eV are required. These are the so-called 'hard'
electrons which must be continuously injected into the plasma,
within which they will not be deflected by the weak electric field
present. Their paths are controlled by the external electrostatic
fields or, to a greater extent, by applied magnetic fields. The
mean free path of these electrons will be much larger than the ion
source dimensions, and on collision with the walls secondary
electrons are produced. It is these soft electrons which create
the plasma, and very important features of an ion source are the
ways in which this plasma is stabilized and located with respect to
the extraction electrode.

One obvious way of increasing the ionization probability is to
increase the path length of the electrons, and figure 2 shows two
ways of achieving this. In 2(a) the electrons are caused to
oscillate by one or more negatively charged repeller plates. In

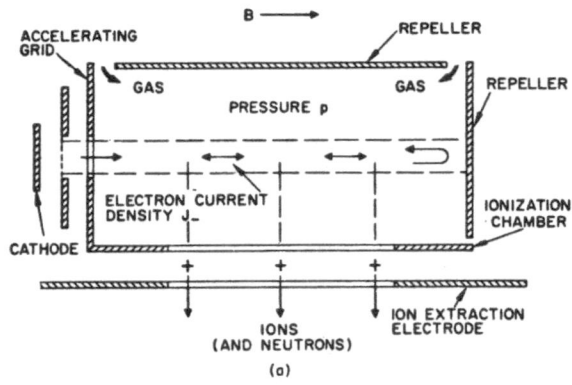

Fig. 2 (a) the oscillating electron   (b)   the electron impact
         ion source   ionization source
                                             (after Wilson & Brewer ref.3)

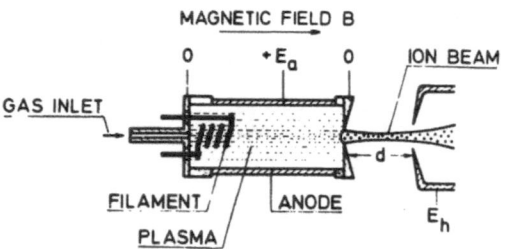

Fig. 3 The Nielsen ion source (schematic)

fig. 2(b) the electrons are caused to spiral by an applied axial
magnetic field, and in this case the ions are usually, though not
always, extracted along an axial direction.

Perhaps one of the simplest types of ion source, developed by
Nielsen for use in isotope separators, is shown in figure 3.
Extraction takes place axially through a single circular aperture,
and a magnetic field concentrates the plasma near the axis of the
source. Instead of a gas feed, various arrangements a furnace may
supply vapour to the ionization region.

Figure 4 illustrates three ways in which gas or vapour can be
fed into an ion source, and it is sometimes found advantageous to
supply gas or vapour from many points symmetrically disposed around
the arc chamber. The flow of cold or unexcited gas into a plasma
will affect the distribution of ion and electron energies, a
feature made use of in Tokamak fusion equipment in the process
known as 'gas puffing'. As a result, the extraction conditions
from an ion source may be influenced by the effects of gas fed in
to the plasma boundary. For this reason, perhaps, the feed
arrangements in figure 4 are all symmetrical with respect to the
extraction electrode.

Two types of oscillating electron discharge sources are shown
in figure 5, one for an axial, circular extraction geometry and the
other making use of a slot. For a given 'brightness' of the
source, it is clear that the extracted current will be greater in
the latter case. (Here the term brightness of an ion source is
defined to be the emitted current density per unit solid angle).

Nielsen has developed a widely-used version of the oscillating
electron ion source for use with solid elements. As shown at the
centre of figure 6, a miniature furnace is arranged to supply
vapour to the discharge region of the source below. A concentric
coil generates a strong axial magnetic field.

By contrast, in the hollow cathode design of ion source, due
to Sidenius, the optimum performance is obtained without a magnetic
field in the cathode region. Figure 7 shows the principles of
operation, and a magnetic field is applied in the immediate
vicinity of the anode, to constrict the plasma as illustrated
schematically by a shaded area. The plasma density at the anode
may be so high that a plasma expansion cup, shaped within the
anode as shown, may be necessary for efficient extraction of the
ions. This is a situation which will be discussed further below.

A different method of ionization is used in the radiofrequency ion source, which is a widely adopted means of obtaining
ions from gaseous materials. The more common arrangement, shown
in figure 8, is to surround an insulating vessel (preferably of
pyrex glass) with a coil which forms part of the tuned circuit of

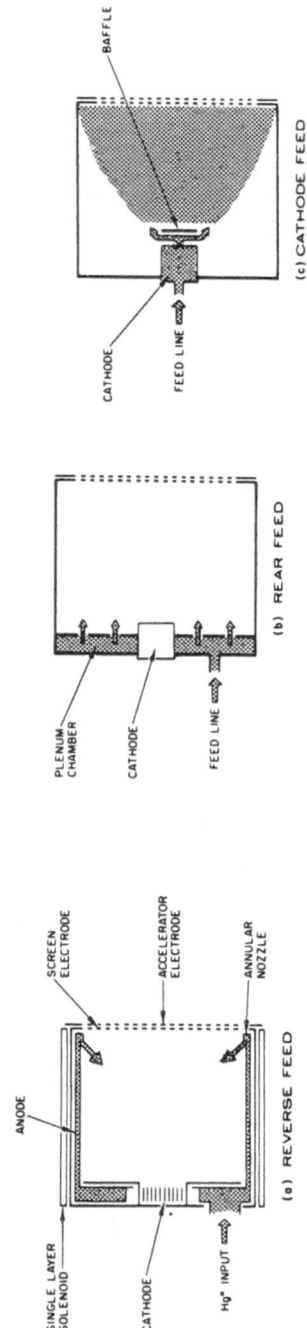

*Fig. 4 Three methods of introducing gas into an ion source (after
Wilson & Brewer, ref. 3)*

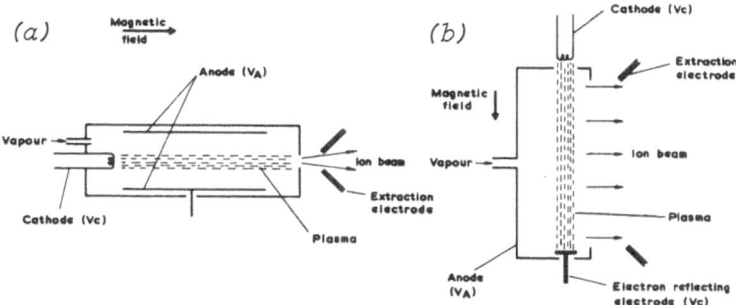

Fig. 5 *Oscillating electron discharge sources.*
(a) Beam extraction from a circular aperture in the end of
the discharge chamber.
(b) Beam extraction from a slit in the side of the dishcarge
chamber.

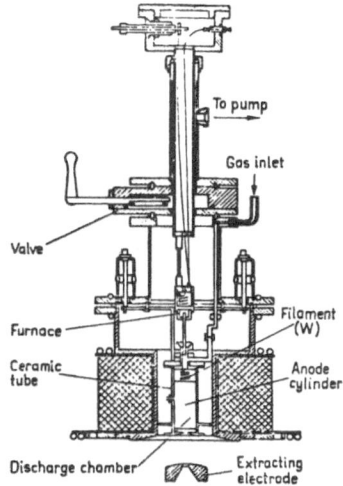

Fig. 6 *Nielsen's version of the oscillating electron source for*
solid elements.

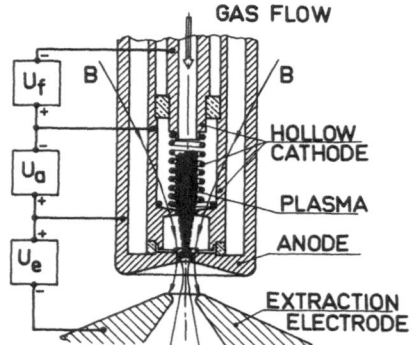

Fig. 7   The Sidenius ion source

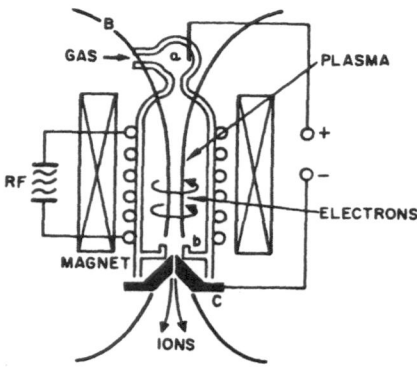

Fig. 8   The R.F. ion source

an oscillator generating a few hundred Watts of R.F. power in the 10 MHz to 30 MHz range. There are always a few electrons initially present which are accelerated by the electromagnetic field and cause impact ionization in gas at $\sim 10^{-3}$ torr pressure. An electrode 'a' is inserted at the top and a cylindrical extraction electrode at the base, and an electric field is applied between them so as to extract ions from the plasma. An axial magnetic field increases the electron path lengths and increases the available ion current. As shown in the diagram, it is necessary to protect the metal electrodes from the possibility of ion sputtering which tends to coat the surfaces of the glass envelope with metal and this causes an increased rate of recombination at the walls, besides removing energy from the plasma by the ionization of high atomic number impurities. Further details of a typical radio frequency ion source are shown in figure 9. This type of source is often fitted to electrostatic Van de Graaff accelerators, and reasonable success has been achieved in feeding vapours into such ion sources in order to produce boron or phosphorus ions for implantation. It is not, however, a particularly versatile source, and tends to be used in cases where it would be difficult to substitute some other type of ion source.

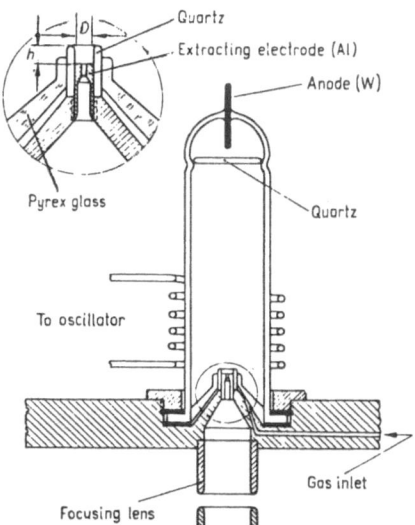

Fig. 9 A practical radio-frequency ion source

G. DEARNALEY

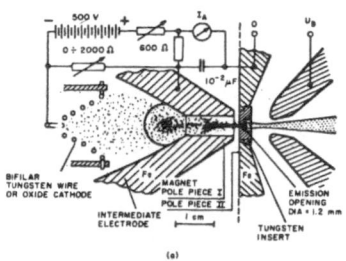

(a)

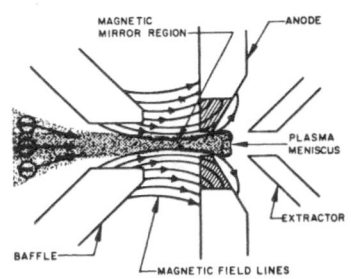

Fig. 10   The duoplasmatron ion
          source

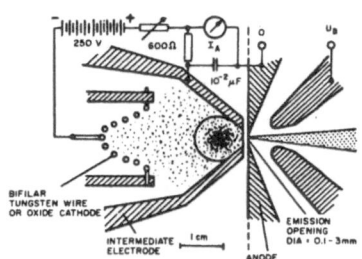

Fig. 11   Two variants of the
          duoplasmatron source

Next, we turn to a group of ion sources which have been given
the name of plasmatron. These are developments of the simple type
of arc discharge source mentioned at the beginning: electrons from a
heated filament are accelerated through the gas to be ionized. Near
the anode however the plasma is constricted by a conical electrode
with an aperture a few mm in diameter. This electrode is held at a
potential intermediate between that of the cathode and anode, and it
is designed to focus the fast electrons into the region of the
extraction aperture in the anode. The most common source of this kind
is called the 'duoplasmatron' (figures 10-12) in which a strong
magnetic field in an axial direction constricts the plasma to an
intense discharge at the centre of the anode. The local power dissipation is very high, so that high melting-point materials must be used.
The plasma density is so high that efficient ion extraction is
impossible: the applied extraction field cannot penetrate the plasma.
Therefore the plasma is allowed to pass through the anode aperture
into a larger 'expansion cup', most clearly illustrated in figure 12.
An attempt was made by Masic et al. to adapt this source for solid
materials by vapourizing them and feeding the vapour into the
expansion cup. Ionization there takes place by charge exchange
mechanisms and by excitations due to energetic electrons emerging
through the anode aperture.

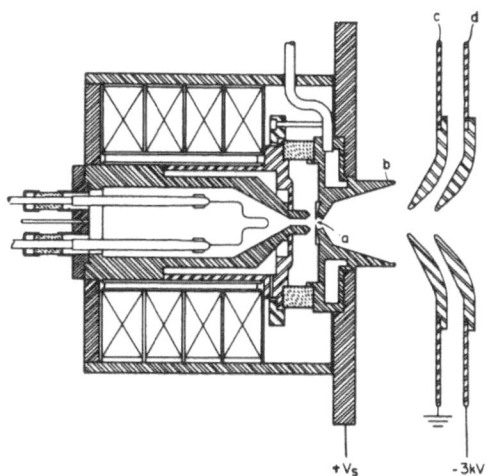

The schematic diagram of duoplasmatron ion source, extraction system and charge-changing target: (a) anode aperture, (b) plasma expansion cone, (c) extractor electrode, (d) suppression electrode,

Fig. 12

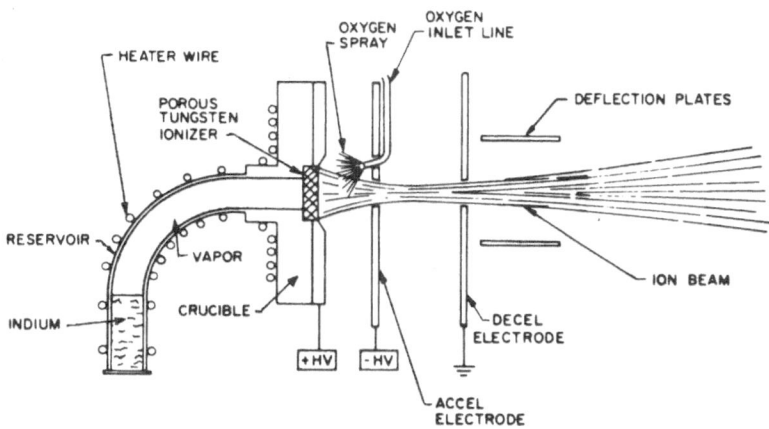

Fig. 13 A surface ionization source (after Daley, ref. 5)

   A type of source which is particularly useful for elements
from the left-hand side of the Periodic Table makes use of the
process of surface ionization, in which atoms may lose an electron
in contact with a suitable hot metal surface. The ion flux is
given by the Saha-Langmuir equation and is exponentially dependent
upon the difference between the work function of the surface and
the first ionization potential of the atom. Thus it is most
efficient for surfaces of high work function such as tungsten,
iridium and osmium or (for lower melting point materials) platinum
or nickel. Figure 13 shows the arrangement used by Daley in which
a vapourizer is attached to a porous tungsten ionizer, and in this
case the work function of the tungsten was increased to over 6 eV
by a controlled oxygen spray. Sources of this pattern are very
promising for the large scale ion implantation of metals which are
reasonably fusible and possess low ionization potentials.

   One of the best established ion sources for implantation purposes
has been termed the magnetron source because a straight wire or
ribbon is mounted in a cylindrical anode and a very heavy heating
current is used. The magnetic field due to this, together with an
applied field parallel to the filament causes complex electron
paths and a high density of ionization. Figure 14 shows the design
due to Freeman, sometimes known as the Harwell ion source, in
which an oven supplies vapour to the cylindrical arc chamber.
Extraction is carried out through a slit. This source provides
an exceptionally stable plasma, and for this reason a very high
quality beam for isotope separation and precise ion implantation
work. The good separation possible is very useful when implanting
radioactive species which require to be selected in the presence
of a large fraction of the stable isotopes. It is also highly
versatile, and a good deal of experience has been gained and
published regarding its use.

   Another very versatile design of ion source was developed by
Hill and Nelson specifically for ion bombardment using species
which are not readily vapourized. As shown in figure 15, atoms
are released by sputtering into a gas discharge in which ionization
occurs by charge exchange and by electron impact. An arc discharge
is created in argon, or some other gas with high sputtering
coefficients, and a 'sputtering electrode' is exposed to this, at
a potential of about -1kV. The detailed construction is fairly
complex, but Goode has recently reported a simplified version
which incorporates some design improvements. A most useful
feature of this source is the ease with which different species
can be generated simply by changing the sputtering electrode. The
absence of an oven leads to a low power consumption. Druaux and
Bernas have described a sputtering source with lateral extraction,
through a slot, which gives large beam currents (> 1mA). The
material in this case is sputtered from the entire wall of a
cylinder.

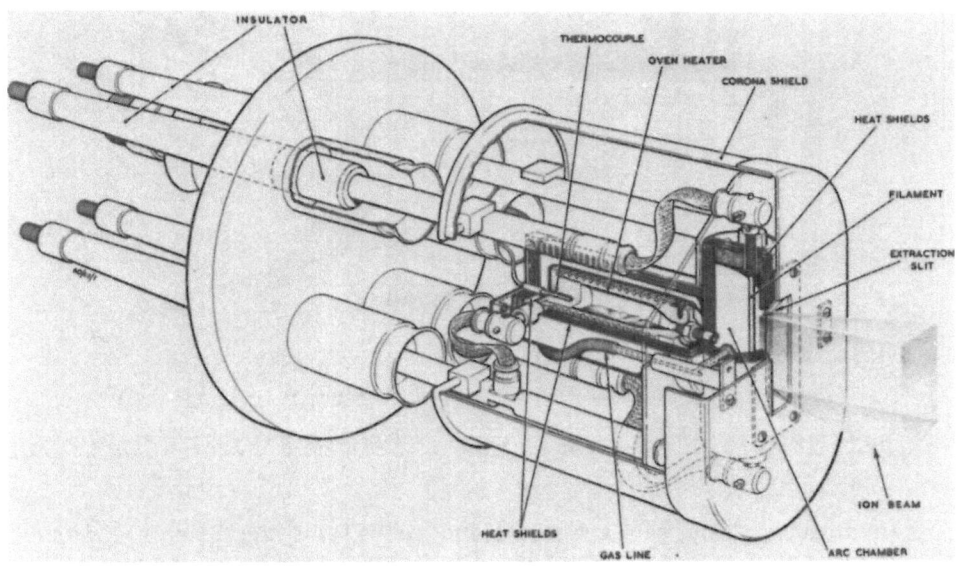

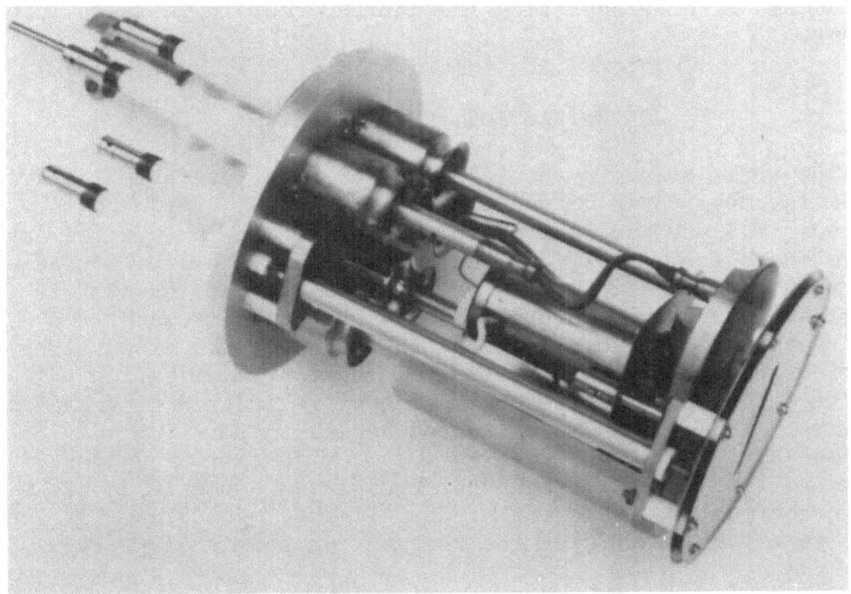

Fig. 14   The Freeman type of magnetron ion source

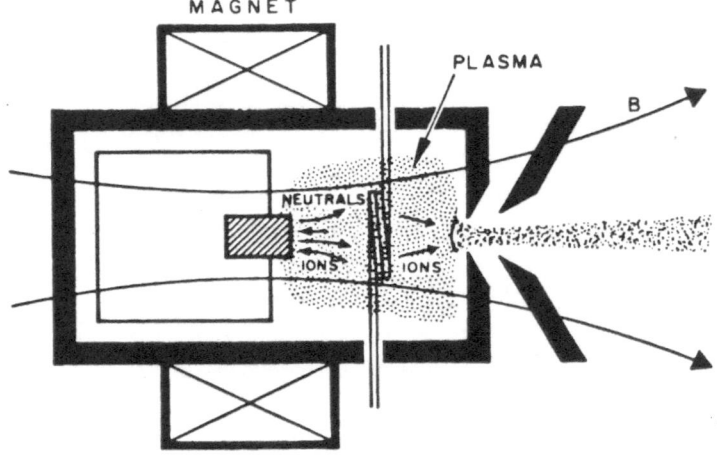

Fig. 15 The sputtering ion source of Hill and Nelson (Schematic)

In summary, the table below (due to Wilson and Brewer) lists the main features of the various ion sources we have discussed, together with a few other earlier types. Several other factors, such as the ease of operating the source, need to be taken into account.

ION EXTRACTION AND ION OPTICS

The first stage in the injection of ions into the acceleration system involves their extraction from the plasma in which they are created. The surface ionization source differs in that the ions are released from a well-defined surface, but a plasma source will take up a configuration which is affected by the flux of ions from the plasma boundary. Figure 16 shows what happens under two conditions (a) a dense plasma and (b) a dilute ion cloud. The behaviour under increasing values of the ion acceleration potential that a curvature or 'meniscus' is created in the plasma boundary. Since ions are extracted normal to this surface one may have a virtual focus (figure 17) from which ions appear to emerge. It is clear that an unstable plasma boundary causes a severe aberration. The stable and well-developed plasma meniscus is an important feature of the Freeman ion source (fig. 14).

The next stage in the system is usually to accelerate the ions to some required high energy. There are several popular forms of accelerator tube (figure 18) but of these the best arrangement is illustrated on the right. An electrode shape is chosen which prevents secondary electrons from arriving directly at the insulator sections and so charging them up. The insertion of

TABLE I   Ion Source Characteristics

Source Type	Energy Spread V	Total Ion Current A	Ion Beam Current Density, A/cm^2	Power W
Field Ionization	20 to 40	10^{-10} to 10^{-8}	1 to 10^2	1000
Duoplasmatron	10	10^{-3} to 10^{-1}	10^{-2} to 1	100-300
RF	30 to 500	10^{-4} to 10^{-2}	10^{-3} to 10^{-1}	500
Spark	10^2 to 10^4	10^{-6} to 10^{-2}	10^1 to 10^2	100
Surface Ionization	0.2 to 0.5	10^{-5} to 10^{-2}	10^{-4} to 10^{-2}	100
Low-Voltage Arc	0.2 to 2	10^{-4} to 10^{-3}	10^{-3} to 10^{-2}	200
High-Voltage Arc	10^4	10^{-3} to 10^{-2}	10^{-2} to 10^{-1}	10
Primary Electron Collision	~1	10^{-10} to 10^{-8}	~10^{-7}	100
Oscillating Electron Discharge	10 to 50	10^{-3}	~10^{-2}	
Hot Cathode Electron Bombardment	1 to 10	10^{-10} to 10^{-3}	10^{-10} to 10^{-3}	50-150
Large-Diameter Penning Discharge	10	10^{-1} to 1	10^{-4}	up to 10^4
Sputtering	10 to 50	10^{-4} to 10^{-2}	10^{-3} to 10^{-2}	200
Electron Bombardment Vaporization	~10	10^{-4} to 10^{-2}	10^{-3}	500
Cold Cathode (Penning)	50	10^{-5} to 10^{-3}	10^{-4} to $10.^{-3}$	20

permanent magnets inside the tube, to deflect secondary electrons
produced by collisions with the electrodes, can have a strong
effect upon the high voltage performance and upon the X-rays
emitted from the positive terminal. Pumping apertures in the outer
regions of the equipotential plates can assist in providing a
reasonable gas pressure in the vicinity of the ion source, where
ion velocities are low: scattering and neutralization of the ion
beam must be minimized.

The ion trajectories through the acceleration system need to
be calculated preferably with the aid of a computer program. The
principles involved have been described in detail by Banford.[11]

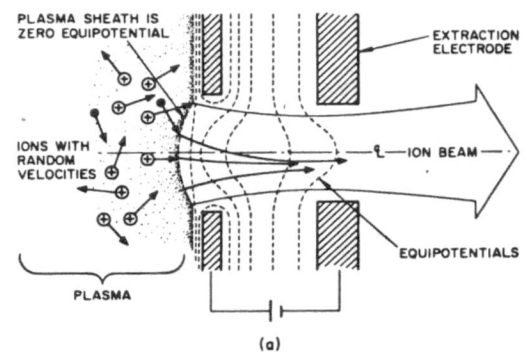

(a)

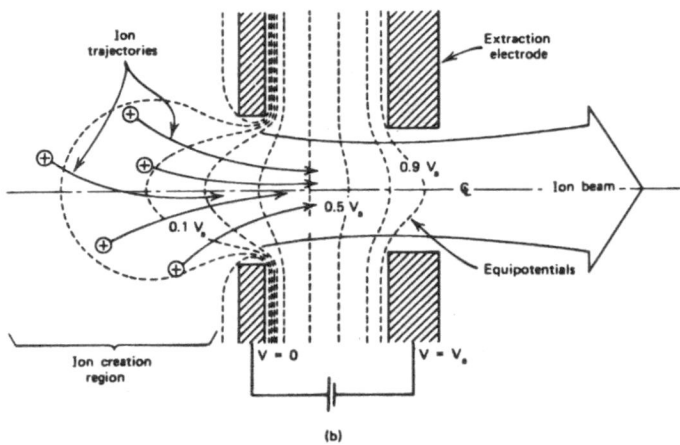

(b)

Fig. 16   Ion extraction from a plasma and from a dilute ion cloud
          (after Wilson & Brewer, ref. 3)

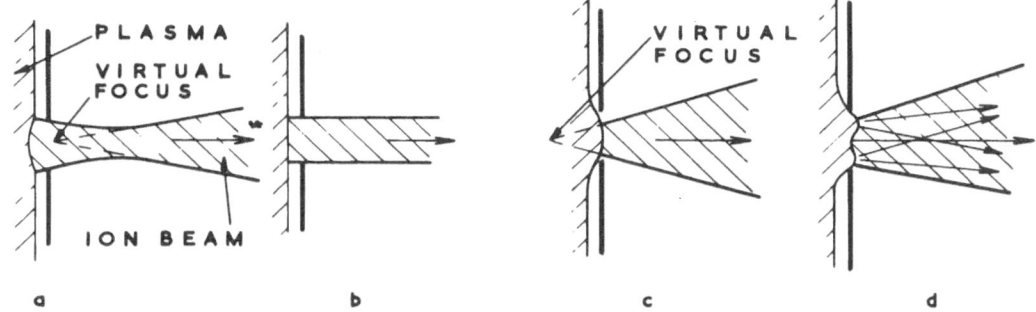

Fig. 17   Curvature of the plasma
surface leads to the formation of
a virtual focus

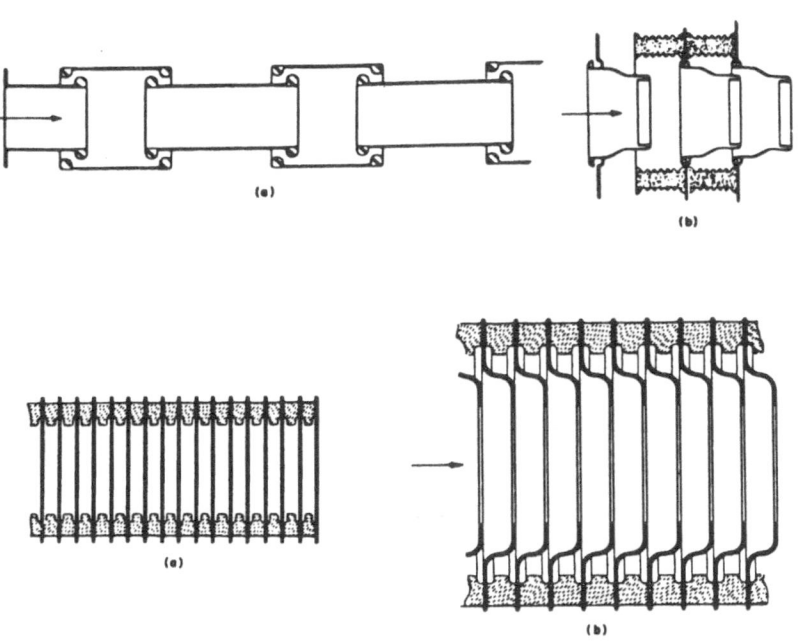

Fig. 18   Several designs of
accelerator tube (after ref. 10)

MASS ANALYSIS

As a rule, the ions extracted from an ion source comprise a
very large number of different molecular and impurity species.
Ion implantation, particularly of radioactive species for HFI
experiments, often demands a high degree of mass discrimination.
The theory of magnetic analysers has been given in detail by
Brewer and Wilson[3] and by Freeman,[12] and here I shall simply outline
the design of the homogeneous field analyzers most commonly used.

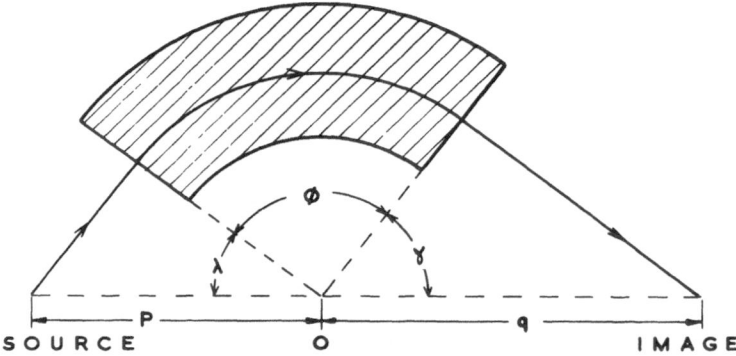

Fig. 19   Barber's rule

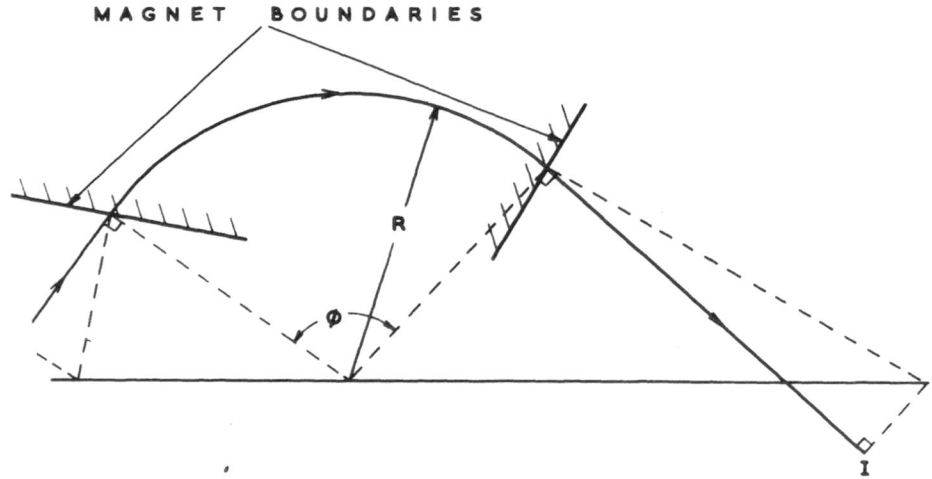

Fig. 20   Cartan's rule

As shown in figure 19, the simplest arrangement is when ions enter a homogeneous, sector-shaped magnetic field at 90° to the pole face. Then it can be shown that Barber's focusing rule applies and ions diverging from a source are refocused to an image at the point such that $\lambda + \theta + \gamma = \pi$. If, on the other hand, the pole faces are tilted (fig. 20) a different construction due to Cartan applies. It is sometimes useful to arrange that the pole faces can be tilted so as to achieve control of the point at which the ion beam is focused: this feature is employed in the Lintott ion implantation system.

After the various ion masses have been dispersed by magnetic analysis, a mass defining slit is used to select the particular species required. It is usually necessary to scan or distribute these ions over a target, and figure 21 shows three ways in which this can be done, by combining electrostatic deflection with a magnetic analyser. By sweeping the magnetic field a mass spectrum is obtained, and in a high resolution system a remarkably large number of ion species can be observed.

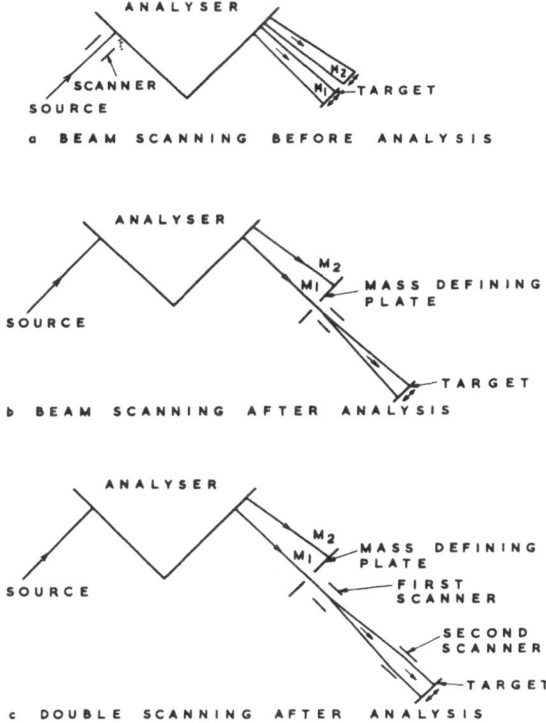

Fig. 21 Three methods of combining beam scanning and mass analysis

  Another method of mass analysis, providing lower mass resolution, is by use of a crossed magnetic and electric field. This is sometimes called a Wein filter or a velocity analyzer. Assuming zero fringing fields, there will be a unique particle velocity $V_O = E/B$ for which paraxial ions will emerge directed along the axis of the system (figure 22). For this condition the electrostatic defection eE balances the magnetic defection veB.

BEAM SCANNING

  In order to provide a uniform distribution of implanted ions over the area of a target specimen it is necessary to arrange some relative movement of the ion beam and target. This can be done in several ways, by electrostatic deflection of the beam, by mechanical manipulation of the target or by some combination of the two.

  Figure 23 shows two forms of electrostatic deflection system using orthogonal pairs of plates to which are applied asynchronous linearly time-dependent waveforms typically of a few kilovolts in amplitude. Neutralization of a small fraction of the ions due to collisions with residual gas is likely to occur along the path towards the target: such neutron atoms will not be deflected and will give rise to a central peak in the lateral distribution. For

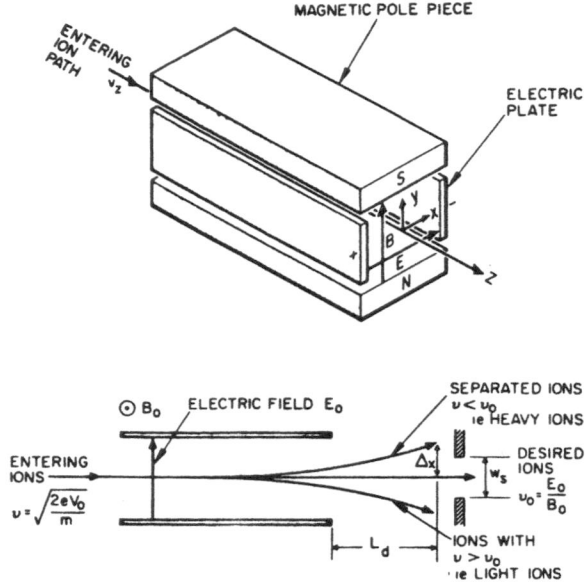

Fig. 22   The Wein filter or E x B analyser

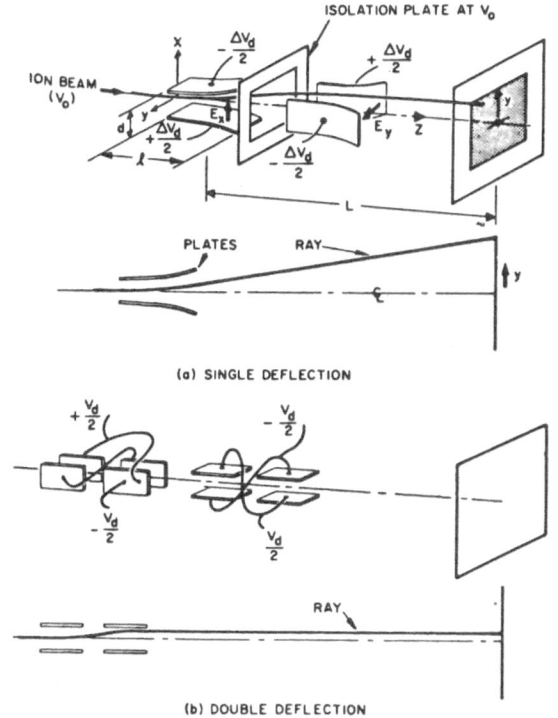

(a) SINGLE DEFLECTION

(b) DOUBLE DEFLECTION

Fig. 23 Two methods of electrostatic beam scanning

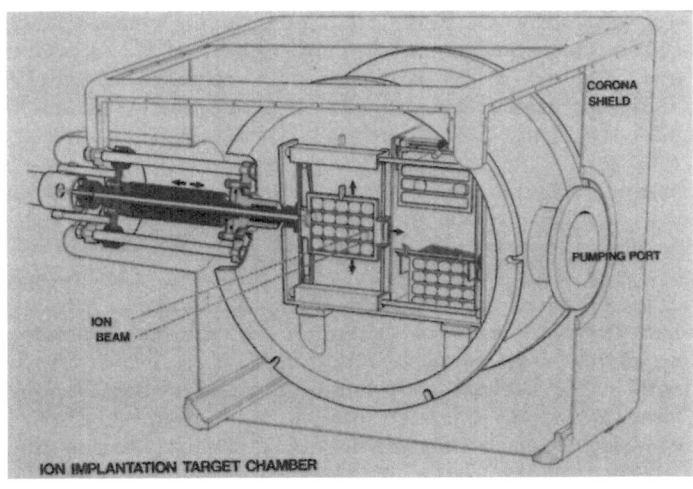

Fig. 24

this reason it is often arranged that there is a deliberate net
deflection of the ion beam as shown in fig. 23(b), and if the
path length between the scanner plates and the target is short
further neutralization can be minimized. Consideration must also
be given to the ratio of beam diameter to the scan area in order
to ensure an adequate overlap of subsequent scans.

An alternative approach is particularly useful in the case
of high intensity (> 0.5 mA) ion beams in which a space-charge
neutralization takes place and an electron cloud surrounds the
column of ions. Electrostatic deflection would separate the ions
and electrons and render control of the beam more difficult. This
is overcome by a mechanical movement of the target (figure 24), the
ion beam being stabilized both in space and time.

A somewhat more complicated arrangement, due to Robertson[13] involves
a rotating target plate which is traversed at a linearly changing
rate to compensate for radial dependence. For precise implantations it is desirable to modulate these mechanical movements in
accordance with the instantenous ion current, and in addition to
scan over the target area many times to smooth out beam intensity
fluctuations. The need for a stable plasma in the ion source now
becomes very apparent, and for this reason sources which have been
highly developed for precise separation of isotopes have proved
very useful for ion implantation under conditions of maximum
uniformity.

TARGET CHAMBER DESIGN

Experience has shown that the correct design of a target
chamber is most important for successful ion implantation. The
requirements for research and for industrial applications are rather
different, the emphasis in research being upon versatility and the
ability to alter the conditions of implantation e.g. temperature,
dose rate, specimen orientation, while in commercial application
the process will be much more repetitive and the need is for a high
rate of sample processing combined with reliability.

Some common features of a target chamber for research on small
samples of material is shown in figure 25. The target stage can
be heated electrically or cooled by a supply of liquid nitrogen.
The pipes carrying this are flexible to allow a rotational
adjustment of the stage for studies of ion channelling. A
secondary electron suppression cylinder is mounted between the
target and an adjustable beam collimator. A helical liquid
nitrogen cooled trap serves to condense contaminant vapours near
the target, and a turbomolecular pump was chosen to give a high
pumping speed for heavy hydrocarbon constituents which cause
surface contamination problems by being cracked or decomposed
under ion bombardment. The specimen magazine and sachet selector
are shown in figure 25.

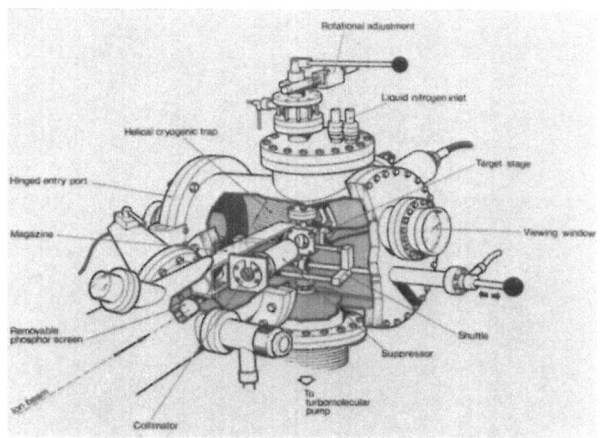

Fig. 25   Target chamber for research involving ion implantation

Accurate target current measurement presents several problems because both secondary electrons and secondary ions can be released from bombarded surfaces. Secondary electrons can be suppressed electrically and by a transverse magnetic field, but if secondary ions are numerous it may be necessary to measure the current by inserting a well designed collector of the Faraday cup variety. A good design may have a series of fine blades (razor blades) set parallel to the beam and with a small magnet to deflect electrons. Particles generated within the stack of blades are unlikely to escape.

If the target specimens are insulators special care is needed in order to measure the ion current. If the beam area is arranged to extend over the specimen surface up to a metallic plate or

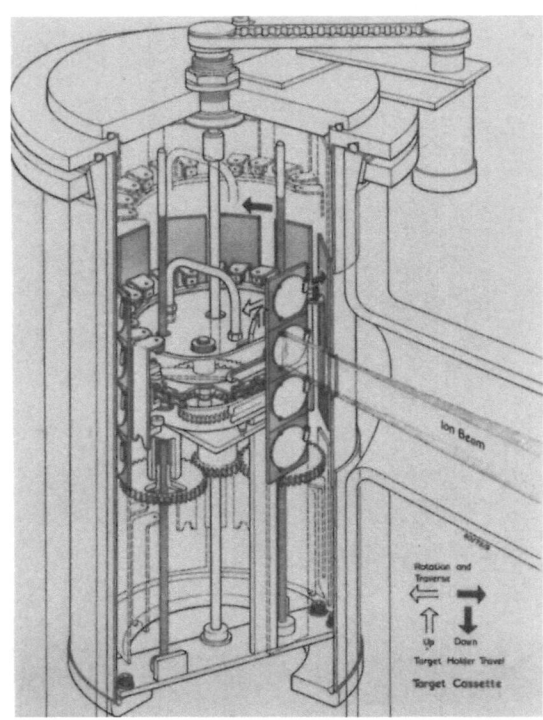

Fig. 26 Target chamber designed for the implantation of silicon
         wafers

ring the bombardment-induced conductivity in the surface layers
(due to charge carrier generation) will usually provide adequate
conduction. Alternatively the surface may be kept in a neutral
condition by a supply of electrons from a nearby heated filament,
but care must be taken that metal is not vapourized from this to
contaminate the specimen.

Carbon contamination from hydrocarbon present can be minimized
by the use of metal sealing gaskets wherever possible, by the
correct choice of pumping system and the use of polyfluorinated
diffusion pump oil. The best protection is probably by cryogenic
cooling at liquid He temperatures in the immediate vicinity of
the target.

For industrial work, and here the chief requirement so far
has been for the implantation of silicon wafers, target chambers
are no less sophisticated. Figure 26 shows one example in which
many specimens are mounted on easily handled plates which can be
rotated on a cassette assembly through the stationary ion beam.[14]
Note that the movement through the beam is planar so that the angle
of the specimen with respect to the beam does not change. The
cassette spirals up and down continuously and the movement speeds
are controlled by the ion beam current which is itself stabilized.
Better than a 1 per cent uniformity and reproducibility of the
implanted dose can be achieved in this way, with a capability of
processing up to 1000 wafers per day.

Newer applications of ion implantation to engineering components
and tools bring some different problems because these may be
relatively heavy and will usually have a more complicated geometry
than that of silicon wafers. Manipulation of such workpieces in
vacuum is necessary both for adequate coverage and so as to avoid
overheating due to the high beam power densities used.

Target temperature measurement may not be easy because of the
poor thermal contact between components in vacuum, and infra-red
techniques require some knowledge of the emissivity. Calculations
of the temperature of target specimens as a function of beam
power density have been made by Freeman, and the results for an
emissivity of 0.5 are shown in fig. 27(a). It can be seen that
liquid nitrogen cooling of a nearby surface (conduction being
ignored) has little effect. The rate of rise of temperature is
shown in fig. 27(b) and from this can be chosen the correct speed
at which to move a specimen through an ion beam in order to
lessen the temperature rise. The cooling of the components is
then arranged by an appropriate ratio between the area under
bombardment and the total specimen area from which radiation
takes place. The more massive engineering components will not
require such rapid movement, because their thermal capacity
limits the temperature rise.

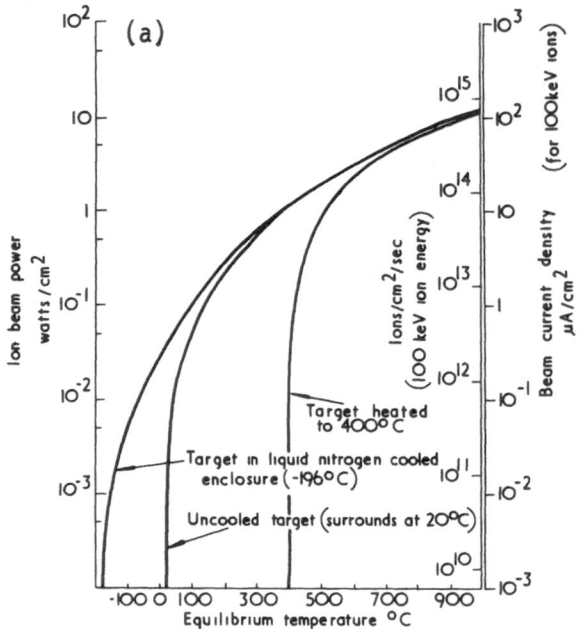

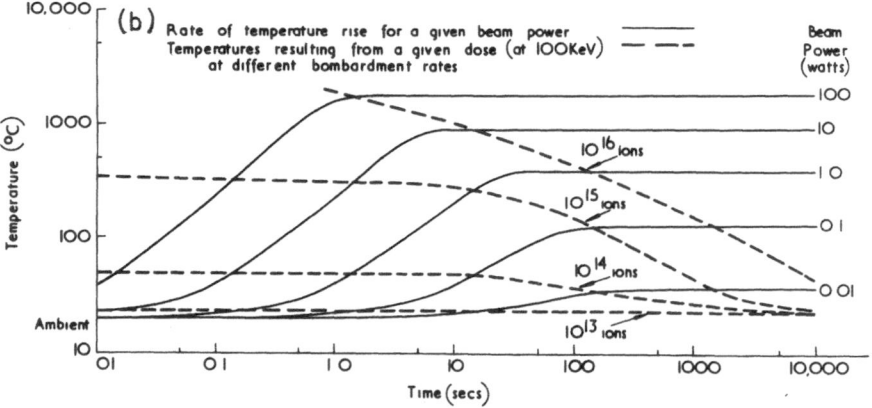

Fig. 27(a)   Equilibrium temperature of ion bombarded silicon wafers.

(b)   Their transient temperature rise (ref. 12).

IMPLANTATION SYSTEMS

We have considered the various component parts of an ion implantation system, and must now examine how these may be assembled.   There are three basic arrangements as illustrated schematically in figure 28.

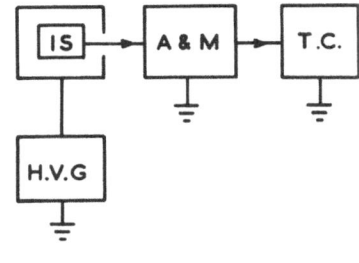

Convenient layout , access restricted to ion source only. Large analyser required for high energy beams, machine performance commonly deteriorates at low energies. Energy programming of ion beam requires analyser adjustment .

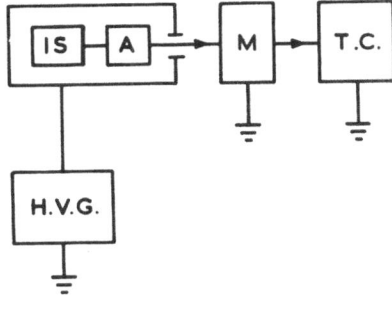

Analysis at constant low energy.
Small analyser adequate. No deterioration of performance over wide energy range.
Beam energy programming by H.T. adjustment only.
Analyser or beam analysis flight tube must be at high voltage .

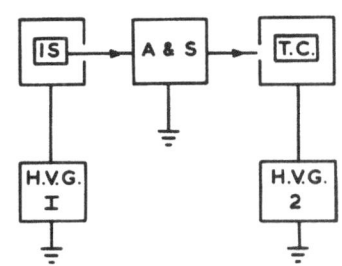

Restricted access to target but convenient method of increasing accelerator energy range. Target does not require high stability H.T. supply.
Beam energy programming by target H.T. adjustment only. Wide energy range with small analyser.
High resolution and intense beam currents readily obtained.

Fig. 28   Three alternative arrangements of the components of an ion implantation system

In the conventional system the ion source is at a high positive potential while mass analysis is carried out at ground potential. Since the particles have the full accelerated energy the magnet may need to be rather large and expensive.

This is avoided in the second design, in which mass analysis is carried out at a constant low energy, say 20–25 keV, with a small magnet operated at high potential. The symbol A in the diagram indicates the acceleration system and M the magnet, while T.C. represents the target chamber.

In the third part of the figure is shown a less common arrangement which has certain advantages for industrial applica-

tions of ion implantation. It is not appropriate for on-line
experimental measurements which require access to the target.
Both ion source (IS) and target chamber are at high potential,
and only the acceleration, mass analysis and beam scanning components are at or near ground potential. This design is particularly
suitable for high beam currents because the second high voltage
generator (H.V.G.2) needs to supply only the current appropriate
for the mass-analysed beam.

There are many instances in the industrial application of
ion implantation when it may not be necessary to apply any mass
analysis at all. Thus, if a gaseous ion species (eg. nitrogen)
is to be implanted into a metal and if small amounts of possible
contaminant species can be neglected, then the arrangement shown
in figure 29 may be used. Here the entire output of the ion source,
comprising atomic and molecular ion species, together with
neutrals, is directed towards the workpiece perhaps with only a
single-gap acceleration lens. The transmission properties are
excellent and the arrangement is attractively simple if the
requirements of the ion source can be met. It is, for instance,
much more difficult to achieve a beam of metal ions by vapourizing
the material and ionizing it: in most cases such high temperatures
are avoided by use of a volatile compound of the metal. Only when
the other constituent of the compound can be tolerated in the
target can this be adopted in the system shown here.

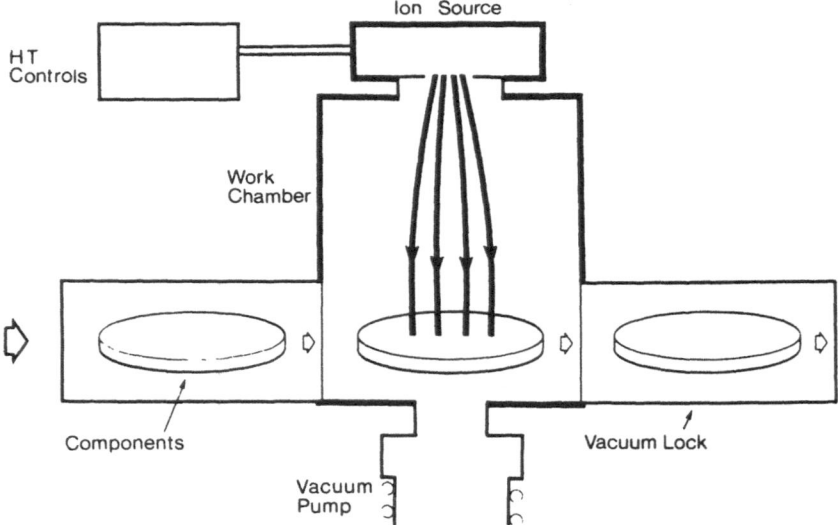

*Fig. 29   A simple arrangement for use when mass analysis is
          unnecessary*

A few examples of complete installations are illustrated in the following few figures. A research facility may be based upon a relatively high voltage accelerator such as a Van de Graaff or Cockcroft-Walton accelerator as shown in figure 30.

Alternatively, for work at lower energies the design may be based upon an isotope separator.[12] Because of their high current capability these have tended more and more to form the basis for industrial implanters supplied for the requirements of the semiconductor device industry[14] (figure 31).

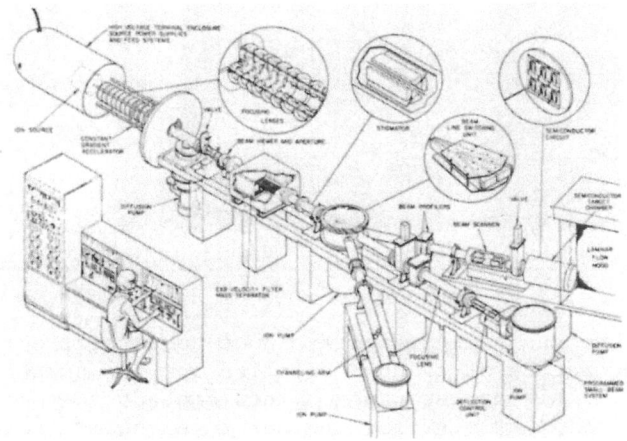

Fig. 30 A 300 keV research facility for ion implantation, with three experimental beam lines

Fig. 31 A high-current ion implanter developed for semiconductor applications

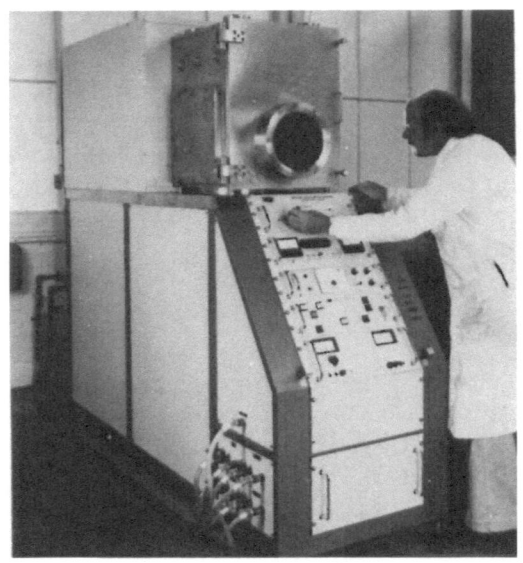

Fig. 32 A prototype implanter for engineering applications

This has become an important field of commercial activity, with over 500 accelerators now installed for the purpose of ion implantation: this number probably now exceeds that of all other particle accelerators e.g. for nuclear research or isotope production, and it will certainly increase. A simplified machine of the type illustrated schematically in figure 29 and designed for the treatment of engineering tools and components[15] is shown in figure 32. It provides up to 5 mA of nitrogen ions at energies up to 100 keV and may be considered as the prototype of a family of larger machines which are shortly to be commercially available. We expect them to prove just as versatile as have implanters for semiconductor device fabrication.[16]

REFERENCES

1. G. Sidenius, Proc. Conf. on Low Energy Ion Beams, Salford, Inst. of Phys. Conf. Series 38 (1978) p. 1.

2. A. von Engel, "Ionized Gases" (Oxford Univ. Press, Oxford 1965).

3. R.G. Wilson and G.R. Brewer, "Ion Beams" (J. Wiley & Sons, N.Y. 1973).

4. C.J. Cook et al., Rev. Sci. Instr. 33, 649 (1962).

5. H.L. Daley, J. Perel and R.H. Vernon, Rev. Sci. Instr. 37, 473 (1966).

6. J.H. Freeman, Nucl. Instr. & Meth. 22, 306 (1963).

7. K.J. Hill and R.S. Nelson, Nucl. Instr. & Meth. 38, 15 (1965).

8. P.D. Goode, Proc. Conf. on Low Energy Ion Beams, Trento, 1978
    (to be published).

9. J.Druaux and R.H. Bernas, "Electromagnetically Enriched Isotopes",
    M.L. Smith (Ed.) (Butterworth, London, 1956).

10. A. Galejs and P.H. Rose, "Focusing of Charged Particles"
    A. Septier (Ed.) (Academic Press, N.Y. 1967).

11. A.P. Banford, "The Transport of Charged Particle Beams" (Spon
    Ltd., London, 1966).

12. J.H. Freeman, in "Ion Implantation" (North-Holland, Amsterdam,
    1973).

13. G. Robertson (Bell Laboratories) U.S. Patent application.

14. Lintott Engineering Co. Ltd., Horsham, U.K.

15. G. Dearnaley, Trans. Inst. of Metal Finishing, 56, 25 (1978).

16. G. Dearnaley, Proc. 7th IVC & 3rd ICSS, Vienna (1977) p.1405.

RADIATION DAMAGE AS PROBED BY IMPURITIES

R. Coussement

University of Leuven
Instituut voor Kern- en Stralingsfysika
Celestijnenlaan 200 D, B-3030 Heverlee, Belgium

I would like to explain, at the beginning of this school, why and how implantations combined with hfi studies can be very useful to obtain complementary information on radiation damage studies. When an impurity is implanted a defect cascade is produced. The impurity lands in the lattice, most of the time in a substitutional site with only a small probability to have a defect on his nearest neighbours. The predominant landing can exceptionally be an interstitial depending on the host impurity combination.

Let us consider the case where the substitutional landing is predominant. The result of the implantation procedure will not always show up the landing probabilities, because after the landing defects can be trapped or detrapped (by annihilation by an anti-defect or dissociation from the impurity).

We will now discuss the trapping of a defect by an implanted impurity. We therefore consider the case of an oversized impurity, (for example Xe in iron, Mo, Wu, etc.). The volume of free Xe atoms is about 70 A^3 while the volume of iron is only 25 A^3. Implanting a Xe atom in iron is really a brute force method and results in a situation where the Xe atom is strongly compressed. This large compression induces an increased s-electron density at the nucleus; as indicated by the very large isomer shifts[1]. The lattice around the impurity must be strongly deformed or compressed. We can even assume that the deformation will not be isotropic. We can expect the deformation along the <111> directions to be different than the one around <100> directions.
Due to this deformation, isotropic or not, the presence of a vacancy can release the pressure and thus the total lattice energy will be decreased by trapping a vacancy in the nearest neighbourhood of

the impurity.
We can expect that a trapping volume exists around that impurity,
and that a vacancy entering that volume will be trapped further
to the nearest neighbours without further thermal activation. We
can imagine a potential well extending to the 3rd, 4th or 5th lattice dimensions and in which the defect is trapped. The result of
this trapping volume is that only the configuration of impurity
with monovacancy is the one with the vacancy in a nearest neighbour lattice point and never in the second, third, etc. This is
a very fortunate situation as the number of possible configurations
of impurity-vacancies is strongly reduced.
We can therefore speak of a reduced number of sites.

    1. The substitutional without any vacancy around, we notice
       IV_1
       I for impurity, V_1 for the missing host atom where the
       impurity is situated.
    2. A substitutional with a monovacancy trapped, we notice IV_2
       V_2 for one vacancy associated at the first neighbours of
       the impurity and for one host atom missing.
       We expect that the impurity strongly relaxes to the vacancy.
    3. Substitutional with two vacancies in the first neighbours,
       we will notice IV_3
       We expect that the impurity will relax to about the gravity
       center of the three missing host atoms.
    4. Etc..., IV_4...

    The question raises how strong the vacancies are bound and
how much vacancies can be trapped. Calculations, using interatomic
potentials have been performed to estimate the binding energy of
the last vacancy of the impurity-vacancies combinations. Fig. 2
shows the binding energy of the last vacancy of a IV_m combination,
calculated for Kr in tungsten. The large binding energy in KrV_1
means the large energy to move the substitutional Kr atom. In KrV_2

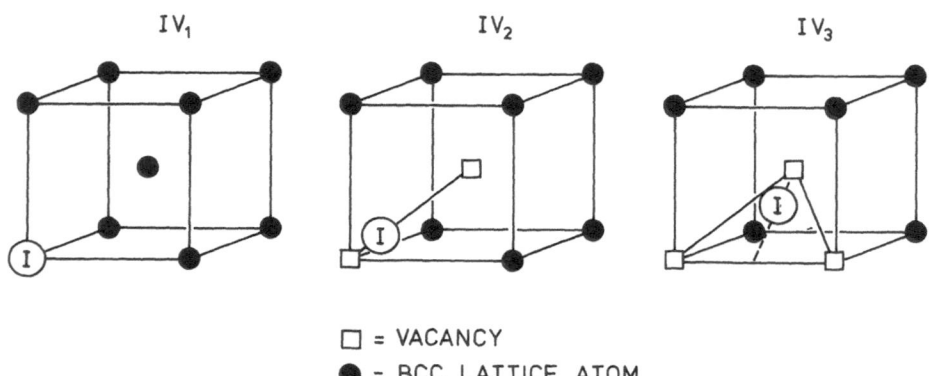

$\square$ = VACANCY
⬤ = BCC LATTICE ATOM

Figure 1

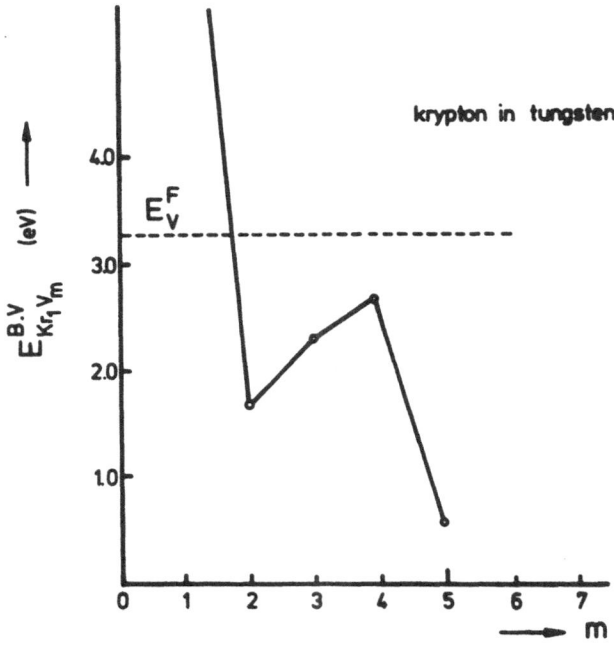

Figure 2

we have the binding energy of a monovacancy to Kr, etc. It is somewhat surprising to note that KrV$_3$ and especially KrV$_3$ shows stronger
stability than KrV$_2$ and KrV$_5$. This can be due to symmetries of the
combination. In fact KrV$_3$ and KrV$_4$ shows threefold symmetry while
KrV$_2$ has only twofold symmetry.

     The different sites are differentiated by the hyperfine inter-
action parameters (magnetic hyperfine field, fieldgradient-tensor,
isomershift). As the contribution to the hyperfine fields de--
creases very strongly with distance ($1/r^3$), only changes in the
first neighbourhood can change the hyperfine interaction parameters
and thus the hfi methods only probe the nearest neighbourhood of
the implanted impurity. The results will not be influenced by defects far away in the lattice, but will be very sensitive to show
trapping or detrapping of defects.

     We can distinguish two mechanisms for the defect to enter the
trapping volume. First of all defects can be mobile if temperature is high enough and can in that way enter the trapping volume.
We call it trapping by thermal diffusion [3]. Secondly a defect
can be produced by a subsequent implantation in the trapping volume.
We call it trapping by overlap of damage cascades[4]. The two mechanisms can be competitive. However, in most cases one is pre-

dominant. In each mechanism defects anneals when they enter the
trapping volume of the antidefect. Therefore the defect density
can reach an equilibrium. Defect density (n_d) change is given by
the relation

$$\frac{dn_d}{dt} = - \lambda n_d + k\Phi$$

where λn_d reveals the annihilation, $k\Phi$ is the production of defects,
which is proportional to the dose rate.
k is the mean number of defects pair per implanted ion.
λ depends on the reaction order. We notice

$$\lambda = k_1 n_d^{\gamma-1} e^{-A_i/kT}$$

A_i is the diffusion energy for the antidefect, and γ is the reaction
order :

　　　$\gamma = 1$ means recombination of correlated pairs of defects
　　　$\gamma = 2$ means annihilation only by uncorrelated defects.

We assume an equilibrium state is reached in which the defect density remains constant.
Then

$$\lambda n_d = k\Phi$$

or

$$n_d^\gamma = \frac{k}{k_1} \Phi\, e^{A_i/kT}$$

The rate equations for the populations of different sites can then
be given as :

$$\frac{dn_1}{dt} = P_1\Phi + \sum_{a=1}^{N} W_{a1} n_a - \sum_{b=1}^{N} W_{1b} n_1$$

$$\frac{dn_2}{dt} = P_2\Phi + \sum_{c=1}^{N} W_{c2} n_c - \sum_{d=1}^{N} W_{2d} n_2$$

$$\vdots$$

$$\frac{dn_N}{dt} = P_N\Phi + \sum_{q=1}^{N} W_{qN} n_q - \sum_{r=1}^{N} W_{Nr} n_N$$

P_i are the landing probabilities on site "i"
W_{ai} is the probability that a site "a" is transformed by a trapping
mechanism in a site "i".
Solutions of this set of equations have been given by Odeurs [3,4] et
al. and some general conclusions can be drawn.

At low doses the relative site populations reflect the landing probabilities. The implanted atom will not have the possibility to
trap a defect. At higher doses the trapping mechanism erases the
landing situation and a saturation situation is reached. Then the
relative site populations reflect the trapping and detrapping mechanisms. These saturation values are temperature dependent in the
case of trapping and detrapping by thermal diffusion of defects,
while they will be independent of temperature if the overlap mechanism is predominant.
At intermediate doses the relative populations can depend on all
parameters, dose, dose rate and temperature. In this region implantation results can only be compared if the implantation parameters are well controlled and known.

   We can conclude that by changing the implantation parameters
one can change the relative populations to some extent, and that
the final resultant relative populations are strongly affected by
trapping and detrapping of defect or defect clusters.
The final resultant relative populations can be changed by the subsequent treatements. One can postimplant ions, postirradiate with
electrons or neutrons. The defects produced during postirradiation
can change the relative populations.
Thermal annealing procedures can change the relative populations
too, and reveal the annealing stages. These results can be connected to resistivity measurements. As the hfi parameters can
reveal the nature of the site, these measurements can be of great
help to identify the nature of the annealing stages in metals.
Once the selectivity to trap vacancies of interstitials is known,
annealing traps can easily be identified.

REFERENCES

1) I. Dézsi, R. Coussement, G. Langouche, H. Pattyn, S. Reintsema,
   M. Van Rossum and J. De bruyn, Proceedings of Tokyo Conference,
   Sept 1978, to be published.
2) A. Van Veen, A. Warnaar and L.M. Caspers, 1978, unpublished.
3) J. Odeurs, R. Coussement and H. Pattyn, Hyp. Int. 3 (1977) 461.
4) J. Odeurs, H. Pattyn and R. Coussement, accepted in Radiation
   Effects.

Part II

Characterization by Nuclear
Methods of Isolated Implanted Atom Sites

MAGNETIC HYPERFINE INTERACTIONS IN IMPLANTED SYSTEMS

N. J. Stone

Clarendon Laboratory, Oxford OX1 3PU, U.K.

The overlap of interest between ion implantation and studies of hyperfine interactions (hfi) in solids has been clear for a decade. Established lines of research involving magnetic inter-actions have centred on the ferromagnetic metal hosts Fe, Co, and Ni into which ions of many elements have been implanted both by recoil following nuclear reactions and using electromagnetic separators. Interest in radiation damage effects studied through hfi in these metals is enhanced since one is bcc (Fe), one fcc (Ni) and Co can be obtained in both hcp and fcc forms.

Experience shows that normally soluble implants, for local doses of order 1 at.% and implanted beyond any surface oxide layer generally show substitutional fractions in excess of 80%, and for such systems implantation has been primarily valuable as a source preparation technique which is fast (stopping times $\sim 10^{-12}$ s) and without chemical complication. Studies of the 'transient fields' experienced by nuclei during slow down have been tied directly to detailed aspects of electronic stopping, but that field will not be covered here. By contrast to the soluble ions, insoluble ion implants have generally shown evidence, in both channelling and hfi studies, of a multiplicity of sites even at the lowest implant doses.

In keeping with the activities of a summer school an attempt has been made to outline a derivation of magnetic hfi in solids from first principles. However emphasis has naturally fallen on those aspects most relevant to investigation of implanted systems. The most concentrated efforts have been directed to studies of the sp impurities, of which the halogens and inert gases are the most clearly insoluble. A second area of activity concerns implanted

rare earth ions. Although these ions have strong hfi which are
highly sensitive to site symmetry and the chemical state of the
ion, the interaction is more complex and to date no clear systematic behaviour has emerged.

The text emphasises current theoretical ideas, however at the
outset it should be clear that a priori calculations of magnetic
hfi in these systems are not yet possible. The ideas serve to
correlate experimental systematics and to assist in relating hfi
measurements with other implant location techniques. The starting
point of the paper is magnetic hfi in free atoms, followed by a
discussion of crystal field effects in solids, conduction electron
polarisation, and overlap interaction terms at non-magnetic ions
in magnetic metals. The main thrust is aimed to give an apprec-
iation of the origins of magnetic hfi at sp impurities in the 3d
magnetic metals and a survey is made both in experiment and theory
of the sensitivity of magnetic hfi in these systems to ionic site
and surroundings. Theoretical complications and experimental
experience concerning rare earth implants .in ferromagnetic metals
are summarised in the concluding section.

MAGNETIC hfi IN FREE ATOMS

(a) Interaction with the Electron Orbital Motion

An electron current density $\underline{J}_e dr$ at point $\underline{r}$ in a co-ordinate
system centred at the nucleus will produce (by Ampere's Law) a
magnetic field at the nucleus

$$\underline{B}_{dr} = \frac{\mu_o}{4\pi} \frac{\underline{r}_\wedge \underline{J}_e dr}{r^3}$$

Writing $\underline{J}_e dr = \underline{v} dq$, where $\underline{v}$ is the velocity of the charge element
dq, and noting that $\underline{r}_\wedge \underline{v}$ is related to the electron angular momentum
$m_o(\underline{r}_\wedge \underline{v}) = \underline{\ell}$, we can write, for each electron

$$\underline{B}_{\ell_i} = \frac{\mu_o}{4\pi} \int \frac{\underline{r}_\wedge \underline{v}}{r^3} dq = -\frac{\mu_o}{4\pi} \frac{\underline{\ell}_i \hbar}{m_o} e\langle r_{\ell_i}^{-3}\rangle$$

$$= -\frac{\mu_o}{4\pi} 2\beta\underline{\ell}_i \langle r_{\ell_i}^{-3}\rangle$$

where $\beta = \hbar e/2m_o$ is the Bohr magneton, and e and β are taken as
positive numbers. Electrons in the same atomic n,ℓ subshell will
have the same value of $\langle r_\ell^{-3}\rangle$ so that, for a closed shell the

summation

$$\underline{B}_{\ell} = -\frac{\mu_o}{4\pi} 2\beta <r_{\ell_i}^{-3}> \sum_i \underline{\ell}_i = 0$$

and only unclosed shells contribute. In simple LS-coupling we can then write the interaction energy with the nuclear dipole moment $\underline{\mu}_I$ as

$$W_L = -\underline{\mu}_I \cdot \underline{B}_L$$

$$= \frac{\mu_o}{4\pi} 2\beta (\underline{\mu}_I \cdot \underline{L}) <r_{\ell}^{-3}>$$

$$= \frac{\mu_o}{4\pi} 2\beta (\frac{\mu_I}{I}) <r_{\ell}^{-3}> \underline{I} \cdot \underline{L}$$

(b) Interaction with the Electron Spin

To evaluate this contribution we must consider separately the interactions with electrons 'outside' and 'inside' the nucleus.

(i) Outside term   This gives rise to a simple dipole sum field

$$\underline{B}_{s_i} = -\frac{\mu_o}{4\pi} \int \frac{\underline{M}_{s_i} - 3(\underline{M}_{s_i} \cdot \underline{r}_o)\underline{r}_o}{r^3} d\tau$$

where $\underline{M}_s$ is the spin magnetisation density at $\underline{r}$ and $\underline{r}_o$ is a unit vector along $\underline{r}$. Writing $\underline{M}_s = - g_s\beta \sum_i \underline{s}_i \rho_{s_i}$ where $\underline{s}_i$ is the spin of the ith electron and ρ_{s_i} is its spin density $|\psi(\underline{r})|^2$ gives

$$\underline{B}_{s_i} = \frac{\mu_o}{4\pi} g_s\beta \sum_i \left[\underline{s}_i - 3(\underline{s}_i \cdot \underline{r}_o)\underline{r}_o\right] <r_s^{-3}>_i$$

The nuclear dipolar interaction energy is then

$$W_s = -\frac{\mu_o}{4\pi} g_s\beta \frac{\mu_I}{I} \sum_i \{\underline{s}_i \cdot \underline{I} - 3(\underline{s}_i \cdot \underline{r}_o)(\underline{r}_o \cdot \underline{I})\} <r_s^{-3}>_i$$

This interaction vanishes when the spin-density distribution is spherical, i.e. for a closed or half-filled sub-shell (in LS coupling), or for any s electron.

(ii) The 'Inside' term   Classically the easiest way to visualize this interaction is to regard the nuclear moment as a current loop in a medium of uniform spin magnetisation density $\underline{M}_o$. Such a magnetisation produces a magnetic field

$$B_c = \frac{2}{3}\mu_o \underline{M}_o = \frac{\mu_o}{4\pi} \cdot \frac{8\pi}{3} \underline{M}_o$$

Since we are considering a point nucleus at the origin, only s electrons contribute to this magnetisation density, and we may write

$$M_o = -g_s\beta\underline{s}|\psi_s(0)|^2, \quad B_c = -\frac{\mu_o}{4\pi} \cdot \frac{8\pi}{3} g_s\beta|\psi_s(0)|^2\underline{s}$$

where $|\psi_s(0)|^2 = \rho_o$ is the electron spin density at the nucleus. The energy associated with this interaction is then

$$W_c = \frac{\mu_o}{4\pi} \cdot \frac{8\pi}{3} g_s\beta|\psi_s(0)|^2\underline{\mu}_I\cdot\underline{s}$$

and it is known as the Fermi 'contact' term.

TOTAL MAGNETIC INTERACTION IN FREE ATOMS

The preceeding terms form the framework in which free atom magnetic dipole hfi is discussed; however there are two further considerations, namely the importance of multielectron interactions and the separability of the various contributions to the total.

CORE POLARISATION

So far we have considered a single electron theory, however its inadequacy is shown by the experimental observation of large magnetic hfi for, in particular, Mn^{++} ($3d^5$, L=0, s=5/2 but spherical) for which, taking only the 3d electrons, all terms above are zero. The origin of the observed splittings was established following a suggestion by Sternheimer that interactions between the spin of the outer 3d electrons and inner, paired, core electrons could produce a net unpaired spin density at the nucleus and hence a magnetic split-ting through W_c. The mechanism is the exchange interaction between electrons of parallel spin. Quantitative multielectron calculations of any sophistication tend rapidly to exceed computer capabilities however inclusion of exchange interactions between all pairs of electrons, and relaxation of the assumption that, for example, both ns electrons should have the same radial wavefunction, have confirmed the importance of this exchange polarisation of inner electrons in atomic and solid state hyperfine interactions[1]. Physically it is found that the spin polarised outer unpaired electrons tend to attract core electrons of parallel spin towards the outer regions of the atom, leaving, at the nucleus, an excess of antiparallel spin and an appreciable contact interaction. This effect is known as 'core polarisation'.

HYPERFINE ANOMALY

For a simple point nucleus the algebraic sum

$$W_T = W_L + W_s + W_c$$

is the only measurable, however this is not the case for a real, finite, nucleus. W_L and W_s are insensitive to nuclear volume since they involve magnetic fields uniform to high order over so small a volume at the origin. However W_c is more correctly given, in terms of the nuclear magnetisation $\underline{\mu}_I(\underline{r})$ and the s electron wavefunction $\psi_s(\underline{r})$, by

$$W_c = - \int \underline{B}_c(\underline{r}) \cdot d\underline{\mu}_I(\underline{r}) d\tau$$

where the integral is over the nuclear volume. The variation of B_c is determined primarily by the nuclear charge distribution through $|\psi_s(\underline{r})|^2$, and for a single element different isotopes and excited nuclear states may have quite different radii and charge distributions. Furthermore the nuclear moment distribution is isotope and level dependent. Thus we have, for two isotopes $\int d\underline{\mu} \cdot dr = \mu_{1,2}$ and the ratio of the hyperfine splittings

$$\frac{W_{c_1}}{W_{c_2}} \cdot \frac{I_2}{I_1} = - \frac{\int \underline{B}_c(\underline{r}) \cdot d\underline{\mu}_I(\underline{r}) d\tau_1}{\int \underline{B}_c(\underline{r}) \cdot d\underline{\mu}_I(\underline{r}) d\tau_2} \cdot \frac{I_2}{I_1} = \frac{\mu_1}{\mu_2} \cdot \frac{I_2}{I_1} (1 + {}^1\Delta^2)$$

where ${}^1\Delta^2$ is known as the 'hyperfine anomaly' between the two isotopes and is typically of order 0.1-1%. Δ is different from zero only for s-electron contact interactions and its measurement for the same pair of isotopes in different environments can give information regarding the relative magnitudes and signs of W_c and the sum $W_L + W_s$ [2].

MAGNETIC hfi IN SOLIDS

Although the great majority of free atoms exhibit magnetic hfi, most solids do not. Chemical bonding, primarily electrostatic in origin, tends to produce systems with zero resultant spin and orbital angular momentum. In those systems which do show magnetic properties in the solid state, normally involving atoms with electrons in unfilled d or f electron subshells, the effect of neighbouring ions is first and foremost to replace the spherical symmetry of the free ion potential with a potential which reflects the particular symmetry of the local surroundings. The effects of this

potential vary widely with the magnitude of its non-spherical components and the extent to which the 'magnetic' sub-shell electrons
are involved with chemical binding or are localised at the magnetic
ion site.

When the magnetic electron levels are not strongly broadened by
admixture with other states, the effect of its surroundings on a
magnetic ion may be expressed in terms of a 'crystalline electric
field'. Depending upon the relative strength of the L.S. coupling
to resultant J (which varies from $\sim 10^2$ cm^{-1}, or 10^{-4} eV, for 3d
ions to $\sim 10^3$ cm^{-1} for 4f ions, the splitting of the various LS
terms ($\sim 10^4$ cm^{-1}), and the crystal field, the resulting states may
be described in terms of J (for 4f electrons, which being well
screened in the ion have small crystal field effects $\sim 10^2$ cm^{-1}), or
L and S (when the crystal field is $>$ L.S coupling as for the more
exposed 3d electrons) or may require a fresh coupling of single
electron ℓ, s states. The 4f ion case is considered later in the
paper, however concentrating on the 3d ions, when L and S remain
good quantum numbers, the crystal field mixes the various L_z states
resulting in linear combination states, generally non-degenerate, in
which, because their spatial distributions are identical, $\pm L_z$
states have equal amplitudes so that the expectation value $\langle L_z \rangle = 0$.
Furthermore in many cases the ground state crystal field multiplet
is either singlet or does not contain elements differing in L_z by
one unit. This means that an applied magnetic field has no effect
in first order, and the orbital magnetism of the free ion is lost,
or 'quenched', by the action of the crystal field.

The electrostatic crystal field does not act in spin space and
thus changes in the orbital states do not affect the electron spin
contribution to the electronic magnetism. However the spin-dipolar
magnetic hfi is strongly dependent upon the spatial distribution of
the spin density, which the crystal field does affect. In particular the spin-dipolar hfi is zero if the spin distribution has
cubic symmetry.

When the magnetic electrons are more strongly involved in chemical bonding their reduced localisation generally leads to a reduction in orbital magnetism. However the direct admixture into wavefunctions involving neighbouring 'non-magnetic' ions can be respons-
ible for the transfer of magnetisation to the neighbours and transferred magnetic hfi.

In the 3d ferromagnetic metals the degree of localisation of
the 3d electrons is not clearly established. Both a full band
structure and an ionic picture indicate full orbital quenching and
magnetism based on electron spin polarisation. As in the free ion
situation, exchange interaction effects are most important, both
for determining the hfi at nuclei of the magnetic ions and more
fundamentally, for giving theoretical understanding of the onset of

magnetic ordering at several hundred ^{o}K through exchange interactions between neighbouring ions. The fact that ferromagnetic, as
opposed to antiferromagnetic, ordering occurs predominantly in
metals indicates the importance of conduction electron polarisation
by exchange with the magnetic ions. This polarisation is also of
fundamental importance to hfi at non-magnetic ions implanted into
Fe, Co and Ni since the conduction band is predominantly s-electron
like, giving a direct contact hfi, including core s-electron
exchange polarisation, at the implant nucleus. With orbital
quenching, and in cubic symmetry, this contact interaction, with
the possible addition of direct polarisation of implant ion electron
states by admixture with neighbour magnetic electron states, is the
dominant source of hfi at non-magnetic implant nuclei. Conduction
electron polarisation and volume overlap effects are discussed in
the following sections.

CONDUCTION ELECTRON POLARISATION

     Basic scattering theory is the formalism for describing the way
in which a regular array of magnetic ions in a metal affects the
properties of the conduction band. Only the bare bones are given
here as many more complete treatments are available[3].

     Consider first a single scattering centre, at the origin and of
radius R, and look for an asymptotic description of the effect, at
distance >> R, on a beam of particles of momentum p. Spherical
symmetry of the scattering centre indicates a partial wave analysis
and the highest angular momentum involved is given by $\hbar\ell = pR = \hbar kR$.
Taking the simplest case kR >> 1 the scattering will be described in
terms of $\ell = 0$ (s wave) only. The incident beam wave function
$u_k(r,\theta,\phi) = e^{ikz} = e^{ikr\cos\theta}$ has an $\ell = 0$ component given by the
overlap integral

$$u_{ks}(r) = \frac{1}{4\pi} \int u_k(r,\theta,\phi) Y_o^o(\theta,\phi) \, d\Omega$$

$$= \frac{1}{4\pi} \int_o^{2\pi} d\phi \int_o^{\pi} e^{ikr\cos\theta} \sin\theta \, d\theta = \frac{e^{ikr} - e^{-ikr}}{2ikr}$$

a linear combination of incoming and outgoing waves of equal amplitude. Scattering will modify the outgoing part, so the s-wave
scattered wave function must have the form

$$u_s(r) \sim \frac{e^{ikr} - e^{-ikr}}{2ikr} + f \frac{e^{ikr}}{r} \sim \frac{Se^{ikr} - e^{-ikr}}{2ikr}$$

where S contains all the information about the scattering. The form
of S is found by requiring a steady state solution, i.e. incoming
flux equal to outgoing flux. This is equivalent to writing

$|e^{-ikr}|^2 = 1 = |Se^{ikr}|^2 = |S^2|$, thus we have $|S^2| = 1$ and hence may

take $S = e^{2i\delta}$, from which $f = \dfrac{e^{i\delta}\sin\delta}{k}$. We thus have the result

that, for $r \gg R$, the effect of the scattering centre is to shift
the phase of the outgoing wave relative to the incoming wave with a
single parameter, the phase shift δ, containing all the dependence
upon the scattering centre.

This form of analysis is quite general and, for any given
potential, the effect of scattering on the wave form, for $r \gg R$,
may be described by a complete set of phases shifts δ_ℓ, one for each
ℓ-wave, limited by $\ell_{max} = k_{max}R$. The form of the scattered wave
within the scattering region can be obtained only for specific
potential wells, e.g. spherical Bessel functions for square wells,
with solutions joined at $r = R$. Results have the general form shown
in figure 1 and the general theory is given in many quantum mechanics
texts.

This calculation was for a single scattering centre. Its relev-
ance to magnetism in metals and to impurities in iron metal in particular is two-fold.

(1) RKKY conduction electron polarisation. These ideas were first
associated with nuclear moment scattering of conduction electrons in
metals by Ruderman and Kittel[4], with extensions to ordered electronic moment systems by Kasuya[5] and Yosida[6]. In iron, s-like
conduction electrons of spin $s_\uparrow$ parallel to the excess iron ion $d_\uparrow$
electron polarisation experience, through the exchange interaction

$E_{ex} = -\dfrac{2S}{N} J(k,k)\underline{S}_1 \cdot \underline{S}_2$, a deeper scattering potential than the anti-

parallel $s_\downarrow$ electrons. Thus $s_\uparrow$ and $s_\downarrow$ have different phase shifts
and the result of summing the two scattered amplitudes is to give a
net polarisation of the $s_\downarrow$ electrons ($s_\uparrow - s_\downarrow$), which is parallel to
the iron $d_\uparrow$ at the iron site but which oscillates with distance from
that site (figure 2). For a full band of electrons a summation over
k must be carried out, but the asymptotic region result can be shown
to depend only upon $\delta_\ell(k_F)$ at the Fermi level.

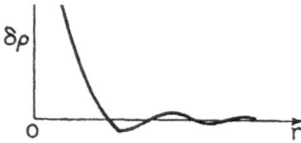

Fig. 1: Oscillations of conduction electron density electron distribution around an impurity atom[9].

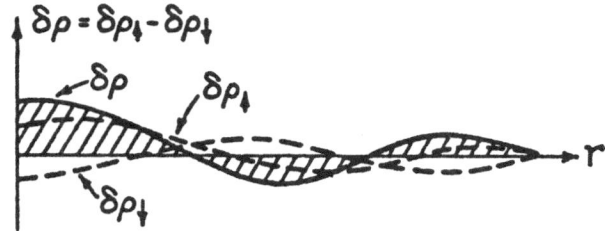

Fig. 2: Oscillations of conduction electron polarisation arising from scattering by a magnetic moment at the origin[9].

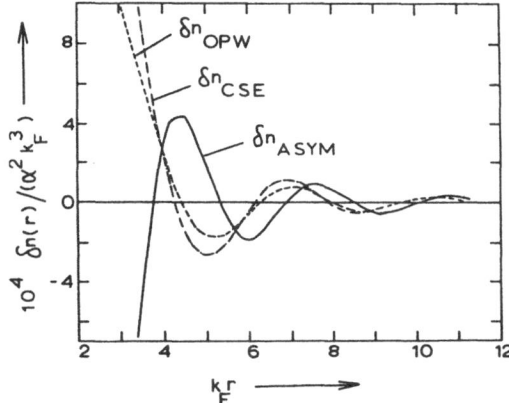

Fig. 3: Results from ref. 7 of detailed electron polarisation density distribution calculations (OPW - orthogonal plane wave, CSE - constant self energy) for a single iron impurity, compared with the asymptotic calculation. $\delta n = n_{s\uparrow} - n_{s\downarrow}$.

Quantitatively this theory has many problems which have been widely discussed in the literature. These include the use of realistic, k dependent, exchange potentials on the Fe sites and of other than free electron conduction band wavefunctions. These modifications serve to alter the effective phase shifts $\delta_{\ell}(k_F)$. In particular Jena and Geldart[7], figure 3, have been able to obtain the same sign of polarisation at the first nearest neighbour site as at the origin, in iron - as is required for a ferromagnetic coupling of ionic spins via this interaction.

(2) <u>Impurity scattering - charge screening effects</u>. Friedel[8] pointed out that an impurity in a metal generally represents a charge anomaly which will be screened by, and will scatter,

conduction electrons. Following the phase shift analysis and summing the difference between the scattered and unscattered solutions
he obtained an expression for the excess charge as a function of
distance from the impurity and derived his sum rule for the phase
shifts $Z_{eff} = \sum_{\ell}(2\ell+1)\delta_{\ell}(k_F)/\pi$, where Z_{eff} is the effective impurity
charge, for each direction of spin in the conduction band.

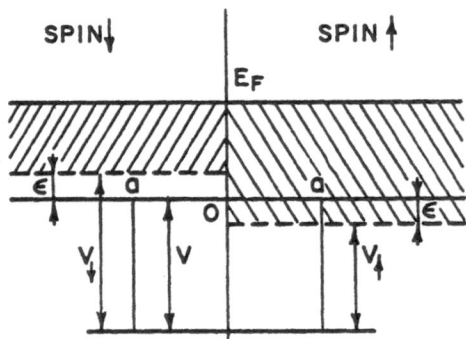

Fig. 4:   Effective potentials for conduction electrons at a non-
magnetic impurity dissolved in iron[9].

     Crudely Fe is considered to give up its single 4s electron to
the conduction band, and have an effective charge Z of 1, so that
$(sp)^n$ impurities in iron, considered to lose all n sp electrons,
have $Z_{eff} = n-1$. What this means for magnetic interactions at the
impurity is that the conduction electron polarisation oscillations
will be modified as a function of Z_{eff}, since the effective
potentials (figure 4) seen by conduction electrons $s_{\uparrow}$ and $s_{\downarrow}$ at the
impurity differ, but the difference, which is caused by the exchange
interaction, becomes less important as the Z_{eff} dependent term in
the potential (which affects $s_{\uparrow}$ and $s_{\downarrow}$ equally) increases.

                CONDUCTION ELECTRON hfi AT sp IMPURITIES IN IRON

     Earlier attempts (Friedel and Daniel[9]; Campbell[10]) to
incorporate impurity charge shielding into hyperfine interaction
theory calculated the effect of the charge dependent impurity potential upon the electron distribution of a plane wave conduction band
with uniform polarisation, in effect neglecting RKKY oscillations.
To remedy this Campbell and Blandin[11] used a model in which the
charge impurity was placed at the origin with a spherical potential
$V_0(r)$ (= 0 for r > r_0). Starting from an unpolarised conduction
band, instead of considering a magnetic neighbour as localised at
some position R they showed that the conduction electron

polarisation at the origin (i.e. at the impurity nucleus which is
what we want) was the same if the magnetic exchange interaction was
distributed uniformly on a spherical shell of radius R. This
spherically symmetric problem was solved using plane waves and
asymptotic phase shifts $\delta_{\ell\uparrow}$ and $\delta_{\ell\downarrow}$ to give the expression

$$m_{R_0} (r=0) = \frac{k_F}{2\pi^2} J\Omega_0 <S_z> |\phi_{k_F} (0)|^2 \cos(2k_F R_0 + 2\delta_0^{\ F}) / R_0^{\ 3}$$

for the conduction electron magnetisation density at the origin.

   The hfi follows from this expression under the following
assumptions:
(i)   that the s-electron contact hfi is dominant. This holds for
      s-p impurities but not for 'magnetic' 3d or 4f ions.
(ii) that polarisation effects from successive neighbour shells are
      additive. This appears reasonable since the addition of terms
      of the form A $\cos(\phi + 2\delta_0^{\ F})$ changes A and ϕ but not the cosine
      form. Also the total spin polarisation is of order a few %.
(iii) that at the impurity ion site the $\ell = 0$ conduction band elec-
      tron states are in effect states in the first unfilled s elec-
      tron shell for that impurity ion.

Under these assumptions, and with measured (free ion) and estimated
values of the hyperfine field strength A_z associated with one electron in the s shell involved as a function of impurity (e.g.
Campbell[9]) we have

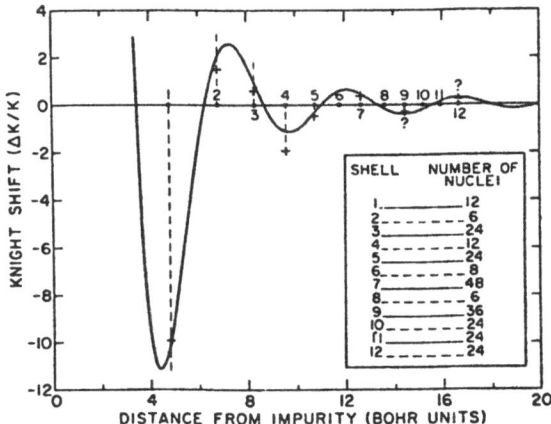

Fig. 5: CuMn Knight shifts ΔK/K vs distance from the Mn impurity;
experiment and theory[12]. The Knight shift is directly
proportional to the conduction electron polarisation.

$$B_c = (n_\uparrow - n_\downarrow) A_z = \frac{m(r=0)}{2\mu_B} A_z$$

After summation this is

$$B_c = \frac{J\Omega_o k_F}{8\pi\mu_B} \cdot \frac{\mu_o}{4\pi} \left\{ \sum_i (S_z(R_o^i) \frac{\cos(2k_F R_o^i + 2\delta_o^F)}{(R_o^i)^3} \right\} A_z$$

TESTS OF THE MODEL

(1) RKKY Oscillations.   For an isolated magnetic ion, Slichter et
al[12] using high sensitivity NMR have studied dilute CuMn alloys.
The Mn ion has Z_{eff} = 0 in this lattice so that charge shielding
is unimportant. Identification of very weak signals from Cu nuclei
in specific neighbour shells relative to the dilute (min 5 x 10^{-4}
at. %) Mn ions showed a spatial distribution of hfi at these Cu
nuclei as shown in figure 5. The figure also shows a theoretical
fit to this oscillatory dependence, however this is obtained only
after considerable development of the simple outline theory given
here.

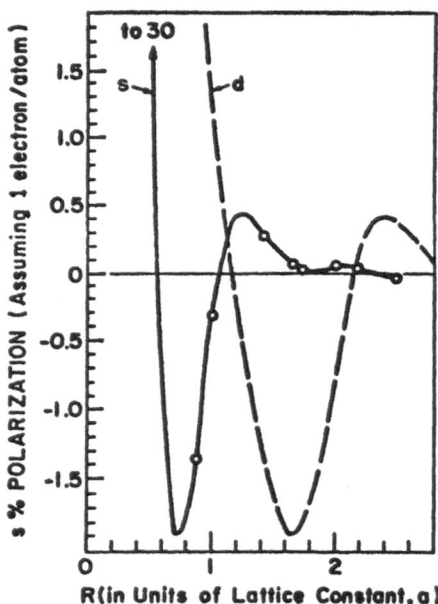

Fig. 6:   Polarisation of 4s-like electrons in iron as deduced from
NMR measurements on ordered $Fe_{1-x}Si_x$ alloys[13].

For an ordered ferromagnet Stearns[13] has studied the FeSi
system, which forms an ordered alloy with well defined numbers of
magnetic Fe ions in various neighbour shells. The hfi contributions
associated with the different shells are shown in figure 6, reveal-
ing a very similar oscillatory form to that in CuMn.

(2) Impurity charge dependence.   The Z_{eff} dependence of the conduction electron hfi at an sp impurity follows intuitively from
figure 4. The effective potential $V_{\downarrow}$ for conduction electrons of
spin opposite to the magnetisation is deeper than that for parallel
spin electrons $V_{\uparrow}$. However there are more $\uparrow$ conduction electron
states below the Fermi surface. Thus for small Z_{eff} the deeper $s_{\downarrow}$
potential leads to $\psi(0)_{\downarrow}^2 > \psi(0)_{\uparrow}^2$, i.e. B_c - ve, whereas for larger
Z_{eff} the potentials are both deeper and the spin dependent differ-
ence 2ε is less, so the hfi directly reflects the polarisation of
the conduction band and $\psi(0)_{\uparrow}^2 > \psi(0)_{\downarrow}^2$.

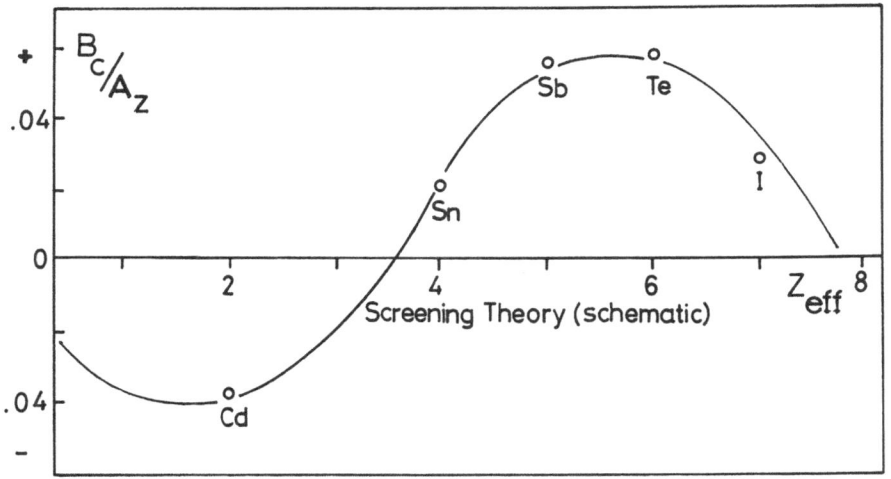

Fig. 7:   Variation of hyperfine field B_c/A_z with impurity effective
charge Z_{eff} for 5sp impurities in dilute Heusler alloys $Pd_2MnSb_xX_{1-x}$:
screening theory and experiment.

Both Daniel and Friedel[9] and Campbell and Blandin[11] predict
a change of sign of B_c in the region $Z_{eff} \sim$ 3-4, and a reduction of
B_c/A_z for $Z_{eff} > 6$. The best systematic experimental test to date
is found in hfi measurements for the 5sp elements in the Heusler
alloys Pd_2MnSb. The variation of B_c/A_z with Z_{eff} is shown in figure
7, with both the sign change at $Z_{eff} \sim 3.5$ and the fall at high Z_{eff}
being observed.

OVERLAP EFFECTS

The second important effect leading to transfer of electron polarisation in solids is admixture of neighbouring atom wavefunctions because of spatial overlap. The formalism is given by Freeman and Watson[14]. Shirley considered this effect for large sp impurity ions in iron, particularly Xe[15]. Considering only first order admixtures of outer (5s)Xe states with the excess spin up $3d_{\uparrow}$ host electrons, and noting that to maintain orthogonality these mix only with spin up impurity electrons, the perturbed 5sXe electron states are

$$|5s_{\uparrow}>' = \{|5s\uparrow> - <3d|5s>|3d\uparrow>\}/\{1-<3d|5s>^2\}^{\frac{1}{2}}$$

$$|5s_{\downarrow}>' = |5s_{\downarrow}>$$

where $<3d|5s>$ denotes the 'two centre' overlap integral of the 3d electron with origin on the Fe atom and the 5s electron with origin on the Xe atom.

The contact hfi due to the 5s electrons at the Xe nucleus is then

$$B_c = -\frac{\mu_o}{4\pi} \cdot \frac{8\pi}{3} g_s\beta_s \{(|5s_{\uparrow}>')_0^2 - (|5s_{\downarrow}>')_0^2\}$$

$$\approx -\frac{\mu_o}{4\pi} \cdot \frac{8\pi}{3} g_s\beta_s <3d|5s>^2(|5s_{\uparrow}>_0^2)$$

and after summing over the 3d electrons on all i surrounding iron atoms and multiplying by Δn_{3d} the excess of $3d_{\uparrow}$ over $3d_{\downarrow}$ electrons on the Fe ion (or more correctly after summing over all the occupied 3d states in the Fe conduction band) we have[16]

$$B_{hf}(\text{overlap}) = -\frac{\mu_o}{4\pi} \cdot \frac{8\pi}{3} g_s\beta_s (|5s_{\uparrow}>_0^2) \Delta n_{3d} \sum_i <3d|5s>_i^2$$

Recently Das[17] has further emphasised the need to evaluate such overlap interaction terms using Fe conduction band wavefunctions.

EVIDENCE FOR OVERLAP EFFECTS

Shirley has argued that, for high Z_{eff} the Daniel-Friedel – RKKY model must ultimately break down. This is because, as Z_{eff} increases and the total conduction electron density at the impurity rises to screen it, so the importance of the polarisation term in the impurity potential falls. Increasing Z_{eff} means adding bound electrons to the impurity site and as they sink into the potential they are more weakly polarised. Shirley found that to produce $B_c \sim 1000$ kG through $\rho_{\uparrow}^{5s} - \rho_{\downarrow}^{5s}$ required a total 5s density at the Xe nucleus 2.5 times that of the filled $5s^2$ shell.

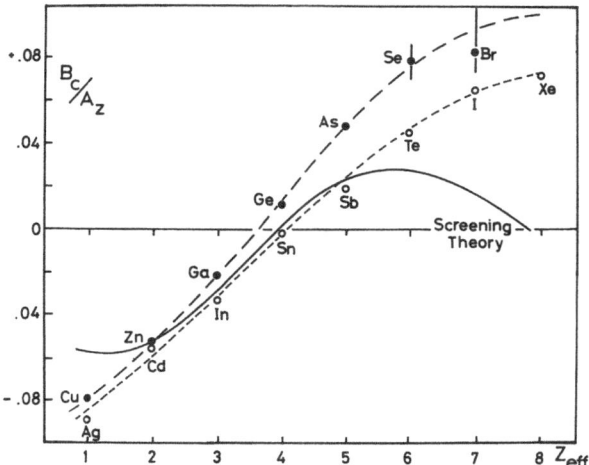

Fig. 8:   Variation of hyperfine field B_c/A_z with impurity effective charge Z_{eff} for 4sp and 5sp series in iron: screening theory
and experiment.

As discussed above, impurity charge screening theories predict
both a change of sign of B_c in the region $Z_{eff} \sim 3-4$, and a falling
off of B_c/A_z for $Z_{eff} > 6$. Experimentally in Fe, in contrast to the
Heusler alloys, the observed fields increase to Xe. Figure 8 shows
that the discrepancy between experiment and theory increases for
higher Z_{eff} in both the 4sp and 5sp series.

Thus there does appear to be a need for a second contribution
to B_c for just those impurities which have large ionic volumes,
(which rise sharply for I, Xe and Br, Kr) after the effects of
impurity charge compensation have been considered. Theoretical
estimates of this term[16] have ranged from Shirley's 3 Mg to
~ 0.1 Mg, the overriding problem being that of estimating the overlap integral in a metal. The result cannot depend upon ionic
volume alone, since the lattice constant of Fe is very close to that
of the Heusler alloys although the overlap term in the alloys is far
smaller.

IMPLANTED LANTHANIDE IONS IN FERROMAGNETIC METALS

A brief review of early work on implanted rare earths in
ferromangetic hosts was given by Campbell[18], a more extensive
recent account being that by Niesen[19]. Bleaney[20] has reviewed
hfi in lanthanide ions and metals.

As outlined earlier, in 4f ions the dominant hfi arises from orbital fields B_L. The 4f electrons are not strongly perturbed by chemical bonding interactions so that a crystal field treatment of the environment is appropriate and L, S, and J can be taken as good quantum numbers. The effect of the crystal field will be a Stark splitting of the (2J+1)-fold degenerate ionic ground state. A general treatment of these splittings exists[21] and although the strength of the interaction cannot be calculated a priori, there is considerable understanding of its variation through the rare earth series.

In a magnetic medium the 4f ion is also subject to exchange fields, introducing a term $B_{ex}\underline{m}.\underline{S}$ where B is a coupling constant, m is the host magnetisation and S the 4f ion spin. This term can be written[22] as a field $Bm(g_J-1)$ acting on J and in the absence of strong or comparable crystal field terms the ionic ground state will have $\langle J_z \rangle = J$ and close to free ion hfi.

Core polarisation and conduction electron polarisation terms are generally small and there has been no improvement on the early prediction of Freeman and Watson[14]

$$H_{cp} = -90(g_J-1)\langle J_z \rangle kOe$$

Conduction electron contributions are estimated from measurements on Eu^{++} and Gd^{+++}, both of which have half filled 4f shells, and on La and Lu, $4f^0$ and $4f^{14}$. The latter indicate $H_{cep} \sim \pm 100$ kG, however complications arise in the $4f^7$ cases because the overall cep is perturbed locally by the ionic spin which, for rare earth ions in 3d host metals, is antiparallel to the host magnetisation. Estimates have been given for cep terms in Eu metal and Gd metal consistent with a total contribution of order 100 Kg[20].

The use of rare earth implants to study hfi at substitutional and non-substitutional sites in 3d ferromagnetic metals has the initial advantage over sp implants that an 'upper limit' free ion interaction strength is known, corresponding to $\langle J_z \rangle = J$. However the fact that both cubic and non-cubic sites can yield this free ion result makes even tentative hfi-site combination assignments difficult. Mossbauer measurements have provided a greater part of the data available, as summarised by Niesen[19].   To date, channelling studies have not yielded detailed analysis of the character of non-substitutional sites, nor has NMR/ON been observed for rare earth implants. There has been a strong controversy[22,23] over the role of oxide formation at the implant site (internal oxydation), although this would appear to be a problem which is avoided if implant energies of 100 keV or greater are used.

EXPERIMENTAL IMPLANT SITE SENSITIVITY OF sp IMPURITY hfi IN Fe

Conduction electron polarisation induced hfi is dominated by the first four or five neighbour shells, and volume overlap interactions are only significant for the first two neighbour shells. Thus hfi is a sensitive probe of the local implant environment.

NMR/ON signals from both recoil implanted and separator implanted systems have shown linewidths no broader than thermally prepared sources for several soluble implants. Remarkably, the smallest fractional linewidth yet observed, 750 kHz at a resonant frequency of 682 MHz was observed for the insoluble implant I in Fe. This is direct evidence of a substitutional site for the impurity in a lattice undisturbed over at least four neighbour Fe shells, occupied by at least 40% of the total I implant dose (see chapter p. 177).

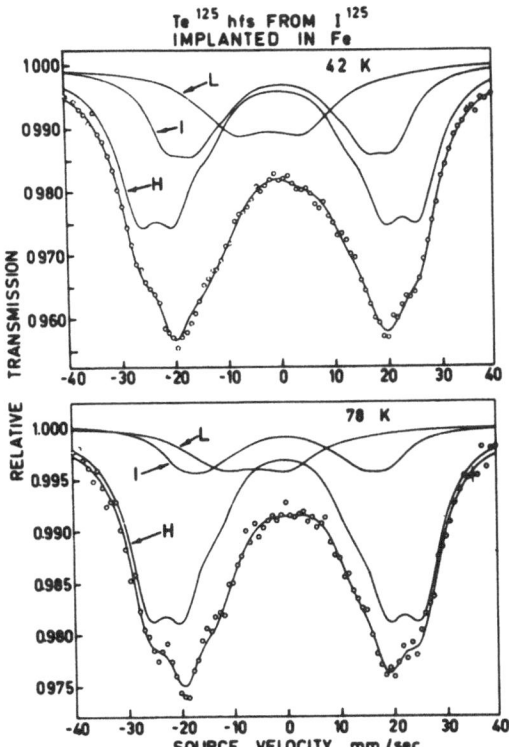

Fig. 9: Mossbauer data on ^{125}TeFe produced in the decay of implanted ^{125}I.

Extensive study of implanted Sb^+, Te, I^+ and Xe^+ in Fe has also given details of hfi at non-substitutional sites and considerable data exists for the As, Se^+, Br^+, Kr systems. The systems indicated + have also been studied by Rutherford backscattering.

As an example, Mossbauer data obtained by the Groningen group on ^{125}TeFe produced in the decay of implanted ^{125}I is shown in Fig. 9. Analysis showed at least three components present in spectra from implanted I and Xe sources, although the same Mossbauer transitions observed following decay of implanted Sb and Te showed a single dominant hfi. Further details of Mossbauer work in the 5sp series implants are given in chapter p. 193.

This evidence, and work on Br, suggests that the 'volume misfit' implants have an increased non-substitutional site occupation pro-bability, with the possibility of attracting associated vacancies. De Waard et al.[16] have proposed that the intermediate and low field sites may be characterised by such associations. Channelling results are consistent with displacement of considerable fractions of the implanted ions into the lattice channels by $\sim$ 25% of a nearest neighbour distance. The most direct hfi evidence for vacancy association with the implant in IFe is based on electric quadrupole interaction measurements as described in the chapter on NMR/ON.

MODEL hfi SITE SENSITIVITY

Despite the empirical nature of the magnetic hfi models presented here, it is of interest to consider their predictions for non-substitutional sites.

Table 1. Relative conduction electron polarisation term contri-butions from first four neighbour shells calculated, using the Stearns[13] distance dependence, for (a) a substitutional impurity in bcc iron, (b) an impurity displaced from $(\frac{1}{2},\frac{1}{2},\frac{1}{2})$ to $\frac{3}{8},\frac{3}{8},\frac{1}{4}$, (c) as (b) but with a vacant near neighbour site along the displacement direction. Lattice relaxation is neglected.

Site	nn	2nn	3nn	4nn	total
Subst.	−11.2	−1.8	+3.1	+0.6	−9.3
displ.	−6.0	−2.3	+2.7	+1.9	−3.7
displ. + nn vacancy	−4.8	−2.3	+2.7	+1.9	−2.5

Table 2.   Hyperfine field ratios for FeXe, FeTe and FeI compared
with overlap integral summation ratios based on single ion wavefunctions, allowing for lattice relaxation[16].

Configuration	Ratio $\Sigma s_i^2 / \Sigma s^2_{subst.}$	Experiment $B_i / B_{subst.}$		
		FeXe	FeTe	FeI
Subst. + 1 nn vac	0.80	0.72(10)	0.80(4)	0.92
Subst. + 1 nnn vac	0.96			
		Ratio of lower component to subst field		
Subst. + 1 nn + 1 nnn vac	0.41			
		0.20(7)	0.30(15)	
Subst. + 2 nn + 1 nnn vac	0.22			

(a) Conduction electron polarisation term.   Campbell and Blandin[11]
indicate that each 1st and 2nd neighbour Fe gives rise to ∿ 5-7% of
the total B_c.   However small movements of the implant towards a
neighbouring vacancy can greatly increase this sensitivity.   As an
example Table 1 shows a calculation of B_c, summed over the first
four neighbour shells, for a Br ion displaced to $\frac{3}{8},\frac{3}{8},\frac{1}{4}$ from a $\frac{1}{2},\frac{1}{2},\frac{1}{2}$
substitutional site, both with or without a host atom at 0,0,0.   For
this calculation the cep-distance dependence curve of Stearns[13]
(Fig. 6) was used, and lattice distortion was neglected.   We see a
reduction in the total B_c by more than a factor of 3 for the
vacancy associated site.

(b) Volume overlap term.   De Waard et al.[16] have estimated the
change in the volume overlap term for different implant-vacancy
configurations as shown in Table 2.   They allowed for lattice
relaxation and implant movement toward the vacancy, and calculated
the ratios of the summed overlap integral to that for a substitutional site assuming a distance dependence $e^{-(r-r_0)}/\rho$, with
$\rho = 0.52$ Å.   We see a marked reduction of the interaction strength
in all sites associated with near neighbour vacancies.

In principal the calculated reduction factors, compared with
experiment, would give the relative contributions of cep and volume
overlap to the total B_c.   Although the theory is too tenuous for
such quantitative use, it is clear that the experimental variations
of B_c with site are of the expected order of magnitude.

CONCLUSION

This review has concentrated on work in ordered magnetic metals. Other applications of hfi to implant-host interaction studies, notably rare earths in noble metals are also of growing significance, but have not yet yielded such clear-cut results.

To date magnetic hfi studies for sp impurities in Fe have proven ability to

(a)   give highly specific evidence of substitutional implantation within a lattice 'undisturbed' up to ~ 5 nm.

(b)   detect small displacements from lattice sites towards associated vacancy sites.

The variation of site population with implant energy, dose, and thermal treatment can also be deduced from hfi measurements.

Finally it must be re-emphasised that magnetic hfi studies can seldom alone characterise a non-substitutional implant site.   In conjunction with other location techniques, such measurements are however capable, from implant doses as low as $10^{11}/cm^2$, of giving very clear evidence of the local environment within the first 1-2 atomic spacings of the injected ion.

REFERENCES

1.   Watson R.E. and Freeman A.J., Phys. Rev. Lett. $\underline{14}$ 695 (1965); and in Hyperfine Interactions, eds. Freeman A.J. and Frankel R.B. (Academic Press) 1967 p.52.

2.   Stone N.J., Journal de Physique Colloque C4 p.69 $\underline{34}$ (1973).

3.   For basic theory see e.g. Watson R.E. in Hyperfine Interactions, eds. Freeman A.J. and Frankel R.B. (Academic Press) 1967 p.413.

4.   Ruderman M.A. and Kittel C., Phys. Rev. $\underline{96}$ 99 (1954).

5.   Kasuya, T. Prog. Theor. Phys. $\underline{16}$ 45 (1956).

6.   Yosida K., Phys. Rev. $\underline{106}$ 893 (1957).

7.   Jena P. and Geldart D.J.W., Phys. Rev. $\underline{B7}$ 439 (1973).

8.   Friedel J., Advanc. Phys. $\underline{3}$ 446 (1954).

9.   Daniel E. and Friedel J., J. Phys. Chem. Solids $\underline{24}$ 1601 (1963).

10. Campbell I.A., J. Phys. C $\underline{2}$ 1338 (1969).

11. Campbell I.A. and Blandin A., J. Magnetism and Mag. Materials, $\underline{1}$, 1 (1975).

12. Cohen J.D. and Slichter C.P., Phys. Rev. Letters $\underline{40}$ 129 (1978).

13. Stearns M.B., Phys. Rev. $\underline{B8}$ 4383 (1973).

14. Freeman A.J. and Watson R.E. in 'Magnetism' Vol. IIA (eds. Rado G.T. and Suhl H.) Academic Press, 1965.

15. Shirley D.A., Physics Letters $\underline{25A}$ 129 (1967).

16. de Waard H., Cohen R.L., Reintsema S.R. and Drentje S.A., Phys. Rev. 10B 3760 (1974), and de Waard H., Physica Scripta $\underline{11}$ $\overline{157}$ (1975).

17. Das T.P., Physica Scripta, $\underline{11}$ 121 (1975).

18. Campbell I.A., Proc. Roy. Soc. $\underline{A311}$ 131 (1969).

19. Niesen L., Hyperfine Interactions $\underline{2}$ 15 (1976).

20. Bleaney B. in 'Magnetic Properties of Rare Earth Metals' ed. Elliott R.J. (Plenum Press) (1972) p.383.

21. Lea K.R., Leask M.J.M. and Wolf W.P., J. Phys. Chem. Solids $\underline{23}$ 1381 (1962).

22. Elliott R.J. in 'Magnetism' Vol. IIA eds. Rado G.T. and Suhl H., Academic Press (1965).

23. Cohen R.L., Beyer G. and Deutch B.I., Phys. Rev. Letters $\underline{33}$ 518 (1974).

24. Thomé L., Bernas H., Chaumont J., Abel F., Bruneaux M. and Cohen C., Phys. Letters $\underline{A54}$ 37 (1975) and Thomé L., Thesis, Centre d'Orsay, Université Paris-Sud (1978).

QUADRUPOLE INTERACTION

Ekkehard **Recknagel**

Universität Konstanz, Fachbereich Physik

775 Konstanz - Germany

INTRODUCTION

For many years the quadrupole interaction was mainly a field of interest to nuclear and atomic physics, because in many cases it was the only way to determine nuclear quadrupole moments, a quantity of great importance to nuclear structure theory. Apart from a few exceptions, this interaction only recently has become an important tool to study solid state properties and phenomena. From the solid state part of the quadrupole interaction - the electric field gradient - more information about the electronic charge distribution, especially in metals, can be extracted.
Furthermore it can be used as an indicator for dynamical properties in solids as phase transitions or migration of defects. Examples will be given in the contribution on perturbed angular correlation. In this paper, some basic aspects of the interaction, between quadrupole moments and electric field gradients will be discussed.

I. HYPERFINE INTERACTION

1. Classical Electrostatic Hyperfine Interaction

The electric quadrupole splitting and the isomeric shift are due to the electrostatic interaction of the atomic nucleus with its surrounding. If the nuclear charge distribution described by $\rho(\vec{r})$ is placed in an external potential $\Phi(\vec{r})$, caused by the atomic electrons and the charges of the lattice atoms, the electrostatic energy of the system is :

$$(1) \quad W = \int \rho(\vec{r}) \Phi(\vec{r}) d\tau$$

with the definition of the nuclear charge $Z.e = \int \rho(\vec{r})d\tau$.
Assuming the potential $\Phi(\vec{r})$ to range slowly over the region, where
$\rho(\vec{r})$ is non-negligible, it can be expanded in a Taylor's series
around $\vec{r} = 0$:

(2) $\Phi(\vec{r}) = \Phi(o) + \sum_{\alpha=1}^{3} (\frac{\partial \Phi}{\partial x_\alpha})_o x_\alpha + \frac{1}{2}\sum\sum_{\alpha\beta} (\frac{\partial^2\Phi}{\partial x_\alpha \partial x_\beta})_o x_\alpha x_\beta + \ldots$

When this expression and the nuclear charge is inserted into (1),
the energy takes the form :

(3) $W = \Phi_o Z.e + \sum_{\alpha=1}^{3} (\frac{\partial \Phi}{\partial x_\alpha})_o \cdot \int\rho(\vec{r})x_\alpha d\tau +$

$\frac{1}{2} \sum\sum_{\alpha\beta}(\frac{\partial^2\Phi}{\partial x_\alpha \partial x_\beta})_o \int\rho(\vec{r})x_\alpha x_\beta \, d\tau + \ldots$

The first term is the Coulomb energy of a point charge, which only
contributes to the potential energy of the crystal lattice and is
therefore of no interest in the discussion of hyperfine interactions.
The second term is the electric dipole interaction, which vanishes
because atomic nuclei possess no electric dipole moment (parity
conservation).
The third term V contains the isomeric shift and the quadrupole interaction. It can be rewritten by introducing $r^2 = \sum_\alpha x_\alpha^2$

(4) $V = \frac{1}{6} \int \rho(\vec{r})r^2 d\tau \sum\sum_{\alpha\beta}(\frac{\partial^2\Phi}{\partial x_\alpha \partial x_\beta})_o + \frac{1}{6} \sum\sum_{\alpha\beta}(\frac{\partial^2\Phi}{\partial x_\alpha \partial x_\beta})_o \cdot \int\rho(\vec{r})(3x_\alpha x_\beta - r^2\delta_{\alpha\beta})d\tau$

The quantities $(\partial^2\Phi/\partial x_\alpha \partial x_\beta) = q_{\alpha\beta}$ form a symmetric (3 x 3) – matrix,
which can be diagonalized by a rotation of the coordinate system.
Applying this rotation to the first term in (4) and since the
electrostatic potential $\Phi(r)$ satisfies Poisson's equation
$\Delta\Phi = -4\pi\rho_e(\vec{r})$, it especially holds for $\vec{r} = 0$

(5) $(\Delta\Phi)_o = \sum_\alpha q_{\alpha\alpha} = 4\pi e |\Psi(o)|^2$

where $-e.|\Psi(o)|^2$ is the charge density of the atomic electrons at
the nucleus.
The electrostatic hyperfine interaction is therefore :

(6) $V = \frac{2\pi e}{3} |\Psi(0)|^2 \cdot \int\rho(\vec{r})r^2 d\tau + \frac{1}{6} \sum\sum_{\alpha\beta}(\frac{\partial^2\Phi}{\partial x_\alpha \partial x_\beta})_o \int\rho(\vec{r})(3x_\alpha x_\beta - r^2\delta_{\alpha\beta}) \, d\tau$

$\equiv V(I)$ $+ V(Q)$

The isomeric shift V(I) reflects the interaction between the ex-

tended charge of the nucleus and the charge density of the electrons at the nucleus. Utilizing the definition of the nuclear
charge, the integration of V(I) can be carried out and V(I) takes
the final form

(7) $V(I) = \frac{4\pi}{3} Z \cdot e^2 |\Psi(o)|^2 <r^2>$

The isomeric shift does not explicitly depend on the angular momentum of the nucleus, but only on the mean square radius of the
nucleus. This only leads to a shift of the nuclear energy levels
and not to a splitting of the degenerate m-substates. A further
discussion of the isomeric shift will be given in the contribution
by H. de Waard.

The quadrupole term V(Q) shows the interaction between the electric field gradient tensor and the non-spherical part of the nuclear
charge distribution, the quadrupole moment tensor. In the following
the quadrupole interaction will be discussed in a more general form.

2. Quantum Mechanical Treatment of the Electrostatic Hyperfine Interaction

Since the energy levels or states of a nucleus are described
by the quantum numbers of the total angular momentum I and its projection M along the z-axis, it is useful to calculate the electrostatic interaction in the framework of the nuclear wave functions.
The interaction energy is then defined as the matrix element of the
electrostatic Hamiltonian :

(8) $E_M = < IM|H_{el}|IM >$

with

(9) $H_{el} = \sum_{c,p} \frac{\rho_p \rho_c}{|\vec{r}_p - \vec{r}_c|}$

where the sum extends over the nuclear coordinates p and the electronic coordinates c (fig. 1). Expanding (9) into Legendre polynomials

(10) $\frac{1}{|\vec{r}_p - \vec{r}_c|} = \sum_k \frac{r_p^k}{r_c^{k+1}} P_k [\cos(\vec{r}_p, \vec{r}_c)]$; $|\vec{r}_c| > |\vec{r}_p|$

and applying the addition theorem for spherical harmonics we find

(11) $H_{el} = 4\pi \sum_{k=0}^{\infty} \frac{1}{2k+1} [\sum_q (-1)^q (\sum_p \rho_p r_p^k Y_{k,q}(\theta_p, \phi_p)) (\sum_c \frac{\rho_c}{r_c^{k+1}} Y_{k,-q}(\theta_c, \phi_c))]$

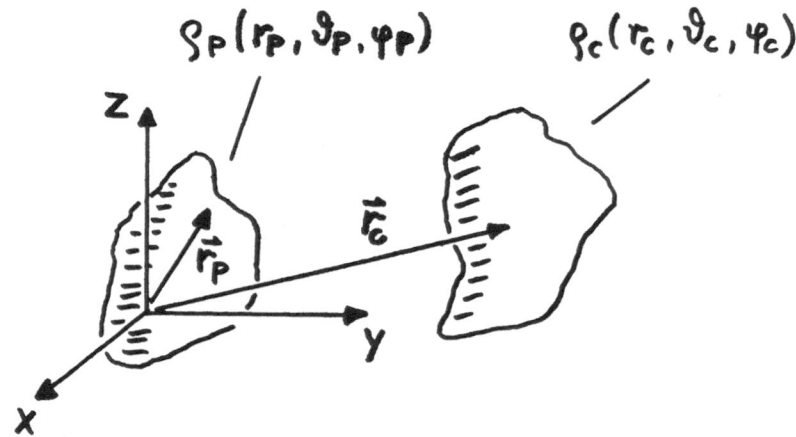

Figure 1.   Illustration of the interaction between the charge dis-
            tributions of the nucleus (index p) and the electronic
            surrounding (index c)

From this expression, one directly gets the tensor operators for the
electric moments $T^{(k)}$ and the corresponding fields $V^{(k)}$:

(12) $T_q^{(k)} = \sum_p \rho_p r_p^k Y_{k,q}(\theta_p, \phi_p)$

       $V_q^{(k)} = \sum_c \rho_c \dfrac{1}{r_c^{k+1}} Y_{k,-q}(\theta_c, \phi_c)$

Since the scalar product of two tensor operators of the same rank
k is defined as

(13) $T^{(k)} V^{(k)} = \sum_{q=-k}^{+k} (-1)^q T_q^{(k)} V_{-q}^{(k)}$

the Hamiltonian can be written

(14) $H_{el} = \sum_{k=0}^{\infty} \dfrac{4\pi}{2k+1} T^{(k)} V^{(k)}$

       $= 4\pi [Z.e_p \cdot \dfrac{1}{4\pi} V(0) + \dfrac{1}{3} \sum_{q=-1}^{+1} (-1)^q T_q^{(1)} V_{-q}^{(1)} +$

$$\frac{1}{5} \sum_{q=-2}^{+2} (-1)^q T_q^{(2)} V_{-q}^{(2)} + \frac{1}{7} \sum_{q=-3}^{+3} (-1)^q T_q^{(3)} V_{-q}^{(3)} + \ldots]$$

The first term is the Coulomb term.
The expectation values of $T^{(1)}$ and $T^{(3)}$ vanish because of parity conservation. Higher moments can be neglected. We are therefore left
with the Hamilton operator of the quadrupole interaction

$$(15) \quad H_q = \frac{4\pi}{5} T^{(2)} V^{(2)} = \frac{4\pi}{5} \sum_q (-1)^q T_q^{(2)} V_{-q}^{(2)}$$

The *electric field gradient (efg) tensor* can be expressed in an
arbitrary cartesian coordinate system (x',y',z'). Taking into account the symmetry of the tensor

$$V_{\alpha'\beta'} = \partial^2 \Phi / \partial_{\alpha'} \partial_{\beta'} = V_{\beta'\alpha'}$$

and Laplace's equation $\Delta\Phi = 0$, only five independent components remain :

$$V_o^{(2)} = \frac{1}{4} \sqrt{\frac{5}{\pi}} \cdot V_{zz'}$$

$$(16) \quad V_{\pm 1}^{(2)} = \mp \frac{1}{2} \sqrt{\frac{5}{6\pi}} (V_{x'z'} \mp iV_{y'z'})$$

$$V_{\pm 2}^{(2)} = \frac{1}{4} \sqrt{\frac{5}{6\pi}} (V_{x'x'} - V_{y'y'} \pm 2iV_{x'y'})$$

The number of components is further reduced by transformation to
principle axis.

$$V_o^{(2)} = \frac{1}{4} \sqrt{\frac{5}{\pi}} V_{zz}$$

$$(17) \quad V_{\pm 1}^{(2)} = 0$$

$$V_{\pm 2}^{(2)} = \frac{1}{4} \sqrt{\frac{5}{6\pi}} (V_{xx} - V_{yy}) = \frac{1}{4} \sqrt{\frac{5}{6\pi}} \cdot \eta V_{zz} \; ; \quad \eta = \frac{V_{xx} - V_{yy}}{V_{zz}}$$

Therefore the electrical field gradient can be described for most
applications by two parameters, V_{zz} and η. Using the convention
$V_{xx} \leqslant V_{yy} \leqslant V_{zz}$, the asymmetry parameter is defined for $0 \leqslant \eta \leqslant 1$. In
the special cases of axialsymmetric fields, i.e. hexagonal crystals,
the V_{xx}- and V_{yy}-components are equal ($V_{xx} = V_{yy} \neq V_{zz}$; $\eta = 0$) and

the efg is given by

$$(18) \quad V_o^{(2)} = \frac{1}{4}\sqrt{\frac{5}{\pi}}\, V_{zz}$$

For this special case the quadrupole Hamiltonian reduces to

$$(19) \quad H_Q = \frac{4\pi}{5} T_o^{(2)}\, V_{zz}$$

and the quadrupole interaction energy takes the form :

$$(20) \quad E_M = <IM|H_Q|IM> = \sqrt{\frac{\pi}{5}}\, V_{zz} \cdot <IM\,|\,T_o^{(2)}|\,IM>$$

The quadrupole moment of a nucleus is usually defined as the I=M component of the quadrupole tensor

$$(21) \quad eQ = <II\,|\,\Sigma\rho_p(3z_p - r_p)|II> = 4\sqrt{\frac{5}{\pi}}\, <II|T_o^{(2)}|II>$$

applying the Wigner-Eckart theorem to equations (20) and (21) we find

$$(20a) \quad E_M = (-1)^{I-M}\sqrt{\frac{\pi}{5}}\, V_{zz}\begin{pmatrix} I & 2 & I \\ -M & 0 & M \end{pmatrix} <I\,\|\,T^{(2)}\|\,I>$$

and

$$(21a) \quad eQ = 4\sqrt{\frac{\pi}{5}}\begin{pmatrix} I & 2 & I \\ -I & 0 & I \end{pmatrix} <I\,\|\,T^{(2)}\|\,I>$$

Inserting (21a) into (20a) and evaluating the 3-j symbols leads to the final result for the quadrupole interaction energies

$$(22) \quad E_M = \frac{3M^2 - I(I+1)}{4I(2I-1)} \cdot eQ \cdot V_{zz}$$

From the properties of the 3-j symbols it directly follows, that $E_m = 0$ for $I < 1$ and $E_m \neq 0$ for $I \geq 1$. Due to the M^2-dependence of the quadrupole interaction energy, there is still a twofold degeneracy of each energy level. The splitting is not equidistant.

3. Example of a Quadrupole Hyperfine Interaction

The energy splitting between two substates M and M' is given by

$$(23) \quad E_Q(M) - E_Q(M') = \frac{3eQ \cdot V_{zz}}{4I(2I-1)}\left|M^2 - M'^2\right|$$

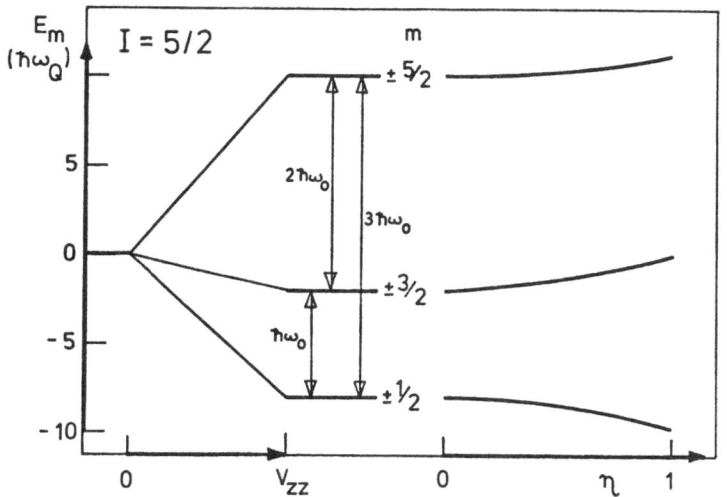

Figure 2.   Quadrupole splitting of a nuclear state I=5/2 with
            and without η.

The differences $M^2 - M'^2$ are integral numbers.   Therefore one defines the quadrupole frequence to be

(24) $\omega_Q = \dfrac{eQV_{zz}}{4I(2I-1)\hbar} \sim Q \cdot V_{zz}$

from which follows

(25) $E_Q(M) - E_Q(M') = 3\hbar\omega_Q |M^2 - M'^2|$

Thus the transition frequencies ω_n are $\omega_n = \begin{cases} 6n\omega_Q \text{ for I half integer} \\ 3n\,\omega_Q \text{ for I integer} \end{cases}$

n = 1, 2, ...

In fig. 2 the quadrupole splitting for a nuclear state with I=5/2
is schematically illustrated.   The ratios of the transition frequencies are $\omega_1 : \omega_2 : \omega_3 = 1 : 2 : 3$

On the right hand side of fig. 2 the level splitting is shown for
a non-axial symmetric field gradient as a function of the asymmetry
parameter η.

    Generally the strength of the quadrupole interaction is expres-
sed by a quantity, which is independent of the nuclear spin, the
quadrupole coupling constant ν_Q :

(25) $\nu_Q = \dfrac{eQV_{zz}}{\hbar} = \dfrac{2I(2I-1)}{\pi}\omega_Q$

II. QUADRUPOLE MOMENTS

To measure quadrupole moments of ground or excited states of
nuclei nearly all experimental methods are based on equation (23),
except the determination from transition probabilities of pure quadrupole transitions (B(E2)-values) which will not be considered in
this context. Therefore one is always confronted with the difficulty, to extract the moments from the product of the nuclear quantity times the corresponding electromagnetic field quantity. In contrast to the magnetic dipole interaction where it is quite easy to
extract the nuclear magnetic moment because the interacting magnetic
field can be produced and measured to high precision by laboratory
means, one has to rely in the case of the quadrupole interaction on
electric field gradients which occur in nature. Artificially produced efg's are eight to nine orders of magnitude too small to generate a measurable interaction. Thus the ambiguity in the determination of efg's (normally more than 10 %) is directly reflected
in the precision with which quadrupole moments can be measured.

Furthermore, as can also be seen from equation (23), only
nuclear states with $I \geqslant 1$ have measurable quadrupole moments.
Since most ground states of stable nuclei have smaller spins, the
number of accessible moments is much less than the number of magnetic moments. Therefore also the number of nuclear quadrupole probes
is limited, which can be used for application purposes.

Ground state moments are measured on free atoms by observing
the hyperfine structure of atomic spectra either via the splitting
of optical lines or by direct ratio frequency transitions within
the hyperfine structure. In these cases the efg on the nucleus produced by the atomic shell has to be calculated from the atomic configuration. On atoms embedded into solids the moments are determined via the nuclear quadrupole resonance (NQR), where the efg has
to be calculated from the atomic and lattice electron configurations.
Both methods can be extended to groundstates of instable nuclei
with lifetimes of about more than 10 min.

**Quadrupole moments of excited nuclear states are accessible by
using nuclear physics methods. One method is the Mossbauer Effect,
which directly measures the energy splitting of nuclear transitions.
Here the number of measurable moments is restricted by the well-known**
limits set by the applicability of the ME. Other methods, which
were very successful within recent years are the perturbed angular
correlation (PAC) on radioactive sources and the perturbed angular
distribution (PAD) which follows nuclear reactions. Both methods
will be presented in more detail in a forthcoming lecture. The problem to determine the efg in these measurements is similar to that
in NQR.

III. ELECTRIC FIELD GRADIENTS

1. Basic Theoretical Approach to Calculate EFG's

Since this summer school has its emphasis more on solid state problems than on nuclear physics ones, the efg's in metals should be discussed in more detail.

The understanding of efg's in non-cubic metals is still poor, because theory as well as experiments are confronted to a rather complex problem. This stems from the fact, that the field gradients are produced by different charges present in the metal : the lattice ions, the localized shell electrons and the conduction electrons which, of course, are not independent from one another. In spite of this intercorrelation, the first approach to calculate efg's is to consider the different contributions independently. Thus the lattice part is calculated separately as is the conduction electron part, while the inner shell contribution is accounted for by the Sternheimer-factor $(1 - \gamma_\infty)$. This leads to the well-known equation

$$(26) \quad eq = (1 - \gamma_\infty) \, eq_{latt} + (1 - R) \, eq_{el}$$

Since no better calculations of Sternheimer factors for atoms embedded into lattices exist, one usually takes the free atomic one's. The factor $(1 - R)$ is introduced in analogy to $(1 - \gamma_\infty)$, describing the influence of the conduction electrons on the atomic shells. The first part of eq. (26) can be easily calculated as a sum over all point charges of the neighbouring lattice atoms.

$$(27) \quad eq_{latt} = e.Z. \sum_{i \neq 0} \frac{3z_i^2 - r_i^2}{r_i^5}$$

This sum depends on the lattice parameters c and a, which for their part depend on external parameters as temperature pressure etc. To calculate the second part

$$(28) \quad eq_{el} = e \int_\Omega \rho_c(\vec{r}) \, \frac{3z^2 - r^2}{r^5} \, d\tau$$

is much more complicated and – up to now – was only possible in a few special cases of light elements (1). To determine the charge density one needs exact wave functions of the conduction electrons $\rho_c(\vec{r}) \sim \sum_k |\Psi_k(\vec{r})|^2$, which are usually unknown. Turning on the conduction electron part means also screening of the ion charges, so that Z has to be replaced by $Z_{eff.}$ in eq(27).

Therefore the theoretical situation so far is not very encouraging. Concerning the experimental side at least a large and still growing amount of data are available, from which certain principal statements can be deduced.

2. Emperical Systematics of efg's in non-cubic metals

2.1. Correlation between q_{ion} and q_{el}.   Today nearly 200 quadrupole coupling constants are determined for different metallic systems, i.e. pure non-cubic metals, impurities in non-cubic metals and alloys. A first remarkable systematic concerning the relation between the ionic part and the conduction electron part was found in 1975 by Raghavan et al. (2). If one extracts the electronic part from the measured values and the calculated ionic part, a plot of eq_{ion} $(1 - \gamma_\infty)$ versus eq_{el} shows that most of the data are centered along a curve with the slope K $\sim$ -3, suggesting a strong correlation between the two contributions (Fig. 3). Thus the empirical relation can be written as

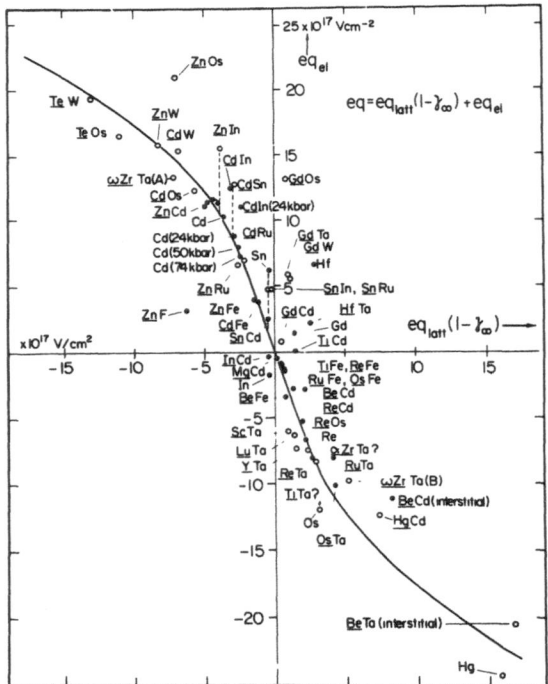

Figure 3. "Universal correlation" between lattice and electronic contributions to the electric field gradient (2)

(29) $e_q = eq_{latt} \, (1 - \gamma_\infty)(1 - K)$

The particular experimental coupling constants were evaluated by
taking the Sternheimer factor of the probe atoms and the Z_{eff} of
the host atoms.   In many cases there exists an ambiguity concerning
the sign of the interaction (open circles in Fig. 3), since the determination of the sign usually requires more complicated experiments.
Full circles correspond to values with definite sign.   Deviations of
this "universal correlation" curve can be attributed to specific
properties of the impurity-host combination.

This general behaviour suggests, that the original separation in
particular contributions might be an artificial one.   A theory where
all charges are treated simultaneously and self-consistent might be
essential.   A task even more difficult than to calculate the elec-

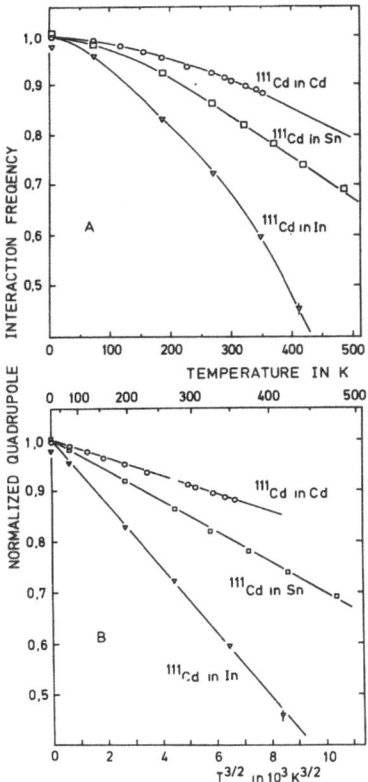

Figure 4.   Temperature dependence of the normalized quadrupole
            coupling constants of the systems ^{111}CdCd, ^{111}CdIn and
            ^{111}CdSn. (3)

tronic part by itself, but strongly favoured by the experimental
evidence.

2.2. Temperature dependence of the efg's.   Another emperical
finding, which beside its own importance might be a decisive step
forward in the understanding of efg's is a simple temperature dependence measured in nearly all metallic systems up to now.   In 1974
Christiansen et el. noticed that in cases, where the temperature variation was known, it could be explained by the simple expression

(30) $q(T)/q(0) = 1 - B \cdot T^{1,5}$

This relation holds for the whole temperature range up to the melting point. The change in temperature is unexpectedly large, typically
25 % and in some cases more than a factor of two.   It is opposite in
sign to what one would expect from lattice expansion considerations.
The $T^{1,5}$ rule was observed in pure metal systems as well as in dilute
binary alloys and in some cases also in stoichiometry alloys.   In
Fig. 4 a few examples of the $T^{1,5}$ temperature dependences are shown.
Most of the experiments have been carried out using PAC techniques,
but a few data exist from NQR measurements ([115]InIn) (4) which are
taken with high precision.

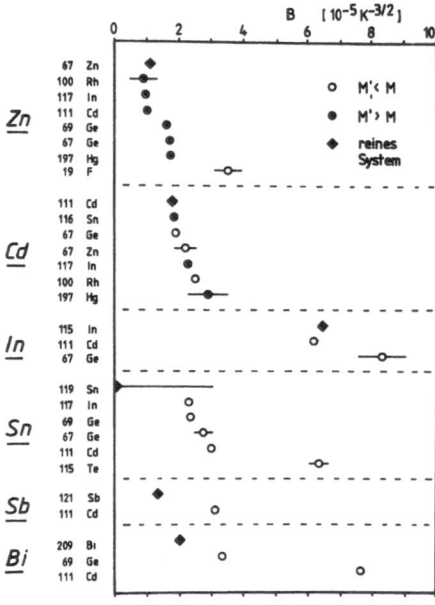

Figure 5.   Systematics of the strenght parameter B of the tempera-
             ture dependence for different host metals. (5)

In most cases the constant B is positive. It seems, that the strength
of the temperature dependence B is mainly governed by the host material as is shown in Fig. 5, where the results for different impurity probes are plotted for the same hosts, though a more careful
evaluation done by Keppner et al. (5) (see contribution to the poster
session) suggests the B-values to be correlated to characteristic
magnitudes of the host as well as the impurity atoms. The characteristic properties taken into account are the spring constants,
the ionic volumes and the valency differences.

3. Theoretical Approach to Explain EFG's and the Temperature Dependence

The concept on which eq. (26) is based leads to the assumption
that the temperature dependence may arise from lattice expansion.
This concept was rather successful to explain the temperature effect
in non-metals (Bayer 1951), but failed in the case of metals. Therefore in a next step lattice vibrations were taken into account, by
this reducing the efg phenomena to the well establised phonon model
of the solid state. A first approach to explain the temperature
dependence by thermally activated lattice atoms vibrating in the
Coulomb potential of the neighbouring ions gave only a small effect
on the efg because of the $1/R^3$ dependence. The utilization of
pseudo-potentials by Nishiyama et al. (6) which are produced by a
combination of ion cores and screening electrons, was more successful (Fig. 6). The efg can be written as

$$(31) \quad V_{zz}(t,T) = (1-\gamma_{eff}) \sum_{i \neq o} \{Z.e \frac{3z_i^2(t)-R_i^2(t)}{R_i^5(t)} - $$

$$e \int \rho_{sc}(\vec{r} - \vec{R}_i(t)) \frac{3z^2 - r^3}{r^5} d\tau \}$$

$(1-\gamma_{eff})$ describes the effective amplification of the external efg
by the polarization of the screened probe ion.
The term $\{ \quad \} \equiv F(R_i(t))$ is determined by the pseudo-potential.
The resulting $V_{zz}(T)$ is the time average of $V_{zz}(t, T)$ which can
easily be proved by considering the characteristic times :

Quadrupole interaction		vibration time of ions		vibration time of electrons
10^{-8} sec	$\ll$	10^{-12} sec	$\ll$	10^{-15} sec

The screening electron could adiabetically follow the ion vibrations.
Taking a realistic pseudo-potential, in which one assumes the Coulomb-

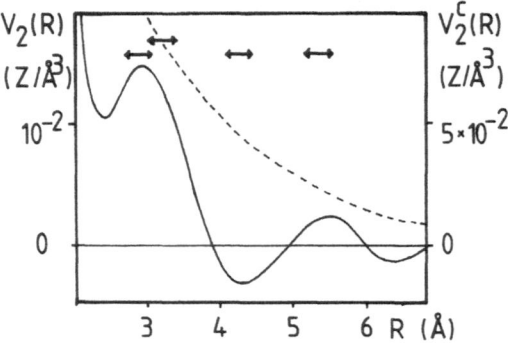

Figure 6.   Influence of the ionic vibrations on the radial part of
the EFG in Cd metal.   The dashed curve describes the EFG derived from
a Coulomb potential (right-hand scale).   The arrows indicate the vibration amplitudes of the first-to fourth-neighbouring ions at the
melting point. (6)

potential of the ions to be screened by a free electron gas leads to

$$(32) \quad V_{zz}(T) = (1-\gamma_{eff})V_{zz}^{sc}(T) \cdot \exp\left(-\frac{4}{3} k_F^2 <u^2> (T)\right)$$

Here the temperature dependence is mainly given by the Debye-Waller
factor, while the factor $V_{zz}^{sc}(T) = \sum_{i \neq o} F_i(R_{io})$ is the "pseudo-potential"
without vibrations, but allows for lattice expansions.
This of course, does not result in the analytical $T^{1,5}$ dependence,
if one calculates $<u^2>$ from Debeye theory.   On the other hand, if
one evaluates $V_{zz}(T)/V_{zz}(0)$, by introducing experimental values for
the temperature dependence of the lattice expansion as well as experimentally determined u^2 values, one finds
$<u^2> (T) - <u^2> (o) \sim T^{1,5}$.   Thus at least a semi-empirical approach
excists to explain the $T^{1,5}$ rule (Fig. 7).

4. EFG's in Distorted Cubic Metals

So far only non-cubic metals have been considered because in
undistorted cubic metals the efg vanishes, even if the probe ions
are placed on an interstitial side.   This changes if localized distortions are introduced by defects, i.e. vacancies, interstitials,
impurity atoms, or clusters of them (7).   Recently quadrupole interactions studies to investigate radiation damage in cubic metals have
found a broad application.   In the context of the lecture on perturbed

angular correlation, experiments of this kind will be discussed in
more detail. At this point, only a few essential remarks should be
made. If a defect is situated next to a probe atom, it produces a
definite efg, the strength of which can be determined by the quadrupole coupling constant. This allows to distinguish between different
defects and to recognize again a particular type of defect. Single
crystal experiments in addition give insight into the orientation of
probed-defect configurations. Furthermore the dynamical behaviour
of defects can be studied, thus supplying information about migration energies as well as trapping and detrapping mechanism.

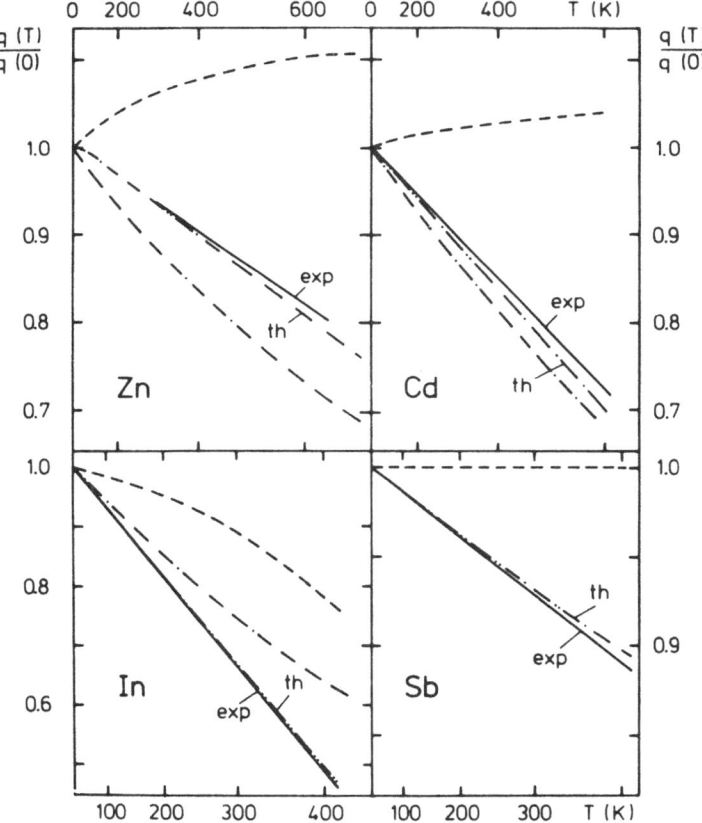

Figure 7. Comparison of the temperature dependence of the EFG
measured for Zn, Cd, In, and Sb with theoretical results (th). Solid
line (exp), experimental $eq(T)/eq(0)$ data (see Ref. 2 and references
therein); dashed line, q_{ion}^{sc}; dot-dashed line, Debye-Waller factor.
All curves are separately normalized to unity at $0°$ K. (6)

REFERENCES

1) T.P. Das, Phys. Scripta 11 (1975) 121.
2) R.S. Raghavan, E.N. Kaufmann and P. Raghavan, Phys. Rev. Letters 34 (1975) 1280.
3) J. Christiansen, P. Heubes, R. Keitel, W. Loeffler, W. Sandner and W. Witthuhn, Z. Physik B24, (1976) 177.
4) R.R. Hewitt and T.T. Taylor, Phys. Rev. 125 (1962) 524.
5) P. Heubes, W. Keppner and G. Schatz, to be published. W. Keppner, Diplomarbeit Universität Konstanz, 1978.
6) K. Nishiyama, F. Dimmling, Th. Kornrümpf , and D. Riegel, Phys. Rev. Letters 37 (1976) 357.
7) A. Weidinger, O. Echt, E. Recknagel, G. Schatz and Th. Wichert, Physics Letters 65 A (1978) 247 ; Erratum : Physics Letters 66A (1978) 514.

ISOMER SHIFT AND RECOILLESS FRACTION

H. de Waard

Laboratorium voor Algemene Natuurkunde, University of

Groningen, Netherlands

I. INTRODUCTION

In this paper we will deal in more detail with two quantities
that can only be determined from Mössbauer spectra, the isomer shift
and the recoilless fraction. We shall see that both may provide
valuable information about the nature of the sites of impurities
implanted in solid matter. Extensive literature exists on the
theory and on the applications of both quantities. As regards isomer
shift, we refer in particular to the edited volume: "Mössbauer
Isomer Shifts" which has recently appeared [1]. For the recoilless
fraction, a quantity that can be derived from the dynamics of the
crystal lattice, we refer to the basic papers of Lipkin [2], Housley
and Hess [3] and the book of Wegener [4] for general considerations.
Special attention to impurity recoilless fractions is given by
Visscher [5], Maradudin and Flinn [6] and, more recently, by Mannheim[7].
Within the scope of this summer institute, the phenomenon will be
treated in a relatively elementary manner, using quantum mechanics
sparingly, and some special attention will be given to measurements
of isomer shifts and recoilless fractions of implanted impurities,
though the number of such measurements is still rather small.

2. ISOMER SHIFT

2.1. *The isomer shift formula*

The isomer shift of nuclear gamma rays results from the
scalar (Coulomb) interaction between the nuclear charge and the
electron charge. This interaction corresponds to the zero order
term in a series development of the electrostatic interaction energy

of the nuclear charge distribution and the surrounding electron charge. It is non-zero only for electrons with a finite charge density $\rho_e(\vec{r})$ at the nuclear site, i.e. for s electrons and for relativistic $p_{\frac{1}{2}}$ electrons. The difference of the Coulomb energy of a finite and a point nucleus is given in non-relativistic approximation by:

$$W_c = \int_0^\infty \{\phi_k(r) - \frac{Ze}{r}\} \rho_e(r) \, d\tau \qquad (1)$$

For the simple case of a homogeneously charged spherical nucleus of radius R the nuclear potential is (see Fig. 1)

$$\left.\begin{aligned}
\phi_k(r) &= \frac{Ze}{2R} (3 - \frac{r^2}{R^2}) \qquad r \leqslant R \\[2mm]
&= \frac{Ze}{r} \qquad\qquad\ r \geqslant R
\end{aligned}\right\} \qquad (2)$$

If we further assume $\rho_e(r) = \rho_e(0)$ throughout the nuclear volume we immediately find from Eq. (1) and (2):

$$W_c = \frac{2}{5} \pi \rho_e(0) \, ZeR^2 \qquad (3)$$

For an arbitrary nuclear charge distribution $\rho_k(r)$ the calculation of (1) yields

$$W_c = \frac{2}{3} \pi \rho_e(0) \, Ze \, <r^2> \qquad (4)$$

again taking ρ_e constant throughout the nuclear volume.

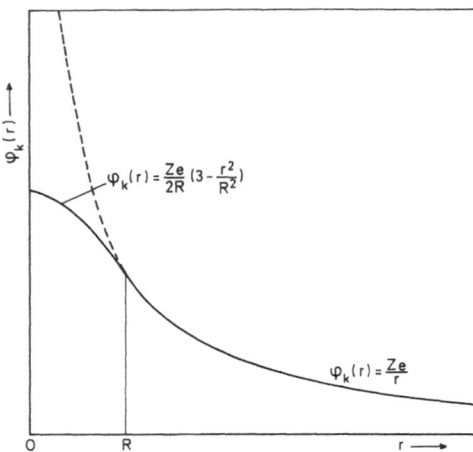

Fig. 1. Potential of a homogeneously charged spherical nucleus of radius R, inside and outside nucleus (——) and Coulomb potential of a point nucleus with equal charge (---).

This is the same expression as has already been given by Recknagel [8] in his more general treatment of the electrostatic interaction between the nucleus and the charges around it.

The mean square of the nuclear charge radius is defined as

$$<r^2> = \frac{\int_0^\infty \rho_k(r)\, r^2\, d\tau}{Ze}$$

Comparison of (3) and (4) shows that, for a homogeneously charged
sphere of radius R, we have the (well known) relation $<r^2> = \frac{3}{5} R^2$.

The non-relativistic expressions (3) and (4) are only good
approximations for very low values of Z. Relativistic calculations
of the electron charge density as first performed by Racah [9] and
by Rosenthal and Breit [10] for the case of a point nucleus with only
one s-electron yield:

$$\rho_e(r) = \frac{2(\rho+1)}{\Gamma^2(2p+1)} (\frac{2Zr}{a_o})^{2\rho-2} \cdot e|\psi_s(0)|^2 \tag{5}$$

$$\rho = \sqrt{1-\alpha^2 Z^2}$$

with $\Gamma(x)$ the gamma function and $e|\psi_s(0)|^2$ the non-relativistic
contact charge density of one s electron. Inserting this expression
in Eq. (1) one finds for a homogeneously charged nucleus of radius R

$$W_c = \frac{24\pi(\rho+1)\ |\psi_s(0)|^2 Ze^2}{2\rho(2\rho+1)(2\rho+3)\Gamma^2(2\rho+1)} (\frac{2Z}{a_o})^{2\rho-2} R^{2\rho} \tag{6}$$

which expression we shall abbreviate to

$$W_c = \frac{2}{5} \pi\, \rho_e(0)\, Ze^2\, R^{2\rho}\, S(Z) \tag{7}$$

in order to bring it into line with Eq. (3).

The relativistic correction factor S(Z) has been tabulated by
Shirley [11]. He also gives an improved factor S'(Z) based on work
of Bodmer [12], who calculated the relativistic contact density for
one s electron in the field of a finite nucleus. A few S(Z) and
S'(Z) values, for important Mössbauer nuclei, are given in Table 1.
In this table we also give the values of ρ, which show that there
exist appreciable deviations from the R^2 dependence of Eq. (3) for
heavier nuclei. This fact has been ignored by many authors.

The next step in refining the relativistic treatment is to
consider a whole atom and to perform self consistent relativistic
calculations of all its electron wave functions. This is known as
a Dirac-Fock treatment. From it, relativistic contact densities for
each of the occupied s shells can be found. Such calculations have
only been published for a small number of selected cases. From them,

TABLE I

Relativistic correction factors for s electron contact densities of outermost shells.

Z	Element	ρ	S(Z)	S'(Z)	Dirac-Fock	
26	Fe	0.98	1.32	1.29		
53	I	0.92	2.75	2.53	2.28	(ref. 13)
79	Au	0.82	8.55	6.84		
93	Np	0.73	19.4	13.6	12.0	(ref. 14)

relativistic enhancement factors are found that are not very different from those given by the S'(Z) factors (see Table I) but the factors do vary from shell to shell.

The energy W_c is a very small term in the expression for the total energy of an atom. It can not be directly observed. In order to see how the Coulomb energy of a finite nucleus gives rise to observable shifts of spectral lines we consider first:
The difference of W_c for two isotopes or for two isomeric states of the same nucleus:

$$\Delta W_c = \frac{2}{3} \pi \, Ze \, \rho_e(0) \left[<r^2>_1 - <r^2>_2 \right]$$

This gives rise to
(i)   a change of gamma transition energy between 2 isomeric states or
(ii)   a change of atomic level energies for two different isotopes.
Both of these are still unobservable directly.
Next, we write down:
The difference of ΔW_c for such nuclei in different electron environments a and b, i.e. with different $\rho_e(0)$:

$$S = \Delta\Delta W_c = \frac{2}{3} \pi \, Ze \left[\rho_e^a(0) - \rho_e^b(0) \right] \left[<r^2>_1 - <r^2>_2 \right] \tag{8}$$

This quantity expresses:
(i)   the isomer shift of a gamma transition for nuclei in different solid state environments a and b,
(ii)   the isomer shift or
(iii)   the isotope shift of an optical transition between two atomic states with different values of $\rho_e^a(0)$ and $\rho_e^b(0)$.
All of these three shifts have been found experimentally.

The order of magnitude of S may be estimated by setting $\rho_e \sim e/(\pi r_o^3)$ (the contact density due to one unscreened 1s-electron with radius r_o):

$$S = \frac{2}{5} \pi \frac{e^2}{\pi r_o^3} Z \, \Delta(R^2) \approx \frac{e^2 Z}{R} \left(\frac{R}{r_o}\right)^3 \frac{\Delta R}{R}$$

The first factor, $e^2Z/R \sim 10$ MeV, for a medium weight nucleus, the second $(R/r_o)^3 \sim 10^{-11}$ and the third $\Delta R/R \sim 10^{-3}$, from experience, so that $S \sim 10^{-7}$ eV. This is a very small fraction of the usual gamma energies (10–100 keV).

The Mössbauer effect, and this effect only, makes it possible to measure such small shifts, because recoilless emission and resonance absorption produce gamma lines with natural width $\Gamma = h/\tau \simeq (6.6 \times 10^{-16})/\tau$ eVs, which, for transitions with lifetimes τ of the order of tens of nanoseconds also amounts to about 10^{-7} eV. If line width and isomer shift are of comparable magnitude the shifts can be found from the Mössbauer spectra and since these shifts yield the changes of contact electron density for nuclei in different environments, they may permit us to draw conclusions about different sites of implanted impurities.

2.2. Isomer shift calibration

One of the main problems in deriving quantitative values of contact charge density changes $\Delta\rho_e(0)$ from isomer shifts S is that the proportionality constant between S and $\Delta\rho_e(0)$ that "calibrates" the isomer shift formula contains the difference $\Delta<r^2>$ of the mean square nuclear radii of the states between which the gamma transitions take place. This quantity is not accessible to direct measurement or accurate calculation.

Much effort has been spent in deriving $\Delta<r^2>$ values from isomer shifts measured for different simple chemical compounds, combined with theoretical values of $\rho_e(0)$. Highly ionic compounds, with the Mössbauer ions in different valence states, are chosen and it is hoped that atomic wave functions are good enough for such ions to determine $\rho_e(0)$. It has become apparent, however, that the use of atomic functions is hardly ever justified.

Recent molecular orbital calculations for some iron compounds give results that inspire some more confidence [15,16]. Most of the $\Delta R/R$ values given in Table 2, where properties of a number of well known nuclides used in Mössbauer spectroscopy are summarized, are derived using atomic valence shell wave functions. They may be in error by as much as a factor 2.

More reliable $\Delta<r^2>$ values are available only for a few cases, namely where the valence shell electron density charge has been derived more directly, either from outer shell electron conversion or from nuclear decay times. A review of these methods and results obtained with them has been given by Pleiter and de Waard [18]. The principle of the conversion electron method for "calibrating" the isomer shift formula is the following. Isomer shifts are measured for two or more sources, taking cases with as large isomer shift

TABLE 2

Measured and derived parameters of isotopes used for Mössbauer spectroscopy.

Isotope	E_γ (keV)	$T_{\frac{1}{2}}$ (ns)	$\Delta R/R$ $\times 10^{-4}$	$\lvert\psi_{ns}^{rel}(0)\rvert^2$ ($\times 10^{26}/cm^3$)		Ref.	S_1 ($\times 10^{-8}$ eV)	$\eta = S_1/\Gamma$
$^{57}_{26}Fe$	14.4	97.8	−8.8	n=4	0.21	5	3.9	8.3
$^{119}_{50}Sn$	23.9	17.9	1.6	n=5	2.2	6	22	8.7
$^{121}_{51}Sb$	37.1	3.5	−11		2.9	6	210	16
$^{125}_{52}Te$	35.5	1.48	0.9		3.6	6	22	0.71
$^{129}_{53}I$	27.8	16.8	3		4.5 4.5 5.0	6 2 a)	95	35
$^{129}_{54}Xe$	39.6	1.01	0.3		5.4 6.0	6 a)	12	0.26
$^{133}_{55}Cs$	81.0	6.3	0.9	n=6	0.64	2	4	0.6
$^{151}_{63}Eu$	21.5	9.7	1.8		1.4	a)	23	5.0
$^{161}_{66}Dy$	25.6	28.2	−0.12		1.7	a)	2	1.3
$^{170}_{70}Yb$	84.2	1.6	0.28		2.5	a)	8	0.3
$^{181}_{73}Ta$	6.24	6800	−7		3.9	a)	330	14×10^5
$^{182}_{74}W$	100.1	1.31	−0.08		4.5	a)	4.5	0.13
$^{197}_{77}Au$	77.3	1.88	1.5		6.3	a)	130	5.3
$^{237}_{93}Np$	59.5	68.3	−1.4	n=7	8.7 7.7	a) 3	200	300

---

a) $\lvert\psi_{ns}^{rel}(0)\rvert^2 = S'(Z)\lvert\psi_{ns}(0)\rvert^2$. $S'(Z)$ is a relativistic factor taken
   from Shirley [1] and $\lvert\psi_{ns}(0)\rvert^2$ are non relativistic densities derived
   from the tables of Herman and Skillman [7].

differences as possible. The isomer shift difference is given by

$$S = \frac{2}{3} \pi \, Ze^2 \, \Delta|\psi_s(0)|^2 \cdot \Delta\langle r^2\rangle \tag{9}$$

We assume
(i)   that the change of $|\psi_s(0)|^2$ is caused only by electrons in
        the valence shell (n): $\Delta|\psi_s(0)|^2 = \Delta|\psi_{ns}(0)|^2$,
(ii)   that the conversion coefficient in a certain electron shell
        (i) is proportional to the contact s density of that shell
        $\alpha_i = (n_{e_i}/n_\gamma) \propto |\psi_i(0)|^2$. This is very nearly true for M1
        transitions, where the conversion is predominantly in the
        s-shells and close to the nucleus.
(iii)   Further, we trust the atomic calculation of the contact s
        density of the n-1 shell, $|\psi_{n-1,s}^{(0)}|^2$.
Now, we measure for all sources for which the isomer shifts have
been found, the ratio $\alpha_{ns}/\alpha_{n-1,s} = |\psi_{ns}(0)|^2/|\psi_{n-1,s}(0)|^2$. Then,

$$\Delta|\psi_s|^2 = \Delta|\psi_{ns}|^2 = \Delta \frac{\alpha_{ns}}{\alpha_{n-1,s}} |\psi_{n-1,s}|^2 \tag{10}$$

Inserting (10) into (8) now yields $\Delta\langle r^2\rangle$, which calibrates the
isomer shift.

        The main difficulty of the method is to obtain very thin
sources that give reasonably well resolved conversion lines in a
high resolution electron spectrometer.

        In Table 3, the results that have been obtained so far are
reviewed. The most reliable value at the moment appears to be that
for the 23.8 keV transition in ^{119}Sn, with $\Delta\langle r^2\rangle = (7.7 \pm 1.5) \times$
10^{-3} fm^2. In this case, there is also available a value derived
from differences of the lifetime of the 23.8 keV level in various
chemical environments: $\Delta\langle r^2\rangle = (7.0\pm1.0) \times 10^{-3}$ fm^2, in good agreement with the conversion electron measurement. Results obtained
from nuclear lifetime changes (both gamma decay and electron capture
decay) are given in Table 4. In the references given in this table
details of the lifetime method can be found.

        The $\Delta\langle r^2\rangle$ value for ^{119}Sn has been related to values for ^{121}Sb,
^{125}Te and ^{129}I by Ruby and Shenoy [27], who compared isomer shifts
of a number of "isoelectric" compounds of the four isotopes. These
values are believed to have an accuracy of about 25%.

        In practice, some cases are much more sensitive for detecting
changes of electronic environment than others. Clearly, $\Delta R/R$ and Γ
are determining factors in the sensitivity of the shift for changes
of the valence shell contact density. In Table 2 some figures of
merit $\eta = S_1/\Gamma$ are given for the isomer shift sensitivity. There, S_1
is the shift calculated from Eq. (8) for a difference of one valence
shell s-electron. Averages of literature values of $\Delta R/R$ were used
and a relativistic value $\rho_e(0) = e|\psi_{ns}^{rel}(0)|^2$, for the density of

TABLE 3

Change of nuclear radius for gamma transitions used in Mössbauer
spectroscopy, derived from isomer shifts and electron conversion
ratios.

Nucleus	E_γ (keV)	$\Delta\langle r^2\rangle$ (10^{-3} fm^2)	$\frac{\Delta R}{R} \times 10^4$	Reference
^{57}Fe	14.4	−13.8(4.4)	−5.4(1.7)	Pleiter [19]
^{119}Sn	23.9	7.7(1.5)	1.8(4)	Emery and Perlman [20]
^{125}Te	35.5	2.4(1.3)	0.55(30)	Makariunas [21]
		2.7(3)	0.63(8)	Martin and Schulé [22]
^{129}I	27.7	18(4)	4(1)	Spijkervet [23]

TABLE 4

Change of nuclear radius for gamma transitions used in Mössbauer
spectroscopy, derived from isomer shifts and nuclear decay times.

Nucleus	E_γ (keV)	Method	$\Delta\langle r^2\rangle$ (10^{-3} fm^2)	$\frac{\Delta R}{R} \times 10^4$	Reference
^{57}Fe	14.4	γ	− 7.9(2.0)	− 3.1(6)	Ruëgsegger and Kündig [24]
		E.C.(^{52}Fe)	−28(5)	−11(2)	Meykens et al. [25]
^{119}Sn	23.9	γ	7.0(1.0)	1.69(22)	Roggwiller and Kündig [26]

one valence shell s electron was obtained from the references. In
general, the values of η are much larger for odd Z nuclei than for
odd N nuclei. This is due to the fact that in an odd Z nucleus a
proton changes its state, leading to a larger change in ΔR than when
a neutron changes its state. Even-even nuclei are very insensitive,
both because the nuclear radius hardly changes and also because the
lifetime of the $2^+ \rightarrow 0^+$ ground state transitions involved is usually
short. The very large value for ^{181}Ta is a mixed blessing because
the very narrow line width places severe conditions on sample preparation. For most isotopes with large η values isomer shift systematics have been established, both for chemical compounds with as
many different valence states as possible and for alloys with
different degrees of electron transfer to or from the gamma emitting
atom. The most extensive data are available for ^{57}Fe, ^{119}Sn, ^{121}Sb,
^{129}I, ^{197}Au and ^{237}Np.

For a successful application of isomer shift systematics, it
is not always necessary to know the calibration constant between
shift and contact electron density. Often a knowledge of the constant
that relates the isomer shift to the valence shell electron occupation number is equally useful. A well developed example of this,
the 27.8 keV gamma transition in ^{129}I, will be briefly discussed
in the next section.

2.3. Characterization of chemical bonds

Isomer shift data are important for characterizing the chemical
state of the recoillessly emitting or absorbing isotope. Clearly, the
oxidation state, which determines the number of electrons in the
valence shell, must have a strong influence on the isomer shift. An
example of this is given in Figure 2, showing the isomer shift of the
77 keV transition in ^{197}Au for Au(I), Au(III) and Au(V) compounds.
The shift is increasingly positive, corresponding to an increase of

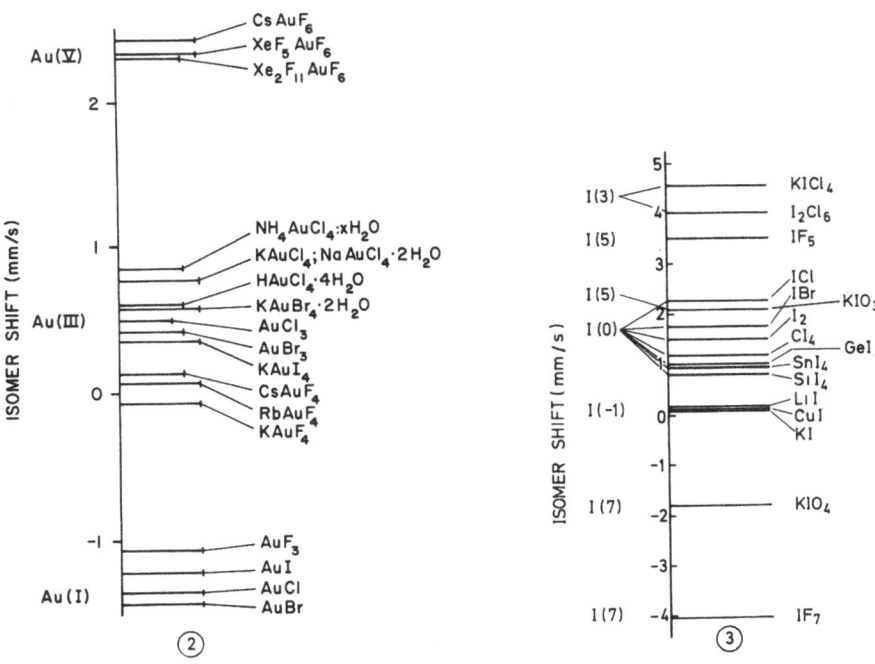

Fig. 2. Isomer shifts of the 77 keV transition in ^{197}Au for Au(I),
Au(III) and Au(V) compounds, showing influence of oxidation
state. Taken from ref. 28.

Fig. 3. Isomer shifts of the 27.7 keV transition in ^{129}I for compounds
of iodine in the 1^-, 0, 1^+, 3^+, 5^+ and 7^+ oxidation state.
Taken from ref. 29.

the contact s density (because $\Delta R/R$ is positive in this case). This trend is explained qualitatively by the removal of electrons from the 5d shell in the higher oxidation states, leading to a decrease of the screening of 5s electrons from the nucleus.

The isomer shifts for compounds of ^{129}I, examples of which are given in Fig. 3, at first sight offer a more confusing picture: within one particular oxidation state a considerable spread may occur, exceeding the boundaries between different states and the shift, first becoming increasingly positive in the direction of 1^-, 0, 3^+ and 5^+ compounds, reverts to negative for the 7^+ compounds. This behaviour can be explained very well, however, by taking into account two other important properties of the chemical bond, viz. ionicity and hybridization.

A variable degree of ionicity, i.e. electron transfer towards (or away from) an iodine ion, explains the spread of isomer shift within one oxidation state. The degree of hybridization depends on the angles between iodine-ligand bonds. The 5^+ compounds have 90° bond angles; there is no hybridization and only 5p electrons are transferred. This explains their large positive shifts. For 7 valent compounds, however, the hybridization is large due to tetrahedral or octahedral coordination of the iodine. Then, both s and p electrons are transferred to the ligand and a negative shift results.

For the case of ^{129}I, a quantitative relation has been established between isomer shift and valence shell electron occupancy [29]:

$$S = -(9\pm1)h_s + (1.5\pm0.1)h_p \quad mm/s \qquad (11)$$

Here, S is given relative to iodine with a fully occupied 5sp shell (I^-; $5s^2 5p^6$) and h_s and h_p are the effective numbers of holes in the 5s and 5p shell, respectively. Knowledge of degree of hybridization and ionicity makes it possible, in principle, to check Eq. (11).

While geometrical considerations based on crytallographic data may yield reasonable estimates of the degree of hybridization, the degree of ionicity is a quantity that can only be roughly guessed, for instance, from the electronegativity difference of the iodine and the ligands.

A powerful source of further information about chemical bonds is the quadrupole coupling strength ΔW_Q, that can be derived from the same Mössbauer spectra as the isomer shift. For the case of an axial iodine-ligand bond (σ-bonding), involving only p_z orbitals, it has been found that ΔW_Q is closely proportional to h_{p_z}, while for planar bonds involving p_x and p_y orbitals, ΔW_Q has opposite sign and is proportional to $h_{p_x} + h_{p_y}$. More general, one has: $\Delta W_Q / \Delta W_Q^{at} = h_{p_z} - (h_{p_x} + h_{p_y})/2$ where ΔW_Q^{at} is the quadrupole splitting for a free iodine atom (one p_z-hole). Evidently, the quadrupole coupling only

involves p-holes, since the s-electrons do not give rise to quadrupole splitting because of their charge symmetry. This makes it
possible to determine h_p independently and then to find h_s from
Eq. (11).

The procedures outlined above for the case of [129]I can be used
as well for other isotopes with 5sp valence electron shells, such
as [119]Sn, [121]Sb and [125]Te, but they fail for transition elements,
in particular [57]Fe, and for the rare earths. Here, the effect of
the crystalline electric field on the electrons in the partly filled
(magnetic) shell is of great importance. These more complicated
cases will not be discussed here, the reader is referred to the
extensive litterature [30]. We now turn to a subject that lies closer
to the program of this institute:

2.4. Isomer shifts in alloys.

By far the most work has been done on alloys containing iron,
measuring the isomer shift of the 14.4 keV transition in [57]Fe. Van
der Woude and Sawatzky [31] have discussed the relation between this
shift and the 3d and 4s shell electron occupation. Combining their
relation (2.12) with the recent isomer shift calibration of Meykens
et al. [25] we may write:

$$S = 2.4h_s - 1.5 h_d + \text{const. mm/s} \qquad (12)$$

with h_s and h_d the numbers of holes in the 4s and 3d shells, respectively. (The signs are reversed compared with Eq. (11) because for
the 14.4 keV transition in [57]Fe the nuclear radius change is negative). Removal of non-s electrons again leads to an enhanced contact
density through reduced s screening while removal of s electrons
directly reduces the total s density.

Van der Woude and Sawatzky have also observed that in ferromagnetic alloys of iron changes of the hyperfine magnetic field
are predominantly due to changes of the 4s density ($\Delta B_{hf} \simeq 360\Delta h_s$ kG).
Combination of hyperfine field and isomer shift data therefore
makes it possible to determine both h_s and h_d changes.

In alloys, two important phenomenological effects can be dis-
cerned that lead to changes of h_s and h_d:
(i)   charge transfer from (or to) Fe to (or from) neighbouring
        non-Fe atoms.
(ii)   compression (or decompression) of Fe-atoms in the alloy
        lattice.
In an empirical approach, Van der Woude and Maring[32] related the charge
transfer to the difference in electrochemical potential $\Delta\psi^*$ between
Fe and non-Fe neighbour, the compression to a volume misfit parameter
$\Delta n_{ws}^{1/3}$. Both parameters were introduced by Miedema [33] in his work on

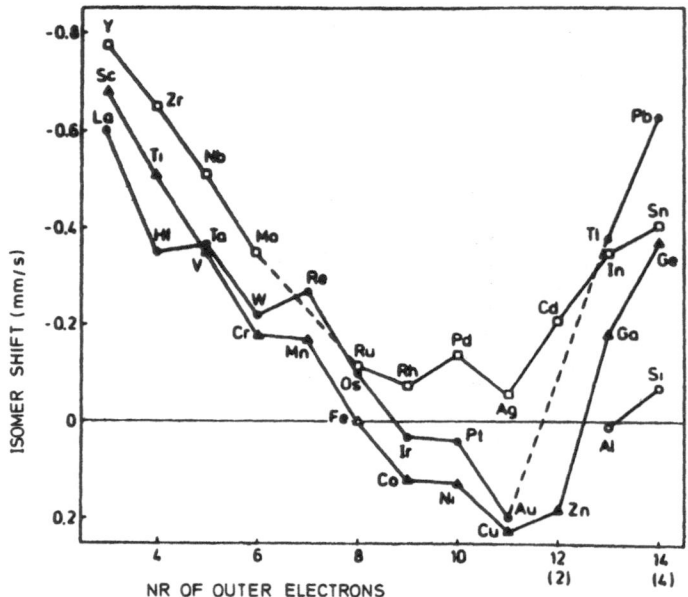

Fig. 4. Isomer shifts of ^{57}Fe in various hosts relative to pure iron, taken from ref. 31.

the thermodynamic stability of alloys; values can be taken from his papers. (Δn_{ws}, the change of electron density at the boundary between two atoms is taken proportional to the ratio of the bulk modulus and the atomic volume of the pure metal). Van der Woude obtained a reasonable fit to the iron alloy isomer shift data with the linear relation

$$S = a\Delta\psi^* + b\Delta n_{ws}^{1/3} \text{ mm/s} \qquad (13)$$

by choosing a = -0.095 and b = 0.262.

Isomer shift data for alloys dilute in Fe are shown in Fig. 4. The shift first increases with atomic number of the transition metal hosts in the 3d-5d series. In this series, the atomic radius is almost constant but the host electronegativity increase from left to right. To the right of the transition element alloys lie the sp element alloys. Here the volume misfit increases and, since its effect is larger than that of further changes of $\Delta\psi^*$, the isomer shift decreases again.

In a few cases, a more fundamental theoretical approach has been tried for calculating isomer shifts in alloys. Roberts et al. [34], for example, did extensive work on alloys of gold in Cu, Ag, Pd and Pt. There is no room to discuss this work any further.

2.5. Pressure dependence of the isomer shift.

A short discussion of this phenomenon is relevant, because if in implanted sources the atomic radius of the impurity is larger than that of the host, one has a situation comparable to that of an atom under very high pressure.

Pressure has influence on the isomer shift through the volume change ∂V of the recoillessly gamma emitting or absorbing atom:

$$\frac{\partial \rho_e(0)}{\partial P} = \frac{\partial \rho_e(0)}{\partial \ln V} \frac{\partial \ln V}{\partial P} = -k \frac{\partial \rho_e(0)}{\partial \ln V} \tag{14}$$

where k is the compressibility modulus.

The most simple assumption is $\rho_e(0)V = const.$, i.e. the contact charge density is inversely proportional to the volume available for the electrons. This assumption works rather well in many cases. It corresponds to:

$$\frac{\partial \rho_e(0)}{\partial \ln V} = -\rho_e(0) \tag{15}$$

More sophisticated calculations using band structure have been made for ^{57}Fe by Ingalls [35], who finds $\partial \rho_e(0)/\partial \ln V = -1.25\rho_e(0)$.

Roberts et al. [36,37] carried out self consistent field Dirac-Fock calculations of atomic wave functions with proper boundary conditions in a Wigner Seitz cell. For gold they find [36] $\partial \rho_e(0)/\partial \ln V = -0.85\rho_e$. Thus, we see that more refined calculations sometimes give results that are rather close to that of a very simple point of view. They are confirmed by experiment. Recent calculations[37] however, also show that the opposite pressure dependence is possible: for elements with atomic numbers just above tin, the calculations yield a decrease of the contact density with increasing pressure. This agrees with experimental results obtained from Mössbauer spectra of ^{119}Sn and ^{121}Sb. In Fig. 5 some results of Roberts et al.'s calculations are shown. For a large enough range of the inverse volume V^{-1} of the Wigner Seitz cell the total contact density $\rho(0)$ always goes through a minimum such that $\rho_{min}(0) < \rho_o(0)$, the density at the normal atomic volume V_o. If this minimum lies at $V^{-1} < V_o^{-1}$, as for Ag, Cd and In, we have the normal case: $\rho(0)$ increases when the V^{-1} increases by compression.

For Te and Sb, however, the situation is reversed, while Sn is a border line case. Roberts et al. have further shown that the calculated curves of Fig. 5 shift somewhat depending on the value chosen for the exchange potential coefficient ξ. By a proper choice of ξ the quantity $\partial \ln\rho/\partial \ln V$ will become positive for Sn, in agreement with experiments on β-Sn.

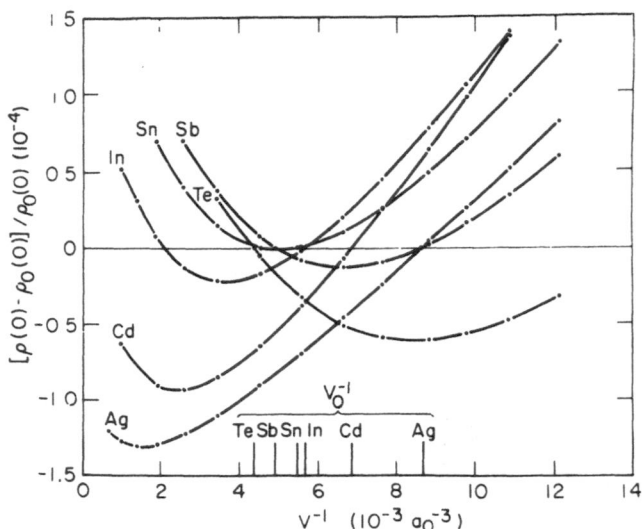

Fig. 5. Contact electron density vs. the inverse of the atomic
volume, V^{-1}, calculated for the 5sp elements Ag-Te. In the
fraction $(\rho(0) - \rho_0(0))/\rho_0(0)$, $\rho(0)$ is the total charge
density as a function of V^{-1} and $\rho_0(0)$ this density at the
zero pressure value V_0^{-1}. According to Williamson et al.[37]

2.6. Isomer shift of implanted samples.

    Two cases can be discerned, implantation of impurities that
are soluble in the host lattice, meaning that they can also be incorporated by alloying or diffusion, and implantation of soluble
impurities, which can not be randomly incorporated by other means.
In the first case, implanted and diffused (or alloyed) samples tend
to have the same isomer shift. This fact has been checked, for
instance, by Sawicka et al. [38] for the case of Fe in different
host elements. It indicates that implanted impurities are in substitutional lattice sites, just as impurities in dilute alloys. A
phenomenological model for the isomer shifts in these cases, in terms
of charge transfer and volume misfit has been briefly discussed in
the previous section.

    A systematic investigation of the isomer shifts of insoluble
impurities in various host metals has so far been reported only
for [129]I and for [133]Cs, using implanted sources of [129m]Te and ^{133}Xe,
respectively [39,40]. Also here, it should be possible in principle
to relate the shifts to volume misfit and charge transfer.

Regarding charge transfer, we may assume complete transfer
of one electron towards the strongly electronegative iodine and of
one electron away from the strongly electropositive cesium in all
metallic hosts. Therefore this effect should not lead to changes in
isomer shift for different hosts. In both cases, the implanted ions
have the $5s^2p^6$ valence shell electron configuration of xenon.

The volume misfit of the implanted ions is large in both cases
and we may expect size differences. A reasonably linear dependence
was found between the low temperature r.m.s. vibration amplitude
$a_O = <x^2>_{T=0}^{\frac{1}{2}}$ of the impurity ion in the host lattice and the isomer
shift. This is shown in Figures 6 and 7.

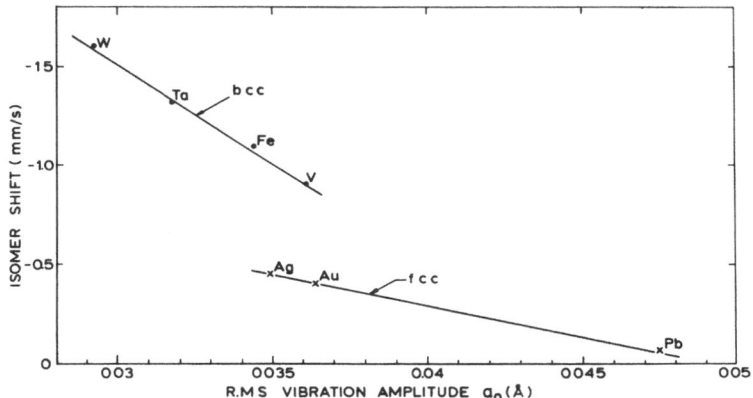

Fig. 6. Isomer shift of 27.7 keV gamma transition in ^{129}I vs. impurity r.m.s. vibration amplitude at T = 0, obtained with
^{129m}Te implanted sources in some b.c.c. and f.c.c. metal
hosts. Shifts are relative to $Cu^{129}I$ absorber.

Values of a_O for the substitutional impurities were obtained
from the following lattice dynamical consideration. In the section
about the recoilless fraction, a relation is given between the mean
square vibration amplitude of an atom in a monatomic lattice and the
Debye temperature, θ_D, assuming a Debye model phonon spectrum (Eq.
(28), which, for T = 0 reduces to:

$$<x^2_H>_{T=0} = \frac{3\hbar^2}{4mk\theta_D} \qquad (16)$$

Here, m is the mass of the lattice atoms, k is Boltzmann's constant.

For an impurity atom of mass $m_i \neq m$, Eq. (16) is no longer
valid. If also the force constants between impurity and host atoms
differ from those between host atoms, no simple relation exists
between a_O and $<x^2_H>^{\frac{1}{2}}$. Disregarding the effect of such force constant
changes, the effective impurity characteristic temperature θ_i can

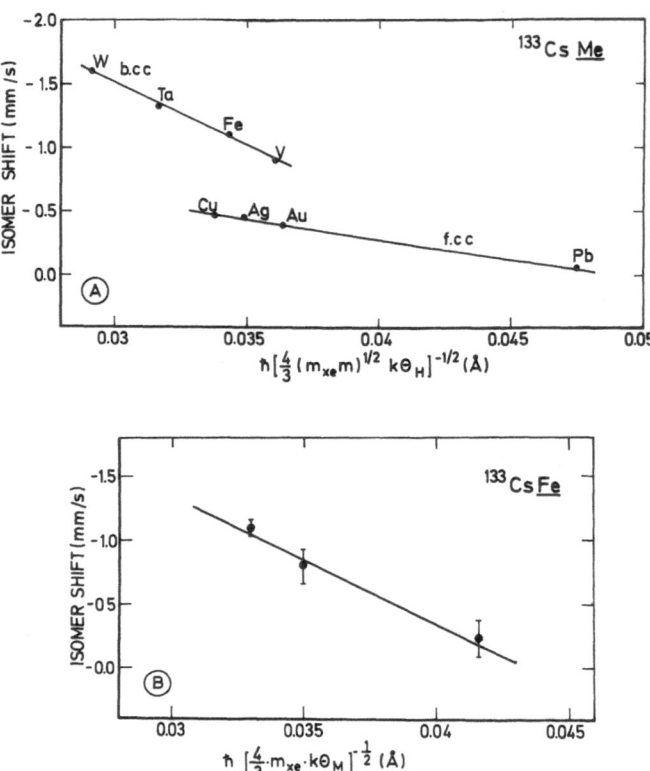

Fig. 7. Isomer shift v.s. impurity r.m.s. vibration amplitude for 81
keV transition in ^{133}Cs, obtained with ^{133}Xe implanted
sources. Shifts are relative to a CsCl absorber. A. Substitutional impurities in various f.c.c. and b.c.c. metals. B.
Substitutional and vacancy associated impurities in iron.

be related to θ_D of the host by Lipkins rule [41]: $\theta_i = \theta_D \sqrt{m/m_i}$ and
thus, the impurity r.m.s. amplitude to θ_D by:

$$a_o = \frac{\sqrt{3}\ h}{2\,(mm_i)^{\frac{1}{4}}\,(k\theta_D)^{\frac{1}{2}}} \tag{17}$$

This is the parameter as a function of which the isomer shift is
plotted in Figures 6 and 7A.

In Figure 7B, isomer shifts are given for substitutional ^{133}Cs
impurities and for two different ^{133}Cs–vacancy clusters. In this
case, the parameter a_o was directly related to characteristic
temperatures θ_i, found for each of the ^{133}Cs fractions as described
in the section on recoilless fractions (p. 23 in this manuscript).
The relation now used is $a_o = (\sqrt{3}/2) \cdot \hbar/(m_i k\theta_i)^{\frac{1}{2}}$.

One may ask, what the quantity a_0 has to do with volume misfit and why small values of a_0 lead to large isomer shifts. Qualitatively, it seems reasonable that a compression of the impurity atom in the host lattice leads to a decrease of its vibration amplitude and, as we have seen in the previous section it will lead to increased contact density of s-electrons in most cases. The degree of compression not only depends on the ratio of atomic radii of impurity and host, but also on the ratio of the compressibilities of impurity and host atoms. Apparently, the parameter a_0 provides a reasonable measure of the degree of compression.

3. RECOILLESS FRACTION

3.1. *Basic recoilless fraction formula*

If atoms in a solid perform harmonic vibrations under restoring forces proportional to their excursions from equilibrium positions, the following general expression holds for the fraction of recoilless emission or absorption of gamma rays with wavenumber k:

$$f = e^{-k^2 <x^2>_T} \qquad (18)$$

Here $<x^2>_T$ is the component of the mean square vibration amplitude of the gamma emitting or absorbing nucleus in the direction of the gamma rays, at a temperature T.

A thorough quantum mechanical treatment of Eq. (18) has been given by Lipkin [2], while Wegener [4] derived this expression in simple classical terms. In the classical approach, which we shall describe here, an emitting nucleus is considered as a transmitter of a continuous frequency modulated electromagnetic wave. Suppose that an atom vibrates with a single frequency Ω_S, so that its position is given as a function of time by

$$\vec{r}(t) = \vec{a}_S \sin (\Omega_S t + \alpha_S).$$

An electromagnetic vibration of frequency $\omega_0 \gg \Omega_S$ emitted by that atom will then show sidebands at frequencies $\omega_0 \pm n\Omega_S$ and the carrier wave amplitude, at frequency ω_0, will be reduced.

In a real crystal with N atoms, any atom is subjected to 3N independent vibrations:

$$\vec{r}(t) = \sum_{s=1}^{3N} \vec{a}_S \sin (\Omega_S t + \alpha_S) \qquad (19)$$

The Doppler shift of radiation emitted under an angle θ relative to the momentary velocity $\vec{v}(t)$ of the emitting atom is

$$\Delta\omega(t) = \frac{v(t)}{c}\,\omega_0\,\cos\theta = \vec{k}\cdot\vec{v}(t)$$

Starting at an arbitrary moment $t = 0$, the phase of the electric field of the emitting atom after a time t is given by

$$\phi(t) = \int_0^t \omega(t)dt = \int_0^t |\omega_0+\Delta\omega(t)|\,dt = \omega_0 t + \vec{k}\cdot\vec{r}(t)$$

and the field itself by

$$E(t) = E_0 e^{-i\phi(t)} = E_0 e^{-i\omega_0 t}\, e^{-i\vec{k}\cdot\vec{r}(t)} \tag{20}$$

Inserting (19) into (20) we may write

$$E(t) = E_0 e^{i\omega_0 t} \prod_{s=1}^{3N} \left[1-i\vec{k}\cdot\vec{a}_s \sin(\Omega_s t+\alpha_s) - \frac{(\vec{k}\cdot\vec{a}_s)^2}{2}\sin^2(\Omega_s t+\alpha_s) + \ldots\ldots \right] \tag{21}$$

Terms of order higher than $(\vec{k}\cdot\vec{a}_s)^2$ may be neglected since it can be shown that $\vec{k}\cdot\vec{a}_s$ is of order $1/\sqrt{N}$. Oscillating terms in the product cause sidebands so that the carrier wave is given by

$$E(t) = E_0 e^{i\omega t} \prod_{s=1}^{3N} \left[1 - \frac{(\vec{k}\cdot\vec{a}_s)^2}{4} \right] = E_0 e^{i\omega t} e^{-\frac{1}{2}\sum_{s=1}^{3N}(\vec{k}\cdot\vec{a}_s)} \tag{22}$$

Normalizing the intensity of the unmodulated carried wave to 1, we can write for the intensity of the modulated carrier wave:

$$f = \left|\frac{E(t)}{E_0}\right|^2 = e^{-\frac{1}{2}\sum_{s=1}^{3N}(\vec{k}\cdot\vec{a}_s)^2}$$

This is just the recoilless fraction, defined as the residual intensity of the wave at carrier frequency. Since in (19) the phases α_s are randomly distributed, the mean square value of $\vec{k}\cdot\vec{r}$ only contains the squares of $\vec{k}\cdot\vec{a}_s$.

$$\langle(\vec{k}\cdot\vec{r})^2\rangle = \frac{1}{2}\sum_{s=1}^{3N}(\vec{k}\cdot\vec{a}_s)^2$$

so that

$$f = e^{-\langle(\vec{k}\cdot\vec{r})^2\rangle}$$

If the vibration vectors $\vec{r}$ are randomly distributed in space (isotropic f) we have

$$\langle(\vec{k}\cdot\vec{r})^2\rangle = k^2\langle r^2 \cos^2\theta\rangle = k^2\langle x^2\rangle = \frac{1}{3}k^2\langle r^2\rangle$$

and

$$f = e^{-\frac{1}{3}k^2\langle r^2\rangle} = e^{-k^2\langle x^2\rangle}, \tag{23}$$

the desired result.

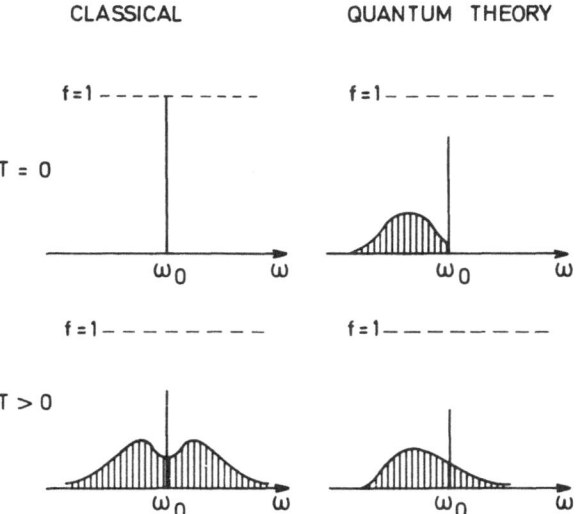

Fig. 8. Classical and quantum mechanical theory of recoilless resonance. At T=0 the resonance line resulting in classical
theory is reduced by zero point vibrations in the quantum
mechanical theory. The symmetric side band spectrum of
classical theory is distorted in quantum theory through the
recoil effect.

A comparison of this classical approach with the correct
quantum mechanical theory shows two -not unexpected- differences,
illustrated in Fig. 8.
(1) In the classical treatment no recoil is associated with the
emission of radiation, whereas in the quantum mechanical theory,
as has been shown by Lipkin[2], the center of gravity of the total
gamma spectrum is displayed just by an amount equal to the nuclear
recoil energy $E_\gamma^2/2mc^2$.
(2) Classical vibrators have no zero point energy so that f would be
one at T=0; the quantum mechanical zero point energy takes care
that a part of the gamma emission is accompanied by recoil even
at T = 0.

3.2. Temperature dependence of f

The quantity f has been a subject of extensive theoretical
lattice dynamical treatment and of experimentation long before the
Mössbauer effect was discovered: it also plays a part in the scattering of x-rays, which, for a certain part, also takes place without
atomic recoil. f is called Debye-Waller factor after the physicists
who first introduced it and derived its temperature dependence on
the basis of the Debye model of lattice vibrations.

Here, we shall discuss the temperature dependence of f for a few different lattice vibration frequency distributions (phonon spectra).

The Einstein model. There is only one phonon frequency $\Omega = \Omega_E$. Vibration energies are given by

$$E_n = (n + \tfrac{1}{2}) \hbar \Omega_E$$

and the probability of finding a vibrator in a state with energy $(n + \tfrac{1}{2}) \hbar \Omega_E$ is

$$P_n = \frac{e^{-n\hbar\Omega_E/\kappa T}}{\sum\limits_0^\infty e^{-n\hbar\Omega_E/\kappa T}} \tag{24}$$

The mean square vibration amplitude $\overline{x_n^2}$ of a vibrator in quantum state n follows from

$$\tfrac{1}{2}m v_0^2 = \tfrac{1}{2}m(\Omega_E x_{on})^2 = m\Omega_E^2 \overline{x_n^2} = (n + \tfrac{1}{2}) \hbar \, \Omega_E \tag{25}$$

Averaging over all quantum states we obtain   from (24) and (25)

$$<x^2>_{TE} = \sum\limits_0^\infty \overline{x_n^2} P_n = \frac{\hbar}{2m\Omega_E} \left(1 + \frac{2}{e^{\hbar\Omega_E/\kappa T} - 1}\right) \tag{26}$$

Even for this quite unrealistic model the agreement with experiment often is not very bad. In the following we shall drop the time averaging bar. Suppose now, that we have a general phonon frequency distribution $G(\Omega)$, normalized to $\int G(\Omega)d\Omega = 1$. Then, we may write

$$<x^2>_T = \int\limits_0^\infty G(\Omega) <x^2>_{TE} \, d\Omega = \frac{\hbar}{2m} \int\limits_0^{\Omega_m} \frac{G(\Omega)d\Omega}{\Omega} \left(1 + \frac{2}{e^{\hbar\Omega/\kappa T} - 1}\right) \tag{27}$$

As a special case of this equation we consider:

The Debye model with $G(\Omega) = 3\Omega^2/\Omega_m^3$ $(0 < \Omega < \Omega_m)$; $G(\Omega) = 0$ $(\Omega \geqslant \Omega_m)$. It is customary to introduce the Debye temperature $\theta_D = \hbar\Omega_m/k$. For this special case, the general expression (27) can be written:

$$<x^2>_T = \frac{3\hbar^2}{4m\kappa\theta_D} \left[1 + 4\left(\frac{T}{\theta_D}\right)^2 \int\limits_0^{\theta_D/T} \frac{x\,dx}{e^x - 1}\right] \tag{28}$$

The integral in this equation has been tabulated by Muir[42].

Real phonon spectra generally deviate widely from the Debye model phonon spectrum except for very low frequencies. All the same, many solids follow the temperature dependence of Eq. (28) very well. Such solids are sometimes called Debye solids.

If the phonon spectrum is known, a better evaluation of Eq. (27) can be obtained by development of $<x^2>_T$ into frequency moments of of $G(\Omega)$, defined as

$$\mu(n) = \int\limits_0^{\Omega_m} G(\Omega)\Omega^n d\Omega \tag{29}$$

At high T eq. (27) can be developed into a series in $\mu(n)$, by using a suitable expansion of the statistical factor. This yields:

$$<x^2>_T = \frac{\kappa T}{m} (\mu(-2) + \frac{1}{12} (\frac{\hbar}{\kappa T})^2 - \frac{1}{720} (\frac{\hbar}{\kappa T})^4 \mu(2) + \dots), \qquad (30)$$

an expression accurate within 2% if $T \gtrsim \hbar\Omega_m/\kappa$.

For T=0, Eq. (27) reduces to

$$<x^2>_o = \frac{\hbar}{2m} \mu(-1) \qquad (31)$$

At intermediate values of T, no simple expansion in frequence moments is possible. The first deviation of $<x^2>_T$ from $<x^2>_o$ can be shown [3] to be quadratic in T. Thus, a T dependence of the mean square amplitude as shown qualitatively in Fig. 9 results.

The frequency moments defined in Eq. (29) can be related to characteristic temperatures

$$\Theta(n) = \frac{\hbar}{\kappa} (\frac{n+3}{3})^{\frac{1}{n}} \mu(n)^{\frac{1}{n}} \qquad (32)$$

This definition ensures that for the Debye model frequency spectrum all $\Theta(n) = \Theta_D$. Values of $\Theta(n)$ can be determined from known phonon spectra. If they lie all close together this indicates that a Debye model temperature dependence of $<x^2>_T$ is a good approximation. In Fig. 10 an example is given of a measured phonon spectrum [43] and of some characteristic temperatures for solid krypton. The Debye frequency dependence is also indicated. Clearly, the phonon spectrum deviates widely from the Debye spectrum, yet the $\Theta(n)$-values still do not differ very much from each other. From Eqs. (30) and (31) we see that, from recoilless fraction measurements at low and high temperatures essentially only two characteristic temperatures, $\Theta(-1)$ and $\Theta(-2)$, are found.

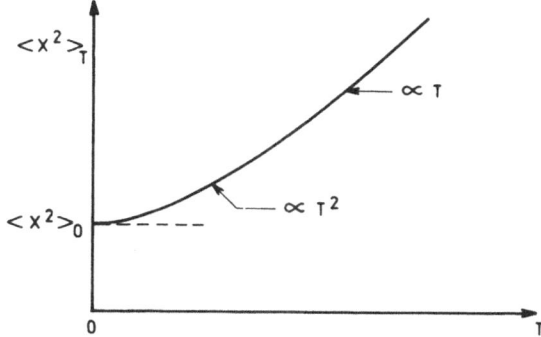

Fig. 9. General appearance of mean square vibration amplitude $<x^2>$ of a lattice atom as a function of T.

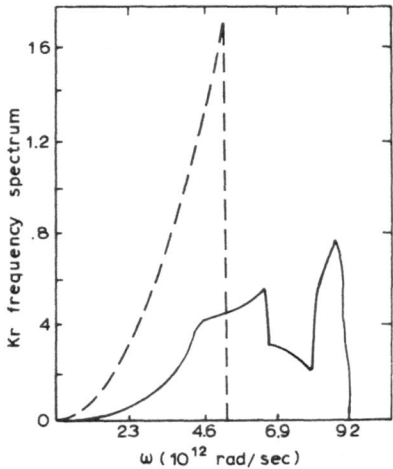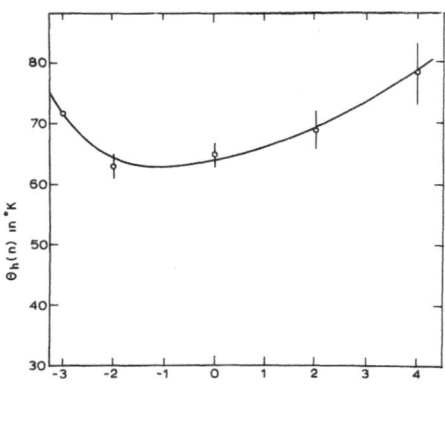

Fig. 10. A. Solid curve: phonon spectrum of solid krypton as given
by Brown[43]. Broken curve: Debye spectrum, using cut-
off frequency determined from recoilless fraction.
B. Characteristic temperatures for solid krypton. $\Theta(-3)$
follows from specific heat data, $\Theta(0)$, $\Theta(2)$ and $\Theta(4)$
from entropy and specific heat, and $\Theta(-2)$ from the
phonon spectrum shown in Fig. 10A.

3.3. *Impurity in monatomic lattice*

So far, a lattice consisting of a regular array of identical
atoms of mass m has been considered. Of course, if we investigate
the recoilless fraction of an impurity of different mass, m', imbedded in a monatomic lattice, we can not use the formulas derived
so far. Among the different approaches to this impurity problem,
that of Mannheim [7] seems to be of most value. Assuming a harmonic
frequency impurity-host interaction and considering only nearest
neighbour central forces, Mannheim derived a general expression for
the phonon spectrum of the impurity, $G'(\Omega)$, in terms of the host
frequence spectrum $G(\Omega)$ and the impurity host mass ratio m'/m and
force constant ratio λ'/λ. This expression makes it possible to
calculate $G'(\Omega)$ and the frequency moments $\mu'(n)$ from $G(\Omega)$. Two
examples of such calculations, carried out by Howard and Nussbaum [44]
for Fe in Pd and in Pt, are shown in Fig. 11. For the case of Fe in
Pt, the calculated spectrum shows an isolated component with frequency ω_L, higher than the maximum of the continuous spectrum.
Such a "localized mode" is a characteristic feature of the lattice
dynamics of impurities. It should give rise to an isolated line at
high velocity in the Mössbauer spectrum. So far, such isolated modes
have not been directly observed.

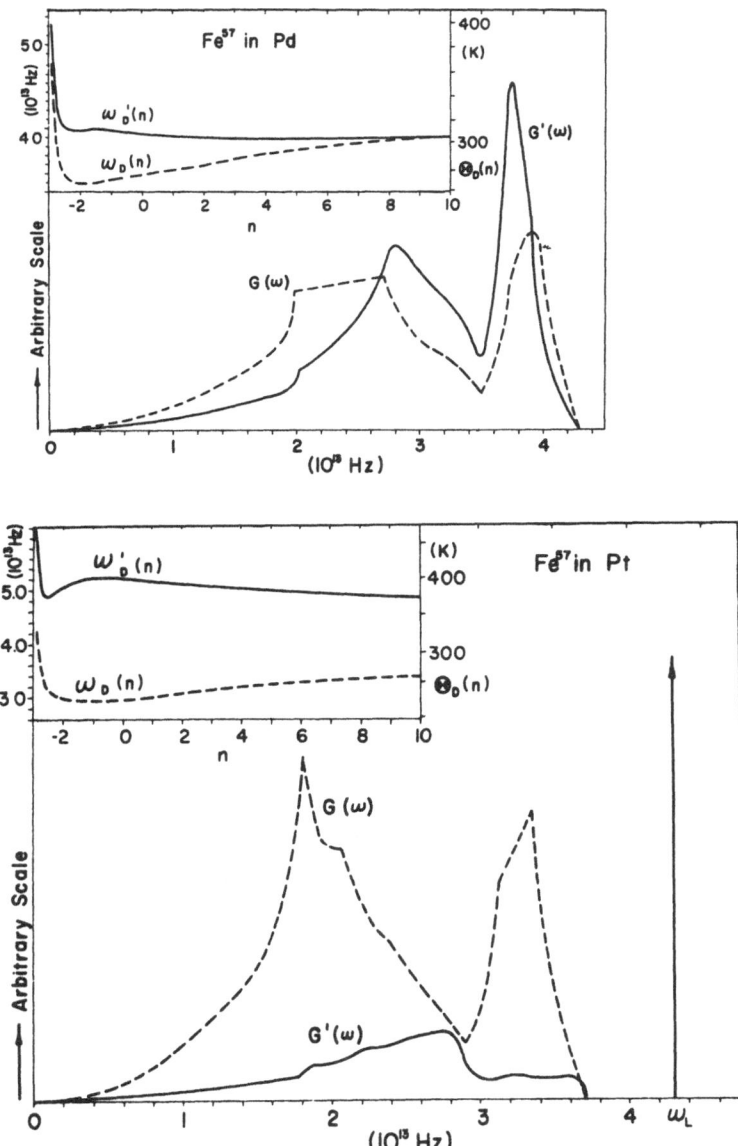

Fig. 11. Impurity response functions $G'(\omega)$ and host phonon spectra $G(\omega)$ for ^{57}Fe in Pd and Pt. Insert: corresponding frequency moments. Taken from ref. 44.

Measurements of the impurity recoilless fraction as a function of temperature yield values of the frqeuency moments $\mu'(-1)$ and $\mu'(-2)$ and from these,values of the impurity-host force constant ratio λ'/λ can be derived when the phonon spectrum of the host is known. Applications of this procedure have been given by J.M. Grow et al. [45] for ^{57}Fe, ^{119}Sn and ^{197}Au impurities in various metals.

3.4. *Recoilless fraction measurements for implanted systems*

Qualitatively, it is clear that $<x^2>$ should depend on the lattice site and on the defect association of implanted impurities. In particular, $<x^2>$ is expected to increase as the impurity traps more vacancies, because this lowers the force constant and thus the impurity vibration frequencies. The basic expression $f = \exp\left[-k^2<x^2>\right]$ further tells us that a fractional change $\Delta<x^2>/<x^2>$ of the mean square vibration amplitude leads to a $\ln(1/f)$ times larger fractional change of the recoilless fraction. Therefore the smaller f, the more sensitive it becomes to changes of $<x^2>$.

There are only a few cases known where f(T) has been studied in enough detail to provide information on impurity location.

In the first place, we mention the work on <u>implanted sources</u> of ^{133}Xe and of ^{131}I in iron [39,46]. In both cases, at least 3 different components, corresponding to at least 3 different sites, are found in the Mössbauer spectra of the 80 keV transitions in the daughter nuclei ^{133}Cs and ^{131}Xe. For each of these components, f(T) was measured for some temperatures between 4 and 80 K. Assuming a Debye model temperature dependence, characteristic temperatures Θ were derived by fitting the measured points to curves given by Eq. (28). Values of $<x^2>_0 = \frac{3}{4}\hbar^2/mk\Theta$ and $f(0) = \exp\left[-k^2<x^2>_0\right]$ were then calculated. The results are given in Table 5.

TABLE 5

Some physical parameters derived from Mössbauer spectra of ^{133}Cs and ^{131}Xe implanted in Fe at room temperature; unannealed. In both cases there are 3 observed components h, i, ℓ of the spectrum, while for ^{133}CsFe the analysis also yields an "invisible" component z with very low recoilless fraction.

	Site	Fraction %	Hyperfine field (kG)	Θ (K)	f(0)	$<x^2>_0^{\frac{1}{2}}$ (Å)
^{133}CsFe	h	29(3)	276(4)	250(15)	0.16(2)	0.033
	i	13(4)	135(15)	220(25)	0.12(3)	0.035
	ℓ	22(4)	< 25	155(10)	0.05(1)	0.042
	z	36(7)			<0.01	
^{131}XeFe	h	45(5)	1540(30)	~300	0.48(24)	0.030
	i	32(4)	1130(70)	260(30)	0.17(4)	0.032
	ℓ	23(6)	270(50)	220(20)	0.12(3)	0.035

   This procedure, in which deviations form a Debye behaviour are
disregarded, is not quite justified.   For ^{131}XeFe, for instance, the
value of f(0) thus derived does not agree with a direct absolute
determination of f(0). For an evaluation of differences between
impurity sites, however, the method is good enough.

   In Table 5 we also give the values of the hyperfine fields at
the different impurity sites. A clear correlation exists between
the magnitude of these fields and the values of $<x^2>_0$. This is
interpreted qualitatively as follows: vacancy association of impurity
atoms leads to an increase of the space around the impurity atoms
and thus to a decrease of the conduction electron density as well
as of the overlap between host and impurity electron orbitals. Both
effects lead to a reduction of the magnetic hyperfine field as well
as to an increase in the effective impurity vibration amplitude.
In section 2.6 the correlation between isomer shift and $<x^2>_0$ observed for ^{133}CsFe has already been discussed.

   An interesting case of anomalous lattice dynamics was discovered
by Vogl et al. [47], who studied the temperature dependence of f for
a source of ^{57}Co dilute in aluminium after electron irradiation at
a temperature of 100 K. At this temperature, irradiation induced interstitials are highly mobile. They are annihilated by vacancies unless
they are first trapped by ^{57}Co impurities. The recoilless fraction
of ^{57}Fe which has trapped single interstitials at a low irradiation
dose, has a very striking temperature dependence, shown in Fig. 12:
f decreases by more than a factor of 4 in the narrow temperature
interval from 13 - 20 K.

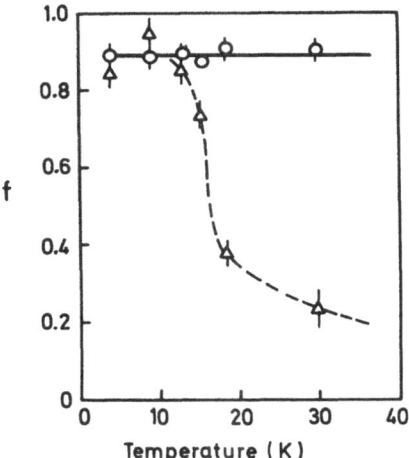

Fig. 12. Temperature dependence of recoilless fraction f of ^{57}Fe in
        Al. Upper curve (o): ^{57}Fe substitutional; lower curve (Δ):
        ^{57}Fe with trapped interstitial. Taken from ref. 47.

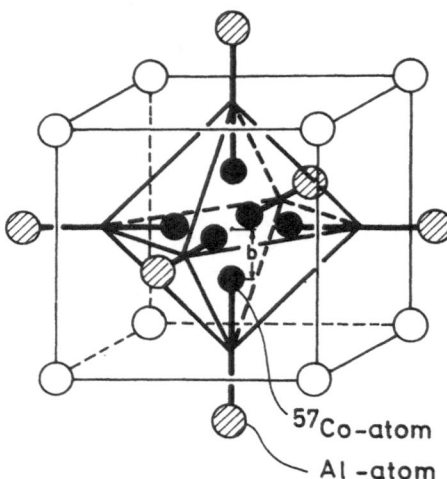

57Co-atom

Al-atom

Fig. 13. Positions of a mixed dumbbell formed by jumps of a ^{57}Co
         impurity in an f.c.c. lattice cell. Taken from ref. 47.

This (reversible) behaviour is quite different from that of purely
substitutional ^{57}Co atoms, which have an almost constant f value in
this temperature range (see Fig.12) as well as from that of larger
trapped interstitial clusters where the decrease of f is much slower.

    The authors offer an elegant explanation of the observed pheno-
menon. They assume that a ^{57}Co impurity forms a mixed dumbbell with
an Al-atom which is stable below 13 K. Above this temperature, the
^{57}Co is activated to jump between six different mixed dumbbell
positions within one f.c.c. lattice cell as shown in Fig. 13, but it
remains confined to this cell. Now its $<x^2>$ value becomes much
larger, but the interstitial associated with the cell holding the
^{57}Co atom is not detrapped before the temperature has risen to about
200 K. Only then does the ^{57}Co atom return to its substitutional
position, where it has a much larger value of f.

    These examples show that both the absolute value and the
temperature dependence of f may provide clues about the microscopic
configuration of impurity-defect clusters.

REFERENCES

1. Mössbauer Isomer Shifts (Ed. G.K. Shenoy and F.E. Wagner, North Holland, Amsterdam, 1978).
2. H.J. Lipkin, Ann. Phys. 9 (1960) 332; 18 (1962) 182.
3. R.M. Housley and F. Hess, Phys. Rev. 146 (1966) 517.
4. H. Wegener, Der Mössbauereffekt und seine Anwendungen, B.I. Bucher, 2/2A (Bibliographisches Institut, Mannheim, 1965).
5. W.M. Visscher, Phys. Rev. 129 (1963) 28.
6. A.A. Maradudin and P.A. Flinn, Phys. Rev. 126 (1962) 2059.
7. P.D. Mannheim, Phys. Rev. 165 (1968) 1011; P.D. Mannheim and S.S. Cohen, Phys. Rev. B4 (1971) 3748; P.D. Mannheim, Phys. Rev. B5 (1972) 745.
8. E. Recknagel, these Proceedings, paper ...... , p. 85.
9. G. Racah, Nature 129 (1932) 723.
10. J.E. Rosenthal and G. Breit, Phys. Rev. 41 (1932) 459.
11. D.A. Shirley, Rev. Mod. Phys. 36 (1964) 339.
12. A.R. Bodmer, Proc. Phys. Soc. (London) A66 (1953) 1041.
13. D.W. Hafemeister, J. Chem. Phys. 46 (1967) 1929.
14. B.D. Dunlap, G.K. Shenoy, G.M. Kalvius, D. Cohen and J.B. Mann, Hyperfine Interactions in Excited Nuclei (Ed. G. Goldring and R. Kalish, Gordon and Breach, New York 1971) p. 709.
15. W.C. Nieuwpoort, D. Post and P.Th. van Duijnen, Phys. Rev. B17 (1978) 91.
16. A. Trautwein, F.E. Harris, A.J. Freeman and J.P. Deschaux, Phys. Rev. B11 (1975) 4101.
17. F. Herman and S. Skillman, Atomic Structure Calculations (Prentice, Englewood Cliffs N.J. 1963).
18. F. Pleiter and H. de Waard, Ch. 5c, p. 253 l.c. ref. 1.
19. F. Pleiter, Thesis, Groningen, 1972.
20. G.T. Emery and M.L. Perlman, Phys. Rev. B1 (1970) 3885.
21. K.V. Makariunas, R.A. Kalinauskas and R.I. Davidonis, Sov. Phys. J.E.T.P. 33 (1971) 848.
22. B. Martin and R. Schulé, Phys. Letters 46B (1973) 367.
23. W.J.J. Spijkervet, Groningen, unpublished result.
24. P. Rüegsegger and W. Kündig, Phys. Letters 39B (1972) 620; Helv. Phys. Acta 46 (1973) 165.
25. A. Meykens, R. Coussement, J. Ladrière, M. Cogneau, M. Boge, P. Auric, Abstracts Int. Conf. Appl. Mössbauer Effect (Kyoto, 1978), p. 5.
26. P. Roggwiller and W. Kündig, Phys. Rev. B11 (1975) 4179.
27. S.L. Ruby and G.K. Shenoy, Ch. 9b, p. 617, l.c. ref. 1.
28. L.D. Roberts, Mössbauer Effect Data Index (Ed. J.G. Stevens and V.E. Stevens (IFI/Plenum, New York, 1973) p. 364.
29. H. de Waard, p. 447, l.c. ref. 28.
30. See, for instance, N.N. Greenwood and T.C. Gibb, Mössbauer Spectroscopy (Chapman and Hall, London, 1971).
31. F. van der Woude and G.A. Sawatzky, Physics Reports 12C (1974) 336.
32. F. van der Woude and K.W. Maring, Proceedings Int. Conf. on Mössbauer Spectroscopy, Vol. 2 (Ed. D. Barb and D. Tarina, Central

Institute of Physics, Bucharest, Romania), p. 133.

33. A.R. Miedema, Philips Technical Journal 36 (1976) 225; R. Boom, F.R. de Boer and A.R. Miedema, Journal of less common metals, 46 (1976) 271.

34. L.D. Roberts, R.L. Becker, F.F. Obenshain and J.P. Thomson, Phys. Rev. 137 (1965) A895.

35. R. Ingalls, Phys. Rev. 155 (1967) 157.

36. L.D. Roberts, D.O. Patterson, J.P. Thomson and R.P. Levy, Phys. Rev. 179 (1969) 656.

37. D.L. Williamson, J.H. Dale, W.D. Josephson and L.D. Roberts, Phys. Rev. B17 (1978) 1015.

38. B.D. Sawicka, J.A. Sawicki and J. Stanek, Phys. Letters 59A (1976) 59.

39. H. de Waard, Proceedings of the I.A.E.A. panel on applications of the Mössbauer Effect (I.A.E.A., Vienna, 1972) p. 123.

40. H. de Waard, P. Schurer, P. Inia, L. Niesen and Y.K. Agarwal, l.c. ref. 14, p. 89.

41. H.J. Lipkin, Ann. Phys. 23 (1963) 28.

42. A.H. Muir, Doc. AI-6699 (Atomics International P.O.B. 309 Canoga Park, Cal.) Jan. 1962.

43. J.S. Brown, Phys. Rev. B3 (1971) 21.

44. D.G. Howard and R.H. Nussbaum, Phys. Rev. B9 (1974) 794.

45. J.M. Grow, D.G. Howard, R.H. Nussbaum and M. Takeo, Phys. Rev. B17 (1978) 15.

46. H. de Waard, R.L. Cohen, S.R. Reintsema and S.A. Drentje, Phys. Rev. B10 (1974) 3760.

47. G. Vogl, W. Manzel and P.H. Dederichs, Phys. Rev. Letters 36 (1976) 1497.

TIME DEPENDENT HYPERFINE INTERACTIONS

D. Quitmann

Institut für Atom- und Festkörperphysik

Freie Universität Berlin, Boltzmannstr. 20
D 1000 Berlin 33, Fed.Rep.Germany

1. INTRODUCTION

Time dependent hyperfine interactions occur in a solid by many mechanisms (phonons, conduction electrons, diffusion...), and they produce changes in the nuclear spin state. When the interactions are statistical, they lead to what is broadly termed "nuclear spin relaxation". Extensive and thorough treatments of the problems involved are available (e.g. /Abr 61/, /Sli 63/, /FreF 67/ and for the detection of hfi by nuclear radiation e.g. /HarB 75/, /GabB 72/).

The present survey is intended to serve at an introductory level only. It will focus on the observation of nuclear spin relaxation using radioactive atoms (Mössbauer effect ME, angular correlations PAC or distributions PAD). For these cases, quite a few papers have appeared in /HypI 75/.

The basic situation to be considered is that of: a hyperfine interaction (hfi) / which changes with time at random / during the time a particular nucleus is being observed.

1. The hfi may be of magnetic or quadrupolar origin. Its main characteristic will be the magnitude of the interaction strength, which typically falls in the range 10^3–10^8/sec.

2. The time dependence will be assumed to be purely statistical, i.e. we shall not be concerned with added rf-fields, pulse sequences etc. In a first approximation it can then be described by a correlation time τ_c (10^{-3}–10^{-12} sec). The value of τ_c is often the essential information obtained; how the hfi changes in detail is usually much less certain.

139

3. The time during which a nucleus is being observed coherently, unperturbed (if there were no time dependent hfi) can be judged from the linewidth for radioactive nuclear states: the dynamic hfi have to compete mainly with the width due to the decay of the state.

Time dependent hfi ensue usually (by not necessarily) a broadening of the line, and lead to drastic changes in the spectrum, if the measurement can be done for a wide range of the characteristic parameters. Effects of time dependent hfi are very common in NMR and also in ESR, and the reader is referred to the classic book of Abragam /Abr 61/.

2. CASES

A few examples may serve to illustrate the situations where up to now the time dependence of hfi has been measured on nuclear states which decay radioactively.

Flips of paramagnetic Fe^{3+} spins in a metal-protein molecule have been observed by Wickman et al. using the Mössbauer effect, see fig.1,sec.4 /WicK 66/. A comparable situation for paramagnetic Yb^{3+} in a metallic Au matrix was studied by Gonzalez-Jimenez et al., again by the ME /GonI 74/. 4f shell flips which are influenced by the Kondo effect have been measured by Mekata et al. in $La_{1-x}Y_x$ containing dilute ^{140}Ce,by the PAC technique /MekK 76/. Fluctuations of the molecular field have been seen on ^{169}Tm as an impurity in Fe by Bernas and Gabriel in a PAC experiment /BerG 73/. The flip times that were accessible to these experiments were of the order of 10^{-7} sec, 10^{-9} sec, 10^{-11} sec, 10^{-11} sec respectively, in the four experiments. A potentially important application is to field fluctuations in a ferromagnet near the Curie temperature as in /GotH 73/.

Changes of sites have so far been studied much less extensively using ME or PAC techniques: The diffusion of ^{57}Fe in FeO_x, $x \approx 1$, was analyzed by Anand and Mullen using ME /AnaM 73/. In Li metal, the diffusion of the atoms could be investigated using β-PAD by Ackermann et al. /AckD 75/. Appearently complicated diffusion and trapping processes were noticed when positive muons were implanted in Nb metal; however, no direct observation of the dynamics, in the sense defined in the introduction, was made /BirC 78/. The ranges of jump times center around 10^{-8} sec, 10^{-6} sec, and 10^{-8} sec, respectively, in the three experiments mentioned. To the knowledge of the author, no observations of time dependent hfi have so far been done on implanted atoms. This talk aims at future use!

3. THE PERTURBATION FACTOR G(t)

Consider an experiment where the angular and time distribution of a γ-(or β-)ray is observed, when the γ-(or β-) emission depopulates a nuclear level with lifetime τ_n. For details concerning the description of angular correlation experiments see e.g. /Ber 79/ and /Rec 79/ and the references given there. Assume that the nuclear level with spin I has been populated at time t = 0 unevenly (i.e. the population p_m of the m-substates is $p_m \neq 1/(2I+1)$) so that

$$\beta_2 = \sum_{m=-I}^{I} p_m (3m^2 - I(I+1)) \neq 0 \quad or \quad \beta_1 = \sum p_m m \neq 0 \tag{1}$$

Here the axis of quantization $\vec{z}$ is given by the process populating the level, e.g. the preceding γ-ray or nuclear reaction. Then the intensity observed under angle ϑ with respect to $\vec{z}$, at time t may be described by

$$I(\vartheta,t) = I_0 e^{-t/\tau_n} [1 + A_k G_k(t) P_k(\cos\vartheta) + \cdots] \tag{2}$$

with k = 2 for γ-, and k = 1 for β-observation (classical NMR **measures** nuclear magnetization and therefore corresponds to k = 1 also). $A_k = \beta_k \cdot F_k$ where F_k is determined solely by nuclear properties of the γ or β transition. Higher k-terms than the first are often negligible.

Eq.(2) describes a very simple case which nevertheless will be sufficient as the starting point, viz. it applies if no preferential direction exists in the sample.

If a **static i.e. time-independent magnetic field** B_0 **is applied perpendicular to** $\vec{z}$**, the nuclear spins precess at the Larmor frequency**

$$\Omega_L = (\mu/I\hbar) B_0 \tag{3}$$

μ is the magnetic moment of the state. When we consider the distribution over the m-states (or density matrix) as determined in the coordinate system which rotates with Ω_L, then the populations are not changed by the homogeneous field $\vec{B}^0$. The simplest description of the angular distribution is by replacing ϑ by $(\vartheta - \Omega_L t)$.

We are here interested only in the so-called perturbation factor $G_k(t)$ which is the correlation function for the orientation of the nuclear spins. It will turn out that some fluctuating hfi's produce just an exponential relaxation with relaxation time $T_{(k)}$

$$G_k(t) = e^{-t/T_{(k)}} \tag{4}$$

Some kinds of static perturbations lead to periodic functions $G_k(t)$.

Very generally the essence of the perturbation factor is always contained in an expression of the form

$$G(t) \propto <e^{i\Omega t}>$$

here <...> means the average over the surroundings in which the nucleus is found, and Ω is the frequency with which the radiation detected is modulated by the interaction.[+) The important generalization is now to time dependent hfi. A time dependent hfi Hamiltonian $\mathcal{H}_1$ adds to the unperturbed Hamiltonian $\mathcal{H}_o$ and thus leads to time dependent eigenfrequencies. Then G(t) becomes

$$G(t) \propto <exp\,[i\int_0^t \Omega(t')dt']> \tag{5}$$

see e.g. /HarB 75/. This is the central ansatz in the present treatment. We shall mostly consider the pure exponential $\exp(i\int \Omega dt')$ for reasons of simplification. It is seen that at each time t, G(t) changes within dt as $<\exp(i\Omega(t)dt)>$. Over the times we then have to integrate the phase $\Omega(t)dt$, and from the time integrated phase we get the phase factor eq.(5).

If one is interested in the frequency spectrum $I(\omega)$ of the radiation emitted or absorbed, as in ME (and in classical NMR), then

$$I(\omega) = \int_0^\infty e^{-i\omega t}\, G(t)\, dt \tag{6}$$

For a static interaction and a ME experiment, Ω is the frequency Ω_o of the center of gravity plus the splitting e.g. due to an effective magnetic hf field B

$$\Omega \to \Omega_o + (m_x\,\mu_x/I_x - m_o\,\mu_o/I_o)\,B/\hbar$$

where X and O designate excited and ground state, respectively. In a PAC or PAD experiment on the very same excited level, Ω contains the modulation frequency

$$\Omega \to \Omega_o + k\,(\mu_x/I_x\hbar)\,B$$

for details, see /Berk 79/ or /Rec 79/ and the references quoted there.

[+) To simplify matters, the lifetime broadening $\hbar/\tau_n$ of the state under discussion will be assumed to be small compared to the hfi. Formally it would introduce a convolution with a Lorentzian, as is seen by setting $f(t) = e^{-t/\tau_n}$ in eq.(14). The original papers on ME, PAC, or PAD spectra which we quote contain this effect regularly. - Also we shall omit the index k from $G_k(t)$; in the actual calculation it usually appears in the expression for Ω.

4. THE PERTURBATION FACTOR AS A FREQUENCY MODULATION PROBLEM

The expression $\exp\left[i\int_0^t \Omega(t')dt'\right]$ is well known in the context of radiofrequency technique where it describes frequency (or phase) modulation /MeiG 68/. An obvious case of frequency modulation occurs when a Mössbauer nucleus moves by lattice phonons or by diffusion /AnaM 73/, thus producing a time varying projection of its velocity on the direction in which the γ-emission is observed.

Let us assume a simple cosine-modulation

$$\Omega(t') = \Omega_0 + \Omega \cos \omega t' \tag{7}$$

The important parameters are:

Ω_0 the average or carrier frequency which typically is very large (10^{20}/sec) in ME, PAC, PAD. Usually, one can split off this very high carrier frequency so that formally $\Omega_0 = 0$. If there is a time independent perturbation, like B_0 in (3), then $\Omega_0 = \Omega_L$.

Ω the magnitude of the frequency shift, i.e. of the hfi frequency which is at times added to, or subtracted from, Ω_0. Ω is therefore a measure of the strength of the fluctuating hfi.

ω is the frequency at which this additional hfi changes.

$M = \Omega/\omega$, the modulation index, determines the general appearance of $G(t)$ and of the frequency spectrum (6).

If (7) is valid throughout the sample, so that the brackets can be omitted,

$$G(t) = \exp\left[i\,\Omega_0 t + i\,M \sin \omega t\right] \tag{8a}$$

which can be expanded in Bessel functions /MeiG 68/

$$G(t) = J_0(M)\exp(i\,\Omega_0 t) + \sum_n J_n(M)\exp\left[i\,(\Omega_0 + n\omega)t\right] \tag{8b}$$

Here $n = \pm1, \pm2\ldots$ numbers the sidebands (which are sharp because of the steady state).

We note some properties of eq.(8) and state them in terms of the frequency spectrum; the situation is illustrated in fig. 2 (derived from /MeiG 68/). These qualitative statements can be transferred to the case of a hfi fluctuating with a characteristic time τ_c; one may then think of a distribution of frequencies $g(\omega)$ extending roughly to $1/\tau_c$, instead of a sharp modulation frequency ω.

1. The carrier frequency Ω_0 retains a finite intensity except at the zeroes of $J_0(M)$ or for $M \gg 1$. For rather small modulation

$M \ll 1$, $J_o(M) \approx {}^o\exp(-M^2)$. We note in passing that this part of the radiated spectrum corresponds to the recoil-free fraction of the ME.

2. The condition for one narrow line at Ω_o, $M \ll 1$, can be achieved by having a perturbation of negligible strength ($\Omega \to 0$) or, alternatively, by very fast fluctuations of a finite hfi ($\omega \gg \Omega$), the famous "motional narrowing" case.

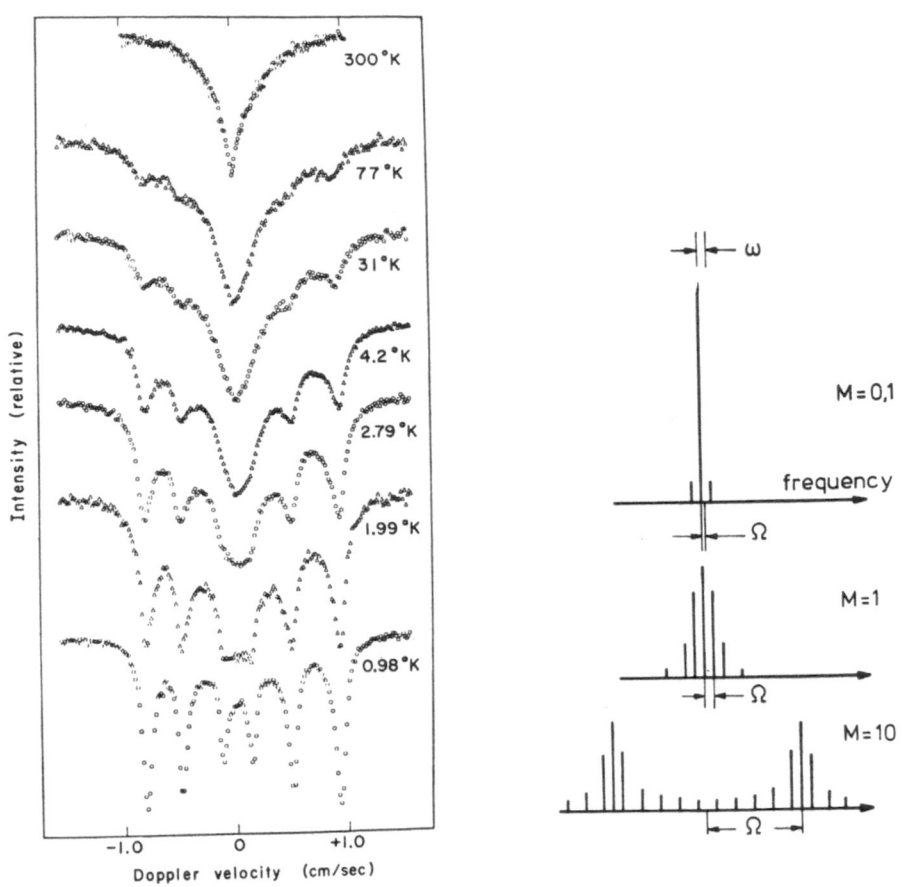

Fig. 1   Fig. 2

Fig. 1 Mössbauer spectra of a Ferrichrome A absorber, from /WicK 66/. The change in spectral shape is essentially due to a drastic increase in τ_c from top to bottom.

Fig. 2 Spectral components from frequency modulation with a square wave, adapted from /MeiG 68/. Note that the modulation index $M=\Omega/\omega$ is varied by varying Ω, not ω.

3. The sidebands occur at integer multiplies of the modulation frequency ($n\omega$), not at the hfi frequency (Ω).

4. When the modulation is not large, $M \lesssim 1$, only the carrier and a
few sidebands around it appear. This corresponds to the broadening
of a line by relaxation processes when $\Omega\tau_c \lesssim 1$.

5. When the modulation is strong, $M \gg 1$, the carrier frequency Ω_o
and the small n-sidebands disappear, and the intensity clusters
around $\Omega_o \pm\Omega$, see fig. 2. This corresponds to the splitting of a
level due to an essentially static hfi and can be discussed in terms
of the development of $J_n(M)$ for large M and large n.

6. For the intermediate case, M of the order of 1, the intensity
spreads over a region wider than either Ω or ω.

   A case often occuring in the study of time dependent hfi is
that the correlation time τ_c decreases by orders of magnitude with
rising temperature T, while the hfi interaction strength Ω is essentially constant. Starting at low T, one then observes a split spectrum (5 above) which blurs, collapses (6 above), and finally con-
tracts towards a single weakly broadened line (2 above). For a hfi
measurement on an excited state, Wickman et al. /WickK 66/ were
among the first to present an example of this transition, see fig. 1.

5. RELAXATION SEEN AS AN ORIENTATIONAL DIFFUSION

   Let us now turn to a rather pictorial classical description of
nuclear spin relaxation, applied to a statistically varying isotropic perturbation which fulfills the small modulation condition

$$\Omega\tau_c \ll 1$$

Imagine a classical angular momentum vector of magnitude I, pointing
in a certain direction at time t = 0. It may then be represented by
a point on the unit sphere. Its motion with time is sketched in fig.3.

   Relaxation means smearing of the probability distribution for
the direction of $\vec{I}$ over the unit sphere and may be discussed in the
following way: the hfi is supposed to change considerably within a
typical time τ_c, the correlation time; more precisely, the hfi is
assumed to have the same magnitude, but another direction, after
τ_c. Let the momentary precession of $\vec{I}$ be characterized by $\vec{\Omega}(t)$.
Then the correlation will be approximated by assuming that on the
average

$$\vec{\Omega}(0)\,\vec{\Omega}(t) = \Omega^2 \cdot \begin{cases} 1 & \text{for } t \leq \tau_c \\ 0 & t > \tau_c \end{cases} \tag{9}$$

Then the vector $\vec{I}$ precesses by an angle

$$\delta = \Omega \tau_c$$

in one direction during one time interval τ_c, in another direction during the next interval etc., making steps of length δ on the unit sphere in random directions. Now we specialize to the case that many steps occur during the time of interest ($n = t/\tau_c \gg 1$). After time t, n random steps will have produced a net displacement

$$\Delta^2 = \delta^2 t/\tau_c = t\Omega^2 \tau_c \tag{10}$$

on the surface of the sphere.

Since the angular distribution of the decay radiation of a radio-active nucleus behaves as if the distribution $A_k P_k(\cos\vartheta)$ had its axis fixed to the direction $\vec{I}$, the spin orientation has to spread over a sizeable fraction of the unit sphere in order to smear out the associated angular distribution. Due to the structure of P_k on the unit sphere, the size of this fractions decreases as $1/k(k+1)$ when k increases. A mean square displacement

$$\Delta^2 \approx 1$$

effectively wipes out the k = 1 distribution. If the corresponding time t, i.e. the relaxation time for k = 1, is called $T_{(1)}$, then from eq.(10)

$$1/T_{(1)} = \Omega^2 \tau_c \tag{11}$$

This is a relaxation of the type anticipated by (4).

The case just discussed was again that of motional narrowing which applies e.g. to liquids, and to paramagnetic ions or to diffusing atoms at sufficiently high temperatures. The most important result of this argument is that - for $\Omega\tau_c \ll 1$ - the relaxation is only sensitive to the product $\Omega^2\tau_c$, a result which will occur again in sections 8 and 9.

The picture allows also an easy visualization of a limiting situation as to the orientation of $\vec{\Omega}(t)$: if the field jumps between "up" and "down" only, $\vec{I}$ cannot spread in up or down direction (see also the discussion of T_2 in section 6). If there is an additional static field B_o, small compared to the time dependent perturbation ($\Omega_L \ll \Omega$), the same diffusion process occurs on the unit sphere while it rotates slowly with Ω_L.

A quantum mechanical discussion would have to start from an "orientational wave packet" on the surface of the sphere. Such a

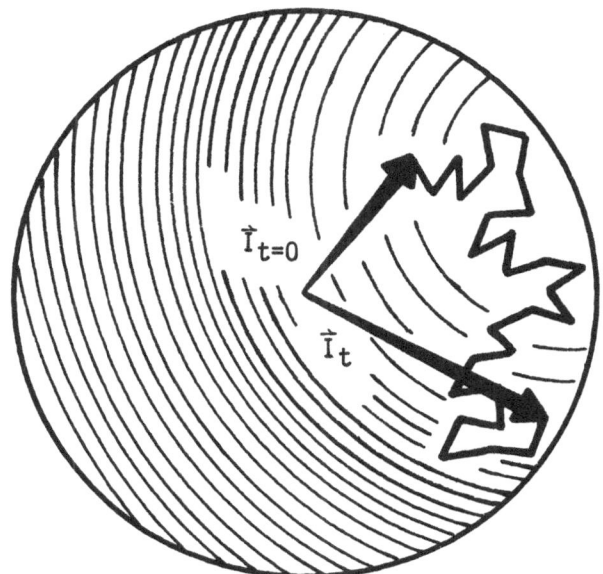

Fig. 3   Representation of a random walk diffusion on the surface of
a sphere. The arrows for the spin direction at t = 0 ($\vec{I}_{t=0}$) and at
t have been drawn for illustration.

packet, constructed from the finite set of m-substates, is comparable to wave packets in $\vec{r}$-space for a free particle in a box with
periodic boundary conditions (for the movement of a wave packet in
a harmonic potential, see /Sch 68/).

6. THE BLOCH EQUATIONS AND TWO TYPES OF LINE BRAODENING

An extremely useful description of the relaxation of a nuclear
spin system is given by the well known Bloch equations. They consider the rate of change of the (macroscopic) nuclear magnetization

$$\vec{M} = \sum_{i=1}^{N} \vec{\mu}_i$$

which in an external magnetic field $\vec{B}$ = (0,0,B_o) is

$$d\vec{M}(t)/dt = (\mu/I\hbar)\vec{M}(t) \times \vec{B} \tag{12}$$

The component $M_z = M_{\parallel}$, if perturbed, will relax towards its thermal equilibrium value $M_{\parallel,0}$ = $N\mu \exp(-\mu B_o/k_B T)$ by exchanging Zeeman
energy with the surrounding (e.g. in the manner discussed in section
5). The Bloch equations assume

$$d M_{\parallel}(t)/dt = (M_{\parallel 0} - M_{\parallel}(t))/T_1 \tag{13a}$$

with T_1 the "spin lattice relaxation time". The resulting width of
the NMR resonance line is a "homogeneous" broadening because it is
assumed to be the same for all nuclei.

There may, however, also be another reason for line broadening,
viz. a distribution of additional small, static hf fields which
lead to a distribution $f(\Omega_0)$ with $\int f(\Omega_0)d\Omega_0 =1$ and $\int \Omega_0 f(\Omega_0)d\Omega_0 =0$. It
may be caused by essentially static nuclear or electronic moments
in the surrounding of a nucleus. In the picture of section 5, the
precession differs in frequency, but not in direction, between the
nuclei, which leads to a diffusional spread of the spins at constant latitude, no energy flowing out of or into the spin system.
According to eqs.(5) and (6) the line shape becomes the convolution
of $I(\omega)$ with $f(\omega)$

$$I(\omega) \rightarrow \iint e^{-i\omega t} f(\Omega_0) e^{i\Omega_0 t} e^{i\int_0^t \Omega(t')dt'} d\Omega_0\, dt \qquad (14)$$

$$= \int e^{-i\omega t} f(t)\, G(t)\, dt = \widehat{f(\omega) I(\omega)}$$

The effect is an "inhomogeneous" broadening. It is possible to distinguish it from homogeneous broadening e.g. by a pulsed NMR experiment or by saturating one subgroup out of the distribution $f(\Omega_0)$,
see /Abr 61/.

In setting up the Bloch equations, the distribution $f(\Omega_0)$ of
precession frequencies is taken into account by a dephasing time
T_2 for the transverse component of $\vec{M}$. Eq.(12) is modified

$$d\,M_\perp(t)/dt = (\mu/I\hbar)(\vec{M}(t) \times \vec{B})_\perp - M_\perp(t)/T_2 \qquad (13b)$$

where $\perp$ is x or y.

Eq.(13a) and (13b), the equations of motion for a classical
magnetization vector $\vec{M}$ in the presence of $\vec{B}$ and two relaxation processes for $M_{\parallel}$ and $M_\perp$, constitute the basic equations first set up
by F. Bloch. (They can easily be extended to include further interactions like rf-irradiation.) From them $M_{\parallel}(t)$ and $M_\perp(t)$ can be
calculated. The connection to $G(t)$ is made by recalling that $G(t)$
is the correlation function for the orientation of the nuclear spins.
Specifically for NMR one needs

$$G(t) = <M_\perp(0)\, M_\perp(t)>$$

After this description of the ansatz, we have to refer the
reader to NMR textbooks like /Sli 63/, /Abr 61/.

7. A FORMAL TREATMENT OF A CHANGING INTERACTION FREQUENCY

The statistical nature of the fluctuations of the hfi with time suggests a treatment using the rate equations concept. P.W. Anderson considered the change of $G(t) = <\exp(i\int_0^t \Omega(t')dt')>$ itself, not of the orientation of $\vec{I}$ or $\vec{M}$, with time in what is called a stochastic model. One starts from the following assumptions:

1. There is a finite number n of different frequencies[+)]

$$\Omega(t') \to \Omega_\alpha = \Omega_0 + \Delta\Omega_\alpha \qquad \alpha = 1,2,\cdots n \tag{15}$$

2. Each nucleus jumps at random between the α's: the probability that a nucleus which had Ω_α at $t = 0$ has Ω_β after a short time Δt, is

$$W(\Omega_\alpha,\Omega_\beta,\Delta t) = \delta(\Omega_\alpha - \Omega_\beta) + \pi(\Omega_\alpha,\Omega_\beta)\cdot\Delta t \tag{16a}$$

3. The transition probability per unit time from α to β, $\pi(\Omega_\alpha,\Omega_\beta)$, depends at most on Ω_α and Ω_β (e.g. there is no memory).

4. The only effect of the transition or jump α→β is an instantaneous change from Ω_α to Ω_β.

Since each nucleus belongs to one of the groups α at each instant of time,

$$\pi(\Omega_\alpha,\Omega_\alpha) = -\sum_\beta{}' \pi(\Omega_\alpha,\Omega_\beta) \tag{16b}$$

the term β = α being excluded on the rhs, and

$$G(t) = \sum_\alpha G_\alpha(t) \tag{17}$$

Each $G_\alpha(t)$ develops within Δt into

$$G_\alpha(t+\Delta t) = G_\alpha(t) + i\Omega_\alpha G_\alpha(t)\Delta t + \sum_\beta \pi(\Omega_\beta,\Omega_\alpha) G_\beta(t)\Delta t \quad \text{or}$$

$$dG_\alpha/dt = i\Omega_\alpha G_\alpha + \sum_\beta \pi(\Omega_\beta,\Omega_\alpha) G_\beta \qquad \alpha = 1,2,\cdots n \tag{18}$$

The rate of change of G for those nuclei which happen to be in group α at time t is due to the characteristic frequency Ω_α of that group, plus the exchange of population to and from all other groups β. The latter term produces the relaxation of G.

Eq.(18) is the desired system of rate equations. It may be considered as a vector equation for the n vector $\vec{G}(t)=(G_1(t)\ldots G_\alpha(t)\ldots$

[+)] In the present section, precessions around other axes than z are not considered and G_α means G for group α.

Gn(t)). As the physical input parameters it contains the set $\vec{\Omega}$ of
n frequencies (not the energy eigenvalues) as well as the n×n matrix
$\overline{\overline{\pi}}$ of the jump rates between these frequencies (not between the
states). The values $\pi(\Omega_\alpha,\Omega_\beta)$ may have to be taken from a model for
the jumps, they may be assumed to follow from one (τ_c) or more fit
parameters, they may themselves obey rate equations (as e.g. when
a nucleus experiences an effective hf field up, α = 1, or down
α = 2, depending on the orientation of its paramagnetic electron
cloud, which itself is acted upon by the random torques exerted by
the thermally activated lattice or by the conduction electrons).
Eq.(18) can be written very concisely as a matrix equation by intro-
ducing the diagonal matrix of eigenfrequencies $\overline{\overline{\Omega}}=\vec{\Omega}\cdot\vec{1}$

$$d\vec{G}/dt = [i\overline{\overline{\Omega}} + \overline{\overline{\pi}}]\vec{G}, \quad \vec{G}(t) = \vec{G}(0)\exp[(i\overline{\overline{\Omega}} + \overline{\overline{\pi}})t] \tag{19}$$

Again, we will not continue to the explicit solution or applications; the reader is referred to /HarB 75/. Suffice it to say
that this treatment has also found very widespread application e.g.
in the work of Blume and his group (e.g. /ClaB 71/, and that it can
be applied for any ratio between the spread of frequencies Ω and
jump rates π.

8. A PERTURBATION APPROACH

If the perturbation is weak, an expansion of the exponential
(5) may be attempted. Let the time dependence of the hf frequency
be

$$\Omega_0 + \Omega(t) \quad \text{with} \quad \overline{\Omega(t)} = 0 \tag{20}$$

Consider now

$$G(t) = \exp[i\int_0^t (\Omega_0 + \Omega(t'))dt']$$

$$= e^{i\Omega_0 t}\{1 + \int_0^t i\Omega(t')dt' - \int_0^t \Omega(t')[\int_0^{t'}\Omega(t'')dt'']dt' + \cdots\} \tag{21}$$

The second term in the curly bracket averages to zero because of eq.
(20). The way in which the double integral is written for the third
term implies the physical argument that the momentary frequency
$\Omega(t')$ can only be influenced by things that have happened at times
t" earlier than t'. From (21) we get

$$dG(t)/dt = \{i\Omega_0 - \Omega(t)\int_0^t \Omega(t')dt' + \cdots\}G(t) \tag{22}$$

Compared to (18) the variation in frequency is completely shifted
into the second term on the right. This term we write

$$\int_0^t \Omega(t)\Omega(t')dt'$$

It is the correlation function of the perturbation Ω, a statistical
measure of the strength of the perturbation Ω and of its correlation
at different times.

If these correlations are independent of t and extend only over
times short compared to the times t˙for which we want to know G(t),
and if they are symmetric in time, G(t) becomes exponentially damped

$$dG/dt = \left\{ \iota\, \Omega_o - 1/T_r \right\} G \qquad\qquad (23a)$$

$$1/T_r = \int_0^\infty \Omega(0)\,\Omega(t')\,dt' \qquad\qquad (23b)$$

For illustration, assume (9). Then $1/T_r = \Omega^2 \tau_c$ is positive and the
connection with section 5 is $1/T_r = 1/T_{(1)}$ for the relaxation time.

The important and very general result is that the relaxation is
determined by the correlation function (23b) of the perturbation.
Thus the correlation function for the nuclear spin orientation, G(t),
has been reduced to the correlation function for the perturbation.
(A simpler model for this one may yield a good approximation for
G(t). Also this reduction can be continued further down on the line
of effect and cause, see comments to eq.(18).)

Again this treatment becomes much more complicated when it is
applied to a real quantum mechanical system. This is mostly done
using the superoperator language[+]. It can then be extended e.g. to
include the case of several lines which are separated in the absence of fluctuations.

For a thorough discussion, we refer to the work of Gabriel and
his collaborators /GabB 72/, /BerG 73/.

9. THE SPECTRAL DENSITY

We turn finally to a way of looking at relaxation processes
which is especially obvious when one starts from thinking in terms
of m-states. Consider again the case of a nuclear state, spin $\vec{I}$,
split by a strong external magnetic field $\vec{B}$ into Zeeman sublevels
$-I \leq m \leq I$ spaced by Ω_L .If there is in addition a fluctuating
magnetic field $\vec{B}'$ which happens to change at the frequency Ω_L and
which has a magnetic field vector perpendicular to $\vec{B}$, it will induce
transitions between the sublevels. As an example, the additional

---

[+]While the usual quantum mechanical operators act on states, the
eigen-energies being the energies of the states, superoperators act
on matrix elements and the associated energies are the energy differences between the states. This may serve as a hint to a   connection
between the treatments discussed in the present and in the preceding
section.

field may be produced by a nucleus with moment μ', so that B' is of the order μ'/r^3, and it may change because each neighbour nucleus with its particular spin component along z changes place by diffusional jumps at an average rate $1/\tau_c$. If the field fluctuations are statistical, as in the example sketched, the net result of the transitions between the sublevels is a relaxation of the spin $\vec{I}$. Qualitatively this is seen from eq.(1), because the $p_m = p_-(t)$ will be **driven toward being almost equal, if $\hbar\Omega_L \ll k_B T$; this time dependence is then expressed through $G_k(t)$ in (2).**

The rate of change of the populations is calculated through time dependent perturbation theory /Abr 61/, /Sli 63/. As in section 8, the rate of change is determined by the correlation function of the perturbation eq.(23b); $\Omega(t)$ is now $(\mu/I\hbar)B'(t)$. To calculate the induced transitions, one introduces the "spectral density" which is the Fourier transform of the correlation function of the perturbation.

The situation is rather simple for a two level system (I = 1/2) in a strong external field $\vec{B}_o$, and for the case that the three components of the perturbing field $\vec{B}'$ jump independently between +B' and −B' at an average rate $1/\tau_c$ with a correlation function

$$<B'(0)B'(t)> = B'^2 e^{-|t|/\tau_c} \tag{24}$$

for all three components $B' = B'_x$, B'_y or B'_z. We then need the spectral density

$$J(\omega) = 1/2 (\mu/I\hbar)^2 \int_{-\infty}^{\infty} <B'(0)B'(t)> e^{-i\omega t} dt \tag{25}$$

(we assume that τ_c is short compared to the time of measurement and, also, $\tau_c \ll T_1$), which gives the nuclear spin lattice relaxation time T_1 as

$$1/T_1(\omega) = 2J(\omega) = (\mu/I\hbar)^2 B'^2 \frac{2/\tau_c}{\omega^2 + 1/\tau_c^2} \tag{26}$$

This result, well known under the name of Bloembergen, Purcell and Pound, displays many features which are rather general. Some of them (points 1,3,4 and 5) are illustrated in fig. 4.

1. T_1 is frequency dependent because of the frequency spectrum contained in (24). Very often, the strongest variation in $J(\omega)$ in a given system occurs through the temperature dependence of τ_c, as e.g. in all the examples quoted in section 2.

2. The frequency integral which contains $\int G(\omega) d\omega = G(t=0)$,

$$\int_{-\infty}^{\infty} 1/T_1(\omega) d\omega = (\mu/I\hbar)^2 B'^2 \tag{27}$$

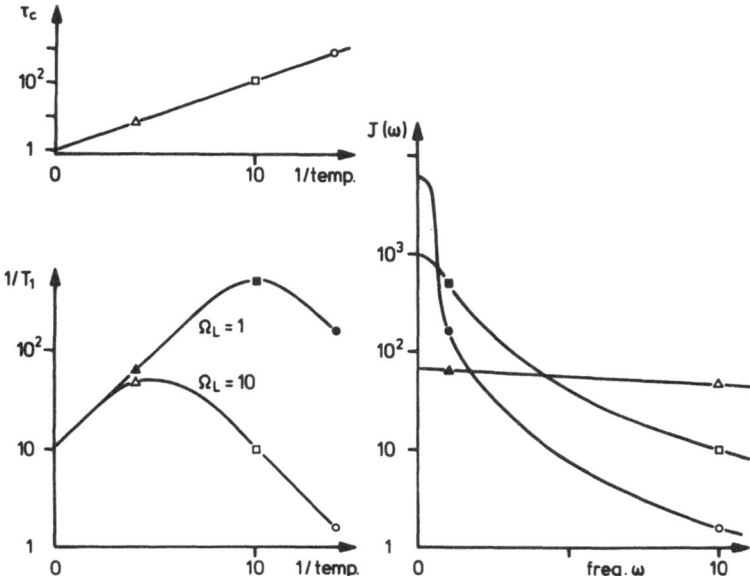

Fig. 4   A relaxation rate $1/T_1$, measured at two Larmor frequencies $\omega = \Omega_L$ as a function of temperature, is represented in the lower left. The right is a reconstruction of $J(\omega)$ for three temperatures from the T_1-data. The corresponding correlation times appear in the upper left. (Schematic)

measures the strength of the perturbation. Note that the measure-ment has to be extended over a frequency range[+)] $\omega \gg 1/\tau_c$.

3. T_1 depends also on the external field applied through Ω_L. Relax-ation is fastest when $\Omega_L \, \tau_c = 1$

$$1/T_1 \, (\Omega_L = \tau_c^{-1}) = (\mu/I\hbar)^2 \, B'^2 \tau_c$$

4. When τ_c is very short we return to the motional narrowing case mentioned several times: the frequency spectrum, the spectral den-sity, spreads its fixed area eq.(27) over a very large frequency range (of order $1/\tau_c \approx 10^{11}$/sec in a normal liquid), leaving very little energy per frequency interval near Ω_L. Here $1/T_1 \sim \tau_c$ and it is independent of frequency as long as $(\Omega_L \tau_c)^2 \ll 1$.

5. When τ_c is very long, the spectral density compresses into a very narrow frequency range (e.g. 1/(diffusive jump time) which falls far below 1/sec in a solid at low temperatures), so that no spectral density is available at the Larmor frequencies at which

---

[+)] See /Elw 78/ for a recent comment on this point.

one usually works $(10^3...10^8/sec.)$. Here $1/T_1 \sim \Omega_G^{-2}\tau_G^{-1}$. Note that $1/T_1(\omega=0)$ gives the average G because $G(\omega=0)^1 = \int G(t)dt$.

6. When $J(\omega)$ is due to a diffusion process, the details of the diffusion process in $\vec{r}$-space have only minor influence on $T_1(\omega)$. For a detailed discussion, see e.g. /MesN 75/, /BarS 76/.

A recent experiment on diffusion which illustrates the discussion given and which uses an excited nuclear state, is /AckD 75/. The effect of vacancy diffusion on $G(t)$ was calculated in /AbrG 77/.

To conclude this qualitative introduction into the spectral density concept, it should be pointed out that the theoretical treatments /Abr 61/, /Sli 63/, /HarB 75/ actually consider a generalization of the occupation probability, the density matrix. It includes the correlations between the probability amplitudes for the m-states (see the discussion at the end of section 5).

10. CONCLUSION

Relaxation is a dynamic process resulting from an interplay between the interaction strength (Ω) and its rate of change (ω). It was the purpose of the foregoing discussion to help to visualize nuclear spin relaxation as a consequence of statistical interactions. The basic similarities between the classical pictures of a precessing magnetization (section 6) or of a diffusing spin orientation (section 5), and the more formal approaches which are suited for a thorough theoretical treatment (sections 7,8), should have come out.

Nuclear spin relaxation is a means of studying the dynamical processes going on around a nucleus. However, one must attempt to cover experimentally a wide range of relaxation times, and one should keep in mind that a "relaxation rate" as observed rarely allows one by itself to identify its cause unambiguously.

The hospitality of the Institute Max von Laue-Paul Langevin is gratefully acknowledged.

REFERENCES

Abr 61   A. Abragam: The Principles of Nuclear Magnetism,
          Clarendon Press, Oxford 1961

AbrG 77   C. Abromeit and H. Gabriel: Hyp. Int. 3 (1977) 231;
          see also P.A. Fedders: Phys. Rev. B14 (1976) 1842

AckD 75   H. Ackermann, D. Dubbers, M. Grupp, P. Heitjans,
          R. Messer and H.J. Stöckmann, phys. stat. sol. b71
          (1975) K91

AnaM 73   H.R. Anand and J.G. Mullen; Phys. Rev. B8 (1973) 3112

BarS 76   W.A. Barton and C.A. Sholl: J. Phys. C9 (1976) 4315

Ber 79   I. Berkes, this volume

BerG 73   H. Bernas and H. Gabriel; Phys. Rev. B7 (1973) 468

BirC 78   H.K. Birnbaum, M. Camani, A.T. Fiory, F.N. Gygax,
          W.J. Kossler, W. Rüegg, A. Schenck and H. Schilling:
          Phys. Lett. 65A (1978) 435

ClaB 71   M.J. Clauser and M. Blume: Phys. Rev. B3 (1971) 583

Elw 78   M. Elwenspoek: Molecular Physics 1978, in the press

FreF 67   A.J. Freeman and R.B. Frankel (eds): Hyperfine Inter-
          actions, Academic Press, New York 1967

GabB 72   H. Gabriel and J. Bosse: Festkörperprobleme 12 (1972)
          505; also in H. von Krugten and B. van Nooijen (eds):
          Angular Correlation in Nuclear Disintegration, Rotter-
          damm U.P., Rotterdamm 1971, p.

GonI 74   F. Gonzalez-Jimenez, P. Imbert and F. Hartmann-Boutron:
          Phys. Rev. B9 (1974) 95

GotH 73   A.M. Gottlieb and C. Hohenemser: Phys. Rev. Lett. 31
          (1973) 1222

HarB 75   F. Hartmann-Boutron: Annales de Physique 9 (1975) 285.
          This review contains extensive references to earlier
          work.

HypI 75   Hyperfine Interactions starting 1 (1975) published by
          North-Holland, Amsterdam. Especially Vol. 4, Proceedings
          of the Fourth International Conference on Hyperfine Inter-
          actions, Madison 1977, and the earlier conferences

referred to there.

MeiG 68   H. Meinke and F.W. Gundlach: Taschenbuch der Hochfre-
             quenztechnik, Springer, Berlin 1968, p. 1356

MekK 76   M. Mekata, Y. Kano, M. Moriya, K. Tsuji and T. Haseda:
             J. Phys. Soc. Japan $\underline{41}$ (1976) 1918

MesN 75   R. Messer and F. Noack: Appl. Phys. $\underline{6}$ (1975) 79

Rec 79   E. Recknagel: this volume

Sch 68   L.I. Schiff: Quantum Mechanics, Mc Graw Hill, New York
             1968, p.74

Sli 63   C.P. Slichter: Principles of Magnetic Resonance, Harper
             and Row, New York 1963

WicK 66   H.H. Wickmann, M.P. Klein and D.A. Shirley: Phys. Rev.
             $\underline{152}$ (1966) 345, see also H. de Waard and R.M Housley
             in /FreF 67/ p. 691

DIRECTIONAL DISTRIBUTION OF
NUCLEAR GAMMA RADIATION

I. Berkès

Institut de Physique Nucléaire (and IN2P3)
Université Claude Bernard Lyon-1
43, Bd du 11 Novembre - 69621 Villeurbanne, France

I. DIRECTIONAL DISTRIBUTION OF AN ELECTROMAGNETIC RADIATION

A nucleus is composed from uncharged neutrons and positive-charged protons. Both have a magnetic dipole moment too. In order to understand the existence of the angular distribution of the nuclear gamma radiation, let us look first only at the charge- distribution of this nucleus.

Let us consider an odd nucleus with an even number of protons and neutrons in a spherical core and an odd proton outside this core (fig. 1a). In an excited state of this nucleus the charges can exhibit individual or collective vibrations, rotations, etc .... The accelerated charges radiate an electromagnetic wave and excitation energy of the nucleus is evacuated via electromagnetic radiation.

If an electric charge q is accelerated with an acceleration $\vec{a}$, we observe at a distance $\vec{R}$ from the charge a time-dependent electric and a magnetic field :

$$\vec{E}(t + \frac{R}{C}) = q \frac{\mu_o}{4\pi} \frac{\vec{R}_o \times \vec{R}_o \times \vec{a}(t)}{|R|}$$

$$\vec{B}(t + \frac{R}{C}) = q \frac{\mu_o}{4\pi C} \frac{\vec{a}(t) \times \vec{R}_o}{|R|}$$

respectively. (Eqs are given in the International MKSA system :
$\mu_o = 4\pi\ 10^{-7}$ Vs/Am , C : light velocity)

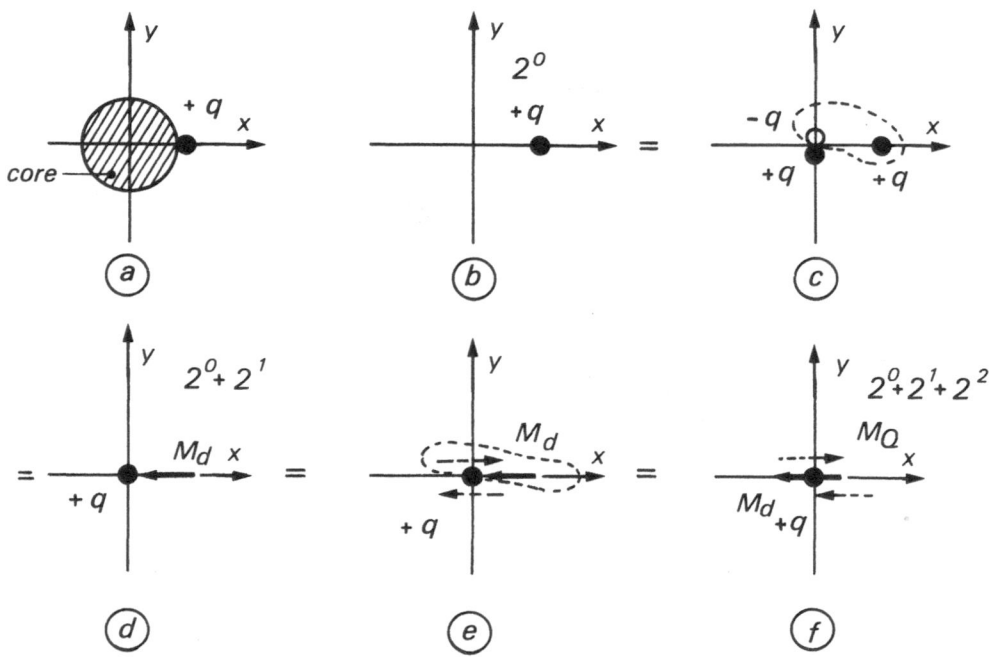

Fig. 1 - Graphical representation of multipole order development
of charge distribution.

The density of radiated energy is expressed by Poynting's
vector $\vec{S} = \vec{E}x\ \vec{H}$. Using $\vec{B} = \mu_o \vec{H}$ and expressing multiple vector
products we find :

$$\vec{S} = q^2\ \frac{\mu_o}{16\pi^2 C}\ \frac{a^2}{R^2}\ \vec{R}_o\ [1 - \cos^2(\vec{R}_o\ \vec{a})] \tag{1}$$

Eq. (1) is time dependent. If the absolute value of $\vec{a}$ varies
in time, $\vec{S}(t)$ gives the instantaneous energy flux, retarded with
respect $a(t')$:

$$t = t' + \frac{R}{C}$$

Usually $\vec{a}$ varies periodically in time. As $\vec{S}$ is quadratic in a,
the time-averaged energy flux, the well known effective value is
just obtained by replacing in Eq. (1) $\vec{S}$ by $\vec{S}_{eff}$ and $\vec{a}$ by $\vec{a}_{eff}$, i.e.

$$\vec{S}_{eff} = \frac{\mu_o q^2}{16\pi^2 C} \frac{a^2_{eff}}{R^2} R_o \lceil 1 - \cos^2(\vec{R}_o \vec{a})\rceil \qquad (1')$$

The qualitative features of the nuclear gamma radiation can be easily understood in the following way : let us assign the radiation to the motion of the only one proton situated outside the core. If this proton didn't move, the static electric potential would be described by U(core) + U(single charge). The first term is central in the coordinate system, the second can be expressed in a multipole order development. The term U(single charge) (fig. 1b) stays unchanged when adding a +q -q pair of charges in the origin (fig. 1c) and the charge-configuration is now equivalent to a charge +q in the origin and an electric dipole which is not centered (fig. 1d) By adding in the origin two opposite dipoles (fig. 1e) we obtain a charge plus a dipole in the origin and an non-centered electric quadrupole (fig. 1f). The static potential can be described in

$$U(core) + U(2^o) + U(2^1) + U(2^2) + U^{(2L)} \quad etc \ldots terms,$$

i. e. L=0 (monopole), L=1 (dipole), L=2 (quadrupole) etc .. terms. If now this charge moves, monopole terms (core + 2°) do not change, but the dipole, quadrupole, etc .. static potentials are to be replaced by the retarded potentials of oscillating 2^L pole configurations. The multipole order L of the radiation is given by the multipole order of the charge oscillation. As static potential does not contribute to the radiation field, the lowest multipole order for an electromagnetic radiation is L =1 (dipole).

Let us consider the radiation emitted by an electric dipole. Two situations can present :

1) Dipole oscillation along the z-axis

In this case the acceleration vector $\vec{a}$ is directed along a constant axis, chosen as the z-axis, so $\cos(\vec{R}_o \vec{a})$ is a constant. Denoting the angle between $\vec{R}_o$ and $\vec{a}$ as θ, the emitted radiation shows an angular distribution expressed in the brackets of Eq. 1 as $W(\theta) \propto 1 - \cos^2(\vec{R}_o \vec{a}) \propto \sin^2 \theta$.

If the oscillation along the z-axis is periodical, the time-averaged mean value of the radiated energy is correctly given by (1) :

$$\vec{S}_{eff} = \frac{\mu_o 2q^2}{16\pi^2 C} \cdot \frac{a^2_{eff}}{R^2} \vec{R}_o \sin^2 \theta = S_o \vec{R}_o \sin^2 \theta \qquad (2)$$

This is the well known dipole-antenna radiation. As the oscillation is always oriented along the z-axis, no angular momentum is carried away by the radiation. The electromagnetic radiation is linearly polarised in the plane [xy]. Optical spectroscopists assign the letter π to this radiation.

2) The dipole rotates around the z-axis in the [xy] plane.

The angle $(\vec{R}_o \vec{a})$ is variable. If φ is the azimuthal angle, $\cos(\vec{R}_o \vec{a}) = \sin\theta . \cos\varphi$. During the rotation φ varies as $\varphi = \omega t$ and the time-averaged mean value of $\cos^2(\vec{R}_o \vec{a})$ enters into (1) :

$$< \cos^2(\vec{R}_o \vec{a}) > \ = \sin^2\theta < \cos^2\varphi > \ = \frac{1}{2} \sin^2\theta$$

an the directional distribution of the emitted radiation becomes

$$W(\theta) \propto \tfrac{1}{2} (1 + \cos^2\theta)$$

As $|\vec{a}| = $ constant, $|\vec{a}| = a_{eff}$, and the time-averaged mean value of the radiated energy becomes :

$$\vec{S}_{eff} = S_o \vec{R}_o \frac{1 + \cos^2\theta}{2} \tag{3}$$

The rotating dipole has an angular momentum, which disappears as the radiation damps the rotation. Thus this dipole radiation carries away an angular momentum which we can consider as ± 1 unity of angular momentum (The sign depends on the rotation sens). The emitted radiation is circularly polarized around the z-axis. Optical spectroscopists assign the letter σ to this radiation.

Eqs. (2) and (3) show, that directional distribution is a classical property of electromagnetic radiation. In atomic or nuclear physics the z-axis is chosen as quantization axis. Its significance will be discussed in chapters 2 and 3. Fig. 2 shows the distributions (2) and (3) in a polar representation. In table I we summarize properties of a dipole radiation.

⎯ · ⎯ · ⎯ · ⎯

Let us consider a nucleus emitting an electromagnetic radiation (γ-ray). The intensity of this radiation can be calculated from Maxwell's equations. If the current-density in the nucleus is $\vec{j}(\vec{r}, t)$ and the charge density $\rho(\vec{r}, t)$ the vector and scalar potentials $\vec{A}(\vec{r}, t)$ and $U(\vec{r}, t)$ are given by

$$\Delta\vec{A} - \frac{\partial^2 \vec{A}}{\partial t^2} = -\mu_o \vec{j}(\vec{r}, t)$$

$$\Delta U - \frac{\partial^2 U}{\partial t^2} = -\mu_o c^2 \rho(\vec{r}, t) \tag{4}$$

and the solutions in terms of retarded potentials give the directional distribution of the intensity of emitted radiation.

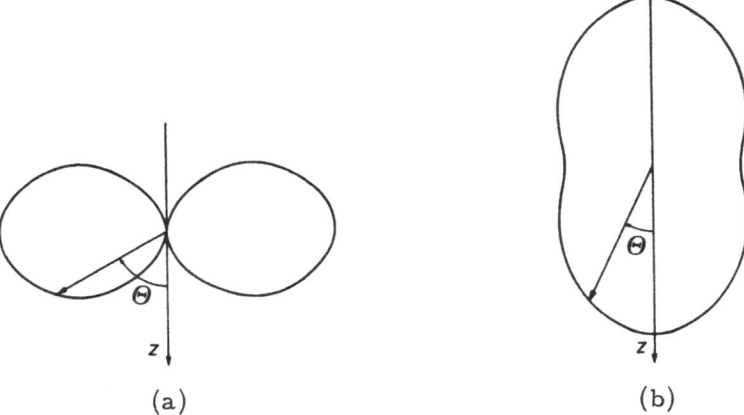

(a)   (b)

Fig. 2 - Polar diagram of the intensity of the dipole-radiation.
(a) Oscillating dipole along the z-axis [(1, 0) mode. The
first number is L, the second $M = m_f - m_i$] ; (b) Rotating dipole around the z-axis [(1, ± 1) mode] .

Table I - Properties of dipole radiation

	Properties	Oscillating (1,0) mode	Rotating (1, ± 1) mode
Classical	Angular distribution	$\sin^2 \theta$	$\dfrac{1 + \cos^2 \theta}{2}$
	Polarization	linear (π)	circular (σ)
	Angular momentum carried away	0	± 1 unity
Quantum mechanical	Changement of magnetic quantum number $(M = m_f - m_i)$	0	∓ 1 unity

The charge density $\rho(\vec{r}, t)$ and current density $\vec{j}(\vec{r}, t)$ are con-
sidered as periodic in time : $\rho(\vec{r}, t) = \rho(\vec{r}) e^{i\omega t}$ and $\vec{j}(\vec{r}, t) = \vec{j}(\vec{r}) e^{i\omega t}$
The space-dependent term can be expanded in multipole order de-
velopment. As we are interested in spherical-wave solutions far
from the nucleus, we develop the solution in terms of the well-
known spherical harmonics.

$$Y_{LM}(\theta,\varphi) = (-1)^{\frac{1}{2}(M+|M|)} \left[\frac{(2L+1)(L-|M|)!}{(L+|M|)!} \right]^{1/2} e^{iM\varphi} P_L^{|M|}(\cos\theta)$$

Introducing the angular momentum operator $\vec{L} = i\vec{r}\times\vec{\nabla}$ the transition electric multipole moment becomes :

$$\vec{\mathcal{M}}(EL, M) = \frac{1}{\omega(L+1)} \int_{\tau'} \vec{j}(\vec{r}')\vec{\nabla}\times\vec{L} r'^L Y_{LM}(\xi,\eta) \, d\tau' \qquad (5)$$

where $\vec{r}'(r',\xi,\eta)$ are coordinates of nucleons inside the nuclear volume τ'. The statique electric multipole operator (see e. g. Recknagel's lecture) is defined in a similar way.

$$\vec{\mathcal{M}}_{st}(EL, M) = \int_{\tau'} \rho(\vec{r}') \, Y_{LM}(\xi,\eta) \, d\tau' \qquad (6)$$

If a radiation is emitted by a configuration of moving electric charges, whose dominating term in (5) is of order L, the emission is called an electric 2^L transition. E. g. $L = 1$ means dipole, $L=2$ quadrupole radiation, etc ...

The protons and neutrons of the nucleus have also magnetic moments, and the internal nuclear motion gives rise to magnetic oscillations. The magnetic transition multipole moment is defined in a similar way as (5) to be : —

$$\vec{\mathcal{M}}(ML, M) = \frac{-i}{L+1} \int_{\tau'} \vec{j}(\vec{r}')\vec{L} r'^L Y_{LM}(\xi,\eta) \, d\tau' \qquad (7)$$

The angular distribution of a radiation, emitted by a nucleus oriented along the z-axis will be described by a periodic function of θ, where θ is the angle between the z-axis and the direction of observation. We calculated these distributions in a classical way for dipole transition. For the general case see detailed discussion in ref. 1. We summarise here some notions.

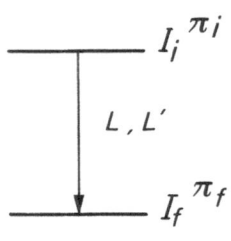

The initial state of the γ-decay is characterized by spin and parity $I_i^{\pi_i}$, the final state by $I_f^{\pi_f}$ (fig. 3). The multipolarity of the transition between I_i and I_f is L. The conservation of angular momentum

$$\vec{I}_i = \vec{L} + \vec{I}_f \qquad (8)$$

Fig. 3 - Spin and parity selection rules

allows several multipolarities L, L' for the γ-transition.

The parity of electromagnetic transition is for an :

Electric transition   (EL) $\pi_\gamma = (-1)^L$

Magnetic transition   (ML) $\pi_\gamma = (-1)^{L-1}$

As for gamma radiation parity is conserved : $\pi_i = \pi_\gamma \pi_f$

The angular momentum and parity conservation rules allow some mixtures between different multipole transition modes. For $I_i^{\pi_i} = 2^+$ and $I_f^{\pi_f} = 2^+$, for example, we can have :   M1, E2, M3, E4.

Intensity rules show, however, that the higher the multipole order of a transition, the lower is its intensity. In general, a radiation of multipolarity L+2 is $\approx 10^4$ times weaker than that of multipolarity L.   Thus we never consider terms other, than the two lowest multipole order transitions (electromagnetic transitions of zero order do not exist)[x].   In our preceeding example the possible mixture is   limited to M1 + E2.

   Eqs. (5) and (7) show that, disregarding scalar and phase factor, the electric multipole operator differs from the magnetic one by replacing $\vec{L}$ by $\vec{rot}\ \vec{L}$. Now Maxwell's equations for periodic vibrations in vacuum $\vec{rot}\ \vec{E} = -\frac{\partial \vec{B}}{\partial t} = -i\omega \vec{B}\ e^{i\omega t}$ and $\vec{rot}\ \vec{B} = \frac{1}{c^2}\frac{\partial \vec{E}}{\partial t} = \frac{i\omega}{c^2}\vec{E}$ show that $\vec{E}$ and $\vec{B}$ have symmetrical roles in the radiation field and the $\vec{rot}$ operation changes only $\vec{E}$ into $\vec{B}$ or vice verse. As $\vec{S} = \vec{E}x\vec{H}$ the directional distribution of an EL transition will be the same as that of an ML transition. The electric or magnetic character of a γ-radiation can be seen only in the linear polarization of the radiation.

   The directional distribution of γ-radiation emitted by an oriented I_i state is developed in terms of Legendre polynomials. Each term will be proportional to product $A_k(L\ L'\ I_i\ I_f)\ P_k(\cos\theta)$ where $P_k(\cos\theta)$ are Legendre polynomials of order k and $A_k(L\ L'\ I_i\ I_f)$ are the distribution coefficients of the electromagnetic radiation of multipolarities L, L'.

   It can be shown (see e.g. ref. 1) that for a γ-transition of multipolarity L, L', A_k is given by a combination of the Ferentz-Rosenzweig coefficients F_k (ref. 2) :

$$F_k(L\ L' I_i I_f) = (-1)^{I_f - I_i - 1}[(2L+1)(2L'+1)(2I_i+1)(2k+1)]^{1/2}$$
$$\begin{pmatrix} L & L' & k \\ 1 & -1 & 0 \end{pmatrix} \begin{Bmatrix} L & L' & k \\ I_i & I_i & I_f \end{Bmatrix} \tag{9}$$

where ( ) and { } are 3j and 6j coefficients, respectively.

---

[x]   A transition between two 0+ levels (E0 transition) can occur via internal conversion

If the selection rules allow more than one multipolarity, in-terference terms rise in A_k. Denoting by α the amplitude of the wave of multipolarity L, by β that of multipolarity $L' = L+1$, the composite amplitude becomes $\alpha + \beta = \alpha(1 + \frac{\beta}{\alpha}) = \alpha(1+\delta)$. δ is the multipole mixture parameter : it is real, positive or negative. The intensity being proportional to the square of the amplitude $Int \propto (1+\delta)^2 = 1+2\delta + \delta^2$. The distribution coefficient $A_k(LL'I_iI_f)$ is proportional to :

$$\alpha^2 [F_k(L L I_i I_f) + 2\delta\ F_k(L\ L+1\ I_i I_f) + \delta^2 F_k(L+1\ L+1\ I_i I_f)]$$

As the directional distribution is normalized to $I(A_o=1)$ we must divide this expression by the total intensity $\alpha^2 + \beta^2 = \alpha^2(1+\delta^2) =$ thus :

$$A_k(L L'I_i I_f) = \frac{F_k(L L I_i I_f) + 2\delta\ F_k(L\ L+1\ I_i I_f) + \delta^2 F_k(L+1\ L+1\ I_i I_f)}{1 + \delta^2} \tag{10}$$

The summation index k satisfies the following selection rule $\vec{L} + \vec{L}' = \vec{k}$ i.e. $\min |L-L'| \leq k \leq \max (L+L')$ $\tag{11}$

The Legendre polynomials of order k are composed from $\cos^k(\theta)$, $\cos^{k-2}(\theta)$, etc .. terms, which all have the parity $(-1)^k$. Odd terms mean asymmetry for angles θ and $\theta + \pi$ (left-right asymmetry). Such an asymmetry can occur in two cases :

i)   Detection sensitive to circular polarization of the emitted electromagnetic radiation (left or right hand polarized, in our example according to the rotation-sens of the dipole)

ii)   Interference between at least two levels $|I_i m_i >$ and $|I'_i m'_i >$
       a) Level spacement between adjacent levels smaller than Γ, the natural linewidth
       b) The $|I'_i m'_i >$ level is a different substate of the same I_i level : $|I'_i m'_i > = |I_i m'_i >$. This situation occurs in a pertur-bed angular distribution or correlation, and in the reorienta-tion of Coulomb excitation.

Items i) and ii-a) don't occur in investigations subject to this school. ii-b) will be discussed in **Prof. Recknagel's** lecture. In this lectu-re we limit the development to k = even values.

II. ORIENTATION OF A NUCLEAR STATE

Let us consider a decaying nucleus with initial state's spin $I_i = 1$ leading to a final state $I_f = 0$ (fig. 4). The populations of the substates of the $I_i = 1$ state are $\rho(-1)$, $\rho(0)$ and $\rho(1)$ respectively. On account of the spin selection rules (8), the transition is dipolar $(L=1)$, thus the energy radiated into the unit solid angle directed to angle θ from the z-axis is :

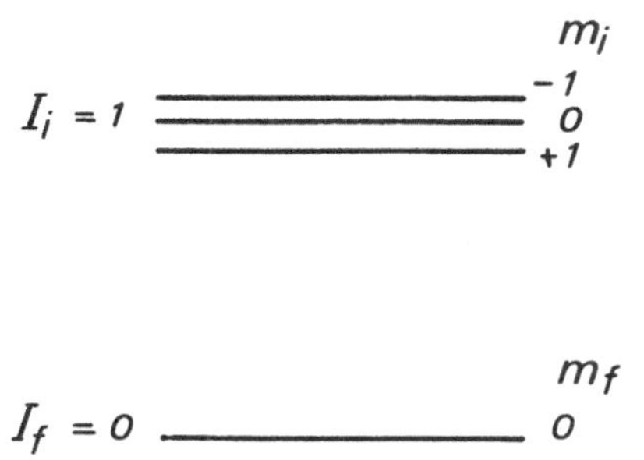

$$m_i$$

$$I_i = 1 \quad \overline{\overline{\overline{\underline{\overline{}}}}} \begin{array}{l} -1 \\ 0 \\ +1 \end{array}$$

Fig. 4 - Orientation of an I = 1 state.

$$m_f$$

$$I_f = 0 \quad \underline{} \quad 0$$

$$\vec{S}_{eff}(\theta) = \sum_{m_i} \rho(m_i) \, \vec{S}_{i\,eff}(\theta) =$$

$$= S_o \vec{R}_o \left\{ \rho(0)\sin^2\theta + [\rho(-1) + \rho(+1)] \frac{1+\cos^2\theta}{2} \right\} \qquad (12)$$

The energy radiated into solid angle 4π is :

$$S_{tot} = \int_{\theta=0}^{\pi} S_{eff}(\theta) \, 2r\pi \sin\theta \, d\theta = \frac{8\pi}{3} S_o [\rho(-1)+\rho(0)+\rho(+1)]$$

As $\quad \sum \rho(m_i) = 1, \quad S_{tot} = \frac{8\pi}{3} S_o$

Eq. (12) shows that the angular distribution of the radiated energy depends on the substate-populations $\rho(m_i)$. If they are equal i. e. $\rho(-1) = \rho(0) = \rho(+1) = 1/3$. $\vec{S}_{eff}(\theta) = \frac{2}{3} S_o \vec{R}_o$ is isotropic. Classically the inequality of the populations of different m_i substates means that the ensemble of nuclei is oriented with respect to the z-axis. In absence of such an orientation every microscopic antenna radiates according to its own angular characteristic but the overall distribution is isotropic.

Without entering into details, three different types of orientation can be considered.

1. Static orientation.

The energy difference between different substates $e_{mm'}$ is not negligible as compared to $\varkappa T$, the thermal energy. The population $\rho(m_i)$ of the $|I_i m_i >$ substates are determined by Boltzmann's law.

2. Dynamic orientation.

The sample-atom is oriented by absorption of electromagnetic radiation (optical pumping, laser-spectroscopy), and the orientation of the atom is transferred to the nucleus via hyperfine interaction. After the switching off of the pumping the dynamic orientation disappears within the spin-lattice relaxation time (see below).

3) Transient orientation

A preceeding nuclear transformation (particle capture or emission, γ-ray capture or emission) can leave the I_i state oriented. The following example is taken from ref. 3. Fig. 5 represents a nuclear γ-ray cascade. We perform a coincidence measurement between $γ_1$ and $γ_2$ rays as a function of the angle between the two γ-rays.

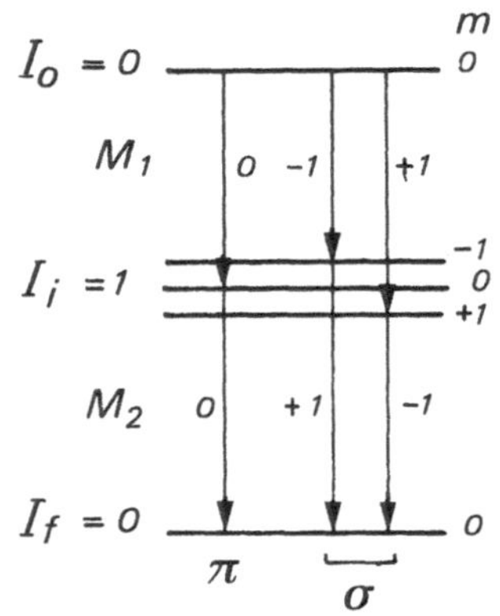

Fig. 5.   0-1-0 spin sequence angular correlation

Let us consider the second transition ($γ_2$). If the observed transition was of a σ type ($M_2 = ±1$) (carrying away ±1 unity of angular momentum) its angular distribution is described by eq. (3) :

$$W(γ_2^σ) \approx 1 + \cos^2 θ$$

If it was of a π type ($M_2 = 0$), its angular distribution with respect to the quantization axis is (eq. 2) :

$$W(γ_2^π) \approx \sin^2 θ$$

Let us choose the direction of the first γ-ray detected by the coincidence system as quantization axis. Obviously this γ-ray could not have been of π-type, because its intensity would have been $\propto \sin^2 0° = 0$, thus not observed (Physicaly this is the only direction to choose as quantization axis). Thus $γ_1$ is of σ type (M =+1 or -1) and in absence of an intermediate-state perturbation $γ_2$ must have been also of σ type (see fig. 5),   and its angular distribution with respect to the first emitted γ-ray (called angular correlation) is given as :

$$W(θ) = 1 + \cos^2 θ$$

This example shows that the nuclear transformation and the measuring device selected intermediate states with $m_i = ±1$ (aligned state).   The $ρ(I_i m_i)$ populations of substates of the intermediate state in this   example are thus $ρ(1, 0) = 0$ $ρ(1, 1) = ρ(1, 1) = 1/2$.

The alignment of the intermediate state is not conserved indefinitely.   Magnetic and electric hyperfine interactions can induce transitions between different m_i substates and thus modify the $ρ(m_i)$ populations.   Classically this means that the hyperfine coupling

changes the degree of orientation of the ensemble of nuclei. The
hyperfine field always contains a time-dependent component due to
thermal agitation. The higher the sample temperature, the greater is the effect of this time-dependent perturbation, which equili-
bes the populations of the substates. This equilization implies an
energy transfer from the nuclear system to the lattice system :
After several spin lattice relaxation times the nuclei will show the
thermal equilibrium density distribution (see i), i.e. at room-temperature practically equal densities in every substate. The degree
of equilization of populations of the intermediate state I_i depends
on the ratio of its lifetime to the spin-lattice relaxation. We must
remember, however, that angular correlation technique is applicable as long as the intermediate states lifetime does not exceed
$\approx 10^{-6}$ s. As room-temperature spin lattice relaxation time is of
the order of 10^{-4} s., during the existence of the intermediate state
the "transient" nature of the orientation is not always evident and
often we can consider $\rho(m_i)$ population time-independent.

-:-:-:-:-:-:-

As to describe oriented nuclei, a multipole order development of the population distribution is used. The orientation parameter of rank k of a state of spin I, $B_k(I)$ is defined (ref. 1) as :

$$B_k(I) = [(2k+1)(2I+1)]^{1/2} \sum_{m=-I}^{+1} (-1)^{I+m} \begin{pmatrix} I & I & k \\ -m & m & 0 \end{pmatrix} \rho(m) \qquad (13)$$

B_k orientation parameters exhibit also k = odd terms if $\rho(m) \neq \rho(-m)$
This is the case for magnetic polarization at low temperature, and
in some nuclear reaction-induced polarizations. Eq. (13) is normalized to $B_o = 1$.

The directional distribution of a radiation emitted by an oriented state will be :

$$W(\theta) = \sum_k B_k(I_i) A_k(L L' I_i I_f) P_k(\cos \theta) \qquad (14)$$

where B_k orientation parameters are defined by the orientation mechanism.

III.   EXPRESSION FOR B_k ORIENTATION PARAMETER IN SOME PARTICULAR CASES

3.1 - Static orientation at low temperature

The population of substates is given by Boltzmann's law :

$$p(m) = \frac{\exp(-e_m/\varkappa T)}{\sum_{m=-I}^{+I} \exp(-e_m/\varkappa T)} \tag{15}$$

3.1.1 - Magnetic_orientation. $e_m = -mg\,\mu_N B$, μ_N being the nuclear magneton $\mu_N = 5.05\ 10^{-27}$ J/T, B the applied magnetic field in T, g the nuclear g-factor. The orientation axis is that of the magnetic field.

Defining $\Delta_M = g\,\mu_N B/\varkappa$ we can write in useful units (mK°) :

$$\Delta_M = 0.366\ g\ B^{(T)} \tag{16}$$

As $p(m) \neq p(-m)$, odd k orientation parameters do not vanish in magnetic orientation.

Numerical values for B_k for $1/2 \leq I \leq 8$ are tabulated in ref. 4.

Let us consider for example, the decay of fig. 4.   Eq.(12) shows immediately that the angular distribution of a magnetically oriented nucleus does not give any information on the sign of the nuclear magnetic moment. Changing g into -g in Eq. (12) leaves Eq.(12) unchanged.

For completely oriented nuclei $(e_m \gg \varkappa T)$ $p(-1) = p(0) = 0$ and $p(+1) = 1$.   The radiation is a σ type dipole radiation, whose directional distribution is given by $W(\theta) = 1 + \cos^2\theta$.

From tables (2) and (4) we find :

$$B_2 = \sqrt{2}/2 \qquad A_2(1\ 1\ 1\ 0) = \sqrt{2}/2 \qquad A_{k>2} = 0$$

$$W(\theta) = 1 + \frac{1}{2} P_2(\cos\theta) = \frac{3}{4}(1 + \cos^2\theta) \propto 1 + \cos^2\theta$$

in agreement with above calculation.

3.1.2 - Electric_quadrupole_orientation. Only axially symmetric electric field gradients are considered : $V_{xx} = V_{yy}$ $(\eta = 0)$. The quantization axis is directed along V_{zz} $(|V_{zz}| > |V_{xx}|)$

$$e_m = \frac{e\,Q\,V_{zz}}{4\,I(2\,I-1)}\,[3\,m^2 - I(I+1)]$$ (17)

where Q is the nuclear electric quadrupole moment

defining

$$\Delta_E = \frac{3\,e\,Q\,V_{zz}}{4\,\chi\,I\,(2\,I-1)}$$ (18)

and using the well-known quadrupole frequency :

$$\nu_Q = \frac{-\,e\,Q\,V_{zz}}{4\,h\,I\,(2\,I-1)}$$ (19)

we find in practical units :

$$\Delta_E\,\frac{mK^\circ}{} = -\,0.144\,\nu_Q\,MHz$$ (20)

As $e_m = e_{-m}$, k = odd terms vanish (alignment). Numerical values for B_k are given in ref. 4.

Considering again our example of the $I_i = 1 \to I_f = 0$ decay, eq. (12) reduces to :

$$\vec{S}_{eff}(\theta) = S_o\vec{R}_o\,[\rho\,(0)\,\sin^2\theta + \rho\,(\pm 1)\,(1 + \cos^2\theta)\,]$$ (12')

In the case of a complete orientation in the   fieldgradient   of a hexagonal crystal $(\eta = 0)$ we find for :

$QV_{zz} > 0$ the lowest state is $m = 0$ thus $\rho(\pm 1) = 0$ $\rho\,(0) = 1$

$QV_{zz} < 0$ the lowest state is $m = \pm 1$ thus $\rho(\pm 1) = 1$ $\rho\,(0) = 0$

The corresponding angular distributions are thus :

for $QV_{zz} > 0$ $W(\theta) \propto \sin^2\theta$

for $QV_{zz} < 0$ $W(\theta) \propto 1 + \cos^2\theta$

The angular distribution is different for positive or negative quadrupole couplings. This difference holds, in general, for other spin and transition multipolarities also.

The table of ref. 4 gives for $QV_{zz} > 0$ $B_2 = -\sqrt{2}$,   for $QV_{zz} < 0$, $B_2 = 1/\sqrt{2}$,   confirming the classically derived angular distribution.

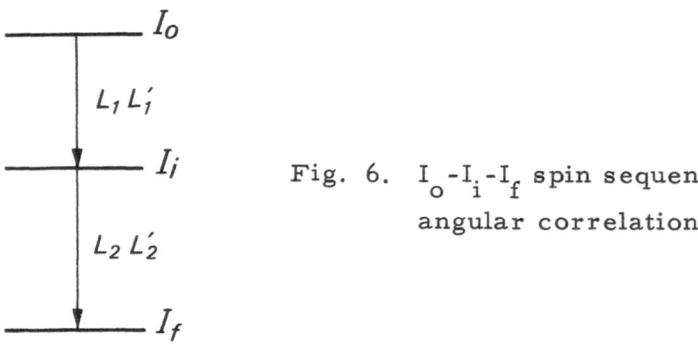

Fig. 6. I_o-I_i-I_f spin sequence

angular correlation

3.2 - Orientation by absorption or emission of a
            γ-ray (angular correlation)

The initial I_o state is not oriented. The directional distribution of
the second γ-ray with respect to the first one is given by

$$W(\theta) = \sum_k B_k(I_i) A_k(L_2 L'_2 I_i I_f) P_k \cos \theta)$$

As shown in parag. 2.3, the I_i state is oriented with respect to the
direction of the first γ-ray. It is shown (ref. 1) that :

$$B_k(I_i) = \frac{F_k(L_1 L_1 I_i I_o) - 2\delta F_k(L_1 L+1 I_i I_o) + \delta^2 F_k(L_1+1 L_1+1 I_i I_o)}{1 + \delta^2}$$
(21)

Comparison of (20) with (10) shows that $B_k(I_i) = A_k(L_1 L'_1 I_i I_o)$ the
distribution coefficient of the first radiation taken in inverse sens
$I_i \rightarrow I_o$ with the sign of interference term inversed.

The angular correlation is often written in the following form

$$W(\theta) = \sum_k A_k(1) A_k(2) P_k(\cos \theta) = \sum_k A_{kk} P_k (\cos \theta)$$

where

$$A_{kk} = A_k(1) A_k(2)$$
(22)

The sign convention of δ in (21) is as accepted in the scientific literature since the Delft-conference on angular correlation
(1970). Older publications use different sign conventions, resulting in ambiguity for the sign of the multipole mixing parameter.

Eq. (21) and (22) hold also if the first γ-ray is not emitted
but absorbed by a non-oriented nucleus (resonant absorption).

IV.   UNOBSERVED INTERMEDIATE TRANSITION

The observed γ-ray is often emitted by a state which is not the directly oriented state.   The radiation leading from the orien-ted state to the initial state of the γ-transition is unobserved, but as it carries away some angular momentum, it modifies (attenua-tes) the orientation.   Two examples are shown in fig. 7

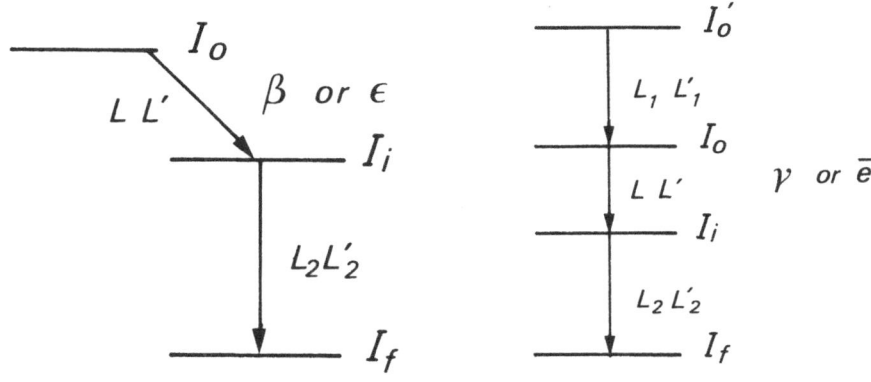

Fig. 7

a) Directional distribution of γ-ray following β (or elec-tron capture) decay of orien-ted nuclei

b) Angular correlation between the first and 3-d γ-ray with intermediate (γ or internal conversion) transition unob-served.

The disorientation coefficient U_k has the same form in case (a or b).   It is given by (ref. 1) :

$$U_k(L\,I_o\,I_i) = (-1)^{I_o+I_i+k+L}\,[(2\,I_o+1)(2\,I_i+1)]^{1/2}\begin{Bmatrix} I_o & I_o & k \\ I_i & I_i & L \end{Bmatrix} \quad (23)$$

where { } is a 6j-coefficient.   Tables for U_k are in annexe of ref. 1.

U_k do not contain interference terms L L'. If any mixture of multipolarities L and L' occurs, the addition is incoherent . With amplitude mixture parameter β we have for the ratio of in-tensities L and L' :

$$\text{Int}\,(L')\,/\,\text{Int}\,(L) = \beta^2$$

$$U_k = \frac{U_k(L\,I_o\,I_i) + \beta^2\,U_k\,(L'\,I_o\,I_i)}{1 + \beta^2}$$

If there are several succeeding unobserved transitions

$$U_{ki} = \prod_i U_{ki}$$

The directional distribution for oriented nuclei take the form (fig. 7a) :

$$W(\theta) = \sum_k B_k(I_o) U_k(L I_o I_i) A_k(L_2 L'_2 I_i I_f) P_k(\cos \theta) \qquad (14)$$

and for an angular correlation (fig. 5 b)

$$W(\theta) = \sum_k A_k(L_1 L'_1 I_o I'_o) U_k(L I_o I_i) A_k(L_2 L'_2 I_i I_f) P_k(\cos \theta) \qquad (22')$$

V. CONSIDERATION OF FINITE SOLID ANGLE OF DETECTORS

If the detector has finite dimensions, the measured directional distribution is somewhat smoothed out with respect to the physical distribution. Let $\theta_\gamma \varphi_\gamma$ be the angle between the quantization z-axis and the γ-ray emitted, θ the angle between the z-axis and the axis of the detector and $e(\beta)$ be the efficiency of the cylindrically symmetric detector, where β is the angle between the detector axis and the impact of the γ-ray given by $\theta_\gamma \varphi_\gamma$, the measured $\tilde{W}(\theta)$ becomes

$$W(\theta) = \frac{\int W(\theta_\gamma \varphi_\gamma) \, e(\beta) \, d\Omega}{\int e(\beta) \, d\Omega} \qquad (24)$$

$\tilde{W}(\theta)$ can be developed in Legendre polynomials. It can be shown that (ref. 5)

$$W(\theta) = \sum_k B_k U_k A_k Q_k P_k (\cos \theta) \qquad (13'')$$

where

$$Q_k = \frac{\int P_k(\cos \beta) \, e(\beta) \, \sin \beta \, d\beta}{\int e(\beta) \, \sin \beta \, d\beta} \le 1 \qquad (25)$$

The attenuation is usually incorporated into the distribution coefficient A_k

$$A_k(\text{measured}) = A_k Q_k \qquad (26)$$

For an angular correlation (involving two detectors) :

$$A_k(1, \text{measured}) = A_k(1) \, Q_k(1)$$

and

$$A_k(2, \text{measured}) = A_k(2) \, Q_k(2)$$

thus :

$$Q_k = Q_k(1) \; Q_k(2)$$

Q_k values are tabulated (e.g. ref. 5). For detection at small solid angle Q_2 does not differ much from unity. On the other hand, for large detectors (great solid angles) the attenuation of the k = 4 term may be important. Q_k is also energy-dependent : the higher the γ-ray energy, the nearer is Q_k to unity.

REFERENCES

1. K. ALDER, R.M. STEFFEN, Emission and absorption of electromagnetic radiation. In D. Hamilton : "The electromagnetic interaction in nuclear spectroscopy" (p. 1), North-Holland Publ., Amsterdam, 1975

   R.M. STEFFEN, K. ALDER, Angular distribution and correlation of gamma-rays., In D. Hamilton : "The electromagnetic interaction in nuclear spectroscopy", (p. 505), North-Holland Publ., Amsterdam, 1975

2. A.H. WAPSTRA, G.F. NIJGH, R. VAN LIESHOUT, Nuclear Spectroscopy Tables (p. 98), North-Holland Publ., Amsterdam, 1959, or

   T. YAMAZAKI, Nuclear Data, Section A, 3, (1967), 1

3. J.M. BLATT, V.F. WEISSKOPF, Theoretical Nuclear Physics, J. Wiley & Sons, New York, 1952

4. Kenneth S. KRANE, Orientation parameters for low-temperature nuclear orientation, Nucl. Data Tables, 11, (1973), 407

5. M.J.L. YATES, Finite solid angle corrections, In In K. Siegbahn, "Alpha, Beta and Gamma-ray Spectroscopy", (p. 1691), North-Holland Publ., Amsterdam, 1965

NUCLEAR ORIENTATION

I. Berkès

Institut de Physique Nucléaire, (et IN2P3)
Université Claude Bernard Lyon-1
43, Bd du 11 Novembre - 69621 Villeurbanne, France

I.   COOLING-DOWN TECHNIQUES

The chapter 3.1 of the lecture "Directional distribution of nuclear gamma radiation " (called below DDGR) shows that the temperature range necessary to observe a measurable anisotropy of γ-rays emitted by oriented nuclei is given by $T \approx e/\varkappa$: $T = 3 - 30\,mK$. Two techniques are mainly developed for these investigations : the adiabatic demagnetization and the ^3He-^4He dilution cycle.

1.1 - Adiabatic demagnetization

A paramagnetic salt is hanged in a cryostat on nylon strings. The salt is cooled down by evaporation of a liquid helium bath to about $T_i \approx 1\,K$. During this cooling-down the demagnetizing magnet, developing a magnetic field B up to 7 T is switched on. The atomic magnetic moments of the paramagnetic salt are oriented at this temperature and the atomic substates are populated according to Boltzmann's law $\rho(m) = \exp(-m\mu_e B/\varkappa T J)$ where J and μ_e are the electronic angular momentum and magnetic moment of the paramagnetic atom, and B the applied magnetic field.

Let us switch off slowly B. As liquid helium is evaporated, there is no more heat contact between the paramagnetic salt and the cryostat. Neglecting small heat losses and eddy current heating during the demagnetization, $dQ \approx 0$, thus the entropy S of the ensemble (salt + sample) stays constant. As $S = \varkappa \ln W$, the occupation of substates must also stay unchanged, and so B/T.

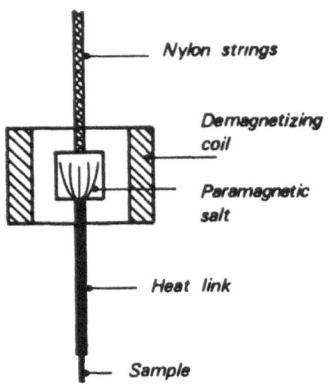

Nylon strings

Demagnetizing coil

Paramagnetic salt

Heat link

Sample

<u>Figure 1</u> - Scheme of an adiaba-

tic demagnetization stage.

The residual local magnetic field $B_r (\approx 0.05\ T)$ is due to the atomic dipoles itself so the temperature falls theoretically to $T \approx \frac{B_r}{B} T_i$. The lowest temperatures achieved with this method are of the order of 3 mK. The warm-up speed depends on the quantity of paramagnetic salt and the quality of the cryostat : typically the temperature is lower than 20 mK for more than 10 hours. There is no continuous cooling–power and, when the sample is warmed-up the cryostat must be refilled with liquid helium and the cycle of cooling-down started again. The heat-link between the paramagnetic salt and the sample is made of some thousands of thin wires or foils.

1.2 - ^{3}He-^{4}He dilution refrigerator

At temperatures lower than 0.8 K liquid ^{3}He is soluble in suprafluid ^{4}He. Evaporation of ^{3}He from dilute solution needs heat ; this heat quantity is higher than that released, if the same quantity of ^{3}He is dissolved in concentrated phase. Thus, it is possible to operate a continuous-cycle dilution refrigerator.

It can be shown, that ^{3}He dissolved in ^{4}He behaves as an ideal gas in the volume of ^{4}He. The passage of ^{3}He through the surface of phase-separation, like a Joule-Thomson expansion, conserves enthalpy. The cooling-power derived from the enthalpy-conservation rule is

$$\dot{Q} = (96.5\ T_m^2 - 12.5\ T_e^2)\ \dot{n} \qquad (1)$$

where $\dot{Q}$ is the cooling power in W, $\dot{n}$ the quantity of circulating ^{3}He in mol/s, T_m and T_e are temperatures of the mixture and the entering concentrated ^{3}He, respectively. The lowest theoretical temperature (base-temperature) is given from $\dot{Q} = 0$ as $T_m \approx 0.4\ T_e$.

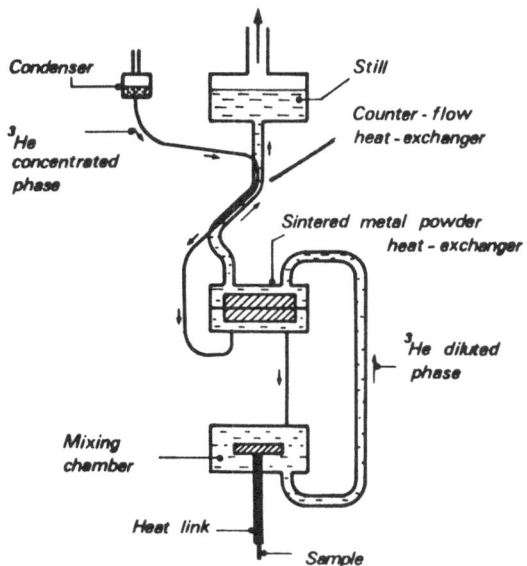

Figure 2 - Scheme of a ^{3}He-^{4}He dilution refrigerator.

As to get low base-temperatures, the concentrated ^{3}He must enter into the mixing chamber at temperatures as low, as possible. Heat of incoming concentrated phase must be transferred to outgoing dilute phase before entering into the mixing chamber. This heat-transfer is realised in the heat-exchangers. On account of the bed low temperature heat transfer between liquids and solids (Kapitza-resistance), great-surface heat exchangers (sintered silver or copper) are emloyed. The heat contact between sample and circulating helium mixture is also realised by means of sintered blocs. If the reduced Kapitza-resistance is denoted by $R_{Kr}(m^2K^4/W)$ and the surface of heat exchangers by σ (m^2), the mixing chamber temperature given in K is given by (cf. ref. 1) :

$$T_m = \left[6.4\ R_{Kr}\ \frac{\dot{n}}{\sigma} + 1.22 \times 10^{-2}\ \frac{\dot{Q}}{\dot{n}} \right]^{1/2} \qquad (2)$$

The lowest temperature achieved - measured inside the mixing chamber - is actually 2 mK. Fig. 2 shows the scheme of the dilution unit of a dilution refrigerator.

It is possible to load samples into the cryostat during operation. The cooling-down time (down to $\approx$ 20 mK) is thus reduced to 3 hours about (ref. 2). Successful experiences allow even direct implantation of radioactive atoms into the refrigerator during operation (ref. 3).

The advantages of the dilution refrigerator and the adiabatic demagnetization can be combined. A paramagnetic salt is hanged with a supraconducting wire, serving as a heat-switch, on the mixing chamber. The salt is cooled down with the dilution refrigerator to 20-30 mK, and the adiabatic cooling starts at this temperature. Recently the temperature of $\approx$ 0.6 mK has been atteined in Grenoble with this method.

2. METHODOLOGY OF NUCLEAR ORIENTATION EXPERIMENTS

Nuclei can be oriented using magnetic dipole or electric quadrupole coupling. In both cases B_k depends on $1/T$ (see DDGR eqs 13 and 15). For this reason measurements are performed in function of the temperature and results are plotted versus inverse temperature.

Usually two detectors are used : one placed in the orientation axis, the other 90° with respect to. The countings are normalized to the high temperature countings, so $W(T \to \infty) = 1$ (T = 1K can be considered as sufficiently high temperature). Figs 3a) and 3b) show a magnetic and an electric orientation experiment, respectively.

In a nuclear orientation experiment β -particles and γ-rays can be both observed. If β -particles are observed in a magnetic orientation experiment, odd terms in the series development $W(\theta) = \sum_k B_k U_k A_k P_k (\cos \theta)$ can also occur. The quantitative measurement of the angular distribution of emitted electrons in a magnetic field is very difficult technically. Observation of the β - particles emitted by oriented nuclei is restricted to detect the sign of the product μB, or to detect the resonant destruction of the orientation pattern. So our following considerations will be limited to the observation of γ-rays.

As long, as T is not really small as compared to $e/\varkappa$, the dominating terms in the series development are the $k = 0$ and $k = 2$ terms. As $P_o(0°) = P_o(90°) = 1$, $P_2(0°) = 1$ and $P_2(90°) = -0.5$, the effect measured at 0° is about twice as strong (with inverted sign) as that at 90°.

In the high temperature limit $e/\varkappa \ll 1$ all orientation parameter B_k with $k > 2$ are negligible. For magnetic orientation B_2 is proportional to $(e/\varkappa T)^2$, for electric orientation to $(e/\varkappa T)$.

For very low temperatures, i.e, $T \ll e/\varkappa$, $W(\theta)$ does not depend anymore on e, only on nuclear parameters $U_k A_k$. Thus, a fit on $1/T$ on a large temperature scale can give both B_k and $A_k U_k$

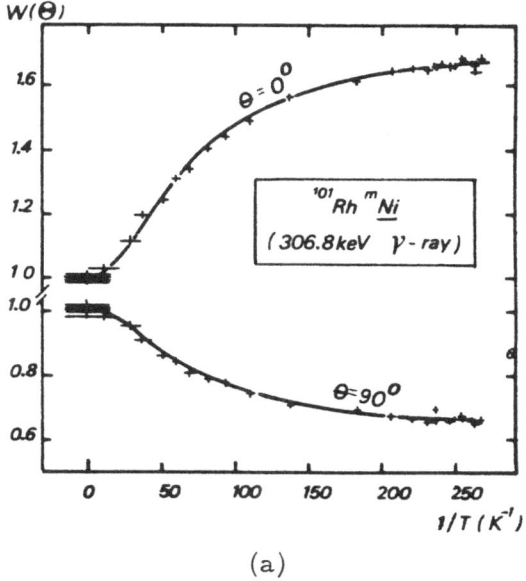

(a)

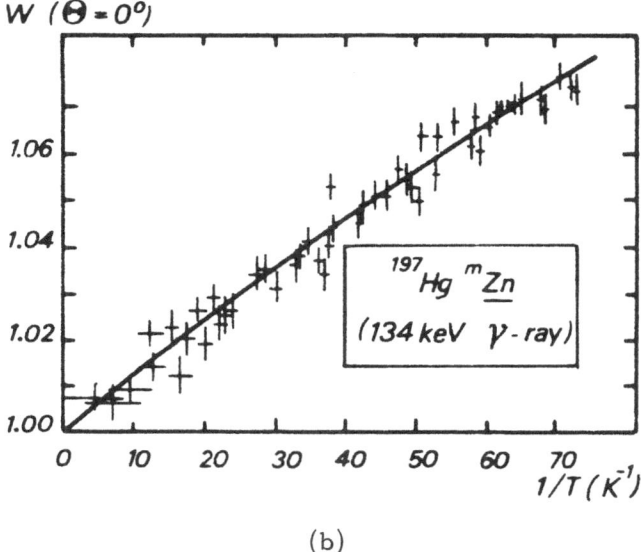

(b)

Figure 3 - Angular distribution of γ-rays emitted by an oriented
nucleus
(a) Magnetic dipole orientation (Ref. 4)
(b) Electric quadrupole orientation (Ref. 5)

as well.   The errors on the two sets of parameters are, however,
correlated.   As to have a sufficiently good reliability on each set,
a great number of points versus temperature are needed.   The
temperature-dependent recording is performed either during the
warm-up period (adiabatic demagnetization technique), or during
the cooling-down (dilution refrigerator). If this period is comparable to the radioactive decay period of the sample, the $W(0°)/W(90°)$
ratio, independent of the decay, can be used to fit orientation parameters and distribution coefficients.

A powerful   method to determine e without the detailed knowledge   of $A_k U_k$ is the resonant destruction of nuclear orientation
It is discussed in Stone's lecture.

The temperature can be measured by means of a "nuclear
thermometer", i.e. with the orientation of a radiactive source,
whose orientation parameter and distribution coefficients are well
known.   The most common nuclear thermometers are [57] Co and
[60]Co, imbedded into Fe, Co or Ni.   Fig. 4 shows the counting
rates versus invers temperature of the 1332 keV or 1173 keV transition of ^{60}Co, imbedded into iron in the direction (0°) and perpendicular to (90°) the external polarizing field.   Other methods, e.g.
the increase of resistivity of silicon resistors or the increase of
the magnetic susceptibility of paramagnetic salts are used mainly
by solid state physicists.   The former is appropriate in the temperature range 0.02-4 K, while the second one is useful even at lower
temperatures.

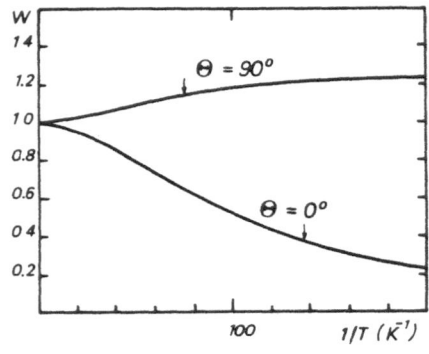

Figure 4 - Calculated
orientation of the 1173 keV
and 1332 keV γ-rays of
[60]Co embedded into iron.
$B_{hf} = 28.75$ T,
$\mu(^{60}Co) = 3.79 \mu_N$

3.   BRUTE FORCE NUCLEAR ORIENTATION

"Brute force" means that the orientation is due to an
external /macroscopic/ field. Macroscopic electric field gradients
are so small, that the only possibility, to orient nuclei with brute
force is the magnetic orientation.

The orientation parameter B_k/I for a magnetic orientation depends on I and $\mu_B/\varkappa\,TI$. The behaviour of B_k is such, as in a rough approximation B_k depends on $\mu_B/\varkappa\,T$ i.e. the magnetic orientation effect depends on μ and not on the g-factor $g = \mu/\mu_N I$.

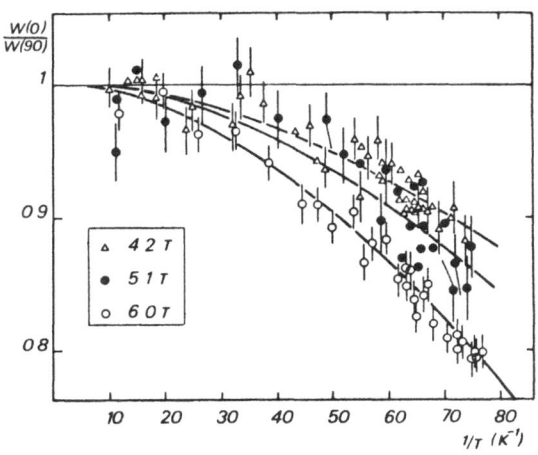

Figure 5 - Brute force magnetic orientation of ^{87m}Y (Ref. 6)

Thus brute force nuclear orientation may be successful for high magnetic moment/usually high spin/nuclei. On the other hand, very low temperatures are required. Fig. 5 show the $W(0°)/W(90°)$ ratio of ^{87m}Y oriented with 3 different magnetic field values. The magnetic moment of the metastable state is $\mu \approx 6\mu_N$. The major part of the curve is in the quadratic region.

Brute force method is advantageous to measure magnetic moment, Knight-shifts, and quadrupole-interactions in polycrystalline sources (cf. par. 4.3 and 4.4)

Thus brute force nuclear orientation may be successful for high magnetic moment / usually high spin / nuclei. On the other hand, very low temperatures are required. Fig. 5 shows the $W(0°)/W(90°)$ ratio of ^{87m}Y oriented with 3 different magnetic field values. The magnetic moment of the metastable state is $\mu \approx 6\,\mu_N$. The major part of the curve is in the quadratic region.

Brute force method is advantageous to measure magnetic moment, Knight-shifts, and quadrupole-interactions in polycrystalline sources (cf. par. 4.3 and 4.4)

4. ORIENTATION WITH MAGNETIC AND ELECTRIC HYPERFINE INTERACTION

4.1 - Orientation in a magnetic alloy

The sample nucleus is alloyed into a magnetic metal (iron, cobalt, nickel, gadolinium, Heussler-alloy, etc..) either in dilute form, or in stoichiometric concentration. The alloy is polarized with an external field to saturation. The hyperfine magnetic

field may be oriented parallel or antiparallel to the external polarizing field.

The effective field acting on the sample is :

$$\vec{B}_{eff} = \vec{B}_{hf} + \vec{B}_{ext} \qquad (3)$$

The choice of B_{ext} is significant, as (3) is valid only if the magnetic alloy is completely saturated. The saturation magnetic field in
iron is about 2 T, but the corresponding excitation H is not defined
unambiguously. This ambiguity has two reasons :

a) The physical curve B(H) is smoothly varying near the saturation.
b) The magnetic field inside a ferromagnetic sample is less than
μ_{rel} times the external field (μ_{rel} is the relative permeability of
the magnetic material). For a rotational ellipsoid (a cylindrical
sample may be assimilated to) of length a and diameter b the effective permeability is given in table 1, for two different relative
permeability materials. The calculation is linear, so valid only
far from the saturation.

Table 1 - Effective permeability giving $B_{int} = \mu_{eff}(B_{ext})$ for an
ellipsoidal sample.

a/b	$\mu_{rel} = 200$	$\mu_{rel} = 1000$
4	7.5	7.75
10	28.5	32
20	68	97

Table 1 shows, that unless for very thin samples, a considerable magnetizing field must be applied to achieve saturation.   On
account of this magnetizing field, B_{ext} in (3) is smaller than the
really applied external field, and it is hazardous to estimate the respective contributions.   Therefore, for precise measurements
(e. g. NMR/ON) B_{eff} is measured in function of B_{ext} and the measured effect extrapolated to $B_{ext} = 0$ to find correct hyperfine
field B_{hf}.

In dilute alloys, especially in implanted ones, nuclei can occupy non-substitutional sites, and experience different $B_{hf,i}$ hyperfine magnetic fields. The orientation parameter B_k becomes :

$$B_k = \frac{\sum_i \alpha_i \, B_{k,i}}{\sum_i \alpha_i} \qquad \text{where} \quad B_{k,i}\left(I, \frac{\mu \, B_{eff,i}}{\varkappa \, T \, I}\right)$$

(Do not confound B_k orientation parameter with the magnetic field
B).

In most cases a two-parameter hypothesis is used to fit the
experimental orientation curve : α proportion of nuclei experience
the "correct" hyperfine field, whereas $1-\alpha$ proportion of nuclei
are localized in a small-field (usually zero field) site. This hypothesis is all the more justified, the nearer α to one is. Resonant
destruction of N.O. allows to determine unambiguously the correct hyperfine magnetic field.

4.2 - Orientation in an electric field-gradient

Electric field gradient is oriented along the cristalline axis,
so single crystal must be used, and the sample activity correctly
incorporated (implanted or alloyed) into the host crystal. The single crystal axis must be mounted with respect to the detector axes.

The orientation parameters B_2 and B_4 depend on the sign
of the quadrupole coupling QV_{zz}. B_4 is always positive, while the
sign of B_2 is inverse to that of QV_{zz}. The absolute values of B_2 an
B_4 are higher for negative than for positive quadrupole couplings.

Electric quadrupole coupling energies are usually 10-100 times smaller than magnetic hyperfine couplings, so, as to get a
measurable orientation, much lower temperatures must be achieved than in the former case. On account of this problem, and single crystal preparation difficulties, the pure electric quadrupole
alignment experiences are rare to this date, but on account of
their considerable interest their number increases.

4.3 - Orientation under the combined influence of oriented
magnetic and randomly oriented electric interaction

Electric quadrupole coupling can act on nuclei in matrices
exhibiting an electric field gradient, but also in cubic matrices, if
impurities vacancies or dislocations are near the sample nucleus.
Localisation of solid state defects is random around the sample
nucleus ; in polycristalline sources the quadrupole interaction is
random too. In absence of any oriented field, no overall orientation occur.

If a magnetic field B polarizes the sample, the nuclei are under the combined influence of the magnetic field and the electric field gradient. The energy of substates becomes $e_{magn.}$ + $e_{el.}$, but the quantization (orientation) axis is different for each nucleus. The superposition of the randomly oriented electric field gradient to the magnetic field yields an attenuation of the orientation parameter B_k. Supposing a unique axially symmetric field gradient V_{zz} the attenuated orientation parameter B_k can be written with the unperturbed orientation parameter B_k and a temperature dependent attenuation coefficient G_k as :

$$B'_k(I, \omega_m \omega_E) = B_k(I, \omega_m) \, G_k(I, \omega_E/\omega_m, T)$$

where $\omega_m = g \, B \mu_N / \hbar$ is the Larmor-frequency,

$$\omega_E = \frac{e \, Q \, V_{zz}}{4I(2I-1)\hbar}$$ the quadrupole frequency.

Calculated G_k attenuation coefficients for $I = 9/2$ are represented on fig. 6.

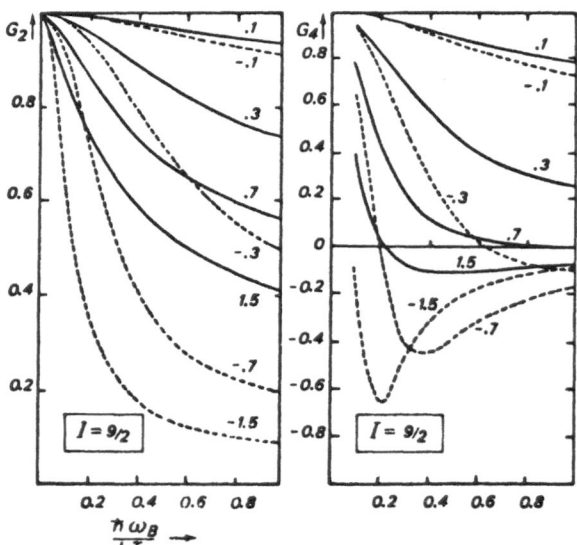

Figure 6 - G_2 and G_4 attenuation coefficients for combined magnetic dipole + random electric quadrupole interaction for $I = 9/2$ Curve parameters are ω_E/ω_m (Ref. 7).

On fig. 7 we represent the angular distribution $W(0)/W(90°)$ of γ-rays of oriented $^{87}Y^m$ nuclei, under 6T polarizing field, perturbed by a random impurity interaction (probably oxygen trapping). The effective field gradients in this case are 10-100 times greater than those observed in a regular lattice site.

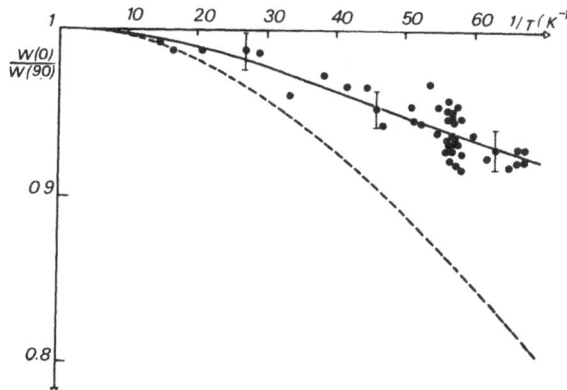

Figure 7 - Attenuation of the orientation pattern of the 381 keV γ-ray of $^{87}Y^m$, due to combined interaction (Ref. 6). Brooken line : Unattenuated pattern (see fig. 5) at B=6 T
Full line : Attenuated pattern. Fit parameter :
$$3\,\omega_E/\omega_m = +7$$

4.4 - Intermediate state reorientation (perturbed nuclear orientation, PNO)

If in a metastable state, other than the initially oriented one, the nucleus experiences a force, whose axis does not coincide with the orientation axis, this interaction mixes up the populations of the substates, thus changing the orientation of the metastable state. The perturbation of the nuclear orientation is time-dependent, and it is stronger, the greater the ratio ω_E/ω_m, calculated on the metastable state is. Orientation measures time-integrated perturbation : the attenuation is significant, as $\omega_E\tau$ is not negligible as compared to one.

Let $B_k(I_o)$ be the orientation parameters of the oriented state, $B'_k(I_i)$ that of the metastable state, and the transition $\vec{I}_o \rightarrow \vec{I}_i$ with multipolarity L is unobserved. We can write :

$$B'_k(I_i) = B_k(I_o)\, g_k(\omega_E, \omega_m, I_i, I_o, L, \tau, T)$$

where g_k contains also the attenuation due to unobserved intermediate transition (cf. DDGR eq. 23). g_k plays the same role in intermediate state reorientation, as $G_k(\infty)$ in the time-integrated perturbed angular correlation measurements.

The usual case is the same as in chap. 4.3. The magnetic polarizing field is oriented and the orientation of the electric field gradient is random. Attenuation due to perturbation of nuclear orientation has been observed for the first time by the Oxford-group (Ref. 8), and estimated theoretically for the intermediate-state's lifetime $\tau \rightarrow \infty$. More detailed calculations for axially symmetric electric field gradient are performed recently by Haroutunian, Meyer and Coussement, but on account of the great number of parameters involved numerical results cannot be tabulated. Fig. 8 shows the counting rate in the magnetic field direc-

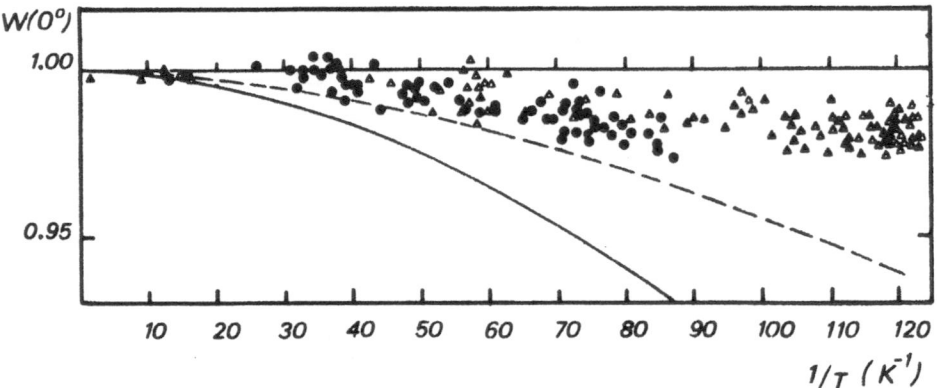

Figure 8 - Counting rate at 0° of the 247 keV γ-ray following ^{111}In decay (Ref. 9)

          6T : • measured points. Full line : calculated non-attenuated curve

          4T : Δ measured points. Broken line : calculated non-attenuated curve

tion of the 247 keV γ-ray of [111]Cd following ^{111}In decay, in a poly-cristalline cadmium lattice, with about 60% of ^{111}In nuclei situated is correct lattice site (The cadmium lattice is hexagonal). Comparison of measured points with the unperturbed orientation curve shows that perturbation of nuclear orientation modifies strongly the angular distribution of γ rays emitted by an oriented nucleus : for a given ω_E/ω_m ratio perturbation of nuclear orientation gives a 10-50 times stronger attenuation, than the combined interaction in the ground state, discussed in chap. 4.3

REFERENCES

(1)   G. Frossati, H. Godfrin, B. Hébral, G. Schumacher and D. Thoulouze, Proc. of the Ultra-Low Temperatures Symposium, Hakone, (Japan), 1977 (or, G. Frossati, Thesis, University of Grenoble, France, 1978)

(2)   I. Berkes, G. Marest and R. Haroutunian, Phys. Rev., C 15, (1977), 1839

(3)   P. Herzog, H.R. Folle, K. Freitag, A. Kluge, M. Reuschenbach and E. Bodenstedt, Nucl. Instr. Meth., (1978)

(4)   K. Kaindl, F. Bacon, H.E. Mahnke and D.A. Shirley,
        Phys. Rev., C 8, (1973), 1074

(5)   P. Herzog, K. Krien, J.C. Soares, H.R. Folle, K. Freitag,
        F. Reuschenbach, M. Reuschenbach and R. Trzcinsky,
        Phys. Lett., A 66, (1978), 495

(6)   G. Marest, R. Haroutunian, and I. Berkes, Phys. Rev.,
        C 17, (1978), 287

(7)   R. Haroutunian, M. Meyer and R. Coussement, Phys. Rev.,
        C 17, (1978)

(8)   M. Kaplan, P.D. Johnston, P. Kitteland and N.W. Stone,
        Nucl. Phys., A 212, (1973), 478   and
        P.J. Johnston and N.J. Stone, J. of Phys., F-4, (1974), 1522

(9)   R. Haroutunian, I. Berkes, M. Meyer, G. Marest and
        R. Coussement,   to be published

NMR DETECTED BY NUCLEAR RADIATION: NMR/ON

N. J. Stone

Clarendon Laboratory, Oxford OX1 3PU, U.K.

Full accounts have been given at this school of the three major techniques whereby hyperfine interactions (hfi) of nuclei can be measured using radiative emissions, namely the Mossbauer effect (MO), nuclear orientation (NO), and perturbed angular correlation (PAC). Stress has been laid on the highly specific signal from particular decay transitions and upon the low concentration and absolute number of nuclei required for a measurement. However typically all three techniques yield data accurate to a few percent and are thus of much lower precision than resonance methods, NMR and EPR, used in studies of stable isotopes present in solids at higher relative concentration and total number. Conventional NMR linewidths in metals are of order kHz at frequencies of hundreds of MHz, i.e. accurate to 1 in 10^5 or better.

Improved precision without loss of sensitivity or specific signal can be obtained by methods combining NMR with radioactive decay detection. The basic idea is simple. NO and PAC depend fundamentally upon setting up unequal populations of the nuclear sub-levels to give a non-uniform radiation distribution. Resonant transitions between the nuclear sub-levels can therefore be detected by the change in angular distribution produced. Extension to the Mossbauer effect is more subtle. Hamermesh (1) has shown **theoretically that the presence of resonant radiation directly** modifies the Mossbauer spectrum, introducing further line splitting. Alternatively if the source or absorber conditions are modified to give a departure from the normal pattern of line intensities through unequal level populations, resonance can be detected. Such modifications can be achieved by polarising source or absorber nuclei at low temperatures or by requiring a correlation between a preceding emission and the Mossbauer transition.

The fact that resonant transitions between the nuclear sublevels must be induced during the lifetime of the nuclear level
whose ordering is being measured introduces a very important practical differential between NMR applied to NO, PAC, and MO. In NMR
on oriented nuclei (NMR/ON) and in NMR/MO on polarised nuclei the
nuclear level concerned is the polarised long lived parent nucleus,
with lifetime >> 1s. In NMR/PAC and NMR/MO (by the Hamermesh
technique) the nuclear level concerned is the intermediate state
with lifetime < 10^{-5}s. The 10^5 fold increase in NMR transition
probability (hence radiofrequency (rf) power) required has greatly
inhibited the development of the latter techniques. NMR/MO has not
been unambiguously demonstrated experimentally (2) and NMR/PAC has
been little used in comparison with pulsed accelerator beam methods
SOPAD and DPAD (3). NMR/ON has been successfully observed in a
wide variety of systems, with nuclear orientation produced by low
temperature means, by optical pumping, by nuclear reaction and
recoil implantation, and by in situ neutron activation by polarised
neutrons. This paper deals primarily with the first of these,
which is limited to temperatures below $\sim$ 50 mK (4). The nuclear
reaction and polarised neutron activation techniques which both
produce non-equilibrium polarisation of a radioactive isotope are
briefly discussed at the end of the paper.

Resonant destruction of low temperature nuclear orientation
was first suggested by Bloembergen and Temmer in 1953, however early
attempts to realise the technique using demagnetised single crystals
of paramagnetic salts failed because of the short nuclear spin-
lattice relaxation times in such salts, which necessitate large rf
fields coupled with their limited enthalpy after demagnetisation.
Thus the system warmed up before resonance was observed.

The first successful experiment was done in Berkeley in 1966 by
Matthias and Holliday (5) and a listing of results prior to 1976 is
given in reference 4.   Success was based on the development of
paramagnetic systems cooling by adiabatic demagnetisation with large
thermal capacity below $\sim$20 mK, and, more recently, of ^{3}He/^{4}He
dilution refrigerators capable of continuous operation against a
several erg s^{-1} heat input at $\sim$ 15 mK.

The technique of NMR/ON requires only minor modification of a
standard NO system. A radiofrequency lead is needed to feed a small
coil (usually two current loops) around the sample to provide a
field B_1 parallel to the (metal) sample surface and perpendicular to
the axis of orientation. The great majority of results have been
obtained on nuclei in ferromagnetic metals and alloys in which the
orientation axis is the axis of an applied polarising field. A full
description of a dilution refrigerator system incorporating NMR and
low temperature implantation recently constructed in Bonn is given
in reference 6.

Since in NO all radioactive decays from the sample will contribute to the counting rate, for NMR/ON in metals the activity must be
confined to the rf skin depth, typically of order 10^{-6}m at 100 MHz.
This can be achieved by using thin foils, by controlled thermal
diffusion, or by ion implantation techniques. The radiofrequency
generator used should produce at least 1 m Gauss at the sample
surface and must have low residual frequency modulation (< 10 kHz)
for ferromagnetic samples. More stringent conditions will apply for
NMR/ON in systems with narrow linewidths, but to date few resonances
have been observed narrower than 500 kHz.

CONDITIONS NECESSARY FOR NMR/ON OBSERVATION

The conditions that a resonance be observable involve several
parameters of the apparatus and of the system under study. Here
these parameters are considered in general, with further particular
reference to ferromagnetic metal systems.

1. The Equilibrium Anisotropy of the Observed Radiation

The mechanisms of nuclear orientation are discussed in
chapter 9.   The maximum change in counting rate with orientation
is generally observed along the orientation axis where

$$W(\theta = 0) \quad = \quad 1 + \Sigma B_{\nu} U_{\nu} F_{\nu} Q_{\nu} \tag{1}$$

Since all nuclei in the sample emitting the gamma ray under study
contribute to the count rate, the resonance signal is a change
$\Delta W(0)$ and typically statistical fluctuations require that this be
at least 1% for detection. The sensitivity of NMR/ON thus depends
directly upon the degree of orientation,'i.e. upon the hyperfine
interaction strength and the sample temperature which can be main-
tained against all non-resonant heat leaks. A multiplicity of
lattice sites or an otherwise complex hyperfine interaction with
several transition frequencies increase the need for a high degree
of nuclear orientation if they are to be fully elucidated.

2, The Nuclear Spin-Lattice Relaxation Time

This time constant T_1 is a measure of the rate at which power
resonantly absorbed by the nuclear spin system is released to the
sample lattice. Thus a short T_1 requires larger rf power to maintain the same resonant disturbance of the nuclear sub-level populations . In metals the nuclear spins relax via the conduction
electrons, as first treated by Korringa. T_1 is determined by the
density of electron states at the Fermi surface $N(\epsilon_F)$, the magnitude
of the hyperfine splitting $h\nu$ (= $\mu_N B_{eff.}/I$ for a simple Zeeman
system), and the lattice temperature T, being given by

$$\frac{1}{T_1} = (\frac{h\nu}{2k_B C_K}) \coth(\frac{h\nu}{2k_B T}), \quad \text{where } C_K \text{ is the Korringa constant,}$$

theoretically $\dfrac{4\pi k_B}{\hbar} \dfrac{N(\varepsilon_F)^2}{(h\nu)}$. For temperatures higher than $\dfrac{h\nu}{k_B}$ this

approximates to the better known Korringa relation $T_1 T = C_K$. For impurities in ferromagnetic metals the resonant frequency dependence is well obeyed and for $^{60}\underline{CoFe}$, resonant at 166 MHz, C_K is close to 2 sK. Thus values of T_1 at $\sim$ 10 mK are 10–1000 s in such alloys. These are much longer than for paramagnetic salts, and are a major factor in the successful observation of NMR/ON in ferromagnetic metals.

3. Γ, the Resonance Linewidth

Contributions from the natural linewidth $1/T_n$, the nuclear dipole-dipole linewidth $1/T_2$, and a power broadening produced by the rf radiation are all present for each nucleus, giving rise to an intrinsic linewidth Γ_i where

$$\Gamma_i = \left[\frac{1}{T_n^2} + \frac{1}{T_2^2} + (\gamma B_1^n)^2 \frac{T_1}{T_2}\right]^{\frac{1}{2}} \tag{2}$$

where γB_1^n is the product of the nuclear gyromagnetic ratio γ and the rf field amplitude B_1^n at the nucleus. However calculated values of Γ_i in ferromagnetic metals are of order a few hundred Hz, two or three orders of magnitude smaller than the linewidth Γ_F typically found in NMR in metals, with NMR/ON linewidths still larger at $\sim$ 1 MHz. These large widths Γ_F are due to inhomogeneties in the local hyperfine splitting over the nuclei in the source.

Such linewidths are far wider than the residual frequency spread of an acceptable rf oscillator. Thus in order to simultaneously influence the orientation of an appreciable proportion of the nuclei in the sample, the frequency must be modulated over a range $\Delta\omega$ of order Γ_F. For a perfectly phase stable oscillator the Fourier transform of a modulated wave is a comb of frequencies with interval equal to the modulation frequency. However provided this is less than the residual F.M. width of the oscillator (usually > 1 kHz at 100 MHz), modulation produces essentially a uniform spread of rf power over $\Delta\omega$, with a consequent power reduction at any one nucleus by $\Gamma_i/\Delta\omega$.

4. Radiofrequency Power

Both resonant and non-resonant rf power absorption are proportional to the square of the applied rf field B_1. Resonant absorption depends upon B_1^n at the nucleus, and in ferromagnetic and paramagnetic systems this may be enhanced over the applied field, through response of the electronic magnetisation, by a factor $(1+B_{hf}/B_e)$ where B_e is the field tending to hold the electronic magnetisation (the external field in a paramagnet, the anisotropy field in a ferromagnet). Enhancement is of order 100–1000 and experimentally a value of B_1 of order 10^{-7} T is found adequate for NMR/ON in ferromagnetic metals. Non-resonant power absorption arises form eddy currents in the sample and other parts of the cooling system and can be a serious source of heat input.

In metals the rf requirement imposes another limitation in that the skin depth, $\delta = (\frac{1}{2}\sigma\omega\mu\mu_o)^{-\frac{1}{2}}$, for iron is of order 1 μm at 100 MHz.

In conventional NMR the resonance saturation condition is given by $\gamma^2 B_1^2 T_1 T_2 \gg 1$. With modulation, and assuming that for very dilute systems $T_2 \approx T_1$, this condition becomes

$$\gamma^2 B_1^2 T_1^2 \Gamma_i / \Delta\omega \gg 1$$

Experimentally non-resonant heating limits the usable rf power.

In summary, resonance will be observed provided the induced change in nuclear level populations, allowing for relaxation, is sufficient to give an observable change $\Delta W(0)$ without non-resonant heating becoming too serious.

LINESTRUCTURE, SIGNAL AND MEASUREMENT TECHNIQUE

Linestructure

An elegant treatment of radiative detection of NMR for a purely magnetic hyperfine interaction has been given by Matthias et al. (8) who consider the effects of varying relaxation time and the degree of radio frequency coherence. The effects predicted by this treatment, of multipole split resonance lines, have been observed in PAC/NMR (9) but not to date in NMR/ON where inhomogeneous broadening Γ_F is so much larger than the power broadening Γ_i that predictions based on a coherent rotation of the angular distribution pattern are not observed. With such broadening, observed resonance lines are Gaussian.

For NMR/ON in ferromagnetic metals the hyperfine interaction hamiltonian may be written

$$\mathcal{H} = -\underline{\mu} \cdot \left[\underline{B}_{eff} + (1+K)\underline{B}_{app} - D\underline{M}\right] + P\left[I_z^2 - \frac{1}{3}I(I+1)\right] \qquad (3)$$

where $P = \dfrac{3eV_{zz}Q}{4I(2I-1)} \cdot \dfrac{1}{2}(3\cos^2\phi - 1)$, K is the Knight shift and D the

sample demagnetisation factor.

This form assumes that the electric interaction has an axis of symmetry, the zz axis, which is at angle ϕ to the resultant magnetic interaction, and that the magnetic interaction is dominant (10). With these simplifications the hyperfine splittings of the parent nuclear level in an NO experiment show a basic Zeeman pattern with a small perturbation proportional to I_z^2. The resonant frequencies for normal $\Delta I_z = \pm 1$ transitions form a series with 2I elements, symmetrical about the Zeeman frequency $\mu B_{tot}/I$ (where B_{tot} is the term in square brackets in (3)) with spacing 2P.

In conventional NMR the resonance signal is proportional to the change of magnetisation M, or $\langle I_z \rangle$. At temperatures > 1 K nuclei are only slightly polarised ($\sim$ 1 in 10^3 or 10^4) and the resulting resonance signal as a function of varied applied field or frequency is a set of equally spaced lines of comparable width and intensity.

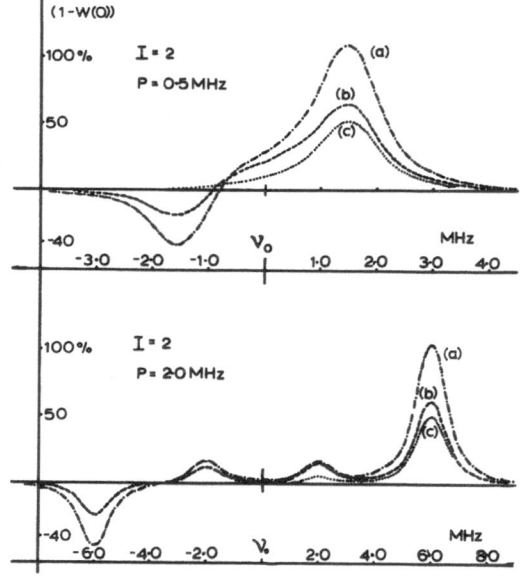

Fig. 1: Computed NMR/ON line shapes for the hamiltonian of equation (3), I = 2 and $P(> \Gamma_F)$ = 0.5 MHz and 2.0 MHz.

In NMR/ON detected by any parity conserving radiation
(typically gamma radiation) the leading term in $\Delta W(0)$ is ΔB_2 where
B_2 is proportional to $<I_z^2>$. Other higher order even powers of I_z
are also involved. Odd terms $<I_z>$, $<I_z^3>$ can only be detected if
parity non-conserving emissions (beta radiation) are observed.
Although beta anisotropies are in general larger than gamma anisotropies, experimental problems in ultra low temperature beta
counting have meant that gamma detection has been much more widely
used. For gamma radiation any resonance which decreases B_2, i.e.
which increases the population of low $|I_z|$ states, is seen as a
reduction in anisotropy, but conversely a resonance which results
in an increased population of high $|I_z|$ states increases the
anisotropy. A second variation from the conventional resonance
signal is that the nuclei involved are quite highly polarised so
that different individual sub-resonances in the series will have
very different amplitude. The result, for $2P > \Gamma_F$ and two
values of $\mu B_{tot}/kT$ is shown in fig. 1.

We see that for a high degree of nuclear polarisation almost
all the interaction strength lies in one of the extreme elements of
the series at the frequency $\nu = \frac{1}{h}(\frac{\mu B_{tot}}{I} + 2IP)$. For a system in
which the quadrupole interaction distribution centres on zero, as in
a lattice with local defects and distortions, the spread of sub-
resonance and the shift at lowest temperatures is a significant
line broadening mechanism. Empirically it is found that linewidths
Γ_F are of order 1% of the resonant frequency. With this limit the
resolution of the technique for a discrete quadrupole interaction
strength is

$$(eQV_{zz})_{minimum} = 0.01 \frac{\mu_n B_{tot}}{I} \cdot \frac{2}{3}(2I - 1) \tag{4}$$

for the case $\phi = 0$.

Much smaller quadrupole interactions have been measured by
NMR/ON using techniques related to adiabatic fast passage in which
the radiofrequency is swept through the resonance region. The
time variation of $W(0)$ can then be analysed to yield quadrupole
interactions which are quite unresolved in the simple lineshape.
These swept and also pulsed frequency techniques, developed by the
Oxford (11) and Duntroon (12) groups, will not be discussed further
here.

Signal and Measurement Technique

As outlined in the previous section the observed signal $\Delta W(0)$
depends upon the radiofrequency power, the resonance line structure
and linewidth, the degree of nuclear polarisation and the relaxation
time. Complete equalisation of populations of the nuclear sublevels corresponds to complete loss of anisotropy. The parameter

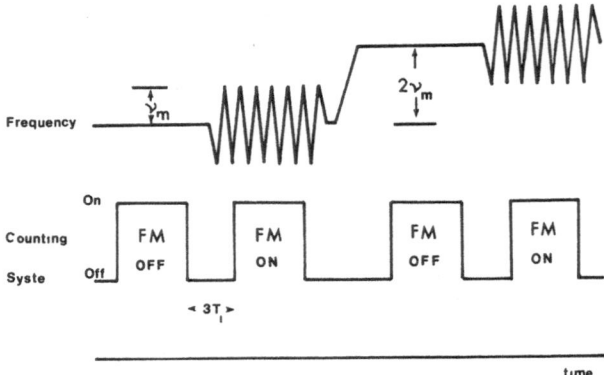

Fig. 2: Time sequence of frequency changes (with and without frequency modulation) and counting periods.

used to measure the resonance effect is the fractional destruction $\frac{\Delta W(0)}{1-W(0)}$. It should be noted that this ratio is not a linear function of nuclear magnetisation, nor of nuclear temperature.

Observation of the resonant destruction requires regular monitoring of the non-resonant count rate. A typical sequence of measurements is shown in fig. 2.   When the system has been cooled to its lowest temperature (see Chapter   ) the radio frequency power supplied to the rf loops is raised until non-resonant heating produces a small increase in the sample temperature. Having decided the resonant frequency search region, possibly on the basis of NO data from the system under study, and the width of the frequency modulation amplitude ν_m to be used (for an initial search $\nu_m \sim 0.2 \times$ the expected linewidth, i.e. $\sim 2 \times 10^{-3} \nu_o$), the search is made as a sequence of modulation ON and modulation OFF counts, the frequency being stepped by $2\nu_m$ between pairs of counts. Provided an interval $\sim 3T_1$ is allowed between ON and OFF counts, the OFF counts monitor the undisturbed anisotropy W(0), including any dependence of non-resonant heating upon frequency.

For an inhomogeneously broadened line of Gaussian shape the fractional destruction at any frequency setting is simply

$$S(\nu) \quad = \quad \int_{\nu-\nu_m}^{\nu+\nu_m} \frac{A}{a\sqrt{\pi}} \, e^{-\left[\frac{\nu-\nu_o}{a}\right]^2} \, d\nu \qquad (5)$$

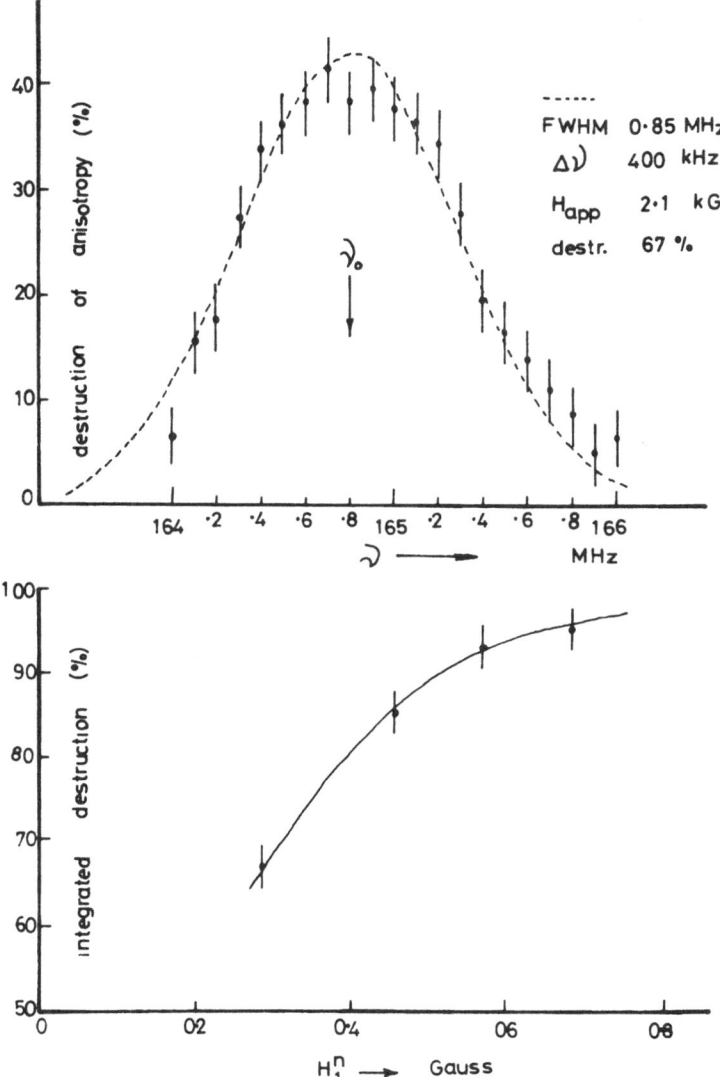

Fig. 3: ^{60}CoFe NMR/ON resonance line (above) and plot of
integrated resonance destruction against rf field at the nucleus.

As an example a ^{60}CoFe resonance line is shown in fig. 3 which also
shows the variation of the integrated resonance destruction A as a
function of the rf field at the nucleus B_1^{nucl} for this system.

This expression for $S(\nu)$ assumes each frequency setting to be independent, i.e. a sufficient time interval between counting periods to allow full relaxation of the partially resonated system to its equilibrium temperature. For relaxation times of more than 100 s it is often simpler to take modulation OFF counts at wider intervals in the search and to make a succession of ON counts. If the nuclear spin-lattice relaxation time is T_1 the signal after n modulation ON settings each of length τ is

$$S(\nu) = \int_{\nu-\nu_m}^{\nu+\nu_m} \frac{A}{a\sqrt{\pi}} e^{-\left[\frac{\nu-\nu_o}{a}\right]^2} d\nu +$$

$$\left[\sum_{k=1}^{n} e^{-\frac{(k-1)\tau}{T_1}} \int_{\nu-\nu_m-2(k-1)\nu_m}^{\nu+\nu_m-2(k-1)\nu_m} \frac{A}{a\sqrt{\pi}} e^{-\left[\frac{\nu-\nu_o}{a}\right]^2} d\nu \right] \left(\frac{1-e^{-\frac{\tau}{T_1}}}{\frac{\tau}{T_1}}\right) \cdot (6)$$

An example of such a sweep in both directions for ^{120}SbFe is given in fig. 4.

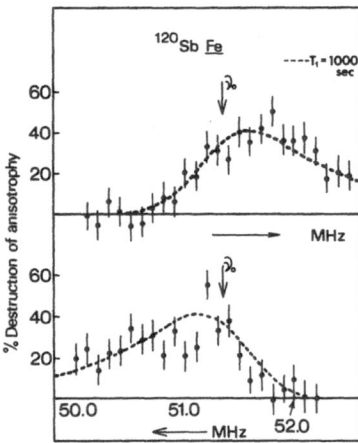

Fig. 4:   NMR/ON sweep through resonance for ^{120}SbFe, a case with long T_1 (22).

Fig. 5:   Quadrupole split NMR/ON resonance as observed for ^{199}Au in iron (11).

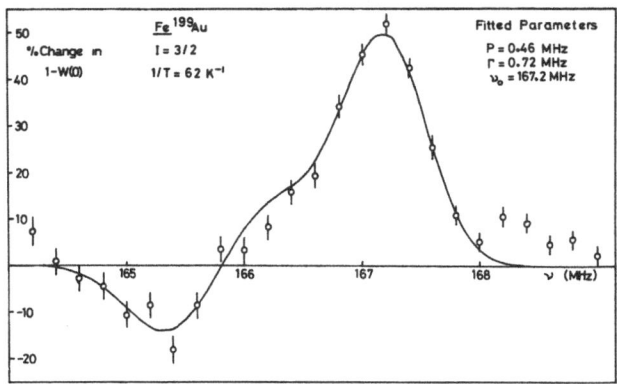

Finally fig. 5 shows an example of a well resolved quadrupole split resonance for ^{199}Au$\underline{Fe}$ in which the quadrupole interaction at all nuclei is collinear with the magnetic interaction (11).

NMR/ON IN IMPLANTED SYSTEMS: FERROMAGNETIC METALS

In ferromagnetic metals the skin depth is a major hindrance to NMR sample preparation and the fact that implantation provides a sample with all nuclei in a layer ($\sim 10^{-7}$ m for 100 keV implant energy) thin compared to the skin depth ($\sim 10^{-6}$ m at 100 MHz) was an early motivation for the study of implanted systems by NMR/ON. Both nuclear reaction recoil and isotope separator implantations have shown successful resonances, including both soluble and insoluble implants (13). Linewidths and relaxation times for implanted samples are generally in line with thermally prepared samples, and in the cases of implanted I (13) and Xe (14) particularly narrow lines have been reported for ions in substitutional sites.

Attempts to observe resonance in non-substitutional sites have only recently been successful. These are of particular importance to this summer school. Following a first report of a broader, lower frequency, second resonance in the ^{129m}Xe$\underline{Fe}$ system by Schoeters et al. (15), work on ^{131}I$\underline{Fe}$ with a single crystal sample by Visser, Niesen, Postma and de Waard has successfully brought the full power of NMR/ON to bear on the problem of the non-substitutional site and its hyperfine interaction. The rest of this paper summarises this work (16).

The NMR/ON of substitutional ^{131}I implanted in single crystal Fe was first observed by James et al. (fig. 6).   They also observed several satellite lines at only slightly lower frequencies which were shown not to be components of a quadrupole split resonance but were assigned tentatively to I ions with vacancies in the 3rd and 4th neighbour shells.   No lower frequency resonances were detected in this work, although searches were made based on Mossbauer evidence (17) that $\sim$ 30% of implanted I nuclei experience a magnetic interaction reduced by $\sim$ 20%.

Visser et al. used three single crystals cut with their longest sides along <100>, <110> and <111> axes. The crystals were implanted with between 0.3 and 2.2 × 10^{14} at./cm^2 110 keV I^{131} ions. The results of their resonance search are summarised in fig. 7. The main substitutional resonance is at 683.8 MHz, only $\sim$ 1 MHz wide, in full agreement with James. In addition weak resonance lines are seen in the 620-660 MHz region with a pattern differing with crystal axis (the crystals were mounted with their longest side parallel to the 0.15 T polarising field B$_{app}$ which determines the magnetic hfi axis).

Taking as a model the non-substitutional I ions being associated with a single near neighbour vacancy, and displaced towards that vacancy, the Hamiltonian of equation   will be applicable, incorporating an axially symmetric V$_{zz}$ along the <111> direction in bcc

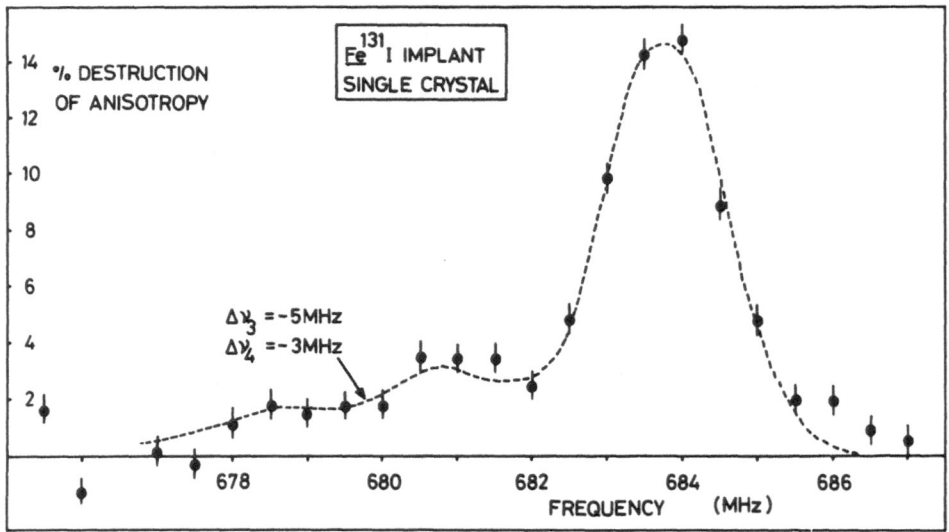

Fig. 6:   Substantial site NMR/ON for ^{131}I implanted in single crystal iron. Smaller low frequency satellites tentatively assigned to sites with vacancies in the third and fourth neighbour shells.

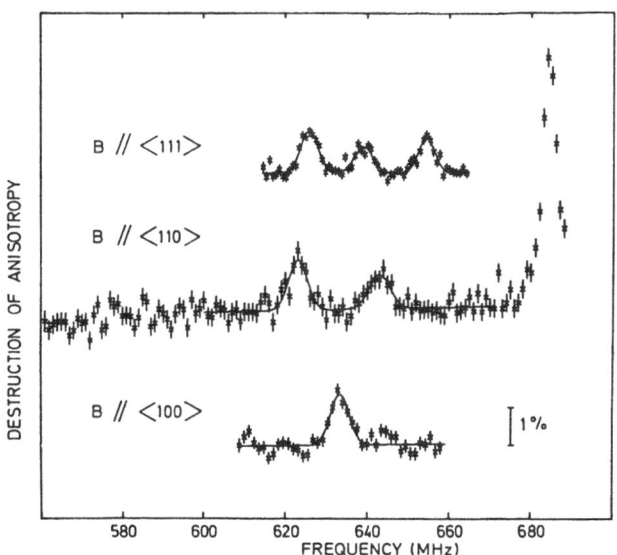

Fig. 7:   Vacancy associated site NMR/ON for ^{131}I implanted in single crystal iron, observed with host magnetised along <111>, <110> and <100> directions.

iron. At the measuring temperature the ^{131}I is highly polarised and, with $I = 7/2$, the resonant signal for this component is highly asymmetric, with the maximum response displaced from the Zeeman frequency by $\dfrac{eQV_{zz}}{4} \cdot \dfrac{1}{2}(3\cos^2\phi - 1)$ where ϕ is the angle between the magnetisation axis and the <111> axis of V_{zz}.

There are eight equivalent <111> directions, and the assumption is made that vacancy-iodine ion pairs are equally likely to lie along any of these directions. For B_{app} along <100> $(3\cos^2\phi - 1)$ is zero for all <100>-<111> angles so that there is no quadrupole splitting and a single resonance is seen. B_{app} along <110> finds $\frac{1}{2}(3\cos^2\phi - 1)$ is $+\frac{1}{2}$ for half and $-\frac{1}{2}$ for half of the <110>-<111> angles. Theory then predicts two systems of sub-resonances with the maximum responses displaced equally above and below the pure magnetic <100> resonance line position. Full agreement with this prediction is found. For B_{app} along <111>, one quarter of the sites have $\frac{1}{2}(3\cos^2\phi - 1) = 1$, and three quarters have $\frac{1}{2}(3\cos^2\phi - 1) = -1/3$. Although resonances are observed at these positions, their relative intensity is not 1:3 and there is an unexplained third component at $\frac{1}{2}(3\cos^2\phi - 1) \approx +1/3$.

These authors have thus been able to use the specific predictions of a simple hamiltonian to fit their results and obtain more detailed evidence for the situation of the implanted I ions in this system by NMR/ON than by other techniques. It is worth emphasising that the largest observed signal in the non-substitutional resonances corresponds to less than 1% destruction of the total anisotropy, and that the site-revealing quadrupole splitting is only 30 Mhz at maximum (for the <111> sample), less than 5% of the dominant magnetic splitting. Thus both the high sensitivity and resolution qualities of the technique are utilised here.

NMR/ON IN OTHER SYSTEMS: NQR/ON

Ferromagnetic metals are advantageous systems for NMR/ON because they are easily cooled, they offer long nuclear spin-lattice relaxation times and high rf enhancement, and the problems associated with the 10^{-6} m skin depth at 100 MHz can be solved in both thin foil, controlled diffused, and implanted systems.

A first report of NMR/ON on nuclei polarised by an external field in a non-magnetic metal has been made (18), but is unconfirmed to date. The loss of rf enhancement increases the power requirement by $\sim 10^4$-10^6 if the same frequency range has to be covered. However, in a sufficiently uniform applied field and a well annealed sample, it may be possible to reduce linewidths by up to 10^3, and for long relaxation time systems the rf power problem may be overcome. Brewer (18) has reviewed the problems in some detail.

Dielectric crystals allow full rf penetration. For paramagnetic ions problems of fast relaxation may be overcome for large applied fields, but thermal contact to the refrigerator is much worse than for metals.

Radioactive NQR detection is also a possible development of interest in dilute alloy systems. Relaxation times are long and frequencies are lower than for most magnetic interactions giving less heating, however lower frequencies require lower temperatures (~ 1 mK rather than 10 mK for NMR/ON) and there is no rf enhancement.

NMR/ON IN OTHER SYSTEMS: NON-EQUILIBRIUM TECHNIQUES

NMR in non-equilibrium polarised nuclear systems allow experiments over a wide temperature range. Two notable techniques for solid state studies exist. The first, developed by Sugimoto at Osaka involves NMR of short lived ($\sim$ ms) light nuclei (^{12}B, ^{12}N, ^{29}P) produced and polarised in a nuclear reaction and recoil implanted into a variety of metals. Relatively long T_1 values ($\sim 10^{-1}$-10^{-2} s for e.g. ^{12}B in Ni up to 600 K) allow observation of NMR by destruction of the asymmetry in the nuclear beta emission. The sites occupied are predominantly interstitial for these light nuclei and the Ni study gave evidence for markedly different B_{hf} at ^{12}B in the octa-hedral and tetrahedral interstitial sites (19).

Secondly the technique of in situ activation with polarised neutrons has been developed at Heidelberg and Grenoble (20). Utilising the longer relaxation times in insulators and semicon-ductors NMR of nuclei of lifetimes 10^{-2}–10^{3} s has been detected by both beta and gamma techniques at temperatures of order 1-50 K. Using this method point defect structures can be studied, as has been done for ^{20}F in CaF_2 (21).

CONCLUSION

NMR/ON can give information concerning the location and inter-actions of implanted ions which is more direct than other hyperfine techniques whilst maintaining their high sensitivity and specificity. There are many questions as yet unexplored before the limits of this combined technique can be set.

REFERENCES

1. M N Hack and M Hamermesh, Nuovo Cimento 19 (1961) 546.
2. E Matthias in "Hyperfine Structure and Nuclear Radiations"(eds.
    E Matthias and D A Shirley) North-Holland Amsterdam (1968) p.815.
3. E Recknagel in "Hyperfine Interactions in Excited Nuclei" (eds.
    G Goldring and R Kalish) Gordon & Breach (1971) p.291.
4. N J Stone, Hyperfine Interactions 2 (1976) 45.
5. E Matthias and R J Holliday, Phys. Rev. Lett. 17 (1966) 897.
6. P Herzog, H-R Folle, K Freitag, A Kluge, M Reuschenbach and
    E Bodenstedt, Nuclear Instruments and Methods (1978) to be
    published.
7. N J Stone in "Hyperfine Interactions in Excited Nuclei" (eds.
    G Goldring and R Kalish) Gordon & Breach (1971) p.237.
8. E Matthias, B Olsen, D A Shirley, J E Templeton and R M Steffen,
    Phys. Rev. A4 (1971) 1626.
9. D Riegel, N Brauer, B Focke and E Matthias in "Hyperfine Inter-
    actions in Excited Nuclei" (eds. G Goldring and R Kalish) Gordon
    & Breach (1971) p.313.
10. P D Johnston, R A Fox and N J Stone, J. Phys. C 5 (1972) 2077.
11. P T Callaghan, P D Johnston, W M Lattimer and N J Stone, Phys.
    Rev. B12 (1975) 3526.
12. H R Foster, P Cooke, D H Chaplin, P Lynam and G V H Wilson,
    Phys. Rev. Lett. 38 (1977) 1546.
13. See tabulation at end of ref. 4.
14. P K James (thesis) Oxford (1975) and in ref. 4.
15. E Schoeters, R Coussement, R Geerto, J Odeurs, H Pattyn,
    R E Silverans and L Vanneste, Phys. Rev. Lett. 37 (1976) 302.
16. D Visser, L Niesen, H Postma and H de Waard, Phys. Rev. Lett.
    41 (1978) 882.
17. H de Waard, R L Cohen, S R Reintsema and S A Drentje, Phys.
    Rev. B10 (1974) 3760.
18. W Brewer, private communcation. See also W Brewer and M Kopp,
    Hyperfine Interactions 2 (1976) 299.
19. Y Nojiri, H Hamagaki and K Sugimoto, Physics Letters 60A (1977)
    77.
20. H Ackermann, Hyperfine Interactions 4 (1978) 645.
21. H Ackermann, D Dubbers, H Grupp, M Grupp, P Heitjans and
    H-J Stöckmann, Physics Letters 54A (1975) 399.
22. P T Callaghan and N J Stone, Physics Letters 40B (1972) 84.

MÖSSBAUER SPECTROSCOPY

H. de Waard

Laboratorium voor Algemene Natuurkunde, University of

Groningen, The Netherlands

1. BASIC CONCEPTS

1.1. Recoilless emission of gamma rays

Rudolf Mössbauer discovered in 1957 that gamma radiation can be emitted and absorbed by nuclei in solid matter without imparting recoil energy to the individual emitting nucleus and without Doppler broadening of the gamma line. At an early stage he gave a theory of the fraction of "recoilless" emission based on considerations of Lamb about elastic neutron scattering in solids.[1] In Chapter 6 a classical approach is used to derive formulas for the recoilless fraction of gamma emission, $f(T)$, as a function of temperature. Here, only the general result is given:

$$f(T) = e^{-k^2 <x^2>_T},$$ (1)

where $k = 2\pi/\lambda$ is the wave number of the gamma ray and $<x^2>_T$ the component of the mean square vibration amplitude of the emitting (or absorbing) nucleus in the direction of the gamma ray. Eq. (1) is valid in the harmonic approximation of lattice vibrations.

In a Debye solid, at $T = 0$, Eq. (1) becomes:

$$f(0) = e^{-\frac{3}{2}E_r/k\Theta_D}$$ (2)

with $E_r = \hbar^2 k^2/2m$ the recoil energy that would be imparted to a free nucleus and Θ_D the Debye temperature.

For useful Mössbauer spectroscopy, we need a recoilless fraction of at least 0.01. With $\Theta_D \sim 300K$ (a value characteristic for many

205

metals) this restricts the gamma-energy to about 100 keV. Higher
energy transitions have occasionally been used. Typical values of
the low temperature recoilless fraction are $f(0) = 0.8$ for the
14.4 keV transition in ^{57}Fe and $f(0) = 0.01$ for the 103 keV transition in ^{153}Eu. Only in a few cases, notably for ^{57}Fe and for
^{119}Sn ($E_\gamma = 23.8$ keV) is it possible to perform Mössbauer spectroscopy at room temperature. In almost all other cases, measurements must be done at liquid helium temperature in order to achieve
sufficiently large values of f.

1.2. Generation of a spectrum

The *source emission spectrum* is determined by the requirement
that the electric field associated with the emission of gamma quanta
decays exponentially as

$$E(t) = E_o e^{-i(\omega_0 - i\frac{\Gamma'}{2})t} \quad (t \geqslant 0)$$
$$E(t) = 0 \quad\quad\quad\quad\quad (t < 0) \tag{3}$$

The constant Γ' is defined here as the inverse lifetime of the gamma
emitting nuclei: $\Gamma' = 1/\tau$. From the Fourier transform of (3),

$$F(\omega) = E_o \int_o^\infty e^{i(\omega-\omega_0+i\frac{\Gamma'}{2})t} dt \propto \frac{1}{\omega-\omega_0+i\frac{\Gamma'}{2}},$$

we immediately find the spectral intensity distribution:

$$I(\omega) = |F(\omega)|^2 \propto \frac{1}{(\omega-\omega_0)^2+(\frac{\Gamma'}{2})^2}$$

or, with $I(\omega)$ normalized to one:

$$I(\omega) = \frac{\Gamma'/2\pi}{(\omega-\omega_0)^2+(\frac{\Gamma'}{2})^2} \tag{4}$$

This is a Lorentz shape with full width at half maximum $\Delta\omega = \Gamma'$.

If the source has a velocity v parallel to the direction of
observation of the gamma rays, the central frequency changes by
an amount $(v/c)\omega_0$ so that

$$I(\omega,v) = \frac{\Gamma'/2\pi}{[\omega-\omega_0(1+\frac{v}{c})]^2+(\frac{\Gamma'}{2})^2} \tag{5}$$

The fraction of gamma rays of frequency ω transmitted through
an absorber of thickness d, containing n identical absorbing atoms
per unit volume is given by

$$D(\omega) = e^{-dn(\sigma_n+\sigma(\omega))}$$

with σ_n the non resonant absorption cross section that depends very
weakly on ω compared with the resonant cross section

$$\sigma(\omega) = \beta f_a \sigma_o \frac{(\Gamma''/2)^2}{(\omega-\omega_o')^2+(\frac{\Gamma''}{2})^2}$$

with

$$\sigma_o = \frac{2\pi}{k^2} \cdot \frac{1}{1+\alpha} \cdot \frac{2I_e+1}{2I_g+1}$$

In this expression β is the isotopic fraction of the resonantly
absorbing nuclei, f_a is the absorber recoilless fraction (which,
of course, depends on temperature), ω_o' is the central absorption
frequency, which may differ from ω_o, α is the total conversion
coefficient of the ground state gamma transition and I_e and I_g are
the spins of excited and ground states, respectively.

The resonance effect can be detected both in transmission and
in scattering geometry. The first method, which is normally used,
is schematically indicated in Fig. 1A. Suppose that with the
absorber removed the detector counting rate is N_o and that all
counts are due to a gamma ray of frequency ω_o that can be resonantly
absorbed. Then, after re-inserting the absorber, we will measure a
rate

$$N(v) = N_o(1-f) \, e^{-dn\sigma_n} + f \int_0^\infty I(\omega,v) \, D(\omega)d\omega$$

when the source moves with velocity v. At very high source velocities, when we are completely outside the resonance absorption region,
we have simply

$$N(\infty) = N_o e^{-dn\sigma_n}.$$

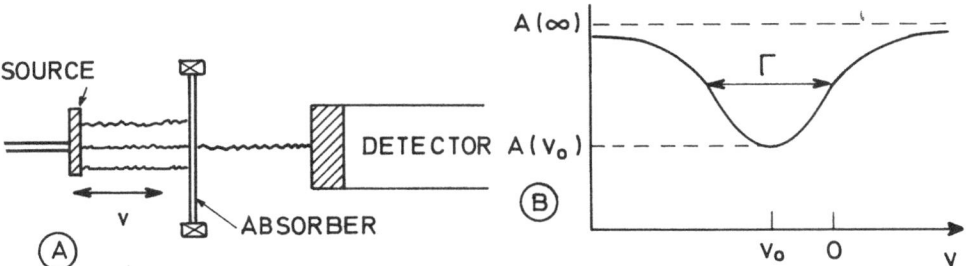

Fig. 1. A. Basic arrangement of transmission Mössbauer spectrometer.
         B. Single line Mössbauer absorption spectrum with velocity
         shift v_0 between source and absorber.

The relative transmission near resonance is

$$A(v) = \frac{N(v)}{N(\infty)} = 1-f + f \int_0^\infty I(\omega,v) e^{-dn\,\sigma(\omega)}\,d\omega \tag{6}$$

The integral in (6) is called transmission integral. For a "thin" absorber, i.e. when the reduced absorber thickness $dn\beta f_a \sigma_o \ll 1$, Eq. (6) reduces to

$$A(v) = 1-f\,\frac{t_a}{2}\,\frac{(\frac{\Gamma'+\Gamma''}{2})}{[\omega_o'-\omega_o(1+\frac{v}{c})]^2+(\frac{\Gamma'+\Gamma''}{2})^2} \tag{7}$$

A comparison of Eq. (7) with the source spectrum, Eq. (4), tells us that the folding of emission and absorption spectra of Lorentzian shape and with widths Γ' and Γ'' results in an absorption line with full width at half maximum: $\Gamma'+\Gamma''$. In Mössbauer spectroscopy it is customary to express frequencies (and energies) in (Doppler) velocities. Therefore, we prefer to write Eq. (7) as

$$A(v) = 1 - f(t_a/2)\frac{(\Gamma/2)^2}{(v_o-v)^2+(\Gamma/2)^2} \tag{8}$$

where the absorption line width is now also given in velocity units. (A velocity v corresponds to a frequency shift $\Delta\omega = (v/c)\omega_o$ or an energy shift $\Delta E_\gamma = (v/c)E_\gamma$). v_o represents a velocity shift between emission and absorption line. This shift usually is about equal to the isomer shift S discussed in Ch. 6. The spectrum given by Eq. (8) is shown in Fig. 1B. For values of t_a not much smaller than 1, Eq. (7) is no longer valid and we must, in general, calculate the transmission integral explicitly. For a single line source + absorber combination, this calculation yields for the minimum transmission[2]

$$A(v_o) = 1 - f[1 - e^{-t_a/2}J_o(it_a/2)] \tag{9}$$

where $J_o(x)$ is the zero order Bessel function of imaginary argument x. The t_a-dependence of the resonantly absorbed fraction of recoillessly emitted gamma rays, $[1-A(v_o)]/f$ is shown in Fig. 2.

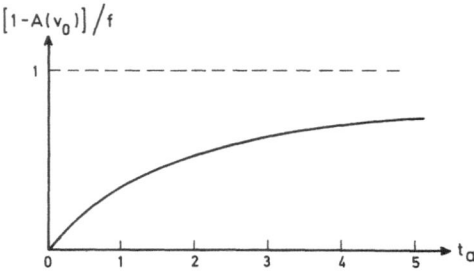

Fig. 2. Absorbed fraction of recoilfree emitted gamma rays vs. reduced absorber thickness.

For not too thick absorbers ($t_a \leqslant 10$) the line shape remains nearly Lorentzian while its width increases linearly with t_a [3]):

$$\Gamma(t_a) = 2\Gamma(1 + 0.135t_a) \tag{10}$$

2. MÖSSBAUER SPECTRA WITH HYPERFINE INTERACTIONS

The nuclear decay Hamiltonian contains terms corresponding to the hyperfine interaction of the nucleus with the surrounding charges and currents. The hyperfine interaction Hamiltonian can be separated into 3 main terms, the Coulomb energy, the magnetic dipole and the electric quadrupole interaction:

$$\mathcal{H} = \mathcal{H}_o + \mathcal{H}_M + \mathcal{H}_Q$$

The first term is always present, it gives rise to the Mössbauer isomer shift that is extensively discussed in Ch. 6.

Magnetic dipole term. If we can attribute a fixed axis to the magnetic field $\vec{B}$ acting on the nucleus this term is $\mathcal{H}_M = -\vec{\mu} \cdot \vec{B}$, which gives the normal Zeeman splitting

$$W_m = -\mu B \frac{m}{I} \quad (-I \leqslant m \leqslant I) \tag{11}$$

where μ is the nuclear magnetic dipole moment.

A gamma transition between sublevels $I^* m^*$ and Im experiences a hyperfine energy shift

$$\Delta W_{m^* \to m} = -\mu B \left(\frac{\mu^*}{\mu} \cdot \frac{m^*}{I^*} - \frac{m}{I} \right) \tag{12}$$

If only magnetic interaction is present, a line is split according to Eq. (12) into a pattern symmetric around $\Delta W = 0$, with a number of components depending on I^*, I and the selection rules for multipole radiation. All Mössbauer transitions have either a dipole character, with selectron rules $\Delta m = 0, \pm 1$ and orbital momentum transfer $L = 1$ to the gamma ray, or a quadrupole character with $\Delta m = 0, \pm 1, \pm 2$ and $L' = 2$, or a mixed dipole quadrupole character ($L = 1$, $L' = 2$, $\Delta m = 0, \pm 1, \pm 2$). In Fig. 3 a well known example is shown: the 6 magnetic dipole transitions between magnetic sublevels of the 14.4 keV excited state and the ground state of ^{57}Fe. The relative intensities of the spectral components depend on: (1) geometric factors determined by the conservation laws of 2^L $(2^{L'})$-pole transitions between levels I^*, m^*, and I, m, the so called Clebsch-Gordan coefficients (C). These have been calculated for many values of I^*, I, m^*, m and L [4]); (2) angular distribution functions $F(\theta, L, L', \Delta m)$ depending on multipolarity and on angle θ between quantization (field) axis and direction of gamma emission [5]).

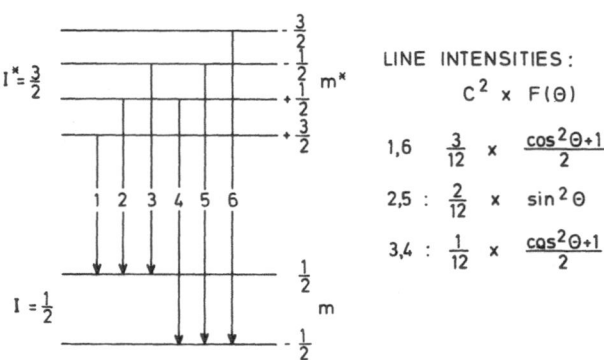

Fig. 3. *Magnetic hyperfine splitting of 14.4 keV state and ground state of ^{57}Fe, gamma transitions between sublevels and relative intensities of these transitions.*

For the simple case of ^{57}Fe the relative component intensities C^2F are given in Fig. 3. If the magnetic field direction is distributed randomly in space, the angular factors all average to 1/3, so that the component intensity ratio's are 3:2:1:1:2:3.

A Mössbauer spectrum, obtained with an absorber of metallic iron and a single line source of ^{57}Co in platinum is shown in Fig. 4. It is an example of the Zeeman pattern just discussed. A clear deviation from the calculated line intensity ratio's is caused by the absorber thickness.

Electric quadrupole term. The nuclear quadrupole moment Q interacts with the field gradient at the nuclear site. In principle, this is a tensor type interaction. It has been discussed at this school in detail by Recknagel [6]. We only give some results here.

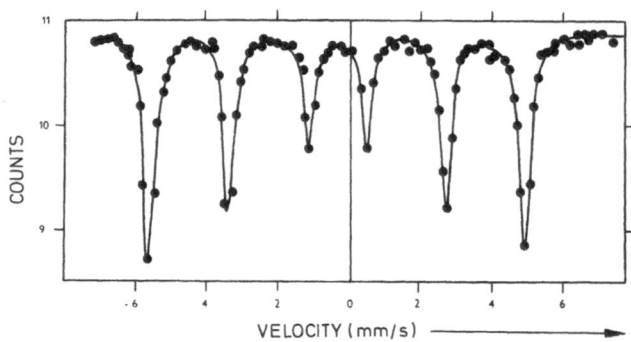

Fig. 4. *Mössbauer spectrum obtained with a source of ^{57}Co diffused in platinum and an absorber of iron metal.*

After choosing a Cartesian coordinate system with x, y, z axes along the principal axes of the field gradient so that $V_{ij} = 0$ for $i \neq j$ and $|V_{zz}| \geq |V_{yy}| \geq |V_{xx}|$; the Hamiltonian becomes

$$\mathcal{H}_Q = \frac{e^2qQ}{4I(2I - 1)} \left[3I_z^2 - I(I + 1) + \frac{\eta}{2}(I_+^2 + I_-^2) \right] \tag{13}$$

with $eq = V_{zz}$ and $\eta = (V_{yy}-V_{xx})/V_{zz}$.

The symbols I_z, $I_+ = I_x + iI_y$ and $I_- = I_x - iI_y$ denote the magnetic quantum number operators. The last two induce non zero off diagonal matrix elements $<Im'|\mathcal{H}|mI>$.

In general, therefore, a calculation of the energy states of Eq. (13) for $\eta \neq 0$ involves a numerical or approximative solution of the secular equation. An analytical solution is possible only for $I < 2$. For ^{57}Fe, the 14.4 keV $I = 3/2$ state splits into two substates, with $m = \pm3/2$ and $m = \pm\frac{1}{2}$ and with energies

$$W_Q = \pm \frac{e^2qQ}{4} \left(1 + \frac{\eta^2}{3}\right)^{\frac{1}{2}}$$

The + sign corresponds with the state $m = \pm3/2$, the − sign to $m = \pm\frac{1}{2}$. For an axially symmetric field gradient, $\eta = 0$, the energy states only involve diagonal matrix elements, $m' = m$ and the Hamiltonian is exactly solved by

$$W_Q = \frac{e^2qQ}{4I(2I - 1)} \left[3m^2 - I(I + 1) \right] \tag{14}$$

showing the $\pm m$ degeneracy of the energy states. Gamma components between sublevels I^*m^* and Im experience energy shifts

$$\Delta W_Q = \frac{e^2qQ}{4} \left[\frac{Q^*}{Q} K(m^*,I^*) - K(m,I) \right]$$

with

$$K(m^{(*)},I^{(*)}) = \frac{3m^{(*)2} - I^{(*)}(I^{(*)} + 1)}{I^{(*)}(2I^{(*)} - 1)} \tag{15}$$

For transitions with unique multipolarity, and for randomly oriented electric field gradient axes the gamma component intensities are given by the same squares of Clebsch Gordan coefficients as in the magnetic case.

Combined magnetic dipole + electric quadrupole terms. In the general case, when both a magnetic field and an electric field gradient occur, with an angle between their axes, matrix diagonalization and solution by computer are required. Procedures for executing such problems have been given, for instance, by Kündig[7].

If $e^2qQ \ll \mu B$ a perturbation treatment shows that the following approximation is valid:

$$W = W_M + P_2(\cos\,\theta)W_Q \qquad\qquad (17)$$

where θ is the angle between $\vec{B}$ and the largest (z-) component of
the field gradient.

3. MÖSSBAUER SPECTROMETERS

3.1. Electronics

Nowadays, electromagnetic transducers with negative feedback
are almost exclusively used to generate the required velocity wave
shape $v(t)$. The most common shapes are:
1.   triangular or saw tooth (constant acceleration)
2.   sinusoidal (especially if high velocities and/or long transfer
     rods are required)
3.   block form (constant velocity)
If spectra are scanned with $v(t)$ shapes 1 or 2, detector counts are
generally stored in a multichannel memory programmed in such a way
that subsequent channels correspond to subsequent small velocity
intervals Δv. The memory itself may form part of a dedicated computer
or it may be directly associated with the drive electronics. In
the latter case, its contents will periodically be transferred to
an on line or off line computer for further processing. A display
is normally provided that allows one to watch the spectrum as it
grows.

In fig. 5 the cross section is shown of a velocity transducer
of a type used in our laboratory. The block diagram of a Mössbauer
spectrometer is given in Fig. 6. A digital waveform generator produces a triangular or sinusoidal reference signal of period T that is
compared in a difference amplifier with the signal from the pick-up

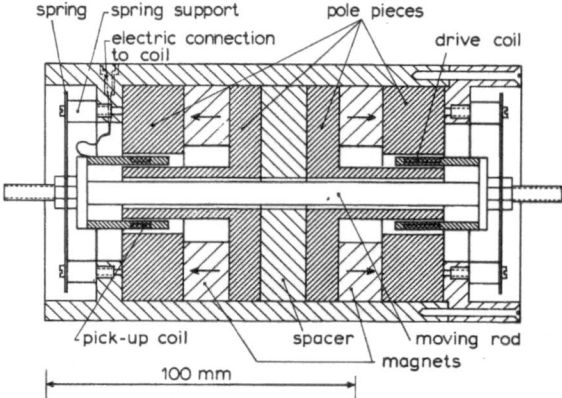

Fig. 5. Schematic drawing of electro-mechanical velocity transducer,
        described by Wit et al.[8].

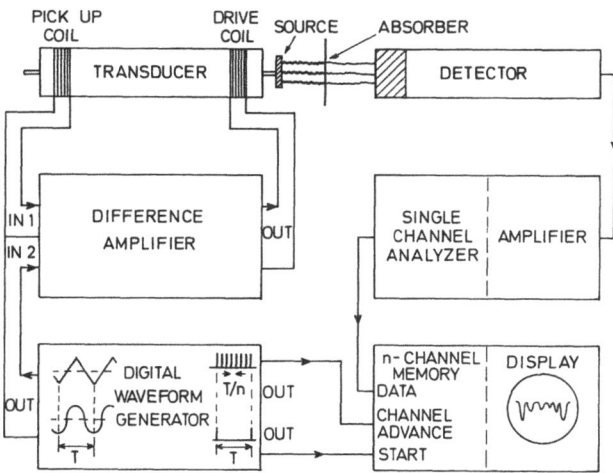

Fig. 6. Block diagram of Mössbauer spectrometer.

coil of the velocity transducer. The amplified difference signal
drives the transducer, minimizing the difference at the amplifier
input by negative feedback. In this way, the pick up signal and,
therefore, the drive velocity can be made to follow the reference
signal within 0.1%. The digital waveform generator at the same
time produces pulses with period T and channel advance pulses with
period T/n, n being the number of memory addresses. The former
pulses start the address at channel zero for each period, the
latter cause the address to advance step by step. In this way a
unique relation is established between velocity and memory address.
In each of the n memory channels pulses are stored during a time
interval T/n for each period. Thus a spectrum is generated. This
can be displayed by transforming the digital contents of each
channel into analog form and using it as the y-deflection singal
or an oscilloscope that receives a voltage proportional to channel
number on its x-deflection plates.

3.2. Calibration

At velocities not exceeding 20 mm/s, a single line ^{57}Co source
+ magnetically split ^{57}Fe absorber, replacing the source-absorber
combination under investigation, is widely used for velocity calibration. At higher velocities, such as used in rare earth experiments, an ^{57}Fe calibration would have to be extrapolated very far,
leading to errors if we do not have perfect linearity. In such
cases special methods, using the Moiré-effect[9,10] or an interfero-
meter[11] may be preferred. A particularly simple device is shown in
Fig. 7. The transducer is connected to an optical transmission grating

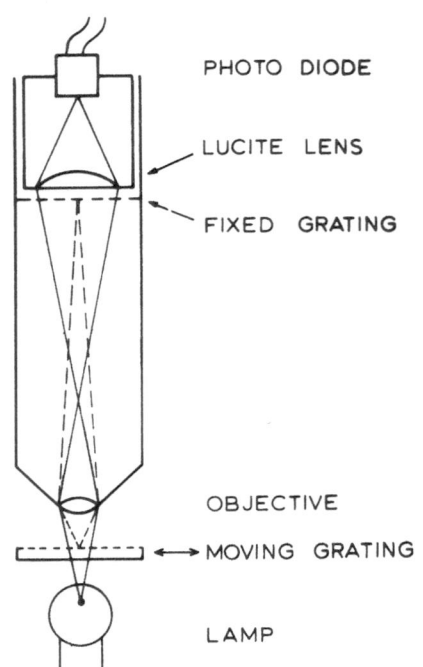

PHOTO DIODE

LUCITE LENS

FIXED GRATING

Fig. 7. Moiré calibrator for Mössbauer spectrometer according to Wit[10].

OBJECTIVE

MOVING GRATING

LAMP

with known grating constant, as closely as possible to the source. A microscope objective projects an enlarged image of this grating on the plane of a coarse transparent grating with constant $d_1 = md_o$, where m is the linear magnification of the objective. Behind the latter grating a photodiode is placed, which measures the intensity variation of the transmitted light as the image of the optical grating sweeps across the plane of the coarse grating. The periodically changing intensity is electronically converted to a train of pulses (2 per period) that are fed to the data input of the n-channel memory. Usually, the time interval between the optically generated pulses is much longer than the channel dwell time T/n. One would therefore expect these pulses to be stored in bunches in certain specific memory channels. This unwanted state of affairs can be prevented by slowly shifting the equilibrium position of the drive. This results in "debunching" of the pulses, so that a smooth calibration spectrum is produced. After running the spectrometer for a precisely determined period t_c, an average number $N = 2vt_c/nd$ of pulses is stored in the channel that corresponds to source velocity v. Thus, the velocity v can be accurately determined for each channel.

3.3. Cryostats

For measurements at low temperatures many different devices, involving the use of bath or flow helium cryostats, have been developed. When both source and absorber are at liquid helium temperature, vertical bath cryostats are mostly used. The drive, vertically mounted above the cryostat is then connected by a driving rod or tube to the source, immersed in the liquid. The absorber is kept fixed below the source. Beryllium windows allow the gamma rays to reach the detector with very little attenuation. If the source temperature must be varied it may be mounted in a small vacuum chamber directly connected to the drive rod. Its temperature can be controlled by a heater-thermocouple-feedback device. If the source (or the absorber) can be held at room temperature, the use of a flow cryostat with temperature control for the absorber (or the source) may be a more simple solution.

A special helium cryostat, allowing source implantation at liquid helium temperature and "on line" Mössbauer spectroscopy was developed by Drentje[12]. With this instrument, source annealing sequences can be performed because the source can be thermally separated from the helium bath by evacuating a helium gas filled thin walled stainless steel tube that connects the source to the bath.

3.4. Conversion electron Mössbauer spectroscopy (C.E.M.S.).

A nuclear level excited by resonant absorption of a gamma ray can decay either by re-emission of a gamma ray or by emission of a conversion electron. A Mössbauer spectrum can be obtained by detecting these electrons instead of the transmitted gamma rays. In this way a much higher ratio of resonant signal to background can sometimes be obtained, but only in a few cases (notably for ^{57}Fe) will the statistical accuracy obtained in a fixed measuring time be better than for a normal transmission experiment.

In Fig. 8A a very simple arrangement for C.E.M.S. is schematically shown. The absorber is mounted inside a gas flow proportional counter. Fig. 8B presents spectra obtained for an ^{57}Fe metallic absorber both with C.E.M.S. and conventional Mössbauer spectroscopy. Due to the high absorption of low energy electrons in solid matter, the absorber thickness accessible to C.E.M.S. is quite small, e.g. about 100 nm for the 8 keV K conversion electrons of the 14.4 keV transition in ^{57}Fe. For this very reason C.E.M.S. is highly depth sensitive. It may be used, for instance, to investigate diffusion in implanted sources. C.E.M.S. has also been used in combination with a magnetic electron spectrometer that measures the conversion electron spectrum with high resolution and thus achieves a certain degree of depth resolution[13].

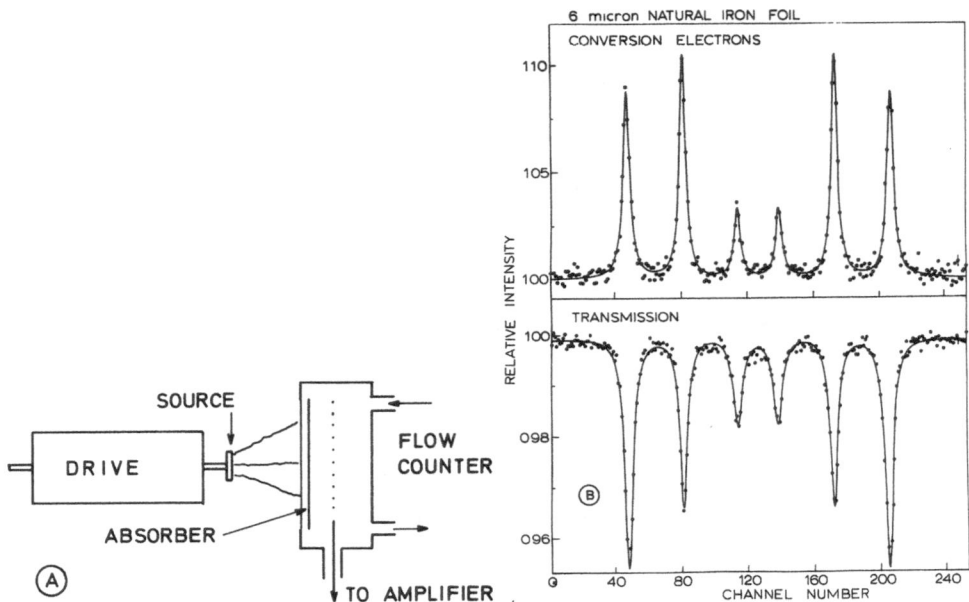

*Fig. 8. A. Schematic arrangement of conversion electron Mössbauer
spectrometer with proportional gas flow counter. B. Conventional transmission and C.E.M.S. spectra.*

4. MÖSSBAUER SPECTROSCOPY OF IMPLANTED SOURCES

After the first experiments with ^{129m}Te sources implanted in
iron, carried out in Groningen in 1965[14], Mössbauer spectrometry
of implanted sources has been performed in various laboratories.
We shall not attempt to present here a complete survey of all
experiments performed so far. We will only briefly discuss a few
cases where rather detailed information on impurity location and
on annealing of defects in metals has been obtained from the Mössbauer spectra. The cases to be considered are those of the heavy
5sp elements Te, I and Xe and of Cs implanted in iron. The results
for these cases have been extensively published in a number of
papers [12,15-19] and in the doctoral thesis of S.R. Reintsema [20].
In Figure 9 a survey is given of the implanted isotopes and daughter
nuclei used in these studies. There are possibilities for systematic
research in this region because each element has at least one
suitable isotope, with a low energy gamma transition to a stable
(or, for ^{129}I, very long lived) ground state. Moreover, the implanted
parent activities have convenient half lives, ranging from a few
days to a few months.

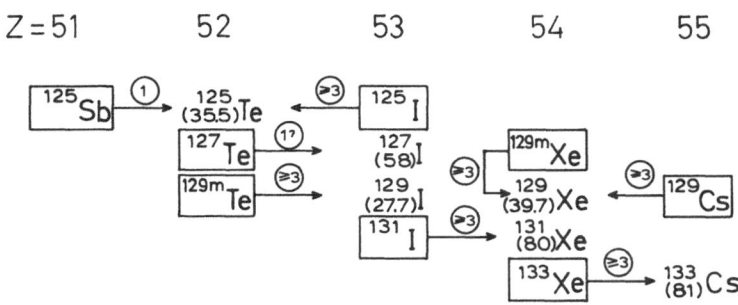

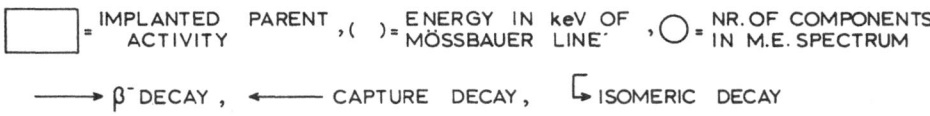

*Fig. 9. Survey of Mössbauer measurements on implanted sources of
isotopes of 5(s,p) shell elements in iron.*

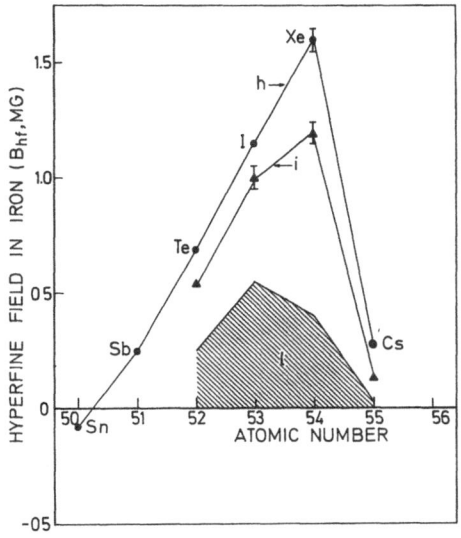

*Fig. 10. Hyperfine fields of the heavier 5(s,p) elements and of Cs
in iron. Different components are given where found. The
low field components are given by a shaded area, indicating
that there may be more than one component for each case.*

The original purpose for studying these implanted systems was
to find the hyperfine magnetic fields of isolated I, Xe and Cs impurities at the iron site. These were unknown at the time, due to
the insolubility of Te, I, Xe and Cs in iron. The results for the
fields are given in Fig. 10. They are of interest for theoretical
considerations of impurity hyperfine magnetic interaction, but they
will not be discussed here any further, except to remark that the
fields are very sensitive to the defect configuration neighbouring
the implanted atom. For this very reason these implants are good
probes for microscopic defect configurations: the resulting differences in the Zeeman splitting of different components of the
Mössbauer spectra facilitate their disentanglement.

Examples of Mössbauer spectra are shown in Fig. 11 for implants
with 4 subsequent atomic numbers: 34 d ^{129m}Te (Z = 52), 8 d ^{131}I
(Z = 53), 5 d ^{133}Xe (Z = 54) and 1.2 d ^{129}Cs (Z = 55). The first
three decay by β^- emission to daughter nuclei with Z' = Z+1, the
last one by electron capture to ^{129}Xe (Z' = Z-1). In each case, the
Mössbauer spectrum is that of a gamma transition in a daughter
nucleus, so that the hyperfine field is that of an impurity with
atomic number Z', but the fractional occupation of different defect
associated sites is characteristic of the parent, with atomic
number Z.

It is clear from the spectra shown in Fig. 11, that (at least)
3 components are needed to fit the ^{129}I, ^{131}Xe and ^{133}Cs spectra
whereas 4 components fit the ^{129}Xe spectrum. In each case, the
component with the highest field (h) is interpreted as due to
substitutional impurity atoms, that with the next lower field (i)
as due to impurity atoms associated with one nearest neighbour
vacancy and that with a small hyperfine field (ℓ) as due to impurity associated with two (or more) vacancies. Besides, the analysis
of the ^{133}CsFe spectra gives evidence of a fourth component with
very small recoilless fraction, interpreted as due to Cs-atoms
associated with at least 3 vacancies.

In Fig. 12 probable impurity-vacancy configurations are shown
with up to three vacancies per impurity. The fractional site occupations of the different components obtained for sources implanted
at room temperature (RT) are given in Table 1. Extensive annealing
studies of site occupations have been performed by Reintsema for
^{129m}TeFe and ^{133}XeFe [20]. The behaviour is markedly different in
both cases: for ^{129m}TeFe, the substitutional fraction first increases when annealing at 300°C, apparently because vacancies are
detrapped from Te-atoms, leading to a disappearance of the i-site.
The substitutional fraction begins to decrease again only at 500°C
annealing. Above this temperature, a new fraction with a low field
appears, presumably corresponding to chemically bound Te-atoms in
Te-vacancy clusters. For ^{133}XeFe, on the contrary, the high field
fraction remains almost constant at 300°C but it starts decreasing

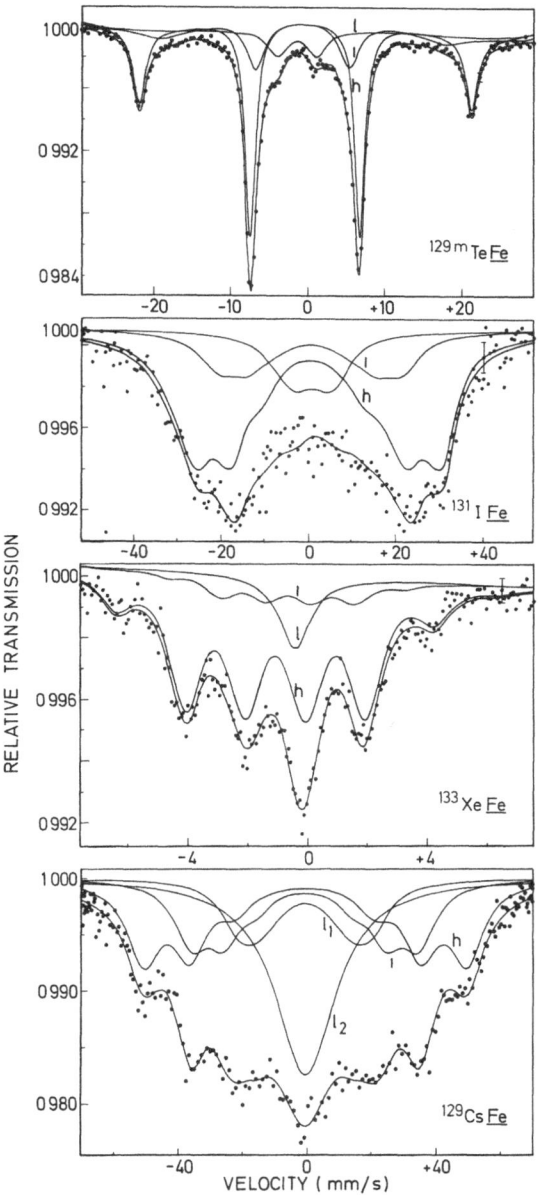

Fig. 11. Mössbauer spectra of implanted sources of ^{129m}Te, ^{131}I, ^{133}Xe *and* ^{129}Cs *in iron. In each case, the spectrum must be decomposed into at least three components.*

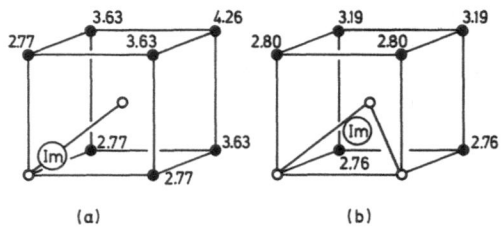

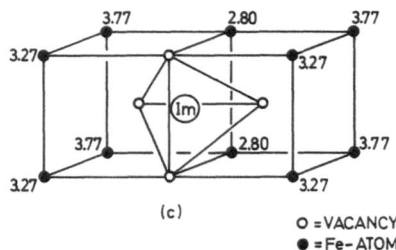

O = VACANCY
● = Fe- ATOM

*Fig. 12. Three possible impurity-vacancy clusters in iron, arising
from initial configurations (a) one impurity + one nearest
neighbour (nn) vacancy, (b) one impurity + one nn + one
next nearest neighbour (nnn) vacancy, (c) one impurity plus
two nn and one nnn vacancy. Numbers near iron atoms give
distances between these atoms and the impurity, taking into
account relaxation as calculated by Drentje and Ekster[21].*

TABLE 1

Summary of hyperfine fields and site populations obtained with implanted sources of ^{129m}Te, ^{131}I, ^{133}Xe and ^{129}Cs in iron.

Source & daughter	^{129m}Te→^{129}I	^{131}I→^{131}Xe	^{133}Xe→^{133}Cs	^{129}Cs→^{129}Xe
	Site populations (%)			
h	65(3)	45(5)	29(3)	30
i	18(3)	32(4)	13(4)	25
ℓ	17(2)	23(6)	22(4)	45
z			36(7)	
	Fields (kG)			
h	1152(10)	1540(30)	276(4)	1540(50)
i	1010(40)	1130(70)	135(15)	1100(100)
ℓ	460(40)	270(50)	< 25	600(125), <100

already at 400°C while at 500°C it has become very small. Moreover, the intermediate field fraction instead of disappearing at 300°C, increases considerably. The growth of both this and the low field fraction occur at the same time that the z-fraction decreases strongly. This behaviour shows that clusters with many vacancies loose vacancies, leading to $z \rightarrow \ell$ and $\ell \rightarrow i$ site transitions, but $i \rightarrow h$ transitions do not take place. At 500°C, almost all Xe-atoms are in ℓ-sites, while at still higher temperature the xenon atoms themselves begin to move, leading to loss of gas from the host and to the formation of small gas bubbles with a low recoilless fraction.

The different behaviour of Te and Xe on annealing as well as the fact that the substitutional fraction of Te directly after implantation is about a factor 2 higher than that of Xe can be explained from the difference of the atomic radii. The large radius of Xe makes it a strong vacancy trapper, whereas Te, almost soluble in iron, tends to detrap vacancies until it becomes chemically active at higher temperatures.

Information on lattice location of implanted impurities in metals is also obtained from the isomer shifts and from the recoilless fractions, as is shown for $^{129m}TeMe$ and for $^{133m}XeMe$ in Chapter p.101.

We end this brief survey by remarking that, though Mössbauer spectroscopy has proved an excellent means for detecting different lattice sites, its resolution is generally insufficient to provide a complete pattern of defective impurity sites. Therefore, its combinations with other methods such as TDPAC and NMRON is of great importance to obtain a more detailed picture.

REFERENCES

1. R.L. Mössbauer, Z. Physik 151 (1958) 124.
2. S. Margulies and J.R. Ehrman, Nucl. Instr. & Meth. 12 (1961) 131.
3. D.A. O'Connor, Nucl. Instr. & Meth. 21 (1963) 318.
4. A.R. Edmonds, Angular Momentum in Quantum Mechanics (Princeton University Press, 1960).
5. D.S. Ling and D.R. Falkoff, Phys. Rev. 76 (1949) 1639.
6. E. Recknagel, these Proceedings, Ch.p. 85
7. W. Kündig, Nucl. Instr. & Meth. 48 (1967) 219.
8. H.P. Wit, G. Hoeksema, L. Niesen and H. de Waard, Nucl. Instr. & Meth. 141 (1977) 515.
9. H. de Waard, Rev. Sci. Instrum. 36 (1965) 1728.
10. H.P. Wit, Rev. Sci. Instrum. 46 (1975) 927.
11. J.G. Cosgrove and R.L. Collins, Nuclear Instruments and Methods 95 (1971) 269.
12. S.R. Reintsema, S.A. Drentje, J. Stavast and H. de Waard, Hyperfine Interactions 2 (1976) 358.

13. U. Bäverstam, C. Bohm, T. Ekdahl, D. Liljequist and B. Ringstrom, Mössbauer Effect Methodology Vol. 9 (Ed. I.J. Gruverman, C.W. Seidel and D.J. Dieterly, Plenum Press, New York 1974), p. 259.
14. H. de Waard and S.A. Drentje, Physics Letters 20 (1966) 38.
15. H. de Waard, R.L. Cohen, S.R. Reintsema and S.A. Drentje, Phys. Rev. B10 (1974) 3760.
16. H. de Waard, Physica Scripta 11 (1975) 157 and references given therein.
17. S.R. Reintsema, S.A. Drentje, P. Schurer and H. de Waard, Radiation Effects 24 (1975) 145.
18. S. Bukshpan, W. Hilbrants, S.R. Reintsema and H. de Waard, Hyperfine Interactions 2 (1976) 356.
19. S.R. Reintsema, H. de Waard and S.A. Drentje, Hyperfine Interactions 2 (1976) 367.
20. S.R. Reintsema, Mössbauer Studies of Implantation Damage in Iron and Nickel, Thesis, Groningen, 1976.
21. S.A. Drentje and J. Ekster, J. Appl. Phys. 45 (1974) 3242.

PERTURBED ANGULAR CORRELATION

Ekkehard Recknagel

Universität Konstanz, Fachbereich Physik,

7750 Konstanz, W. Germany

I. PERTURBED ANGULAR CORRELATION THEORY

1. PAC-Theory for Electric Perturbations

In a preceding lecture Dr. Berkes has described the angular correlation between two successive γ-rays by

$$W(\Theta) = \sum_k A_k^{(1)} (L_1, I_i, I) A_k^{(2)} (L_2, I, I_f) P_k (\cos \Theta) \tag{1}$$

In this expression the coefficients $A_k(1)$ describe the γ-radiation from an initial state I_i to an intermediate state I with multipolarities of the radiation field L_1, while $A_k(2)$ describes the corresponding transition from the intermediate to the final state I_f. The summation extents to a maximum value of k given by $k_{max} = \min(2I, 2L_1, 2L_2)$ and Θ forms the angle between the direction $\vec{k}_1$ and $\vec{k}_2$ of the γ-rays (Fig.1). In equation (1) the summation over all magnetic substates m_i, m, m_f is already carried out which has been led to the simple angular dependence described by legendre polynomials. Using the addition theorem for spherical harmonics

$$P_\ell (\cos \Theta) = \frac{4 \pi}{2\ell+1} \sum_{-\ell}^{+\ell} Y_{\ell m}^* (\Theta', \phi') Y_{\ell m} (\Theta, \phi) \tag{2}$$

equ.(1) can be written in the more general form

223

$$W(\vec{k}_1, \vec{k}_2) = \sum_m \langle m_f | H_2 | m_b \rangle \langle m_a | H_1 | m_i \rangle \delta_{m_a m_b}$$

$$\times \langle m_f | H_2 | m_a' \rangle^* \langle m_b' | H_1 | m_i \rangle^* \delta_{m_a' m_b'} \qquad (3)$$

where the Hamilton operators H_1, H_2 describe the interaction of
the nucleus with the γ-radiation. (The summation extends over all
m-values.) In the absence of an extranuclear perturbation the
final states $|m_a\rangle$, $|m_a'\rangle$ after the emission of the first γ-radia-
tion are identical with the initial states $|m_b\rangle$, $|m_b'\rangle$ of the sec-
ond radiation. In what follows the nucleus is considered to be no
longer isolated but is influenced by an extranuclear perturbation,
i.e. an electromagnetic field. This interaction effects the nucleus
between the emission of the first γ-ray (t = 0) and the emission
of the second γ-ray (t = t). During this time interval, which is
given by the lifetime τ of the intermediate state I, the external
fields act on the nuclear moments of this state, causing a change of
the magnetic substates from $|m_a\rangle$ to different states $|m_b\rangle$. This
interaction is described by the Hamiltonian K while the time evolu-
tion of the states $|m_b\rangle$ is represented by the unitary operator Λ(t),
which satisfies the time-dependent Schrödinger equation:

$$\frac{\partial}{\partial t} \Lambda(t) = -\frac{i}{\hbar} K \Lambda(t). \qquad (4)$$

In the case of static time independent interaction, K does not depend on time and the solution of equ.(4) is simply:

$$\Lambda(t) = \exp(i K t/\hbar). \qquad (5)$$

Fig.1: Schematic illu-
stration of the para-
meter involved in
γ-γ-perturbed angular
correlation.

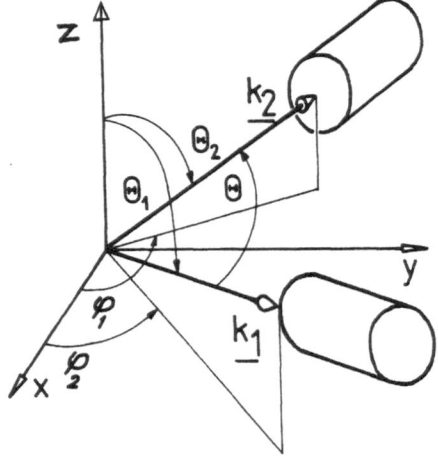

The perturbed angular correlation is therefore

$$\vec{W(k_1,k_2,t)} = \sum_m <m_f|H_2\Lambda(t)|m_a><m_a|H_1|m_i>\cdot<m_f|H_2\Lambda(t)|m_a'>^*$$
$$\times <m_a'|H_1|m_i>^* \tag{6}$$

where the states $|m>$ form a complet set and the state vector $\Lambda(t)|m_a>$ can be expanded in the complete system of the substates $|m_b>$, $|m_b'>$

$$\Lambda(t)|m_a> = \sum_{m_b}|m_b><m_b|\Lambda(t)|m_a>. \tag{7}$$

Introducing equ. (7) into equ. (6) gives

$$\vec{W(k_1 k_2,t)} = \sum_m <m_f|H_2|m_b><m_b|\Lambda(t)|m_a><m_a|H_1|m_i>$$
$$\times <m_f|H_2|m_b'>^*<m_b'|\Lambda(t)|m_a'>^*<m_a'|H_1|m_i>^* \tag{8}$$

A comparison of expression (3) and (8) shows, that the time-dependent angular correlation can be separated into the time independent matrix elements of the unperturbed correlation and a time dependent matrix element describing the influence of the extranuclear interaction on the angular correlation

$$<m_a m_a'|G(t)|m_b m_b'> = <m_b|\Lambda(t)|m_a><m_b'|\Lambda(t)|m_a'>^* \tag{9}$$

Thus finally the angular correlation can be written as

$$\vec{W(k_1,k_2,t)} = \sum_{\substack{k_1 k_2 \\ N_1 N_2}} A_{k_1}(1) A_{k_2}(2) G_{k_1 k_2}^{N_1 N_2}(t) [(2k_1+1)(2k_2+1)]^{-1/2}$$
$$\times Y_{k_1}^{N_1*}(\Theta_1\phi_1)\cdot Y_{k_2}^{N_2}(\Theta_2\phi_2) \tag{10}$$

where we define the perturbation factor

$$G_{k_1 k_2}^{N_1 N_2}(t) = \sum_{m_a m_b} (-1)^{2I+m_a+m_b} [(2k_1+1)(2k_2+1)]^{1/2} \begin{pmatrix} I & I & k_1 \\ m_a' & -m_a & N_1 \end{pmatrix}$$
$$\begin{pmatrix} I & I & k_2 \\ m_b' & -m_b & N_2 \end{pmatrix} \times <m_b|\Lambda(t)|m_a><m_b'|\Lambda(t)|m_a'>^* \tag{11}$$

For vanishing perturbation (t = 0) the evolution matrix $\Lambda(t)$ reduces to the unit matrix and the orthogonal relation of the 3-j symbols gives

$$G_{k_1 k_2}^{N_1 N_2} \equiv \delta_{k_1 k_2} \, \delta_{N_1 N_2} \tag{12}$$

Thus, equ. (10) reduces to the expression given by Dr. Berkes.

2. The Perturbation of a Static Axial-Symmetric Quadrupole Interaction

Since in this lecture we will mainly deal with the PAC caused by a quadrupole interaction, the time differential perturbed angular correlation for the simple case of an axial-symmetric and static interaction will be discussed in more detail.

Taking the Hamilton operator H_{el} from the lecture on quadrupole interaction in the diagonalized form, the perturbation matrix elements in equ. (11) can be easily evaluated, if K in equ. (5) is replaced by H_{el}.

Since H_{el} as well as the eigenvalues E_M are diagonal we find

$$< m_b | e^{-i/\hbar H_{el} t} | m_a > \ = e^{-(i/\hbar) E_{m_b} t} \quad < m_a | m_b > \ = e^{-(i/\hbar) E_{m_b} t} \delta_{m_a m_b} \tag{13}$$

(and a corresponding equation for m_a' and m_b'). The perturbation factor $G(t)$ is only different from zero, if $m_a = m_b = M$ and $m_a' = m_b' = M'$, and reduces to

$$G_{k_1 k_2}^{N N} (t) = \sum_{MM'} \left[(2k_1+1)(2k_2+1) \right]^{1/2} \begin{pmatrix} I & I & k_1 \\ M' & -M & N \end{pmatrix} \begin{pmatrix} I & I & k_2 \\ M' & -M & N \end{pmatrix} \tag{14}$$

$$\times \exp \left[-(i/\hbar) (E_M - E_M') t \right]$$

The 3-j symbols will be equal in their M quantum numbers and therefore $N_1 = N_2 = N$.
Taking $E_Q(M) - E_Q(M') = 3 \, \omega_Q (M^2 - M'^2)$ from the lecture on 'Quadrupole Interaction', the perturbation factor can be written as

$$G_{k_1 k_2}^{N N}(t) = \sum_n S_{n \, N}^{k_1 k_2} \cos(n \omega_0 t) \tag{15}$$

where the smallest non-vanishing energy difference is given by the basic frequency ω_0: for I integer $\omega_0 = 3 \, \omega_Q$; for I half-integer $\omega_0 = 6 \, \omega_Q$. The summation is carried out only for those values which fulfill the relation

$$|M^2 - M'^2| = \begin{cases} n & \text{for I integer} \\ 2n & \text{for I half integer} \end{cases} \tag{16}$$

The coefficients S_{nN}^{k,k_2} are given by

$$S_{nN}^{k_1k_2} = \sum_{MM'} \begin{pmatrix} I & I & k_1 \\ M' & -M & N \end{pmatrix} \begin{pmatrix} I & I & k_2 \\ M' & -M & N \end{pmatrix} [(2k_1 + 1)(2k_2 + 1)]^{1/2} \qquad (17)$$

The influence of the quadrupole interaction results in a rotation of the angular correlation with frequencies ω_o, $2\omega_o, \ldots, n_{max}\omega_o$ where each frequency contributes with the weight factor $S_{nN}^{k_1k_2}$.

As a special case, which in fact is the most common in quadrupole interaction experiments, a statistical distribution of efg's should be considered, i.e. a polycrystalline probe. Then the perturbation factor $G_{k_1k_2}^{NN}$ (t) takes the simple form

$$G_{kk}(t) = \sum_n s_{kn} \cos(n\omega_o t). \qquad (18)$$

In this case the perturbation factor does not depend any more on N and $k_1 = k_2 = k$. (There are only even values of k, if one observes non-polarized γ-transitions.) The coefficients S_{kn} are

$$s_{kn} = \sum_n \begin{pmatrix} I & I & k \\ M' & -M & -M'+M \end{pmatrix}; \quad s_{oo} = 1 \qquad (19)$$

There always remains an unperturbed fraction, given by the hard core value s_{ko}. Numerical values for $S_{nN}^{k_1k_2}$ and s_{kn} are tabulated. As a result, the angular correlation for a polycrystalline probe is

$$W(\theta,t) = \sum_k A_k(1) A_k(2) G_{kk}(t) P_k(\cos\theta) \qquad (20)$$

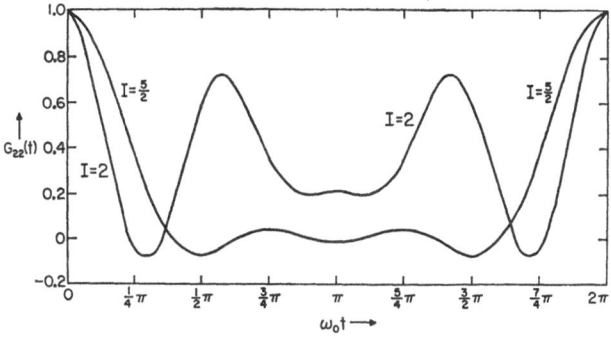

Fig.2: The differential attenuation coefficient $G_{22}(t)$ for an axially symmetric randomly oriented quadrupole interaction involving nuclear states of spin I = 2 and I = 5/2 |ref.1|.

In fig.2 the differential attenuation coefficients $G_{22}(t)$ are shown
for an axial-symmetric quadrupole interaction in polycrystalline
samples (I = 2 and I = 5/2). For non-axial-symmetric field gradients, i.e. rhombohedral structure, the evaluation of $G_{kk}(t)$ is
more complicated. The differential attenuation coefficients $G_{22}(t)$
for non-axial quadrupole interaction in polycrystalline sources as
a function of the asymmetry parameter η are shown in fig.3.

   A general discussion for perturbation factors of quadrupole
interaction is given in ref.|1|.

3. The Perturbation by Magnetic Dipole Interaction

   The perturbed angular correlation formalism for magnetic
dipole interaction is much simpler, because in a static magnetic
field B the energy splitting of the magnetic substates is equidistant and given by

$$E_M(M) - E_M(M') = -\mu B (M - M')/I = \omega_L \hbar \qquad (21)$$

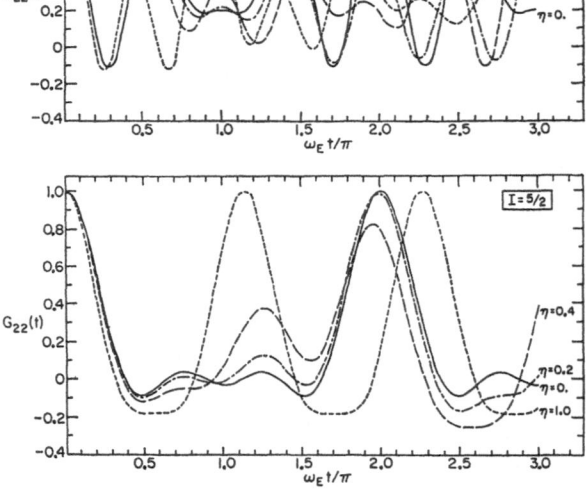

Fig.3: Differential coefficients $G_{22}(t)$ for a randomly
oriented non-axial quadrupole interaction involving nuclear
states of spin I = 2 and I = 5/2. The parameter η is the
asymmetry parameter η = $(V_{xx} - V_{yy})/V_{zz}$ |Ref.1|.

where ω_L is the nuclear Larmor frequency. In the case of B applied perpendicular to the correlation plane, expression (10) for the perturbed angular correlation reduces to

$$W(\theta,t,B) = \sum_k A_k(1) A_k(2) P_k\left[\cos(\theta - \omega_L t)\right] \tag{22}$$

A detailed survey of the theory as well as experimental examples are given in references |1|, |2| and |3|. In ref. 2 also the different times involved in perturbed angular correlation experiments are discussed.

II. EXAMPLES OF A TIME DIFFERENTIAL PERTURBED ANGULAR CORRELATION

1. The ^{111}In Decay

To simplify the discussion, all experimental examples given in this lecture will concentrate on the application of the first excited nuclear state of ^{111}Cd. The nuclear decay scheme is shown in fig.4. In radioactive source experiments, we start from ^{111}In, which decays by electron capture with a half-life of 2.8 days to the 419 keV $7/2^+$ state in ^{111}Cd. This state decays by emission of a first γ-ray of 172 keV to the isomeric $5/2^+$-state with a half-life of 84 ns, the state which is used as a nuclear probe ($Q = 0.8$ barn; $\mu = -.7656$ μ_N). The second γ-ray of the cascade has an energy of 247 keV and depopulates the isomeric state by transition to the ground state $1/2^+$ of ^{111}Cd.

The isomeric state in ^{111}Cd can also be populated by nuclear reactions: ^{110}Cd(d,p) ^{111}Cd or ^{108}Pd(α,n)^{111}Cd. These types of ex-

Fig.4: The ^{111}In decay

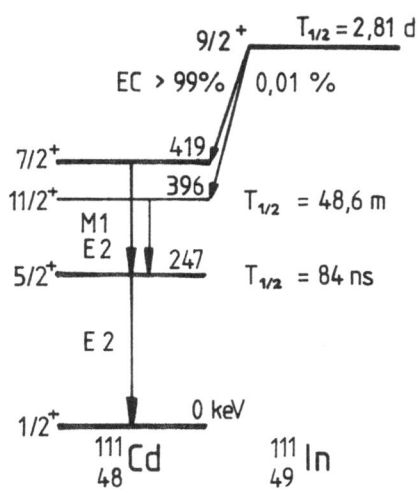

periments are called perturbed angular distribution (PAD) experiments, since only the angular distribution of the second γ-ray is
observed while the first γ-ray is replaced by the reaction.

2. The Quadrupole Interaction of ^{111}Cd in Hexagonal Cadmium

    In fig.5 a typical experiment set-up is schematically shown.
The first γ-ray is observed in detectors A or B and is taken as a
start signal for an electronic time counter, while the second
γ-ray is observed in detectors C or D and stops the counter. The
intensity distribution I(t) of the second γ-rays as a function of
time displays a straight exponential decay curve as long as no perturbation is present. If a perturbation exists, given by the perturbation factor G(t), a modulation of the type shown in fig.2 or 3
is superimposed to the exponential decay (fig.6) |4|. This stems
from the fact that by observing γ_1 in a certain direction the magnetic substates are non-equally populated with respect to γ_2.
(An unequal population of the m-substates is also the result of the
nuclear reaction process.) The intensity distribution is therefore
given by

$$I(t) = I_o e^{-t/\tau} \cdot W(\theta,t) \tag{23}$$

Since only the modulation part is of interest for the evaluation
of the quadrupole interaction, one usually forms the expression

$$R(t) = \frac{I(180^\circ,t) - I(90^\circ,t)}{I(180^\circ,t) + I(90,t)} \tag{24}$$

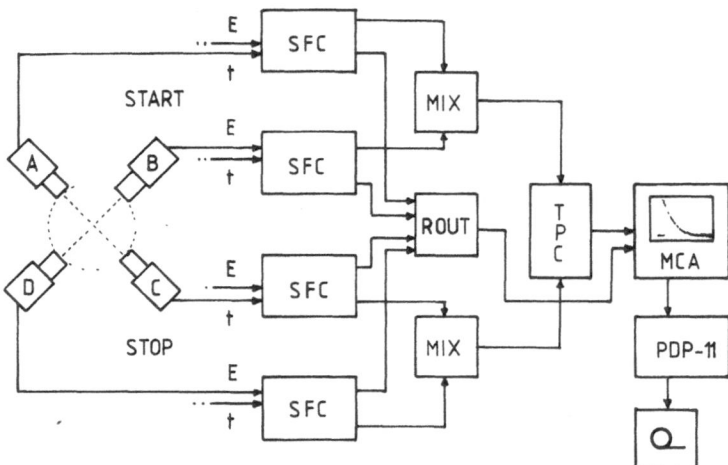

Fig.5: Experimental set-up for a PAC experiment.

where the angles between the detectors are 180° and 90°, respectively. In most cases the coefficients $A_k(1)$ $A_k(2)$ fulfill the relation $A_o(1)A_o(2) = 1 \gg A_2(1)A_2(2) \gg A_4(1) A_4(2) \gg \ldots$ Then the ratio $R^o(t)$ is approximately

$$R(t) = \frac{A_2(1) A_2(2) G_{22}(t)}{1+\frac{1}{4}A_2(1) A_2(2)G_{22}(t)} = \frac{3}{4} A_2(1) A_2(2) G_{22}(t) \qquad (25)$$

In fig.7 this ratio is plotted as a function of time for the cadmium experiment shown in fig.6.
If one carries out this experiment at different temperatures and evaluates the ratio $eqQ(T)/eqQ(0)$ one obtains the temperature dependence of the efg shown in fig.4 of the lecture on 'Quadrupole Interaction'.

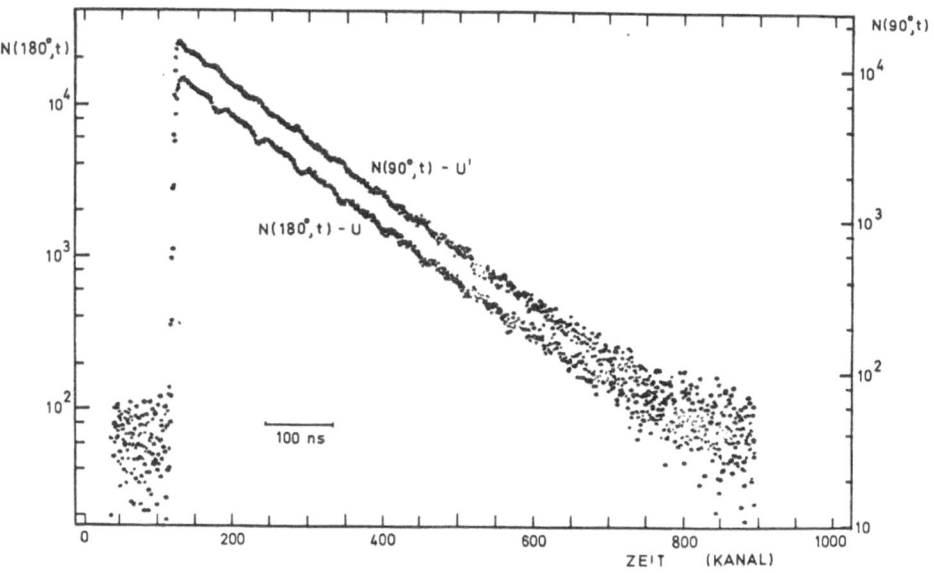

Fig.6: Time differential PAC spectrum for ^{111}CdCd |4|.

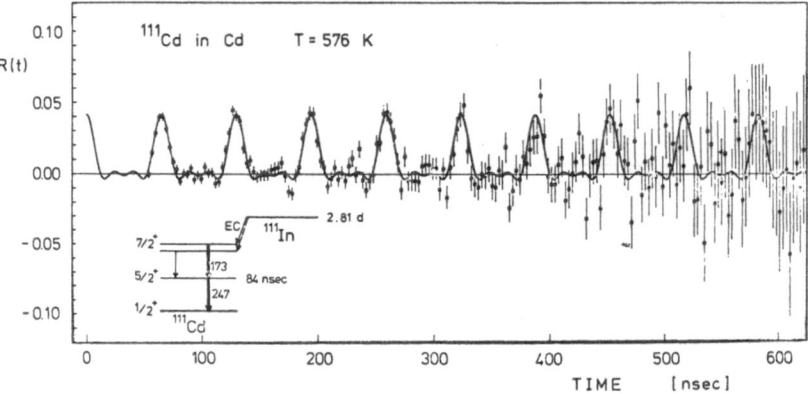

Fig.7: Modulation of the 247 keV γ-radiation of $^{111}CdCd$ after
        evaluating the experimental decay |4|.

III. RADIATION DAMAGE STUDIES IN METALS

The application of PAC to solid state problems started about
15 years ago. In the beginning the interest was much more on the
determination of nuclear moments, because the technique was developed by nuclear physicists. But as we have seen, the nuclear quantity is always connected with a solid state one. The interest soon
shifted to solid state problems, as to the determination of inner
magnetic fields of various impurities in ferromagnets, to transient
fields, or, as mentioned before, to electric field gradients. Recently the research has focussed on the study of defects, especially
in cubic metals. Since this summer school is mainly devoted to
problems of this kind, the second part of this lecture will concentrate on the discussion of a representative example, namely the
investigation of radiation damage in copper by using ^{111}Cd as a
microscopic probe |5|.

1. Classical Damage Studies in Copper

Most results about defects in metals were accumulated by rest-
resistivity-, neutron scattering-, and electron microscopic measurements on samples, where the defects were created by a number of
different methods. Each of them is assumed to produce defects in
certain combinations and configurations. Thus quenching produces
mainly vacancies, electron bombardement ($\sim$ 3 MeV) Frenkel pairs,
while irradiation with ions (H^+, He^{++} and heavier ones) is considered to form damage cascades. To get information about specific

properties of defects as production rate, mobility, cluster formation etc., the samples are normally treated by an isochronal
annealing procedure. Fig.8 shows a typical result of a rest resistivity measurement of an irradiated copper sample. Certain stages
I-V can be distinguished, the interpretation of which is still
a matter of discussion. Conclusions of two different models, the
so-called one- and two interstitial models, are compared in Tab.I.

Without going into a detailed discussion of the different
classical methods, and of course, without having in mind to lessen
the success, those experiments contributed to the understanding
of defects in metals, some disadvantages should be mentioned with
respect to PAC investigations.

In rest-resistivity measurements all conclusions about processes which happen on a microscopic scale have to be drawn from
macroscopic observations. Thus, for example, the migration energy
of a defect is taken from an annealing stage, but there are no
means to decide, whether one or more kinds of defects having the
same migration energy are involved. Furthermore the recognition
of a certain defect is only indirect.

Neutron scattering or electron microscopy work is limited to
a minimum size of defect clusters or minimum concentration.

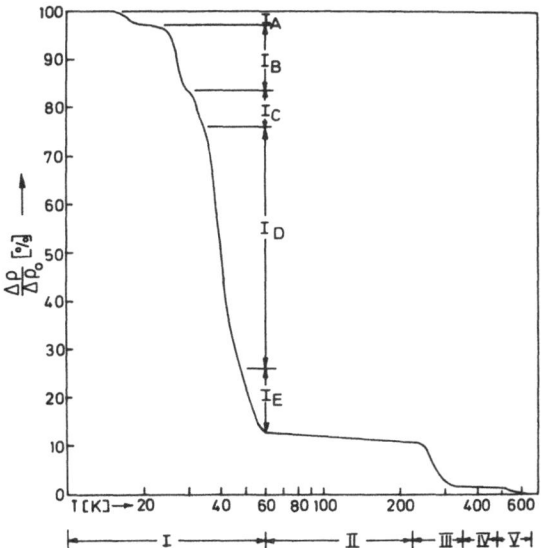

Fig.8: Annealing curve of a damaged copper sample from
a rest-resistivity measurement.

Table 1: Comparison of the one- and two-interstitial model for pure metals.

Annealing stage	1-Interstitial Model	2-Interstitial Model
I	Migration and agglomeration of interstitials.	Migration of a metastable interstitial configuration (Crowdion). Athermal conversion of these interstitials to dumbells and inter-stitial-clusters.
II.	Rearrangement and growing of interstitial clusters.	Thermal conversion of interstitials to dumbells.
III	Free migration of vacancies.	Free migration of dumbells.
IV	Growing of vacancy clusters.	Free migration of vacancies.
V	Dissolving of vacancy and interstitial clusters.	Dissolving of vacancy and interstitial clusters.

Some of these disadvantages can be overcome by PAC experiments. They are therefore a valuable supplement for a complete
understanding of the defect problem.

2. Defects in Copper Investigated by PAC

Before going into the details of the TDPAC-experiment in
copper, the characteristical features of the perturbation function
$G_{22}(t)$ for different static situations within a crystal lattice
should be mentioned. If one replaces in a cubic lattice like copper
one copper atom by a radioactive probe atom like In, the surrounding has still cubic symmetry, i.e. no perturbation will influence
the immediate state. In equ.(25) $G_{22}(t) \equiv 1$, a full unperturbed
anisotropy is observed during the whole measuring time. If defects
are statistically distributed around the probe atoms, i.e. in the
n-nearest neighborhood (n = 1,2,3...), $G_{22}(t)$ will roughly show
an exponential decrease. Only in the case defects as vacancies
or interstitials (or clusters of them) form a definite configuration with the probe atom, a certain efg exists, resulting in
a R(t) pattern like that one shown in fig.7. In general, the
observed spectra are more complex, formed by a superposition of
different contributions.

Having these general features in mind, we can now turn to the
experiment. If one bombards a well annealed copper foil with
heavy ions, i.e. [111]In, with an energy of about 350 keV, it penetrates to a depth of about 1000 Å. On the path the ion looses it's
energy, thereby creating defects in the metal, roughly one
Frenkel pair for 30 eV. Some of these pairs immediately recombine,
but at the end a zone is left which is rich in vacancies and de-
pleated in interstitials. At room temperature irradiation the
TDPAC curve shows an exponential decay, which can be interpreted
as a distribution of efg's around $V_{zz} = 0$. (Fig.9).
Starting now an isochronal annealing procedure, i.e. the foil is
kept for 10 min. at an annealing temperature T_A, while the measurements are carried out at room temperature, a pattern turns
up, which is known to be due to a definite efg. The amplitude of
the structure corresponds to the percentage of ^{111}In probe ions,
which experience the same efg. They all form a distinct probe-
defect configuration. At $T_A \sim 750$ K the structure has been completely disappeared and the R(t) curve corresponds to an unperturbed cubic surrounding of the probe ion. In fig.10 the dependence of the amplitude from annealing temperature is shown. The
increase of the curve is due to a trapping of defects on the probe,
while the decrease reflects the detrapping process.
In this case a maximum amplitude of 35% is reached at $T_A \sim 650$ K,
the quadrupole coupling constant is $\nu_{Q3} = 52(1)$ MHz, which corres-

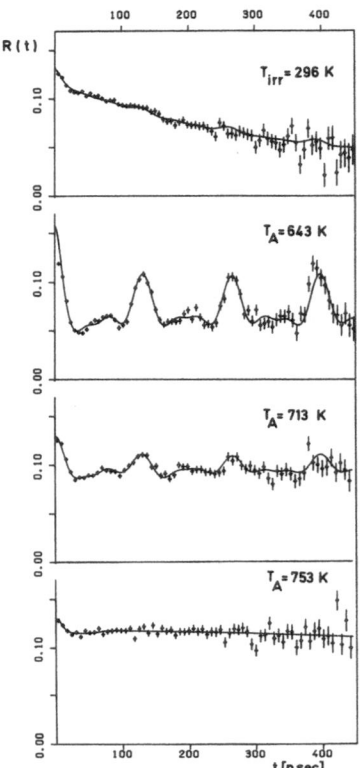

<u>Fig.9</u>: DPAC spectra of ^{111}In in copper measured directly after irradiation and at different annealing temperatures.

<u>Fig.10</u>: Dependence of the relative fraction B_3 of probe atoms in the defect configuration $\nu_{Q3} = 52$ MHz on the annealing temperature

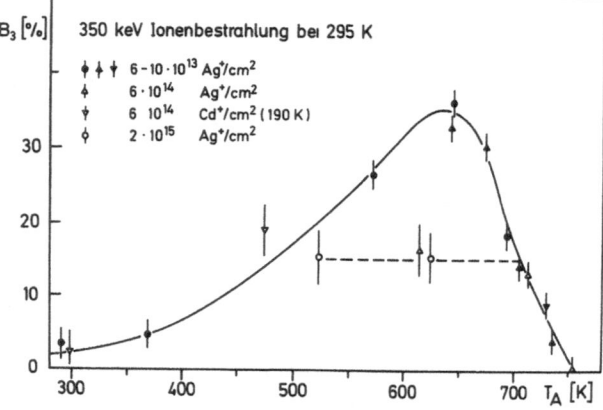

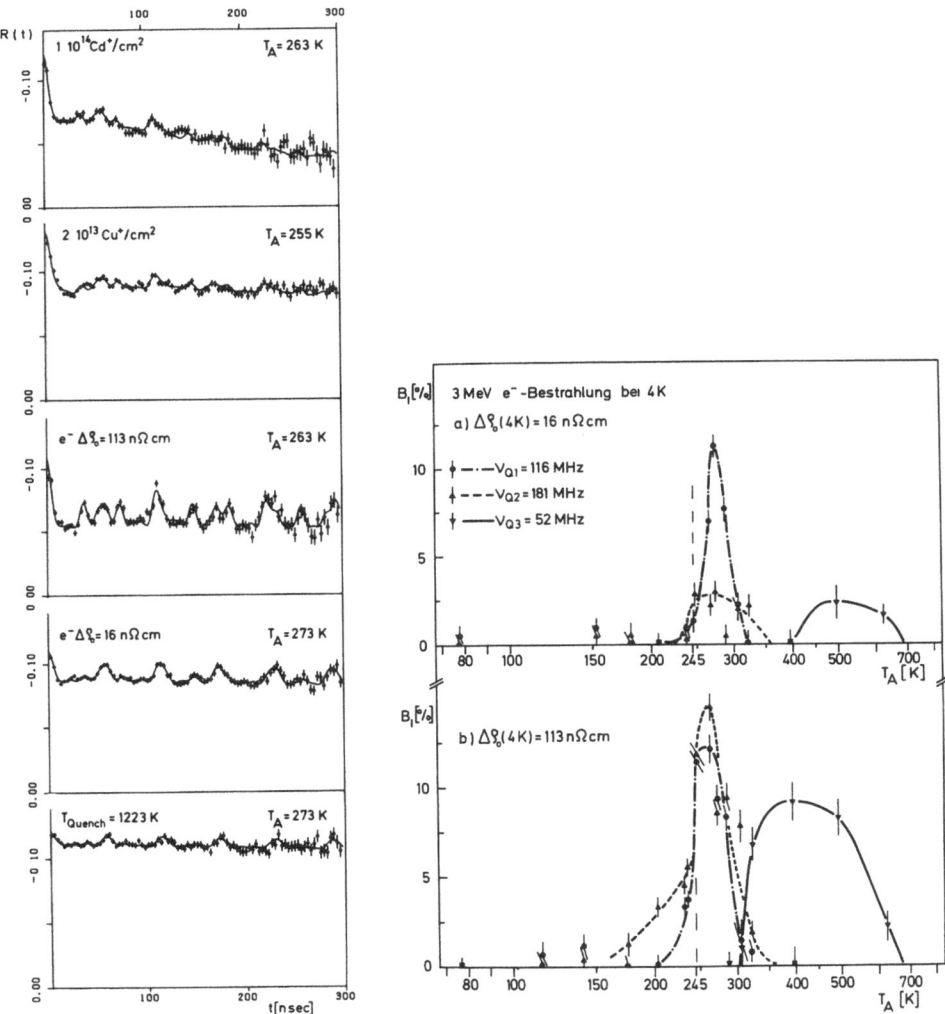

Fig.11: (left) DPAC spectra of ^{111}In in copper
following different damage procedures.

Fig.12: (right) Dependence of the relative fraction B_i
of probe atoms in the defect configurations
ν_{Q1}, ν_{Q2} and ν_{Q3} on the annealing temperature. Defects are produced by two different electron
irradiation doses at 4 K .

ponds to an efg of $eq = 2.9 \cdot 10^{17}$ V/cm^2.

Before an interpretation of this observation is given, other results of damage studies on copper should be summarized. Beside the heavy ion bombardement, different other experiments were carried out, most of them analogous to the rest-resistivity experiments, and, in fact, such measurements were also performed on the same probes to get a consistent picture of macroscopic and microscopic results. Thus copper foils, into which ^{111}In was either implanted or diffused, were well annealed ($T_A \sim 900$ K) and afterwards treated by electron bombardement ($E(e^-) = 3$ MeV), by quenching, or by proton irradiation. All measurements were carried out at liquid nitrogen temperature, while the isochronal annealing process was as described above. Typical TDPAC spectra are shown in fig.11. There are two more quadrupole interactions to be seen, characterized by the coupling constants $\nu_{Q1} = 116(2)$MHz and $\nu_{Q2} = 181(3)$ MHz. The experimental data can be evaluated by the expression

$$A_{22}^{eff} G_{22}(t) = A_{22}^{eff} \sum_{n=0}^{3} S_{2n} \left\{ A + \sum_{i=1}^{3} B_i \cos(n \omega_{Qi} t) + C e^{-n\sigma_o t} \right\} \quad (26)$$

where A denotes the unperturbed fraction, B_i is associated with defects resulting in a unique interaction frequency ν_{Qi} and C is a fraction of nuclei, which are positioned in a nearly unperturbed region and observe a frequency distribution σ_o around zero.

The dependence of the fractions B_i on annealing temperature for electron irradiation experiments is shown in fig.12. The defects characterized by ν_{Q1} and ν_{Q2} are observed in the temperature range of recovery stage III, while the appearance of ν_{Q3} is closely correlated with recovery stage V.

In quenching and low dose electron irradiation experiments only ν_{Q1} has been observed. Therefore this frequency can be attributed to an ^{111}In-next nearest monovacancy configuration. Though rest-resistivity measurements interpret stage III to be due to the annealing of only one defect, the PAC experiment clearly shows that different defects are involved. The second configuration (ν_{Q2}) most probably can be attributed to a divacancy or a small vacancy cluster. The increase of the B_i amplitude is therefore due to the migration of vacancies, and, presumably with a somewhat lower migration energy, to divacancies. The decrease at about $T \sim 300$ K can be interpreted as being due to detrapping (or to the accumulation of more vacancies). Above this temperature ν_{Q3} starts to turn up which can be attributed to the interaction of the ^{111}In probe with planar faulted loops, presumably of vacancy type. This interpretation is supported by another experiment on a single crystal, where the symmetry axis of the efg was shown to point along the crystallographic <111> direction (fig.13).

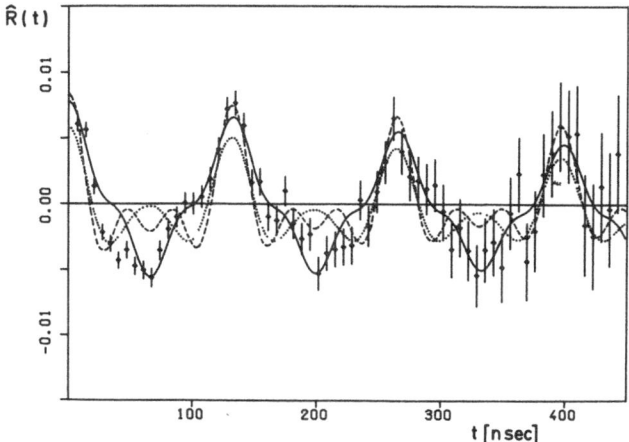

Fig.13: DPAC spectra of ^{111}In in a single copper crystal.
The best fit is obtained if the symmetric axis of the efg
points along the <111> direction (drawn line).
Dashed line: <100>, Dotted line: <110>

The annealing of this defect coincides with stage V, which from
electron microscopy work is known to be connected with the dissolution of clusters.
Generally the observed pattern may depend also on other parameters, like the dose − as already shown in the case of electron
bombardement − or the irradiation conditions. For demonstration
an example should be given for each of these two cases.

Fig.14 shows the results of experiments, where copper foils
were irradiated with doses of $6 \cdot 10^{14}$ Ag$^+$/cm^2 and 10^{14} Ag$^+$/cm^2 (fig.15)
respectively. After annealing to 650 K, the temperature where B$_3$
was at its maximum, the high dose irradiation displayed a more
complicated pattern. The analysis of this TDPAC spectrum exhibited
two frequencies, one similar to ν_{Q3} as observed in the low dose
experiment, the other one is somewhat higher, ν_{Q4} = 53 MHz and
shows an asymmetry of η = .48. Since ν_{Q3} was attributed to a
planar loop, the other field gradient might be produced at
an interconnection of two <111> planar loops.

Concerning the irradiation conditions, one usually has to
distinguish between correlated and uncorrelated damage. In the
former case the probe ion senses defects produced by its own
damage cascade, while in the latter case the probe ion is situated
on a normal lattice site before the sample is irradiated with
other ions, so that it only feels defects created by other damage
cascades. An example is shown in fig.15 which shows essentially
the same pattern in both cases (ν_{Q3}).

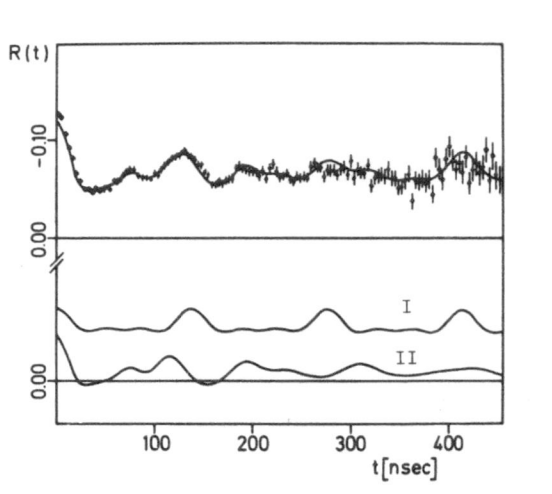

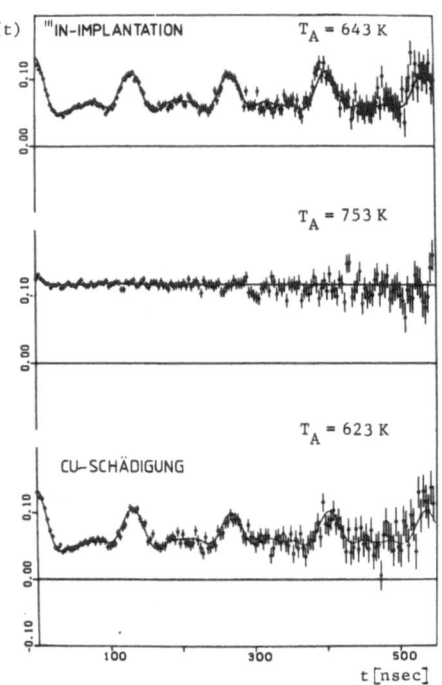

Fig.14: (left) Dose dependence of the DPAC spectra of ^{111}In
in copper (Dose $6 \cdot 10^{14}$ Ag$^+$/cm^2 at E = 300 keV).
Lower curve shows contribution of the two components.

Fig.15: (right) Correlated and uncorrelated damage in copper
(Dose 10^{14} Ag$^+$/cm^2 at E = 300 keV).

REFERENCES:

|1| Frauenfelder, H. and Steffen, R.M. in 'Alpha-, Beta-, and
Gamma-Ray Spectroscopy' (K. Siegbahn, ed.) Vol.2 (1965)
North-Holland Publ., Amsterdam.
|2| Recknagel, E., in 'Nuclear Spectroscopy and Reactions'
(J. Cerny, ed.) Part C (1974) Academic Press, New York.
|3| Recknagel, E., Physica Scripta 11 (1975) 208.
|4| Christiansen, J., Heubes, P., Keitel, R., Klinger W.,
Löffler, W., Sandner, W., Witthuhn, W., Z.Physik B24(1976)177.
|5| Echt, O., Recknagel, E., Weidinger, A., Wichert, Th.,
Z.Physik B32 (1978) 59, and Wichert, Th., Deicher, M.,
Echt, O., Recknagel, E., Phys.Rev.Lett. 41 (1978) 1659.

ATOM SITE CHARACTERIZATION IN METALS USING CHANNELING TECHNIQUES

L.M. Howe and J.A. Davies

Solid State Science Branch
Atomic Energy of Canada Limited
Chalk River Nuclear Laboratories
Chalk River, Ontario, Canada   K0J 1J0

1. INTRODUCTION

Detailed information concerning the crystallographic location of solute atoms in solids is of great importance since many physical properties of materials are determined by the specific location of the solute atoms in the lattice. This information can be obtained using ion channeling by monitoring close impact parameter collision events of the incident ions with the host atoms and the solute atoms. Rutherford scattering, ion-induced nuclear reactions and ion-induced X-rays are the principal close impact parameter collision events which are utilized in ion channeling investigations.

The various solid-state applications of channeling are based on the unique property of a channeled beam that it cannot penetrate closer than approximately 0.1 Å to a lattice site (see references 1-4 for detailed discussions of channeling phenomena). Hence the channeled beam has close-collision interactions only with those atoms which are displaced > 0.1 Å from normal lattice sites. Consequently, ion channeling provides information on geometrical site locations but is insensitive to conditions of local bonding, fields, etc., which is in contrast to the situation which exists for other nuclear techniques. Many atom site location studies were initially performed in ion implanted semiconductors but in recent years considerable data has been obtained on the location of solute atoms in metals. The technique has also been very useful in studying the interaction between irradiation-produced defects and solute atoms in metals. In addition, ion channeling is proving to be a rather novel means of studying the structure of metal surfaces.

241

2.   THE CHANNELING PHENOMENON

Channeling is a gentle steering process arising from a
correlated series of collisions whenever a beam of energetic
charged particles moves through a crystal in a direction almost
parallel to a major axis or plane. Provided the angle ψ between
the beam and the row of atoms is sufficiently small, the increasing electrostatic repulsion between the ion and the screened field
around each successive target nucleus (due to the gradually
decreasing impact parameter at each successive collision) is
sufficient to deflect the particle smoothly away from the row – as
illustrated schematically in fig.1.

Three basic particle trajectories can be distinguished, as
illustrated in fig.2a.

(1) An energetic charged particle, entering a lattice at an angle
less than $\psi_{\frac{1}{2}}$, the predictable critical angle, of an atomic row (or
plane) is steered by successive gentle collision (trajectory A),
and is thereby prevented from entering a forbidden region around
each lattice row. The radius r_{min} of this forbidden region may be
equated roughly to the two-dimensional vibrational amplitude $\rho_\perp$ in
the transverse plane.

(2) When the incident angle is much larger than $\psi_{\frac{1}{2}}$ the particle
has no "feeling" for the existence of a regular atomic lattice,
and so has a random trajectory C.

(3) If the incident angle ψ is only slightly larger than $\psi_{\frac{1}{2}}$, then
the particle trajectory (B) actually has an enhanced probability
of being close to the atomic rows, and hence of undergoing violent
collisions.

A very important consequence of the above model is that all
physical processes requiring smaller impact parameters than

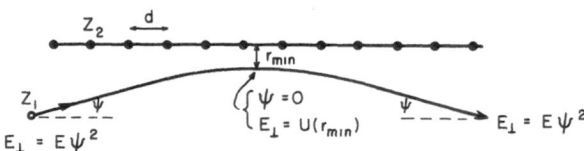

Figure 1 – Schematic diagram illustrating how a correlated
sequence of collisions with an aligned row of crystal atoms
(atomic number Z_2) can gently steer (channel) a particle (atomic
number Z_1). Channeling will occur for particles having trajectories such that the incident angle ψ at the mid-channel plane is
less than a critical angle $\psi_{\frac{1}{2}}$.

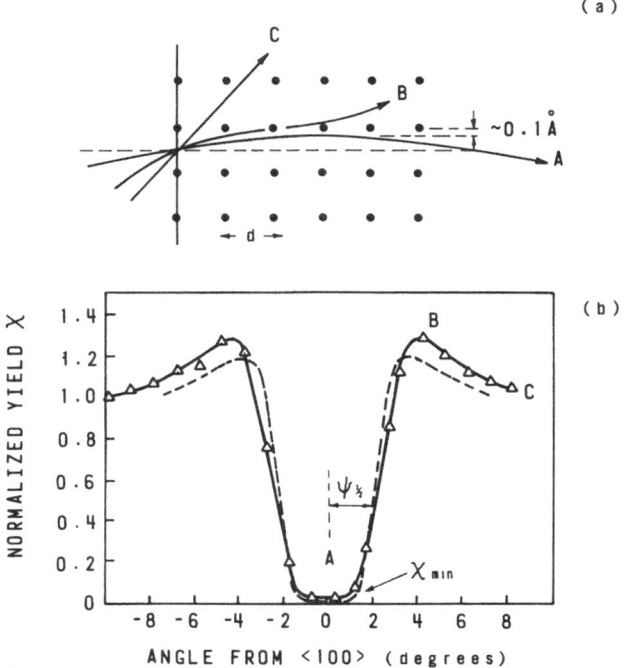

Figure 2   (a) Schematic diagram of three typical charged particle
trajectories in a crystal.   (b) Experimental (Δ) and calculated
(---) angular dependence of the yield of a typical close encounter
process (in this instance Rutherford scattering of 480 keV protons
in <100> W).

r_{min} ($\sim$ 0.1 A) are completely prohibited for a channeled beam.
Consequently, the yield of such a process is a quantitative
measure of the non-cnanneled fraction of the beam, and so provides
a sensitive "detector" for studying the transition between
channeled and non-channeled trajectories.   The experimentally
observed and theoretically predicted orientation dependence of a
typical close encounter process is shown in fig.2b; the yields in
regions A, B and C arise from particles with the corresponding
trajectories in fig.2a.   The yield at ψ = 0 does not fall quite to
zero because there is still a small random fraction ($\sim$ 0.01),
determined by the point of impact on the crystal surface.

        Rutherford scattering has been the most widely used close-
encounter process in channeling studies.   It is particularly useful in atom location studies where the solute atom is heavier than
the host atom, since the solute atom spectrum then appears at a
higher energy and separate from the host atom spectrum, as shown
in fig.3 (see ref.5 for more detailed information on analyzing

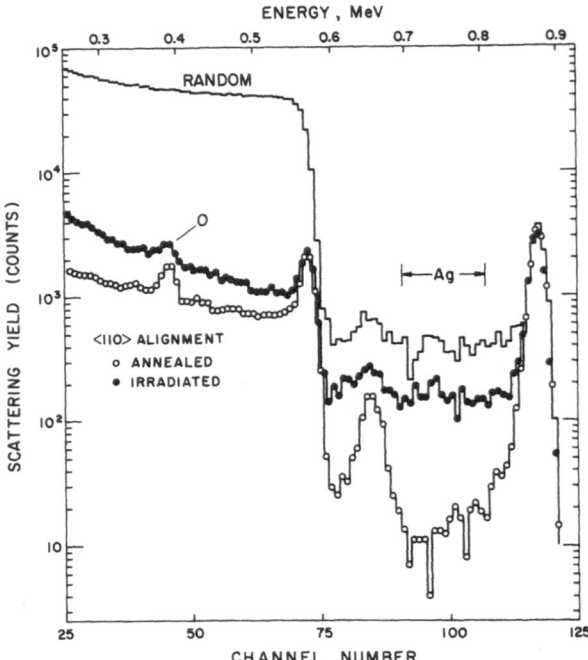

Figure 3: Energy spectra of backscattered 1 MeV He$^+$ ions from an
Al–0.08 at% Ag crystal at 35 K showing the <110> aligned spectra
before and after an irradiation with 3 x 10^{15}–0.3 MeV He$^+$ ions/cm^2
at 60 K. Also shown is the random spectrum which has been normalized to the same integrated current as the aligned spectra. Note
that the irradiation has increased the scattering yield from both
the Ag solute atoms (channels 91–107) and the Al host atoms
(channels < 75). The peak extending from channels 80 to 90 is due
to Cl present from the electropolishing of the crystal (from ref.
22).

Rutherford scattering spectra). For studying solutes which are
lighter than the host, nuclear reactions are generally used.[6]
According to theory, the yields of all processes requiring an
impact parameter less than approximately 0.1 Å should show the
same orientation dependence. Thus, different close-encounter
processes having impact parameters $\stackrel{\sim}{<}$ 0.1 Å can be used inter-
changeably in studying channeling effects.

     The observed minimum critical angle ($\psi_{\frac{1}{2}}$) and minimum yield
(χ_{min}) both depend on the crystal temperature and the depth
beneath the surface at which the measurements are made. The minimum yield χ_{min} is defined as the ratio of the yield in the perfectly aligned direction to that in a random direction; it is there-

fore a direct measure of the unchanneled fraction of the beam.
Theoretical expressions[7] which appear to give fairly reasonable
estimates for $\psi_{\frac{1}{2}}$ and χ_{min} (extrapolated to zero depth) for axial
channeling of projectiles having energies in the MeV range are:

$$\psi_{\frac{1}{2}} = 0.25 \; F_{ax} \; (\rho_{\perp}/a) \cdot (Z_1 Z_2/Ed)^{\frac{1}{2}} \qquad (1)$$

and

$$\chi_{min} = Nd\pi \; (C_1 \rho_{\perp}^2 + C_2 a^2) \qquad (2)$$

In the above expressions Z_1 and Z_2 are the atomic numbers of
incident projectile and crystal atoms respectively, E is the
incident energy of the projectile, d is the distance between atoms
along the channel row, a is the Thomas-Fermi scaling function, F_{ax}
is a function depending upon the appropriate $\rho_{\perp}/a$ ratio, N is the
atomic density and C_1 ($\sim$3) and C_2 ($\sim$0.5) are numerical constants.
Similar expressions also apply for planar channeling. Values of
$\psi_{\frac{1}{2}}$ vary from a few hundredths of a degree to a few degrees. Values
of χ_{min} are typically about 0.2-0.4 for planar channeling and
about 0.01-0.05 for axial channeling. Channeling effects are
strongly enhanced at low temperatures since $\psi_{\frac{1}{2}}$ increases and χ_{min}
decreases as $\rho_{\perp}$ is decreased.

3. BASIC ATOM LOCATION PRINCIPLES

The important solid-state applications of channeling are
based on the fact that a channeled beam has close-collision interactions only with those atoms which are displaced $>$ 0.1 Å from
normal lattice sites.[8] In a lattice location analysis by the
channeling technique, a comparison is made between the interaction
yields from the solute and host atoms with the beam incident along
various different axes and planes. Strong interaction yield
attenuation of the channeled beam with the solute atoms indicates
that the solute atoms lie within the shadow of the aligned set of
atomic rows or planes. For example, as schematically illustrated
in fig.4a, a solute atom occupying a substitutional lattice site
(O) would give channeling behaviour similar to the host lattice
for channeling along the <100>, <110> and <111> directions. A
solute atom in the interstitial site indicated by (Δ) would
exhibit the same channeling behaviour as the substitutional solute
atom for the <100> and <111> directions. However, this
interstitial solute atom does not lie within the shadow of the
<110> rows and would give an enhanced yield along this direction.
The magnitude of the enhanced yield will depend upon the ion flux
distribution in the channel. For a solute atom in an off-lattice
position such as X, an increase in yield would be experienced for
a beam incident along the <100>, <110> and <111> directions.

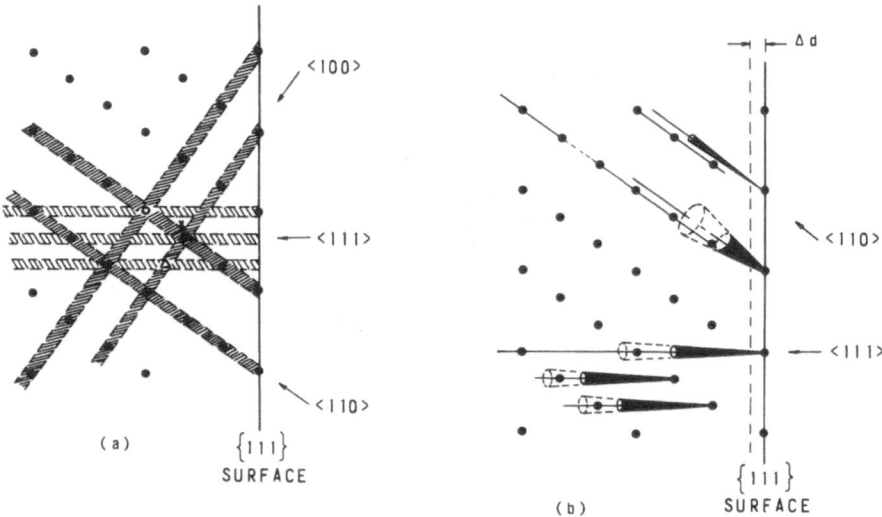

Figure 4: (a) Schematic diagram illustrating the principle of
determining the positions of solute atoms in a fcc lattice by
means of triangulation. The shaded areas represent the regions
which are forbidden for a channeled particle. (b) Atomic configuration near the surface of a {111} fcc crystal illustrating how
the channeling effect may be used to measure the surface relaxation Δd.

     Ion channeling can also be used to study the structural
changes occurring at a crystal surface. In this case, information
is obtained regarding the position of the host crystal atoms in
the surface region. The basic principle involved in using this
technique is illustrated in fig. 4b. The area of the surface peak
in a Rutherford backscattering energy spectrum provides a quantitative measure of the number of unshadowed lattice atoms per cm^2
in the surface region. For perpendicular incidence (<111> in fig.
4b), the second atom in each row is perfectly shadowed and hence
the surface peak area should be equivalent to 1 atom per row at
all energies, provided the 2-dimensional vibrational amplitude is
much smaller than the collisional shadow-cone radius R. A surface
relaxation of Δd has no effect on this <111> shadowing. For non-
perpendicular axes (<110> in fig. 4b), a surface displacement Δd
shifts the shadow cone relative to the underlying row of atoms,
thus causing the surface peak to increase towards a value of 2
atoms/row. Since the cone radius R varies inversely with incident
beam energy E, the surface peak should increase from a value of 1
atom/row at low E to $\sim$ 2 atoms/row at high E. If one or more
surface planes of atoms have been laterally displaced with respect

to the underlying crystal (as in the reconstructed surface region
of crystals), then, even for perpendicular incidence the surface
peak would correspond to appreciably more than one atomic plane.

4. ION FLUX DISTRIBUTION AND SOLUTE ATOM YIELDS

In order to obtain detailed information on the location of
solute atoms in lattices, it is necessary to determine the spatial
distribution of the channeled ions. The flux distribution of the
channeled ions can be calculated either by an analytical model[9-12]
based on the continuum theory of channeling or by Monte Carlo
computer simulation[7,13-15]. An example of the ion flux distribution of channeled particles is shown in fig.5 for 1 MeV He[+] in
<110> Al at 30 K. Note the appreciable enhancement in ion flux
near the center of the channel (and at the potential minimum
position) relative to that for a beam incident along a random
direction.

The yield Y(s) from displaced solute atoms may be obtained

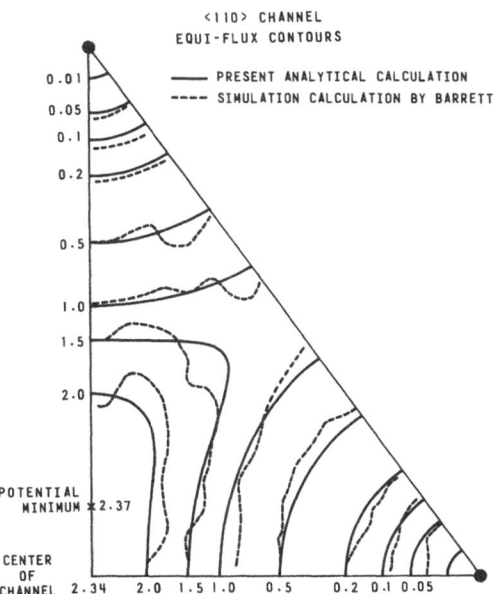

Figure 5: Equi-flux contours of one quarter of a <110> channel
for 1 MeV He[+] in Al at 30 K. The results of an analytical calculation (solid lines) are compared with those from a Monte Carlo
simulation (dotted lines). The flux values have been normalized to
the value for random incidence (from ref.12).

by averaging the ion flux around the equilibrium positions of the displaced atoms,[12] hence

$$Y(s) = \int F(\bar{r}, \psi_{in}) \, P_s(\bar{r}) \, d\bar{r} \tag{3}$$

where s is the magnitude of the displacement of the solute atoms from a lattice site along a <hkl> direction, $F(\bar{r}, \psi_{in})$ is the flux distribution at a position $\bar{r}$ in the channel for an incident angle ψ_{in} and $P_s(\bar{r})$ is the probability for the displaced atoms being at $\bar{r}$ in the channel. $P_s(\bar{r})$ is given by[12]

$$P_s(\bar{r}) = (d\bar{r}/\pi\rho_\perp^2) \exp (-| \bar{r}-\bar{s}|^2/\rho_\perp^2) \tag{4}$$

where $\rho_\perp$ is the thermal vibrational amplitude perpendicular to the string. The calculated yields for an angular scan through a <110> channel are shown in fig.6 for atoms displaced along <100> directions in Al. It can be seen that, for large values of s, the yield in the <110> aligned direction can be more than twice the random yield. On the other hand, in the <100> and <111> aligned directions, these <100> displacements produce a considerably smaller yield.[12]

When both the host and the solute atoms are shadowed along a particular <lmn> direction, the angular scans for the host and the solute atoms will be almost indistinguishable as long as the mean vibrational amplitudes of the two species are of the same order. This behaviour has been observed, for example, in substitutional alloys of Au(Cu), Au(Ni), Au(Ag) and Au(Pd) which were produced by ion implantation.[16] If the mean vibrational amplitude of the solute atom is considerably greater than the host atom, then an angular scan through the shadowed direction will show a narrower yield for the solute atom than for the host atom. Angular scans through the [0001] axis in Mg crystals implanted with deuterium (10^{15} D^+/cm^2, 5 KeV) exhibit this behaviour when analyzed at 50 K.[17] The deuterium is believed to be in a tetrahedral interstitial position at 50 K (thus shadowed along a [0001] direction in the hcp lattice) and to have a vibrational amplitude 2.5 times greater than Mg.[17]

For rather complicated distributions of interstitial solutes, it may be necessary to perform fairly detailed experiments coupled with the appropriate ion flux distribution calculation in order to specify the location of the solute atoms. However, interstitial solute atoms are often located on well-defined interstitial sites in the lattice which greatly simplifies the analysis. For example, consider the lattice location of deuterium implanted into the b.c.c. crystals W and Cr.[18] The angular distributions for scans through the <100> axis are shown in fig.7. The deuterium was dected by the protons emitted from the nuclear reaction $D(^3He,p)^4He$ at 750 keV, and the lattice atoms by backscattering of

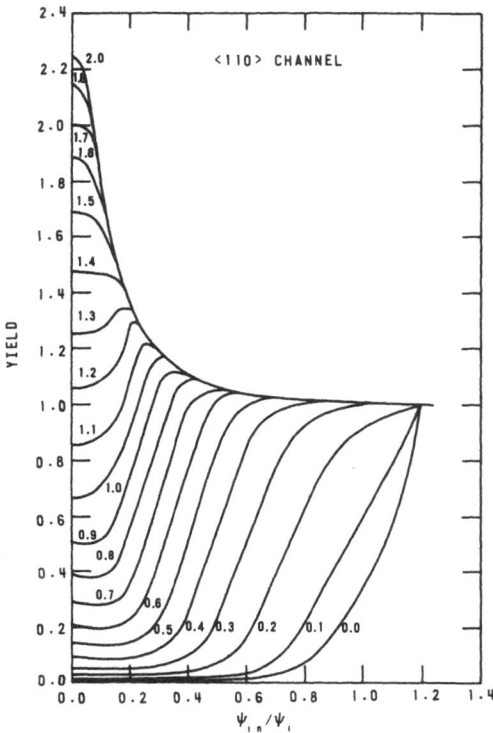

Figure 6: Calculated yields in a <110> channel for atoms displaced along <100> directions as a function of the incident angle
ψ_{in} normalized to the Lindhard[1] characteristic angle ψ_1. The
numbers on the curve designate values of the displacements in Å.
Calculation for 1 MeV He[+] in Al at 30 K at a depth of 1000 Å (from
ref.12).

[3]He ions. For W, the strong enhancement in the D yield suggests
that the solute is located in the tetrahedral interstitial sites,
which are not shielded by the W atoms along the <100> rows. For
Cr, the D yield exhibits a sharp flux peak superimposed upon a
broad dip with an angular width similar to that observed for the
Cr lattice and with an extrapolated minimum yield of about 70%.
Since 1/3 of the octahedral sites are shielded by Cr atoms along
the <100> rows, the minimum yield value and the dip width are
consistent with 1/3 of the D lying along the <100> row whereas the
central flux corresponds to the remaining D atoms in the center of
a <100> channel. Angular scans along the {100} plane were also
used to check this interpretation. The results showed only a dip
for D in Cr but a small flux peak superimposed on a dip for D in W
which is as expected since the octahedral sites lie within {100}

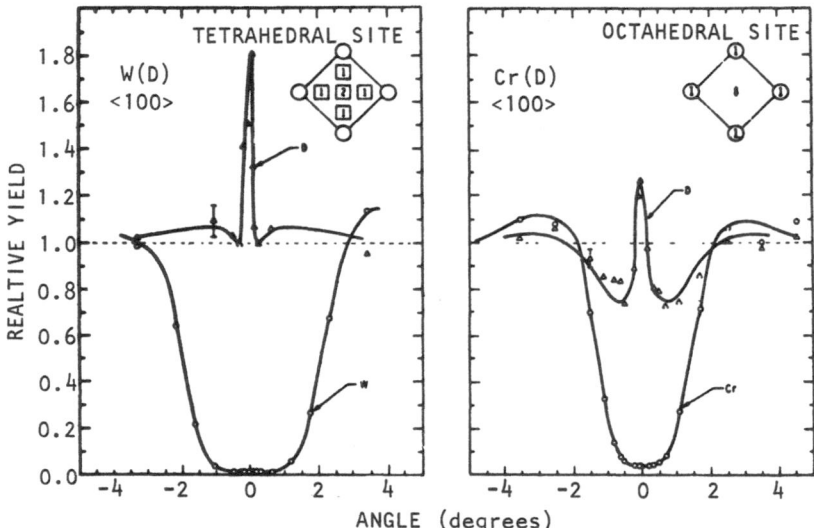

Figure 7: Angular scans through the <100> axis for W and Cr for 3×10^{15}/cm^2 D implants at 30 and 15 keV, respectively (from ref.18).

whereas 2/3 of the tetrahedral sites are in the {100} plane and 1/3 between the planes.

    The main limitations in the use of ion channeling for locating solute atoms are: (a) detection sensitivity ($\sim 10^{-4}$ atom fraction), (b) crystal perfection, (c) radiation damage induced by the analyzing beam, (d) ambiguous site determination for solute atoms lying in low symmetry sites, and (e) difficulties associated with obtaining detailed information on ion flux distributions in certain cases (e.g. depth oscillation of the particle density occurs at small penetration depths ($\lesssim 1000$ Å) since statistical equilibrium is not yet achieved). These problems, as well as more detailed information on the location of solute atoms in various metallic systems are discussed in refs. 10, 18-21.

    5.   INVESTIGATIONS ON THE INTERACTIONS BETWEEN SOLUTE ATOMS AND IRRADIATION PRODUCED DEFECTS

    Defects such as vacancies and self-interstitials which are produced during energetic particle irradiations can be trapped by solute atoms. Channeling measurements can be used to determine accurately the displacement of the solute atoms from lattice sites as a result of this trapping, and thus the trapping configura-tion.[22-28] After low temperature irradiation, annealing measure-ments can be used to indicate the onset of trappping and thus the mobility of the defect which is trapped. Also, from measurements

of the trapping rate as a function of irradiation fluence and
temperature, the trapping efficiency of solute atoms for migrating
defects can be found. In fig.8, for example, the irradiation-
induced displacement of solute atoms into <110> channels is shown
as a function of irradiation fluence for Al(Ag), Cu(Be) and Mg(Ag)
alloys. The fraction of displaced solute atoms is expressed in
terms of an apparent displaced fraction $f_{di}^{<1mn>}$ divided by the
calculated normalized ion flux $F_i^{<1mn>}$ at the position of the
displaced solute atoms. The apparent displaced fraction is given
by:[24]

$$f_{di}^{<1mn>} = (\chi_i^{<1mn>} - \chi_h^{<1mn>})/(1-\chi_h^{<1mn>}) \tag{5}$$

where $\chi_i^{<1mn>}$ and $\chi_h^{<1mn>}$ are the experimental values of the
normalized yields from solute atoms and host atoms, respectively,
for the ion beam incident along the <1mn> channel. When the self-
interstitials are mobile (above $\sim$ 40 K in Al and Cu and already at

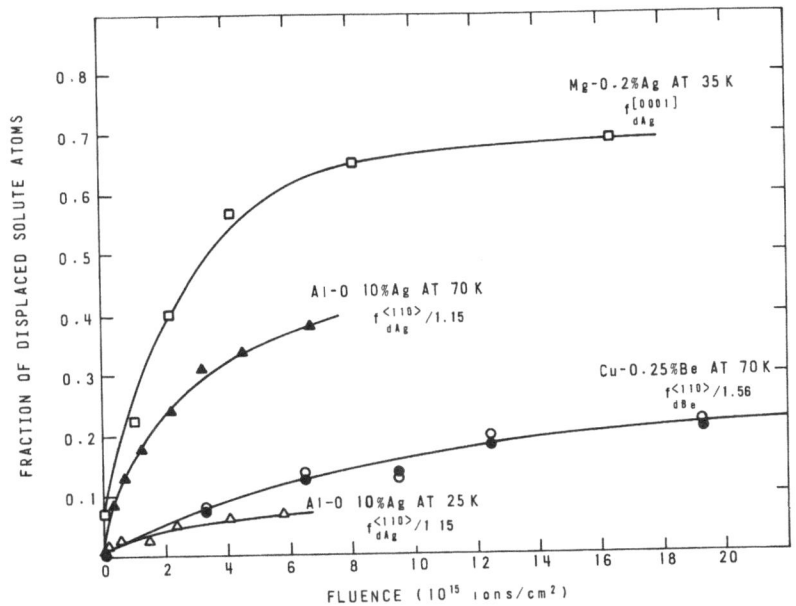

Figure 8: The fraction of solute atoms which are displaced from
substitutional lattice sites as the result of irradiating with 1
MeV He$^+$ (Al(Ag) and Mg(Ag)) or with 0.6 MeV D$^+$ (Cu(Be)) at the
indicated temperatures. Rutherford scattering was used to obtain
the Ag solute atom yields in Al(Ag) and Mg(Ag) crystals (see fig.3
for example) whereas the Be in Cu(Be) was obtained from the meaured
alpha particle yields from the ^{9}Be(d,α_0)^{7}Li (o) and the ^{9}Be(d,α_1)^{7}Li (●)
reactions (based on data in ref.26-28).

35 K in Mg), they become trapped at the solute atoms and displace them from their substitutional lattice sites. The dominant trapping configuration is believed to be a mixed-dumbbell interstitial in which one solute atom and one solvent atom straddle a lattice site.[22-28] The ratio of trapping efficiencies of interstitials at vacancies to that at solute atoms was determined to be $\sim$ 5 for all of the alloys investigated.[27] The 25 K irradiation results for the Al-0.1 at% Ag alloy indicate that when the self-interstitials are not freely migrating, only a small fraction of the interstitials are trapped spontaneously at the solute atoms in the displacement cascade.[27]

For Al-Cu, Al-Mn, Al-Ag and Cu-Be mixed dumbbells, $f_{di}^{<110>} > f_{di}^{<100>}$ and $f_{di}^{<111>}$ (also $f_{di}^{\{100\}} \sim 1/3 \, f_{di}^{<110>}$) which is consistent with the axis of the mixed dumbbell lying along <100> directions of the f.c.c. lattice.[22-24,28] This was confirmed by using flux calculations of the type shown in fig.5 to determine yields as in fig.6. Angular scans through the <110> and <100> axes were also consistent with this interpretation (see fig.9). The magnitude of the displacement s of the solute atom can be obtained from the relationship[12]

$$\chi_i^{<lmn>} - \chi_h^{<lmn>} = Y(s)^{<lmn>} \, c \, g^{<lmn>} \qquad (6)$$

where $Y(s)^{<lmn>}$ is the calculated yield for a displacement s into a <lmn> channel, c is the true fraction of displaced solute atoms and $g^{<lmn>}$ is the geometrical factor for displacement into a particular <lmn> channel. The dependence on c in equation 6 can be eliminated by considering more than one channel, say <lmn> and <l'm'n'>. Then one has:[12]

$$\frac{g^{<l'm'n'>}}{g^{<lmn>}} \frac{(\chi_i^{<lmn>} - \chi_h^{<lmn>})}{(\chi_i^{<l'm'n'>} - \chi_h^{<l'm'n'>})} = \frac{Y(s)^{<lmn>}}{Y(s)^{<l'm'n'>}} \qquad (7)$$

Using experimental values of $\chi_i^{<lmn>}$ and $\chi_h^{<lmn>}$ for the <110>, <100> and <111> channels and calculated $Y(s)^{<lmn>}$ values, the average values for the displacement of the solute atoms from lattice sites along <100> directions in Al-Mn, Al-Cu, Al-Ag and Cu-Be mixed dumbbell interstitials were determined to be 1.43, 1.48, 1.27 and 1.32 Å respectively.[12] The error in this determination was estimated as 0.05 Å.

For Mg-Ag mixed dumbbells, the fractions of Ag atoms displaced into [0001], <11$\bar{2}$0>, <10$\bar{1}$0> and (0001) channels were comparable.[26] Also, angular scans through the [0001] and <11$\bar{2}$0> directions revealed the presence of a small peak in the scattering yield from the Ag atoms, thus indicating that the Ag atoms were displaced far enough into these channels to reach a flux peaking region. As shown in fig.10, Ag atom displacements along <40$\bar{4}$3> which are an

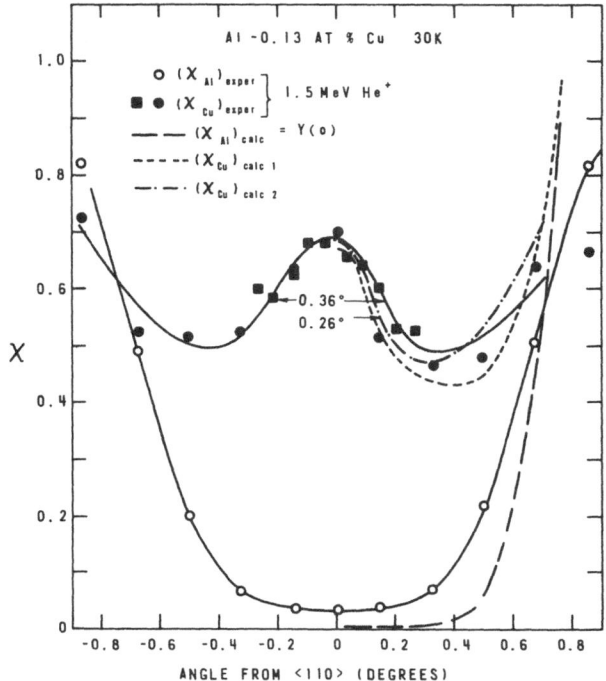

Figure 9: Experimental and calculated yields for an angular scan
through <110> in an Al-0.13 at% Cu crystal for backscattering of
1.5 MeV He$^+$ ions at 30 K. The crystal was irradiated at 70 K with
9.0 x 10^{15}-1.5 MeV He$^+$ ions/cm^2. This resulted in the formation of
Al-Cu mixed dumbbell interstitials in which the Cu atoms were
displaced 1.48 Å along <100> from their initial substitutional
positions (from ref.12).

appreciable distance towards the octahedral site (i.e. about 2/3
the distance) would place the Ag atoms near the center of the
[0001] and <1120> channels, hence where peaking in yield would be
expected. For the same type of displacement the Ag atoms would be
located closer to the edges of the <10$\bar{1}$0> channel which would be
consistent with a fairly large f_{dAg}<10$\bar{1}$0> value but no peak in Ag
yield in the <10$\bar{1}$0> angular scan. Hence, even in the absence of
detailed flux and yield calculations, it appears reasonable to
consider the trapping configuration of the Ag-Mg interstitial atom
complex as a mixed dumbbell lying along the second nearest-
neighbour directions <40$\bar{4}$3> in the hcp lattice.[26]

     The annihilation of the trapped defects during annealing
determines the thermal stability of the trapping configuration and
the mechanism of annihilation. Also, since the rate of dechanneling dχ/dz (where z is the penetration depth of the ion beam) is a

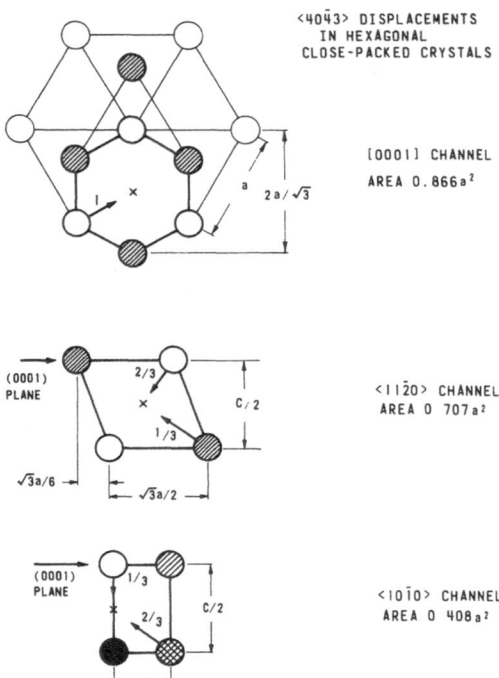

<4043> DISPLACEMENTS
IN HEXAGONAL
CLOSE-PACKED CRYSTALS

[0001] CHANNEL
AREA 0.866a²

<1120> CHANNEL
AREA 0 707a²

<1010> CHANNEL
AREA 0 408a²

Figure 10:   Cross-sections of [0001], <1120> and <1010> channels
in an ideal hexagonal close-packed crystal, drawn to scale. The
nearest neighbour distance is a.   The different shadings represent
atoms in different planes.   The octahedral interstitial site x is
shown.   Displacements of √2a/3 in <4043> directions (one-third of
the second-nearest neighbour distance) are indicated by arrows and
the numbers alongside the arrows represent the fraction of displacements into equivalent positions   (from ref.26).

qualitative measure of the total defect concentration, the
monitoring of $(d\chi/dz)^{<1mn>}$ as well as $f_{di}^{<1mn>}$ can be used to
determine whether a given annealing stage corresponds to defect
migration rather than merely to an alteration of a trapping
configuration.   As shown in fig.11, in irradiated crystals of
Al(Ag), Al(Cu), Al(Mn) and Al(Fe), little change in either the
fraction of displaced solute atoms or the irradiation-induced
dechanneling increment was observed in the region 70-150 K, whereas
both recovered completely in stage III (near 200 K).[24]
Hence the displaced solute atoms in the mixed dumbbell configuration return to substitutional lattice positions when vacancy-
interstitial annihilation occurs in stage III.   The results shown
in fig.12 for Mg(Ag) crystals indicate that there is a fairly good
correlation between the recovery of the fraction of displaced Ag
atoms and the irradiation-induced dechanneling increment in the

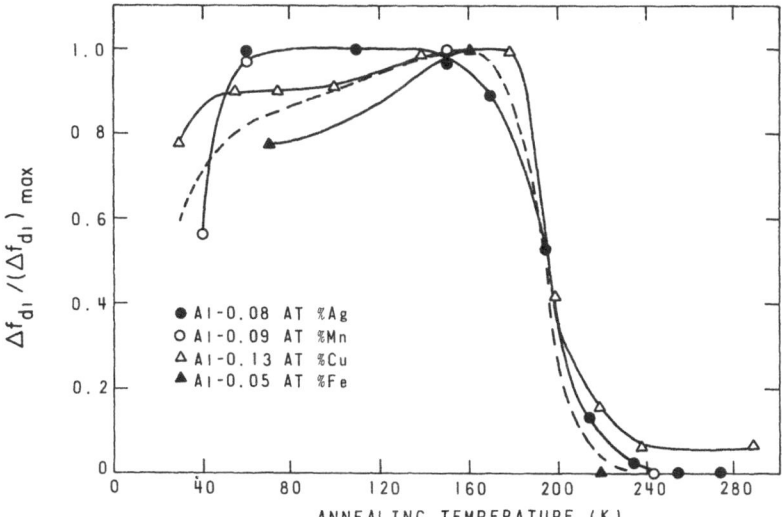

Figure 11:   The effect of isochronal annealing (600 s pulses) on
the displacement of solute atoms in Al alloys after irradiation.
$\Delta f_{di}/(\Delta f_{di})_{max}$ is the fraction of the maximum change in displacement
of solute atoms which remained after annealing at the indicated
temperatures.   The dashed line is the recovery of the irradiation-
induced change in dechanneling, $\Delta d\chi/dz$, for the Al(Cu) crystal
(from ref.24).

two main annealing stages at 80–160 K and 200–280 K.[26] Recently,
Chami et al[17] have combined the lattice disorder and atom location
channeling results to determine the position of H in Mg as implanted at 50 K and then the subsequent migration, trapping and
detrapping of the H during annealing up to $\sim$ 300 K.   Although the
dechanneling measurement provides a useful indication of the
changes in the total defect concentration during annealing, it is
not nearly as sensitive or specific as an electrical resistivity
measurement for this purpose.   Hence, ideally, it is useful to
obtain annealing data using both channeling and electrical resistivity measurements on the same system, as has been done in
various instances.

The ion channeling technique has also been extended by
Picraux et al[29] to obtain depth profiles of disorder in systems
where the principal irradiation produced defects present were
dislocation lines and loops. The density, size and Burgers vectors
of the dislocations were determined independently by transmission
electron microscopy and these results along with dechanneling data
at various ion energies enabled the dechanneling cross-section
of the defects to be evaluated.   The depth distribution of the

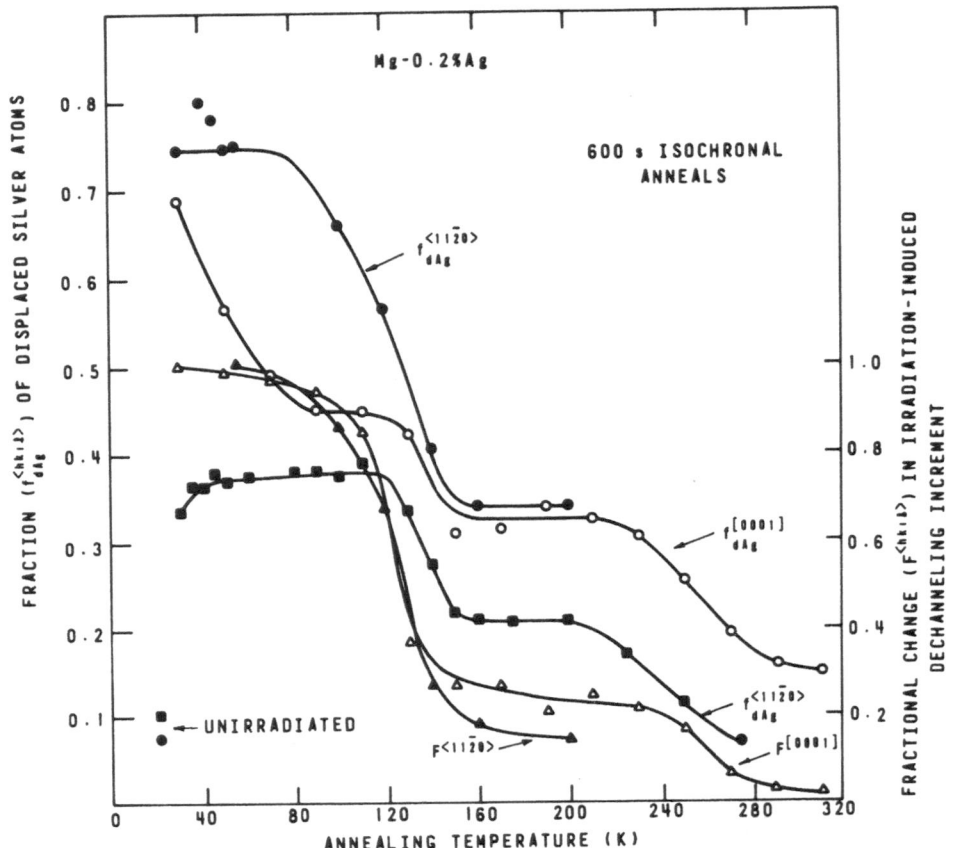

Figure 12:   The recovery of Ag atom displacements $f_{dAg}^{<hkil>}$ and
the fractional change $F^{<hkil>}$ in the irradiation-induced dechanneling increment as a function of annealing temperature for a Mg-
0.2 at% Ag crystal. The irradiations as well as the backscattering
analyses were performed at 30 K using 1 MeV He$^+$ ions. The $f_{dAg}^{<1120>}$
data shown as solid squares ■ was for an irradiation of 1.9×10^{15}
He$^+$ ions/cm^2,whereas all other experimental data pertain to irradiations of $1.65-1.8 \times 10^{16}$ He$^+$ ions/cm^2 (based on data in
ref.26).

disorder was then obtained by differentiating the expression,[29]

$$\int_{o}^{z} n(z') \, dz' = 1/\lambda \, \ln[(1-\chi_D)/(1-\chi_V)] \tag{8}$$

where n is the density of irradiation-produced defects, z is the
depth at which a particular defect is located, χ_D is the channeled
yield in the damaged crystal and χ_V is the channeled yield in the

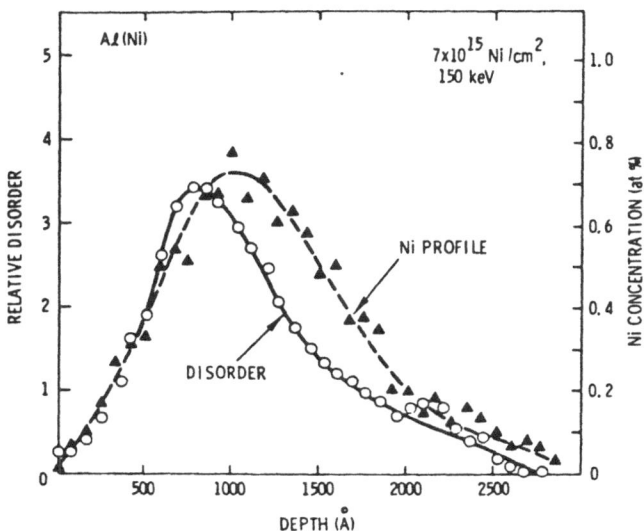

Figure 13: Disorder and Ni depth profile from the analysis of
backscattering spectra for 2 MeV He analysis along the {111}
planar and random directions in unimplanted Al and Al implanted
with 7 x 10^{15}-150 KeV Ni$^+$ ions/cm^2 (from ref.29).

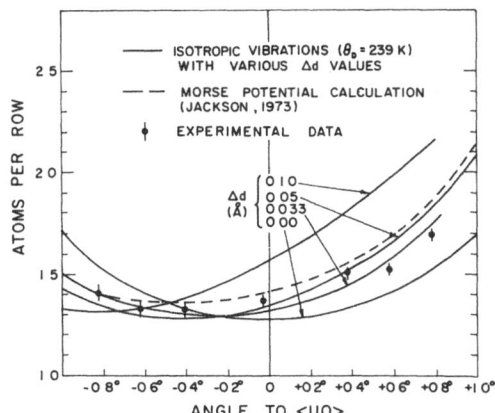

Figure 14: Angular scan of the <110> surface peak on Pt {111} at
78 K, using a 2 MeV He$^+$ beam. The experimental data are compared
with Monte Carlo simulations (solid and dotted curves) which were
obtained using various values of Δd and isotropic surface vibrations characterized by the bulk Debye temperature, θ_D = 239 K
(from ref.35).

unirradiated crystal. The resulting disorder profile as well as the Ni atom profile is shown in fig.13 for an Al crystal implanted with 7×10^{15}-150 KeV Ni^+ ions/cm^2. The disorder–depth distribution is similar to that which would be predicted for the relative damage deposition by the implanted Ni, consistent with nucleation and evolution of point defects into dislocation loops and their subsequent growth. The peak in the disorder profile is observed to be at $\sim 0.8 R_p$ where R_p is the projected range for the Ni ion profile and this is consistent with the value typically obtained for the relative positions of the peaks. This type of analysis should also be quite useful in investigating the interplay between defects and solute atoms when migration of the solute atoms occurs during the implantation process.

6. SURFACE STUDIES

The value of high energy (0.2–2.0 MeV) H^+ and He^+ scattering for studying the structure of crystal surfaces has now been demonstrated successfully in several laboratories.[30-36] An excellent review of this technique and of the basic concepts involved has been given by Bøgh.[30] In studies of this type, it is not only essential to obtain and maintain extremely clean crystal surfaces but considerable attention must also be paid to background subtraction and yield calibration in order to obtain high precision in the determination of the number of unshadowed atoms/cm^2 from measurements of the surface peak area (see Davies et al[35] for more details). Here, only two applications of the technique will be discussed.

One application of particular interest is the determination of the magnitude Δd of any surface relaxation of the outermost layers of a single crystal. The most reliable and sensitive method for investigating surface relaxation is to perform a detailed angular scan through a close-packed axis, as shown in fig.14 for a Pt crystal.[35] In such a scan, any outward (or inward) displacement of the surface plane will produce a significant asymmetry in the observed surface peak area. As shown by the computer simulations in fig.14 an outward displacement or relaxation shifts the surface peak minimum towards the surface normal, i.e. to negative angles. An inward displacement or contraction would shift the minimum away from the surface normal. Comparison of the experimental data with predicted curves for various Δd values (fig.14), suggests that there is an outward relaxation of the surface atoms in the {111} plane of ~ 0.03 Å. In obtaining the above experimental data considerable care was taken[35] to prevent radiation damage efects from introducing any spurious anisotropy into the measurements, since an unexpectedly large temperature-dependent surface-damage phenomenon was observed at low fluences of the He^+ analyzing beam.

Another application of the technique is the investigation

of lateral displacements of atoms in the surface region of
crystals (i.e. of surface reconstruction). In this case, the ion
beam is aligned with the surface normal in order to obtain maximum
sensitivity for detecting laterally displaced atoms. In fig.15,
the results of Norton et al[36] are shown for a study of a $\{100\}$
surface of platinum. The surface was prepared in its reconstructed
state by in-situ cleaning and subsequent cooling to 300 K. Under
conditions where reconstruction occurs, an extra 1.4×10^{15} Pt
atoms/cm^2 (0.55 ± 0.05 atoms/string) are detected by the incident
beam. Hence, approximately one atomic plane (1.28×10^{15}/cm^2) of
Pt atoms has moved out of registry with the substrate. The crystal
was then cooled further (4 K s^{-1}) in the presence of ~ 1 μ Pa H$_2$
and measurements were taken at various temperatures. Above 250 K,
the experimental data fit well to the reconstructed (upper) curve.
At temperatures below 150 K there is apparently sufficient hydrogen
coverage to cause complete reversion to the bulk structure (i.e.
unreconstructed surface) and the data fit well to the bulk-
structure (lower) curve. Angular scans through the <110> axis,
which makes a large angle to the surface normal, showed that the
outward relaxation of the $\{100\}$ unreconstructed surface was
negligible (i.e. $0.5 \pm 0.5\%$).

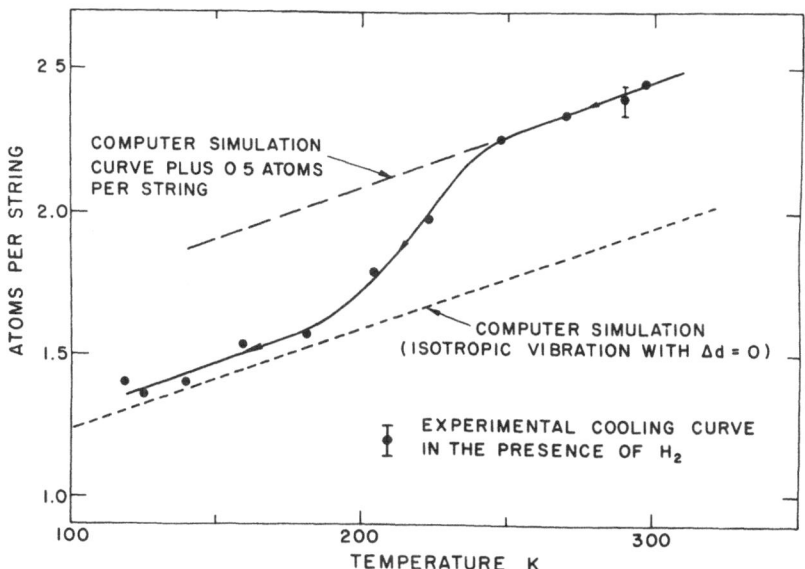

Figure 15:   Temperature dependence of the <100> surface peak on Pt
$\{100\}$ surface, using 2.0 MeV He$^+$ (from ref.36).

7.   SUMMARY

Ion channeling is a very useful technique for atom site characterization in metals and its application is particularly relevant to various problems encountered in ion implantation studies. The virtue of the technique is that, provided the solute atom is sitting in a reasonably high symmetry position in the host crystal, its position can be determined fairly precisely by essentially a geometrical pinpointing procedure. In addition, by combining solute atom location data with dechanneling data from the host crystal, information can be obtained regarding the interaction between solute atoms and irradiation-produced defects (i.e. the trapping configuration, the trapping efficiencies of solute atoms for defects, the thermal stability of the trapping configuration and the mechanism of annihilation of the defect-solute atom complex). The technique has also been applied successfully for determining the position of host atoms at the crystal surface, thus providing information on surface relaxation, surface reconstruction and irradiation damage enhancement in the surface region.

ACKNOWLEDGEMENTS

The authors wish to acknowledge the contributions made by the many scientists in the channeling field, as referred to in the paper. We are also particularly grateful for the collaboration with M.L. Swanson, N. Matsunami, P.R. Norton and D.P. Jackson in recent channeling investigations.

REFERENCES

1.   J. Lindhard, Kgl. Danske Videnskab, Selskab Mat. fys. Medd. $\underline{34}$, #14 (1965).
2.   D.S. Gemmell, Rev. Modern Phys. $\underline{14}$, 129 (1974).
3.   Channeling-Theory, Observation and Applications, ed. D.V. Morgan (Wiley & Sons, London 1973).
4.   J.A. Davies in Material Characterization Using Ion Beams, ed. J.P. Thomas and A. Cachard (Plenum Press, N.Y. 1978) p.405.
5.   J.W. Mayer and M.-A. Nicolet in Material Characterization using Ion Beams, ed. J.P. Thomas and A. Cachard (Plenum Press, N.Y. 1978), p.333.
6.   A. Cachard and J.P. Thomas in Material Characterization using Ion Beams, ed. J.P. Thomas and A. Cachard (Plenum Press, N.Y. 1978), p.367.
7.   J.H. Barrett, Phys. Rev. $\underline{3B}$, 1527 (1971).
8.   J.W. Mayer, L. Eriksson and J.A. Davies, Ion Implantation in Semiconductors (Academic Press, N.Y. 1970).
9.   K. Komaki in Ion Beam Surface Layer Analysis, Vol.2 (Plenum Press, N.Y. 1976) p.517.

10. S.T. Picraux in Ion Beam Surface Layer Analysis, Vol.2, Plenum Press, N.Y. 1976) p.527.

11. N. Matsunami and N. Itoh, Radiat. Eff. 31, 47 (1976).

12. N. Matsunami, M.L. Swanson and L.M. Howe, Can. J. Phys. 56 1057 (1978).

13. J.H. Barrett, in Proc. of the Fourth Conference on Applications of Small Accelerators (Denton, Tex. 1976) IEEE Pub. No. 76CH1175-9 NTS (IEEE, N.Y.).

14. H.-D. Carstanjen and R. Sizmann, Radiat. Eff. 12, 225 (1972).

15. R.B. Alexander, G. Dearnaley, D.V. Morgan, J.M. Poate and D. Van Vliet, in Proc. Europ. Conf. on Ion Implantation (Peregrinus, Stevenage, England 1970) p.181.

16. J.M. Poate, W.J. Debonte, W.M. Augustyniak and J.A. Borders in Ion Implantation in Semiconductors, ed. S. Namba (Plenum Press, N.Y. 1975) p.361.

17. A.C. Chami, J.P. Bugeat and E. Ligeon, Radiat. Eff. 37, 73 (1978).

18. S.T. Picraux and F.L. Vook, Phys.Rev. Lett. 33, 1216 (1974)

19. S.T. Picraux in New Uses of Ion Accelerators, ed. J.F. Ziegler (Plenum Press, N.Y. 1975) p.229.

20. H.-D. Carstanjen in Ion Beam Surface Layer Analysis, Vol.2 (Plenum Press, N.Y., 1976). p.497.

21. P. Mazzoldi in Material Characterization Using Ion Beams, ed. J.P. Thomas and A. Cachard (Plenum Press, N.Y. 1978) p.429.

22. M.L. Swanson, F. Maury and A.F. Quenneville, in Applications of Ion Beams to Metals, ed. S.T. Picraux, E.P. Eer Nisse and F.L. Vook (Plenum Press, N.Y. 1974), p.393.

23. M.L. Swanson and F. Maury, Can.J. Phys. 53, 1117 (1975).

24. M.L. Swanson, L.M.Howe and A.F. Quenneville, J. Nucl. Mat. 69 & 70, 372 (1978).

25. L.M. Howe, M.L. Swanson and A.F. Quenneville, J. Nucl. Mat. 69 & 70, 744 (1978).

26. L.M. Howe, M.L. Swanson and A.F. Quenneville, Radiat. Eff. 35, 227 (1978).

27. M.L. Swanson and L.M. Howe, in preparation for publication.

28. M.L. Swanson, L.M. Howe and A.F. Quenneville, Can. J. Phys. 55, 1871 (1977).

29. S.T. Picraux, D.M. Follstaedt, P. Baeri, S.U. Campisano, G. Foti and E. Rimini, to be published in the Proc. of Int. Conf. on Ion Beam Modification of Materials, Budapest, Hungary, 1978.

30. E. Bøgh in Channeling – Theory, Observation and Applications, ed. D.V. Morgan(Wiley & Sons, London 1973) p.435.

31. D.M. Zehner, B.R. Appleton, T.S. Noggle, J.W. Miller, C.H. Jenkins and O.E. Schow, J. Vac. Sci. Technol. 12, 454 (1975).

32. J.A. Davies, D.P. Jackson, J.B. Mitchell, P.R. Norton and R.L. Tapping, Nucl. Instr. Meth. 132, 609 (1976).

33.   W.C. Turkenburg, W. Soszka, F.W. Saris, H.H. Kersten and
       B.G. Colenbrander, Nucl. Instr. Meth. 132, 587 (1976).
34.   L.C. Feldman, R.L. Kaufman, P.J. Silverman, R.A. Zuhr and
       J.H. Barrett, Phys. Rev. Lett. 39, 38 (1977).
35.   J.A. Davies, D.P. Jackson, N. Matsunami and P.R. Norton,
       Surface Science 78, 274 (1978).
36.   P.R. Norton, J.A. Davies, D.P. Jackson and N. Matsunami,
       private communication (1977).

Part III

Implantation Effects At Higher Concentrations, Aggregation Phenomena

RADIATION EFFECTS IN METALS

J. LETEURTRE

Section d'Etude des Solides Irradiés

C.E.N. de Fontenay-aux-Roses (92260)   France

Radiation creates point defects in metals ; this had been recognized by Wigner in 1946 soon after the first nuclear reactor operation. Seitz had even predicted that the threshold displacement energy is about 25 eV. During the two following decades, nuclear reactor technology was developed without any serious metallurgical problem after the adoption of refractory nuclear fuel (oxides) ; scientific investigations in radiation damage were mainly focussed on studying point defect properties through low temperature irradiation experiments. But about 10 years ago, it was discovered that exposure to fast neutrons in reactors (at $\simeq 600°C$) induce significant changes in the external dimensions of the material cladding the nuclear fuel or forming the reactor structure. The reason for this dimensional instability is void formation and radiation creep. The financial implications of these phenomena have promoted a considerable effort to understand the radiation damage in metals in its entirety.

We shall try to review the current understanding of radiation damage in metals, simplifying the actual complexity of the effects by considering some aspects separately. The production of point defects in metals, the primary damage state will first be studied. The second part of the lecture is devoted to the evolution of this primary damage state as a function of temperature and dose : the steady state concentration of point defects, the nucleation of secondary defects and their growth will be successively considered.

1 - PRODUCTION OF DEFECTS BY RADIATION IN METALS

Energetic particles can affect a metal in two ways :

    i/ lattice atoms are removed from their regular lattice sites,
i.e. displacement damage production ;
    ii/ chemical composition of the target can be changed by ion
implantation or transmutation - noticeably helium production via
(n, α) reactions. This second aspect of radiation damage is particularly important for secondary defect nucleation.
    Let us focus our attention on the displacement damage. An atom
can be displaced from its position if it receives a high enough
energy : a vacancy is left at the place previously occupied by the
recoiling atom which becomes an interstitial atom. The minimum
energy transfer for an atom to be displaced in a bulk crystal is
called the threshold displacement energy, E_d ; in most crystals,
its value is about 25 eV, i.e. several times higher thant the sum
of the formation energies of 1 vacancy plus 1 interstitial

$$(E_f^v \simeq \frac{\mu b^3}{5} \simeq 1 \text{ eV and } E_f^i \simeq \mu b^3 \simeq 5 \text{ eV})$$

because this Frenkel pair formation is highly irreversible.
    What are the collisions responsible for damage in metals ?
Particles travelling through matter lose their energy via elastic
collisions (with atoms of the target) and via inelastic collisions,
i.e. in electronic excitation. Atomic displacements could occur if
any excited electronic state, energetic enough to transfer at least
E_d, also has a lifetime long enough for it to transfer effectively
its energy and cause atomic ejection ($\geq 10^{-12}$ s.). In metals, any
electronic excitation will be dissipated among the conduction electrons in a time of the order of 10^{-15} s., and thus will not be able
to displace atoms. Thus, only elastic collisions are able to induce
atomic displacements in metals.
    If an atom is ejected with a very high energy transfer ($T \gg E_d$),
it becomes an interstitital atom moving in the target and losing
its energy in a series of atomic collisions. Some of its collision
partners are ejected from their lattice sites and can possibly
produce still more displaced atoms. This sequence of collisions
induced by this primary knock-on atom (P.K.A) is called a displacement cascade.
    Now, for a specified irradiation (nature of the particle (mass,
charge)   , energy of the particle (E) and number of these particles
per unit area and unit time Φ), we must estimate the damage rate in
our target (mass of atom : M). We have thus to know :
    i/ how many defects are created by a given P.K.A, of energy
E_r. This damage function $g(E_R)$ is studied in §1.2.
    ii/ the number and energy spectrum of the P.K.A ejected by
each kind of particle. The answer is the differential scattering
cross section, i.e. the probability that a particle of energy E
produces a recoil of energy E_R, with an incertainty in energy ΔE_R

$$\frac{d\sigma (E_R)}{dE_R} = K (E, E_R).$$

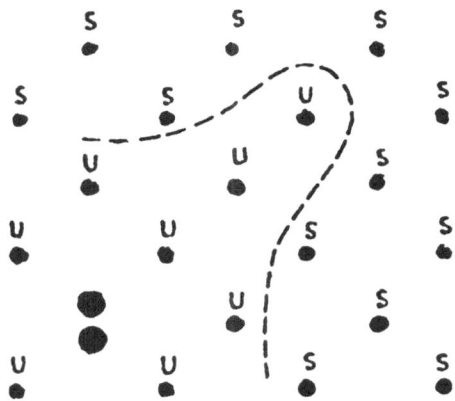

Figure 1

Computer simulation [3] of the annihilation volume : inside this volume, a Frenkel pair is unstable.

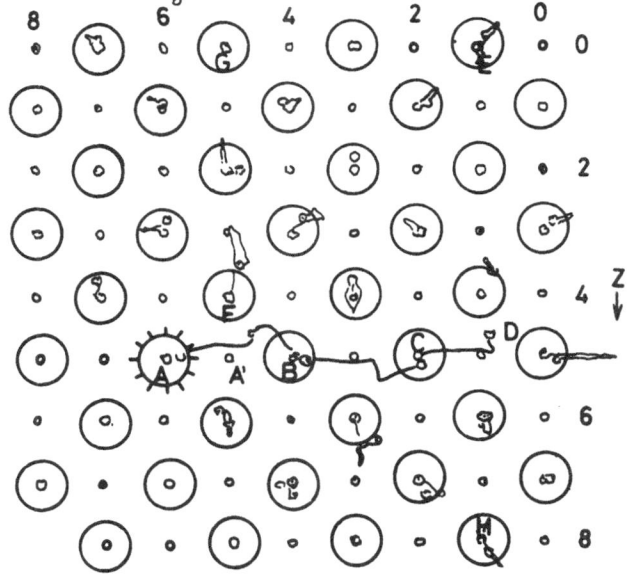

Figure 2

Atomic orbits produced by shot in (100) plane at 40 eV. Knock-on was at A and directed along y axis. Large circles give initial positions of atoms in plane ; small dots are initial positions in plane below. Vacancy is created at A, split interstitial at D [F] .

The damage cross section is $\sigma(E) = \int_{E_R} g(E_R) . K(E,E_R) \, dE_R$, and the damage rate is $G = \sigma . \Phi$; it is expressed in displacements per atom per unit time (d.p.a./s.).

1.1 - Primary recoil spectra [1].

Fast neutrons : Above 2 MeV, collisions between neutrons and nuclei resemble those between hard spheres. Cross sections are constant up to the max. transferred energy $E_R^{max} \simeq \frac{4}{M} . E$.

Ion irradiations: Ions (mass m, charge of the nucleus z; energy E) can transfer to target atoms (M,Z) up to :

$$E_R^{max} = \frac{4\,m\,M}{(m+M)^2} \, E,$$

but the probability of transfer E_R decreases as $1/E_R^2$:

$$\frac{d\sigma}{dE_R} = \frac{\pi(z\,Z\,e^2)^2}{\frac{M}{m}\,E.\,E_R^2}$$

For electron irradiation also, low energy recoils are much more probable than those with energies close to $E_R^{max} = 2\,E(E+2\,m\,c^2)/Mc^2$, where m is the rest mass of the electron ($mc^2 = 511$ keV), and c is the velocity of light. Displacement cross sections are given by OEN [2].

1.2 - Damage produced by an energetic primary ($E_R > E_d$).

1.2.1 - The displacement threshold energy (E_d)

The displacement threshold is defined as the minimum P.K.A. energy to produce a stable Frenkel pair in a bulk lattice. This means that the recoiling atom must travel far enough for the interstitial to come to rest outside the recombination volume, about 100 atomic volumes around the vacancy ; inside this volume (fig. 1), an interstitial is unstable and annihilates athermally with the vacancy. Since the separation of the interstitial from its vacancy proceeds via focussed collision sequences that propagates along close packed rows of the lattice, less energy is necessary to form a stable Frenkel pair if the direction [h k l] of the recoiling atom coincides with some close packed row of the lattice (fig. 2). The dependence of E_d(h k l) has been studied first by computer simulations [3] ; results are schematically shown in fig. 3.

How is E_d measured?   Displacement threshold determinations are essentially obtained with electron irradiations, by observing the changes in some physical property as a function of electron energy ; this has been done either using residual electrical resistivity per unit dose in specimens irradiated at low temperature, or by observing the interstitial dislocation loop formation with very thin specimens irradiated in a high voltage electron microscope.

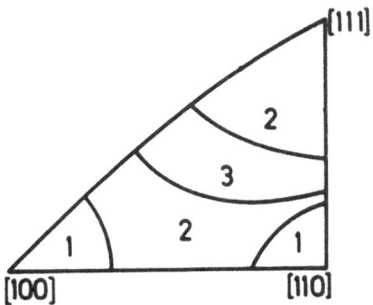

<u>Figure 3</u>
Displacement threshold energy anisotropy ; results of computer simulations.

Whatever the technical method, E_d determination have sometimes been
derived by extrapolation to zero damage ; a much better method consists in first assuming a certain shape of the threshold function,
then trying to fit to the observed damage rate at several electron
energies with computations of the total displacement cross section
$\sigma(E, E_d)$ by varying the value of E_d. Fig. 4 shows some threshold
functions ; and fig. 5 illustrates the proper method for measuring
E_d.
     The determination of the threshold anisotropy is also complex ;
it cannot come out directly from experiments ; the reason is that
the onset of damage is nearly always determined by the displacements
in the easy directions. Experiments only directly give E_d in the
direction where it is minimum (for instance $E_d < 110 >$ for Ni). In
other directions, experiments provide apparent threshold (in Ni,
$E_d < 001 >$ apparent $= 31$ eV and $E_d < 111 >$ apparent $= 28$ eV). True values are computed : they are taken as adjustable parameters for
fitting, according to Von Jan and Seeger [4] , the experimental

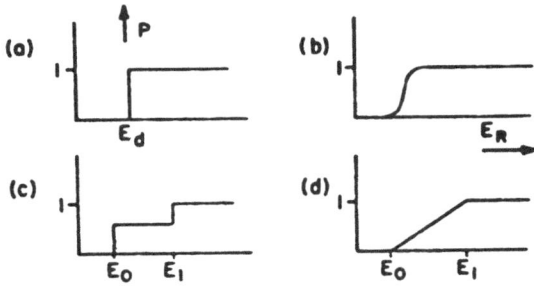

<u>Figure 4</u>
Threshold functions

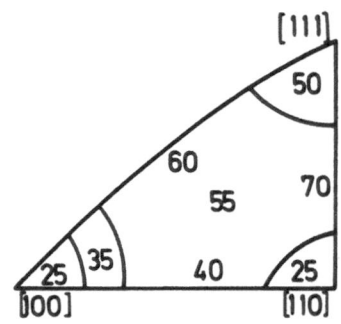 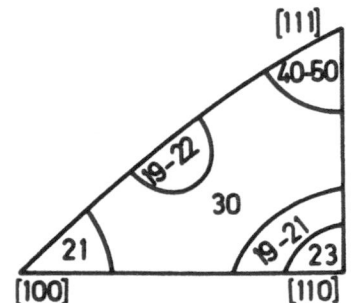

<u>Figure 5</u>
a/ experimental results of E_d (hkl) for austenitic stainless steel.
b/ experimental results of E_d (hkl) for copper.

	Metals							
	Al	Ni	Cu	Ag	Au	Nb	Mo	Ta
E_d^{min}	16	23	19	25	35	36	33	34
E_d^{avg}	27	33	29	39	43	78	70	90

<u>Figure 6</u>
Atomic displacement threshold energies (in eV)

damage rates . The best fits, taken as "true" values are E_d <111> =60
eV and E_d <100> = 38± 3 eV, in Ni [5]. Fig. 5 shows some experimental results. A review paper of E_d measurements is [6].
    For irradiations in which all directions are equally probable,
the relevant threshold energy is the average threshold, E_d^{avg}, over
all crystallographic directions. Some values of threshold energies
are tabulated in fig. 6 ; the average values are typically about
1.4 and 2.2 above the minimum values in f.c.c. and b.c.c. lattices.

   1.2.2 - <u>Number of Frenkel pairs produced by an energetic</u>
           <u>primary (E_R>> E_d)</u>.

    A P.K.A. has received the energy E_R. A rough estimation of the
number of Frenekl pairs created in the displacement cascade is
given by the model of Kinchin and Pease [7] . Atoms are supposed to
behave as hard spheres ; for hard spheres with the same masses, the
incident particle energy is, on an average, shared equally between
collision partners. Thus on an average, the displacement cascade
looks like a chain of collisions, in which the number of displaced
atoms is doubled at each step whereas their mean kinetic energy is
reduced to half of the previous value.

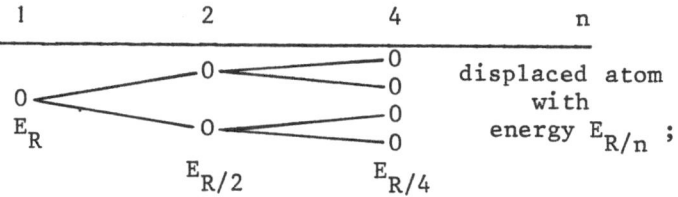

The sequence of collisions is interrupted when the energy $E_{R/n}$ becomes less than $2E_d$. Otherwise, a further division would create displacements with an energy less than E_d, which is impossible according to the definition of E_d :

$$\frac{E_R}{n} \le 2E_d \text{ or } n \simeq \frac{E_R}{2E_d}$$

Actual atomic collisions are not hard sphere collisions, with a realistic interatomic potential, the number of Frenkel pairs created by a P.K.A. of energy E_R is :

$$n = \frac{0.8 \times E_R^*}{2E_d^{avg}} \quad \text{where } E_R^* = E_R - \text{electronic losses suffered in the displacement cascade.}$$

Practically, these electronic losses are negligible when the P.K.A. energy, in keV, is lower than the number of mass of the target atoms.

This modified Kinchin-Pease model had been compared to experimental measurements. Seidman et al.[8] took advantage of the atomic resolution attainable by the field ion microscopy and of the possibility of field evaporation to operate a dissection of displacement cascades in tungsten made by 20 keV or 30 keV ions. Irradiation and field ion microscopy (F.I.M) were performed at 18 K, temperature at which all point defects are immobile in tungsten. Fig. 7 shows typical observations of the depleted zones, i.e. the vacancy rich cores left by displacement cascades created in tungsten by 30 keV W^+, Mo^+ and Cr^+ ions. The number of vacancies produced by each ion is also shown in fig. 7 bis ; it is about 200 in all cases, a little more for Cr^+ ions, a little less for W^+ ions, despite the fact that electronic losses are more important for Cr^+ than for W^+ ions ; this small effect that could be due to more active close pair recombination in hotter, more dense W^+ cascades, does not invalidate the Kinchin Pease formula. The average vacancy concentration within a depleted zone varies from 3-5 % with Cr^+ ions to 15-25 % with W^+ ions. This last value is extremely high ;

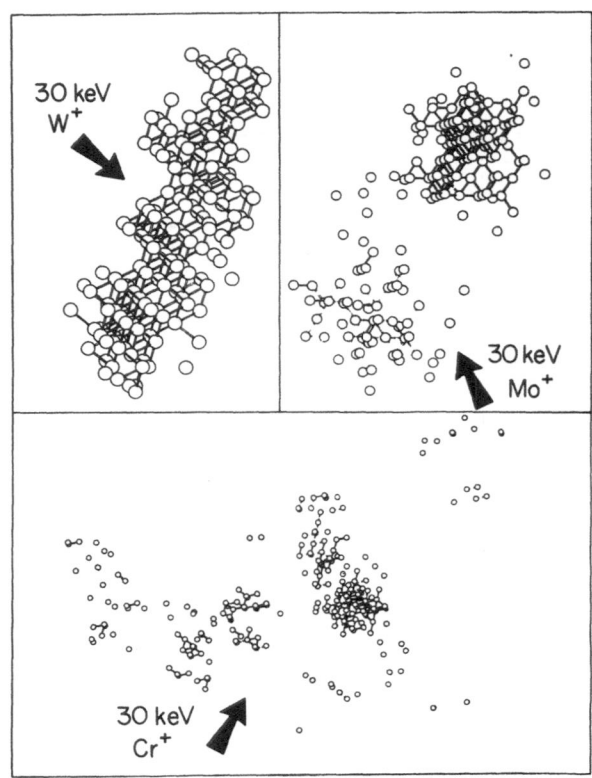

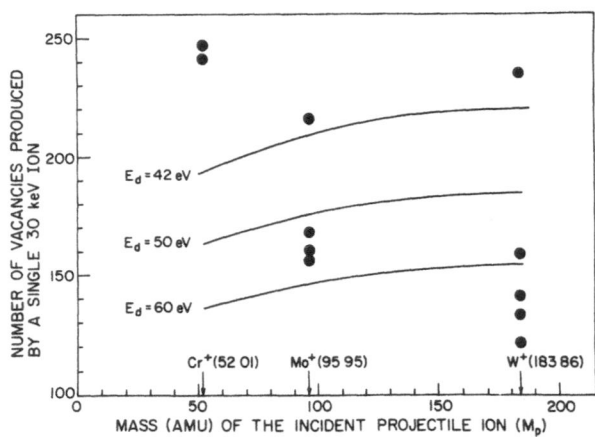

Figure 7

a) Typical observation made by field ion microscopy of the depleted
   zones created in tungsten by 30 keV Cr^+, Mo^+ and W^+ ions.
b) Number of vacancies produced by each of these displacement cascades.

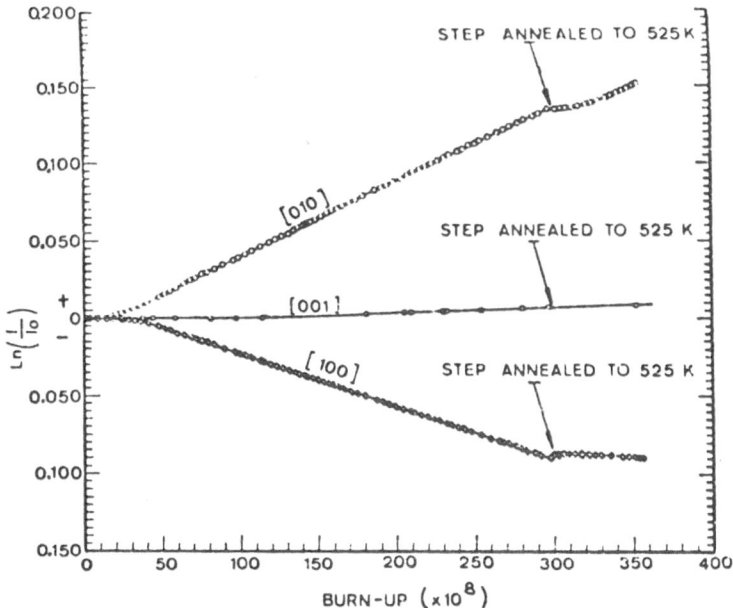

Figure 8
Length changes in the three directions [100], [010] and [001] for a
uranium single crystal irradiated at 4 K. (After Loomis and Gerber,
1968). Phil. Mag. 18, 539.

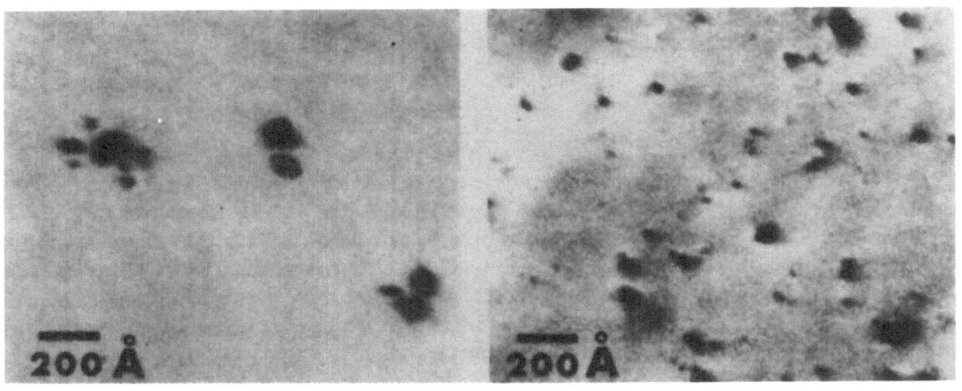

Figure 9
a/ 14 MeV Neutron Bombarded Au at 7 x 10^{15} n/cm^2. Note subcascade
   formation
b/ Reactor Neutron Irradiated Au at 6.2 x 10^{17} n/cm^2 (> 1 MeV).

vacancy clustering however does not occur within the displacement
cascade and its subsequent thermal spike, as proven by fig. 7. But
fig. 8 proves that vacancy clustering occurs at liquid helium temperature when a second displacement cascade is produced in a region
already damaged by a previous cascade [20] . At higher temperature,
clustering occurs within each displacement cascade. Transmission
electron microscopy (T.E.M.) can thus be used to study the cascade
damage.

This technique, as compared to FIM, has a much reduced resolution
but it allows to look at a greater number of cascades ; furthermore, these cascades can be produced by P.K.A. much more energetic
than those it is possible to study by FIM. Merkle [9] observed
cascades in gold irradiated by 500 keV self-ions or 14 MeV neutrons.
Fig. 9 shows the cascade spatial extension, i.e. the splitting of
these energetic cascades into sub-cascades. The number of clustered defects, i.e. the total area of dislocation loops, is lower
by a factor 2 than what is predicted by the modified Kinchin
Pease formula. The damage function, $g(E_R)$, introduced in the beginning of this paragraph can be represented by

$$g(E_R) = \xi(E_R) \times \frac{0,8 \ E_R^*}{2 \ E_d^{avg}} \quad , \text{ where } \xi \text{ is an efficiency factor}$$

starting from 1 for low values of E_R and decreasing to 0,3 when
E_R reaches several keV.

We are now able to know the nascent damage rate $P = \sigma.\Phi$ produ
ced in our target irradiated by particles at the fluence Φ and with
a displacement cross section σ.

How this nascent damage state continuously formed by irradiation
evolves in the material held at the temperature T is the topic of
§ 2.

2 - ACTUAL DAMAGE : THERMAL EVOLUTION OF THE NASCENT DAMAGE

Point defects move through the crystal until they disappear
either by recombination with the opposite type of defect, or by
incorporation into the lattice at sinks such as dislocation; surfaces or grain boundaries. The situation is very complex since the
density of sinks varies as the irradiation dose increases. The
dislocation density increases by the apparition of dislocation
loops formed by point defect clustering ; interstitial loops grow
until they intersect each other and eventually form dislocation
tangles. Surface area also increases in huge proportion by the
formation of voids that are three-dimensional vacancy clusters. A
further complication is point defect trapping at impurities . This
effect changes the point defect mobilities, noticeably for interstitials : in some cases, vacancies can become more mobile than
interstitials ! Trapping of point defects is also thought to govern

the clustering of point defects. Furthermore, the trap concentration can change during an irradiation, for instance if a new phase
precipitates of if the temperature is lowered during irradiation,
some traps, inoperative at higher temperature, may enter now as
reaction partners.
Models have been developed [10] taking due account of all these
reaction possibilities, sketched in fig. 10, and also of the complications arising from the formation of vacancy clusters within displacement cascades.

$$V + v \rightarrow V_2 \qquad\qquad\qquad v + t \rightarrow vt$$
$$i + V_2 \rightarrow V \qquad\qquad\qquad i + t \rightarrow it$$
$$V + V_2 \rightarrow V_3 \text{ etc (clustering)} \qquad vt + i \rightarrow t$$

$$i + i \rightarrow i_2 \qquad\qquad\qquad it + v \rightarrow t$$
$$V + i_2 \rightarrow i \qquad\qquad\qquad V + vt \rightarrow V_2 t \quad \text{etc (clustering)}$$
$$i + i_2 \rightarrow i_3 \text{ etc (clustering)} \qquad i + it \rightarrow i_2 t \quad \text{etc (clustering)}$$

Figure 10

Reactions between point defects, including trapping impurities (t).

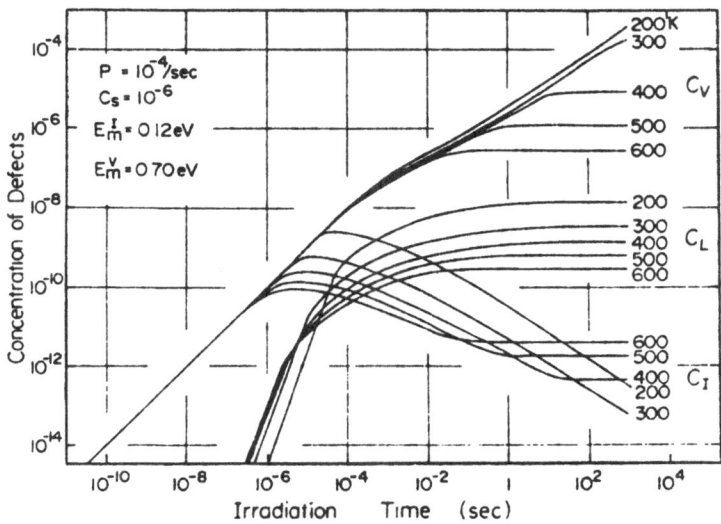

Figure 11

A computer simulation of the variation of point defect concentrations at vacancy—mobile high temperatures.

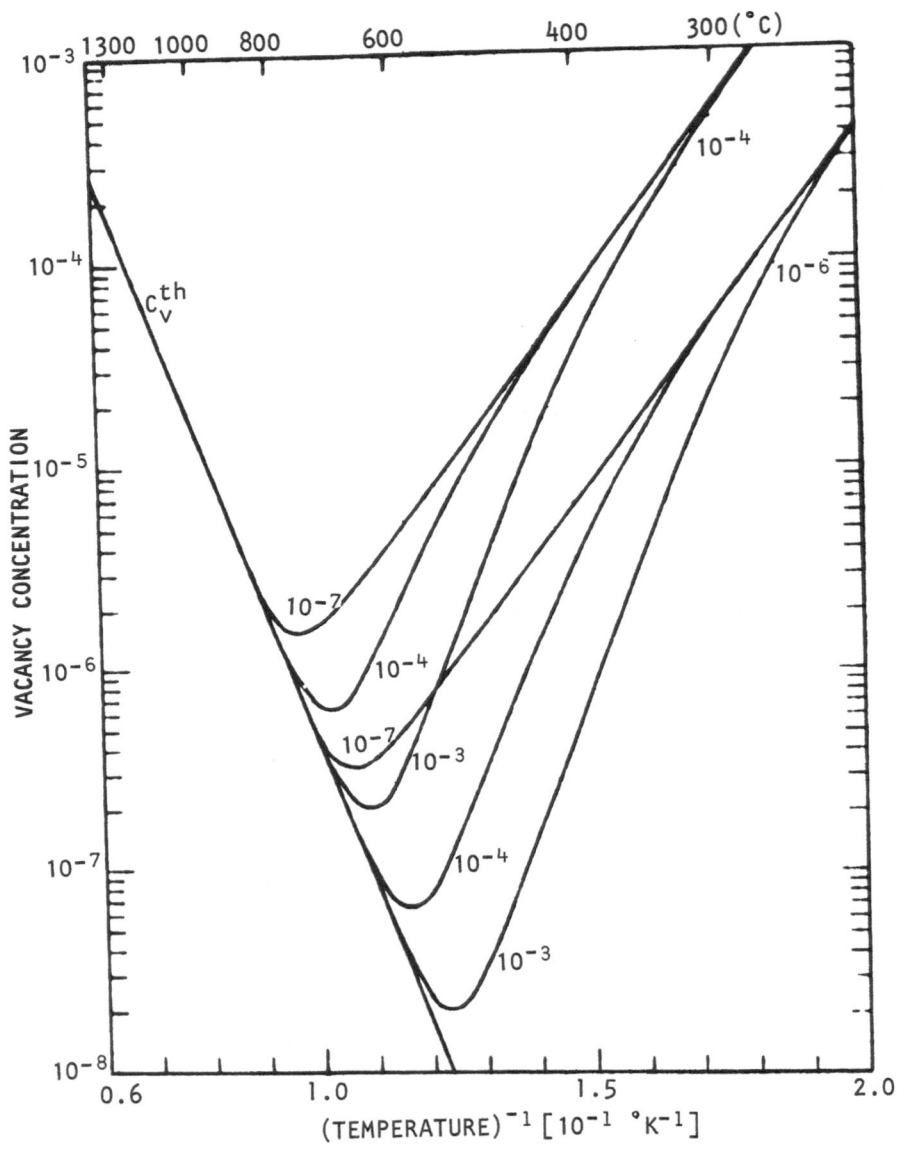

<u>Figure 12</u>

Steady-State Vacancy Concentration as Function of Inverse Temperature. The defect-production rates, P, in dpa/s and sink-annihilation
probabilities, a, are given.

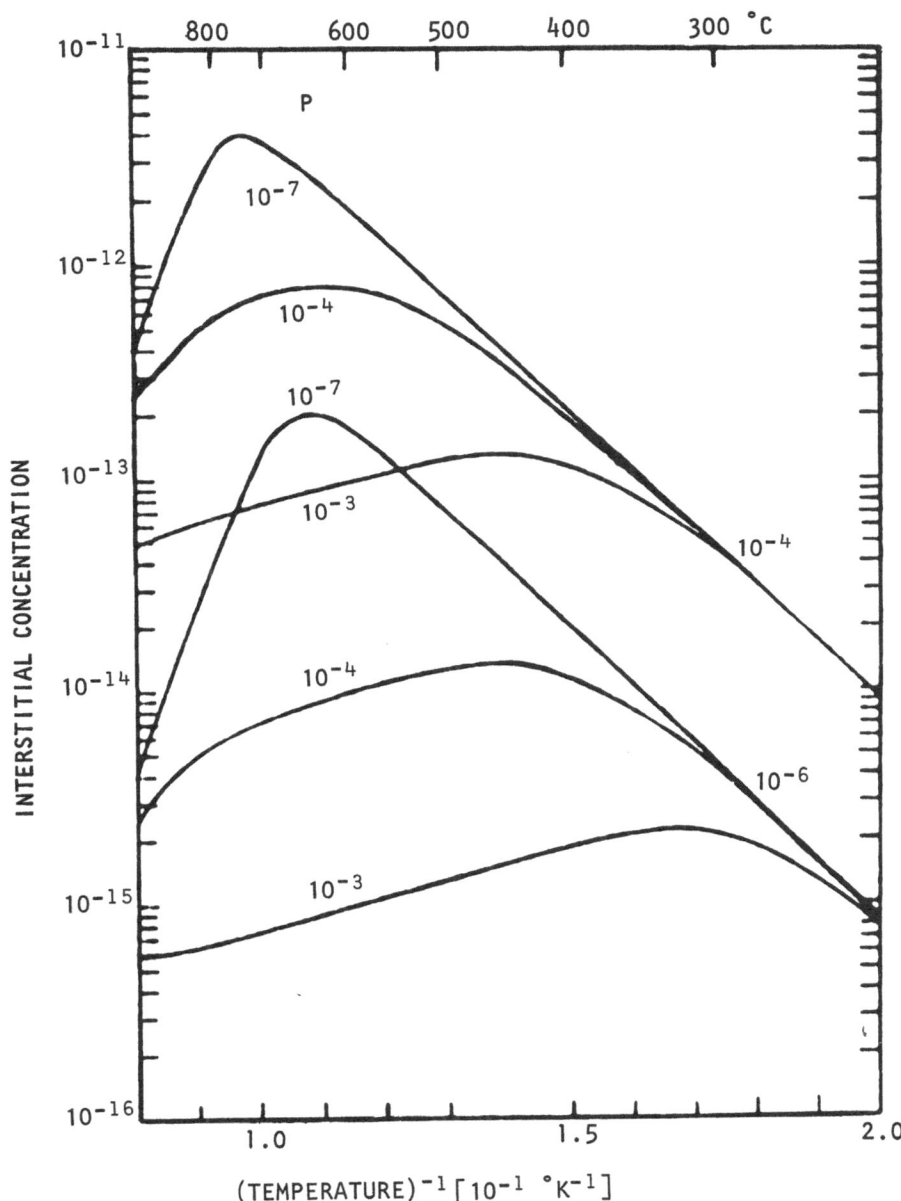

Figure 13

Steady-State interstitial concentration as a Function of Inverse temperature. The parameters are as in Figure 12.

2.1 - Steady state defect concentrations [12]

The rate of change in point defect concentrations (C_V and C_i) follow the equations :

$$\frac{dC_v}{dt} = P_v - (M_v z_v + M_i z_i) C_i C_v - a_v M_v C_v$$

$$\frac{dC_i}{dt} = P_i - (M_v z_v + M_i z_i) C_i C_v - a_i M_i C_i$$

production   annihilation i+v→o   losses at sinks.
rates

M_v and M_i are the vacancy and interstitial mobilities ; a_v and a_i are the probabilities that a jump leads to annihilation at a sink, whatever it is ; z_v and z_i are dimension less parameters corresponding to the recombination volume.
P_v and P_i are the vacancy and interstitial <u>total</u> production rates ; if the radiation production rate $P = 0$, there are thermal production rates P_v^{th} and P_i^{th} such that the equilibrium point defect concentrations $C_v^{th} = \exp(-G^v/kT)$ and $C_i^{th} = \exp(-G^i/kT)$ are reached; G^v and G^i are the Gibbs free energies of formation for vacancy and interstitial.

The basic rate equations for the vacancy and interstitial concentrations during irradiation are

$$\frac{dC_v}{dt} = P - (M_v z_v + M_i z_i)(C_i C_v - C_i^{th} C_v^{th}) - a_v M_v (C_v - C_v^{th})$$

$$\frac{dC_i}{dt} = P - (M_v z_v + M_i z_i)(C_i C_v - C_i^{th} C_v^{th}) - a_i M_i (C_i - C_i^{th}).$$

The interstitial mobility $M_i = \nu_i \exp(-G_m^i/kT)$ is very high compared to the vacancy mobility $M_v = z \nu_v \exp(-G_m^v/kT)$; ν_i and ν_v are the relevant vibration frequencies, Z is the number of nearest atoms around a vacancy and G_m^i and G_m^v are the activation Gibbs energies for migration of respectively interstitial and vacancy point defects. Migration entropies are of the order of 1 or 2 k for vacancy and 8 k for interstitial ; values of activation energies for migration E_m^v and E_m^i are given in fig. 24 and 29.

Interstitial formation energies are rather high ; fig. 30 shows that the maximum intersitial equilibrium concentration is always negligible in pure metals.

The rate equations may be used to obtain the evolution of point defect concentrations with time towards a quasi steady state, since the sink-annihilations probabilities, a, may change with time. Fig. 11 are schematic diagrams of point defect concentration build-up during irradiations, depending on the temperature and the sink density [11]. The final values, i.e. the steady state, are obtained

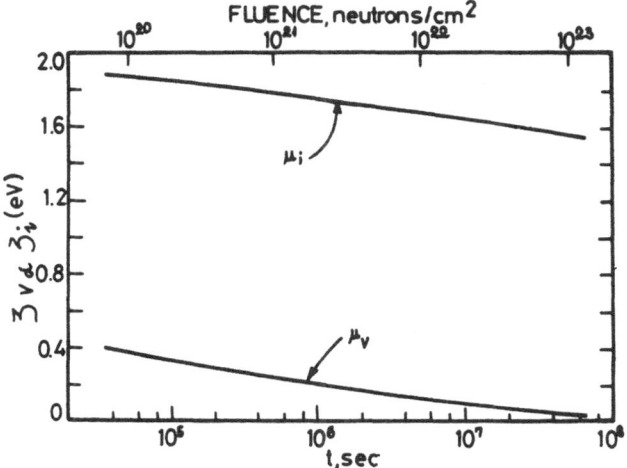

Figure 14

Vacancy and interstitial chemical potentials as a function of fluence during an irradiation [16].

by putting the time derivatives to zero. One gets the relation

$$a_i \, M_i \, (C_i - C_i^{th}) = a_v \, M_v \, (C_v - C_v^{th})$$

which states that deviations of the defect concentrations from thermal equilibrium are inversely proportional to the corresponding mobilities.

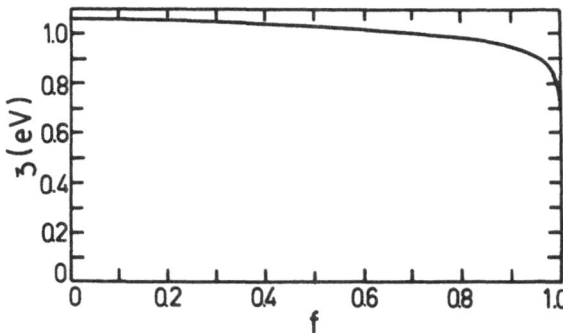

Figure 15

Vacancy chemical potential, as a function of f, the fraction of defects annealed, during isothermal annealing at $T_a = 320°C$ after a quench from $T_q = 1400°C$ in a stainless steel, assuming $E_V^F = 1.65$ eV and $E_V^M = 1.30$ eV [16].

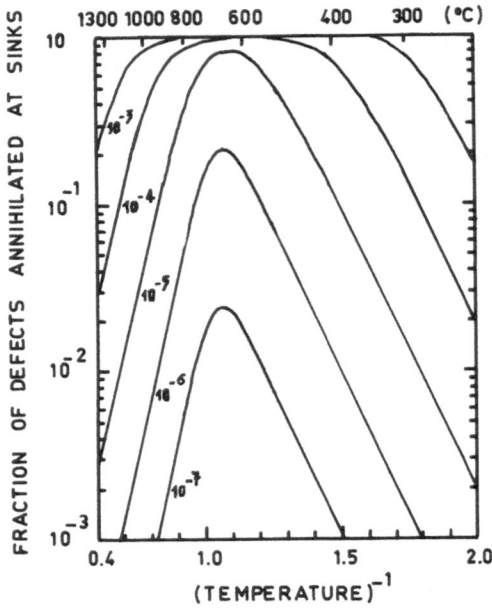

Figure 16

Fraction of radiation-produced defects that are annihilated at sinks during steady state as function of inverse temperature. The parameter is the sink-annihilation probability. Solid lines are for a defect-production rate $P = 10^{-6}$ s^{-1}.

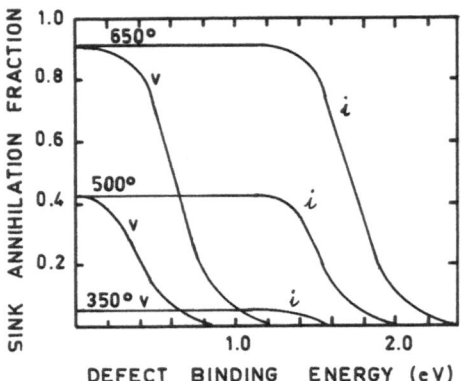

Figure 17

Sink-annihilation fraction as function of the defect-solute binding energy for several temperatures. Curves identified by v and i indicate trapping of vacancies and interstitials, respectively. Trap concentration c_t = 0.01, P = 10^{-2} dpa/s, a = 10^{-3}.

Steady state concentrations are $C_v = C_v^{th} + B\left[(1+F)^{1/2} - 1\right]$

with :

$$B = \left[\frac{M_i a_i}{M_i z_i + M_v z_v} + C_v^{th} + C_i^{th} \frac{M_i a_i}{M_v a_v}\right] \frac{1}{2} \approx \frac{1}{2} \left[\frac{P_i}{z_i} + C_v^{th}\right] \quad \text{and}$$

$$F \approx P \frac{a_i}{z_i} \bigg/ \left[M_v a_v (a_i/z_i + C_v^{th})^2\right] \quad \text{and} \quad C_i = \left[C_v - C_v^{th}\right] \frac{M_v a_v}{M_i a_i} + C_i^{th} \; .$$

Fig. 12 and 13 respectively show vacancy and interstitial steady state concentrations as a function of temperature for two production rates (10^{-6} and 10^{-4} d.p.a./s,) and three values of the sink parameter a (10^{-7}, 10^{-4} and 10^{-3}).

The expressions of C_v and C_i give the values of the chemical potentials $\xi_v = kT \ln(C_v/C_v^{th})$ and $\xi_i = kT \ln(C_i/C_i^{th})$. Fig.14 shows typical values of the chemical potentials, i.e. driving forces for point defect clustering.It is interesting to see that the vacancy chemical potential is rather low during an irradiation, whereas it can be high after a quenching from high temperature (fig. 15).

An other important parameter is the fraction of the point defect production that annihilates at sinks

$$S = \frac{(C_v - C_v^{th}) M_v a_v}{P} = \frac{(C_i - C_i^{th}) M_i a_i}{P} \; .$$

This fraction is plotted in fig. 16 for $P = 10^{-6}$ d.p.a./s. and several values (10^{-7} to 10^{-3}) of a. The point is that most interstitials and vacancies migrate to sinks at high sink densities and intermediate temperatures.

Finally, the effect of trapping of point defects by impurities, neglected up to now, can be incorporated into the rate equations by substituting C_v by $C_v^* = C_v + C_v^t$ and C_i by $C_i^* = C_i + C_i^t$, where C_v^t and C_i^t are the concentrations of complexes formed with traps. Mobilities are modified : detrapping requires an activation energy which is the sum of the binding energy of the point defect to the trap and of the activation energy for the migration of the point defect. Mobilities are thus to be changed into "effective mobility" defined by $M^{eff} = M/(1 + \tau_{diss}/\tau_{free})$ where τ_{diss} and τ_{free} are characteristic times for respectively, dissociation of complexes and free movement of the point defects. Fig. 17 shows how strong is the effect of trapping as soon as the defect binding energy becomes high.

2.2 - Interstitial dislocation loop nucleation (observed by High Voltage Electron Microscopy [13].

Interstitial clusters can only be dislocation loops; on the contrary, vacancies exhibit a great variety of secondary defects (dislocation loops, stacking fault tetraedra, voids). Because of

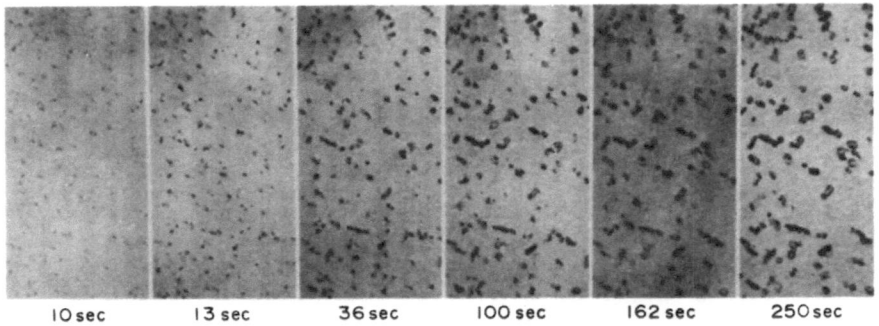

| 10 sec | 13 sec | 36 sec | 100 sec | 162 sec | 250 sec |

Figure 18

Early cessation of intersitital loop nucleation in gold. Only the growth of loops is observed during a prolonged irradiation. 2000 keV, 4.3 x 10^{18} e/cm^2-sec at room temperature (Yoshida and Kiritani 1973).

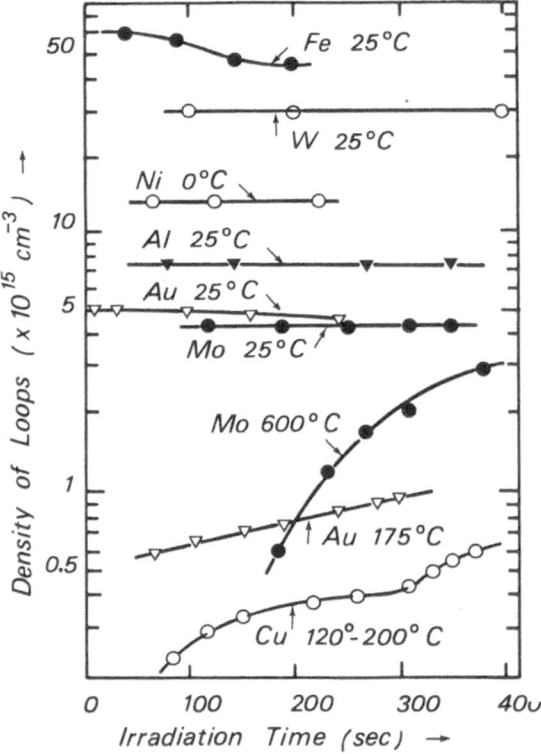

Figure 19

Variation in the interstitial cluster density during steady irradiation.

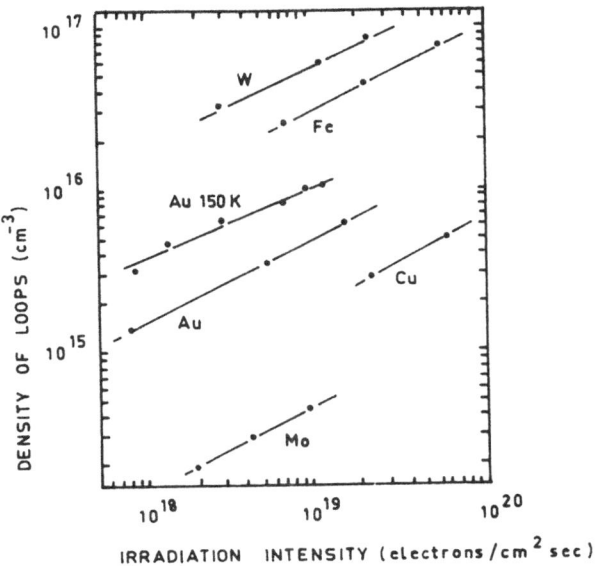

Figure 20

Irradiation intensity dependence of interstitial cluster density.
(Irradiation temperatures are room temperature when not indicated)

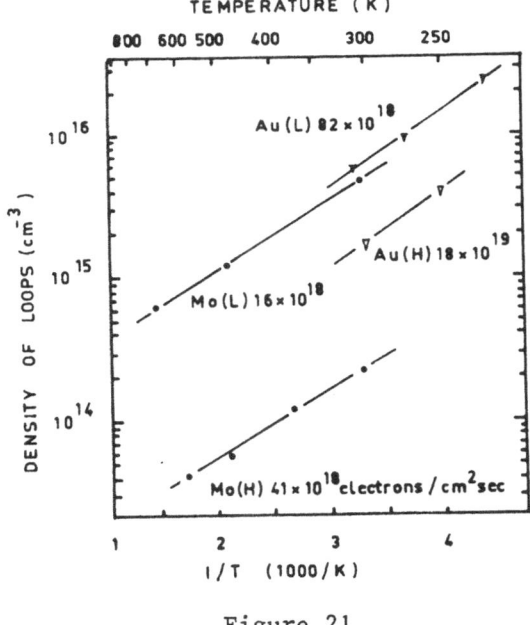

Figure 21

Comparison of temperature dependence of the cluster formation between two purities (Kiritani, 1975).

their high chemical potential (fig. 14) and their high mobility,
interstitials start clustering sooner than vacancies do. Due to the
interatomic potential form, a cluster of 4 or 5 interstitial point
defects has most probably acquired the dislocation loop character
(i.e. has taken a Burgers vector). The di-interstitial, which is
the first step towards clustering, is usually so stable that its
life-time is longer than the characteristic time to receive a
third intestitial, i.e. the di-interstitial is usually the dislocation loop nucleus ; there are thus as many loops as di-interstitials have been formed.

At not too high temperature (so as vacancy mobility is kept
low), the interstitial concentration C_i increases up to a maximum,
after which it goes on decreasing (fig. 11). When C_i becomes too
low, the di-interstitial formation stops, i.e. the dislocation
loop nucleation is achieved. On the contrary, at very high temperature, as shown in the last fig. 11, C_i reaches a steady state
concentration ; than dislocation loop go on forming with irradiation time. These two points about dislocation loop density that
saturates except at very high temperature are well observed (fig.
18-19).

When the dislocation loop density saturates, the rate equation

$$\frac{d\,C_i}{dt} = P - Z_{iv}M_iC_iC_v - Z_{ii}M_iC_i^2 - Z_{is}M_iC_iC_s$$

provides (by integration of the term $Z_{ii}M_iC_i^2$ which describes the
di-interstitial formation, i.e. the dislocation loop nucleation)
the saturation density of the loops :

$$C_L \propto P^{1/2}\, M_i^{-1/2}\quad .$$

This relation is well verified for the dependence with the
square root of the production rate (fig. 20). Then keeping Voltage
and current density of the High Voltage Microscopy constant, and
varying the specimen temperature allows experimental determinations
of the interstitial mobility ; the apparent activation energy found
between the i loop density and the irradiation temperature is one
half the activation energy for mobility of interstitial atoms.

Now the strong effect of impurities on the dislocation loop
nucleation is obvious in fig. 21. What sort of interstitial defect
is actually migrating ?
at medium temperature, where interstitials are deeply trapped at
impurities, the probability of complex formation is very much
higher than the probability of meeting of two free interstitials.
Than all dislocation loops are nucleated at interstitial impurity
complexes.The apparent activation energy is half the activation
energy for the migration of i complexes; i.e. $E^i_m + E_b$ is the
binding energy of the interstitial i to the impurity.

at very low temperature, the interstitial migration is so low
that it takes a long time for the interstitials to reach their

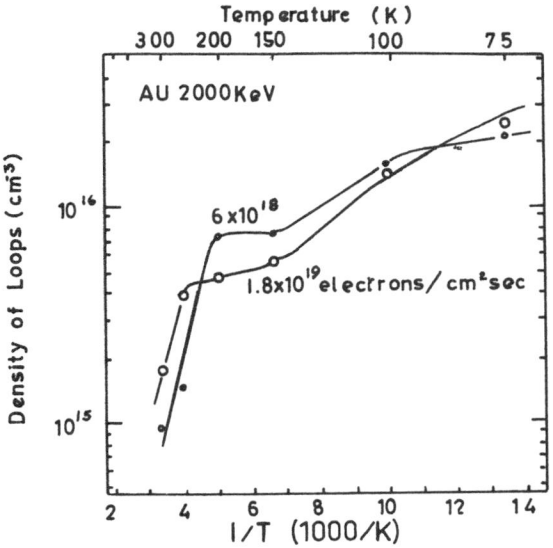

Figure 22

Irradiation temperature dependence of interstitial cluster density
in gold at low temperature (Kiritani, 1975)

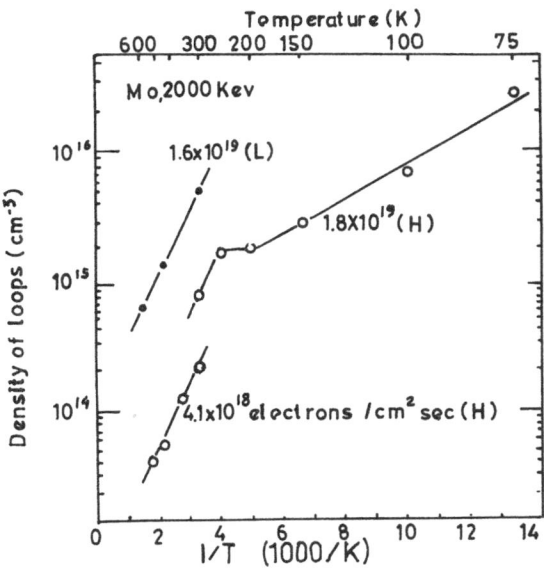

Figure 23

Irradiation temperature dependence of interstitial cluster density
in molybdenum at low temperature (Kiritani, 1975).

Metal	Intermediate temperature $(E_m^i + E_b)$ (eV)	Low temperature E_m^i (eV)
Au	0,19	0,04
Al	0,08	0,03
Fe	0,26	–
Mo	0,18	0,05
W	0,21	–

Figure 24

Interstitial migration energies derived from measured activation
energies for dislocation loop nucleation. Low temperature values
are too low, because of the radiation induced migration.

trapping impurities. C_i and C_v increase monotonically with P.t ;
annihilation of interstitials with vacancies become preponderant to
interstitial-impurity complex formation. Then, at very low T, di-
interstitial formation preferentially occurs via free interstitial
encounters ; impurity governed dislocation loop nucleation is less
and less important as temperature decreases. Plots of loop densities vs 1/T should be a straight line of slope 1/2 E_m^i at very low
temperature ; these ideas have been used recently to derive
E_m^i [13,14] ; the very low values that were found are probably not
correct because of the radiation induced defect migration ($\simeq 1$ jump/s
in a High Voltage microscope) which is not thermally activated and
which becomes more preponderant as the temperature decreases.

at high temperature, i complex is not stable enough : it will dissociate before the arrival of a next interstitial to form a stable
nucleus. It is thus expected that impurity atoms have no influence
on the dislocation loop nucleation at high temperature. The apparent activation energy is thus 1/2 E_m^i. This "high temperature" regime, however, is not easily used to derive E_m^i, because the dislocation loop densities become very low. Specimen surfaces become
very influent. A further reason is that the temperature range of
this regime may be rather short : at too high temperature, dislocation loop nucleation is never achieved ; their density increases
with the irradiation dose.
    Fig. 22,23 show that the preceding estimations are actually
verified by experiments. Some results are collected in fig. 24 [14].
    Studying the dislocation loop nucleation by high voltage elec-
tron microscopy thus provides E_m^i and E_b, the binding energy of i
with the dominant impurity in the specimen.The obvious method is
to very carefully prepare a series of "pure alloys" doped with one
impurity at several concentrations= 0 (reference pure alloy), 30ppm
1000 ppm ; the problem is to get a reference alloy with a residual
impurity level as low as possible ; it is a long but promising task.

The di-interstitial stability can be checked at high temperature. The theory [15] had been applied to Fe, Cu and Ni ; binding energies of di-interstitials were found to be respectively $E_{ii}^{Fe} = 0.20$ eV [15] $E_{ii}^{Cu} = 0.64 \pm 0.14$ eV and $E_{ii}^{Ni} = 0.8 \pm 0.1$ eV [16] .

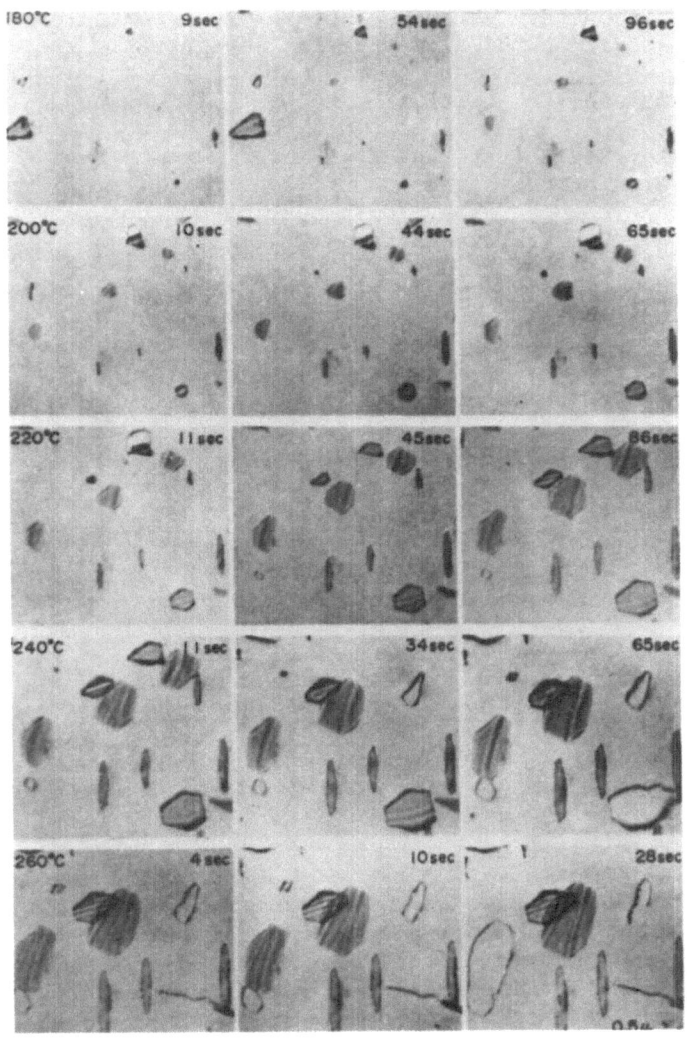

Figure 25

Growth of dislocation loops in copper for a stepwise increase of specimen temperature. 2000 keV, 6.5×10^{18} e/cm^2.sec (Kiritani[13]).

2.3 – Interstitial dislocation loop growth in High Voltage Electron Microscopy irradiations [13].

Fig. 25 shows the growth of interstitial dislocation loops in copper irradiated by 2 MeV electrons, at several high temperatures. The growth speed of some particular loops is easily measured ; it is found to be a linear function of time (Fig. 26).

i loops grow because more interstitial point defects arrive at the dislocation than vacancies do. The long range interaction between a dislocation and a point defect essentially arises from the interaction (E_o) between the hydrostatic (p) stress field of the dislocation and the relaxation volumes of the point defect (ΔV_{local}) = E_o = $3.\frac{1-\nu}{1+\nu}.p.\Delta v_{local}$.The local volume change is much larger for an interstitial than for a vacancy ; that is to say that the absorption cross section of loops for i, Z_{iL}, is larger than for vacancies, Z_{VL}. The growth rate of i loops is the difference between i and v fluxes towards the dislocation.

$$\frac{dR}{dt} \alpha Z_{iL} M_i C_i - Z_{vL} M_v C_v \text{ and because } M_i C_i = M_v C_v = (\frac{P M_v}{Z_{iv}})^{1/2}$$

$$\frac{dR}{dt} = (\frac{Z_{iL} - Z_{vL}}{Z_{iv}^{1/2}}) . P^{1/2}.M_v^{1/2} .$$

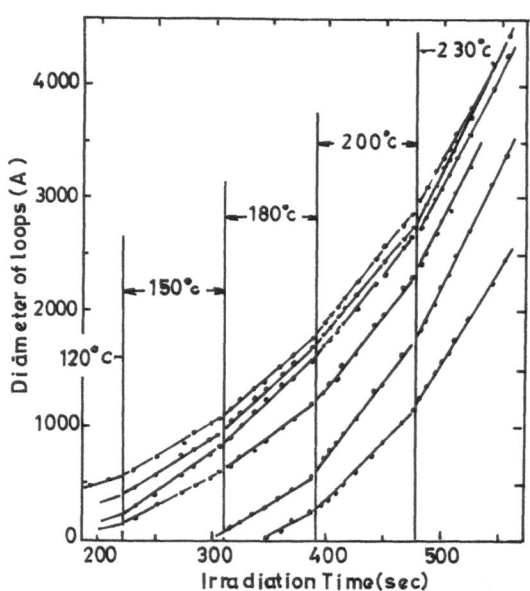

Figure 26

Measured size increase of dislocation loops in copper for the step-wise increase of specimen temperature, 2000 keV, 1.6 x 10^{19} e/cm^2. sec.

The dependence of the growth rate of loops vs beam intensity
(i.e.P) has been verified ; the dependence of dR/dt vs temperature
is more interesting. Whereas the dislocation loop nucleation is
controlled by the i mobility, <u>the growth of these loops is governed</u>
<u>by the vacancy mobility</u>. Measurements of growth rates versus tempe-
rature (fig. 27) give the activation energy for the migration of
vacancies. Once more, it should be possible, with specimens at several well controlled impurity levels, to measure also binding
energies between a vacancy and these impurities. This has not yet
been done. Up to now, binding energies of vacancy impurity complexes have been measured (fig. 29) by electrical resistivity or
positron annihilation techniques. However a result, new and surprising at first sight, comes from a clever use of high voltage
electron microscopy; <u>in iron</u>, and presumably in all B.C.C. metals
<u>the di-vacancy is not mobile</u>! Actually the geometry of this point
defect and its migration configurations are quite different in the
B.C.C. and F.C.C. structures. According to [21] , a di-vacancy
in Fe has to dissociate into its two monovacancies to migrate ;
the monovacancy migration activation energy has been found

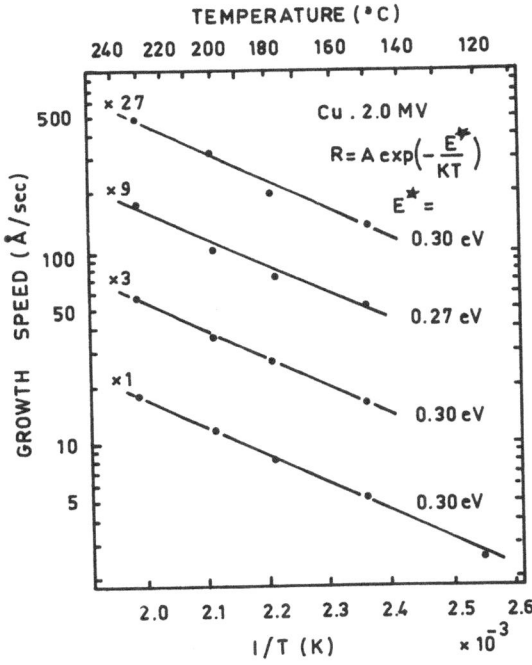

Figure 27
Temperature dependence of the growth speed of interstitial-type
dislocation loops in electron-irradiated cooper. 2000 keV, 1.6 x
10^{19} e/cm^2.sec. The value of E_m^V found here (0.60 eV) should not
be surprising : 2 MeV electron in Cu can produce divacancies!

to be 1.24 ± 0.14 eV, by the method described in this paragraph ;
but studying a phenomenon governed by the migration of many vacancies does not provide $E_m^v = 1.24 \pm 0.14$ eV, because mobile single
vacancies meet each other and form immobile divacancies, then the
activation energy which is found is the same of the energy necessary to dissociate sum divacancy into 2 single vacancies (i.e.
E_b^{2v}) plus the activation energy of migration of vacancies :
$E_a = E_b^{2v} + E_m^v$. Kiritani et al. [21] have studied the annihilation
of vacancies produced either by electron irradiation or by plastic
deformation ; in both cases respectively provided $E_a = 1.47 \pm 0.05$eV.
and $E_a = 1.50 \pm 0.05$ eV. The first estimation of the binding energy of
a divacancy in iron is thus derived $= E_b^{2v} \simeq 0.25$ eV.

Metal	E_{1v}^m	E_{2v}^m	E_{3v}^m	E_{2v}^b	E_{3v}^b
Au	0.83–0.89	0.62–0.79	0.48–0.56	0.25–0.57	1.5 E_{2v}^b
Al	0.65	0.50	0.47	0.20	~0.3
Pt	1.45	' 10	–	0.19	–
	1.38	1.11	–	0.11	–
	–	–	–	0.23 *	–
Cu	0.72	–	–	–	–
Ag	–	0.57	–	–	–
W	1.8	–	–	–	–
M	1.3–1.9	–	–	–	–

Figure 28

Selected values of vacancy defect migration energies and binding
energies obtained from quenching experiments (eV) (R.W. Balluffi).

Solvent	Solute	Valency	B_{Vi}	S_{Vi}^b/k
Al	Cu	1	0.00 ± 0.12	0.0 ± 1.5
	Ag	1	0.05	−0.5
	Ag	1	0.013	
	Mg	2	−0.01 ± 0.04	−0.1 ± 0.5
	Mg	2	0.04 ± 0.01	2.7 ± 0.2
	Zn	2	0.019	
	Zn	2	0.02	−4.0 ± 1.2
	Zn	2	0.10	
	Si	4	0.07	−1.7 ± 0.5
	Si	4	0.03	−2.0
Cu	Al	3	0.15 ± 0.10	
	Ge	4	0.27 ± 0.10	
	Ge	4	0.25 ± 0.10	

Figure 29

Some experimental values of the binding energy B_{Vi} between a vacancy and impurity atoms (Doyama).

2.4 - Vacancy clusters in irradiated metals
Nucleation.
    Most information about vacancy clusters comes from quenching
experiments. Impurities seem to always play an essential role in the
nucleation of these clusters ; the impurity content in metals (even
the purest grade material) is sufficient to account for the density
of vacancy precipitates, which has rarely exceeded $10^{16}/cm^3$, i.e.
0.2 ppm. Since, despite the high chemical potential of vacancies
in metals quenched from high temperature (fig. 15), homogeneous
vacancy cluster   nucleation apparently does not occur during ageing
of quenched metals, it cannot occur in irradiated metals, except
within displacement cascades(cf § 122) or in special circumstances.
    Many experiments prove that void nucleation is not homogeneous:
it needs the presence of gas atoms [17, 18].

Growth.
    Vacancy loops are unstable in irradiated bulk material : their
growth rate :

$$(\frac{dR}{dt})_v = \frac{1}{b} (Z_v D_v C_v - Z_i D_i C_i - Z_v D_v C_v^{th} \exp \{\frac{F_{el}(R)b^2 + \gamma_{sp} b^2}{kT}\})$$

is negative. The first terms describe the vacancy and interstitial
arrivals at the dislocation core ; the last term is the loss of
vacancies from the loop by vacancy evaporation : the activation
energy of this evaporation is the sum of the formation energy of a
vacancy at the dislocation loop (i.e. E_f^v - the help provided by the
dislocation line tension $F_{el}xb^2$ and the stacking fault contribution

Metal	Au	Al	Pt	Cu	Ag	W	Mo
E_{1v}^f(eV)	0.95	0.67	1.51	1.28	1.13	~3.6	~3.2

Figure 30
Vacancy formation energy ; the max. equilibrium concentrations
are a few 10^{-4}.

	Al	Cu	Pt
E_{1i}^f [eV]	3.2 ± 0.5	2.2± 0.7	3.5± 0.6
C_{1i} (T_m) atomic fractions	10^{-18}	10^{-7}	10^{-6}

Figure 31
Interstitial formation energy and maximum equilibrium concentration.

γ_{spb}^2 if the loop is faulted) plus the activation energy for removing the vacancy E_m^V (this term is contained in the diffusivity D_v). A vacancy loop is always expected to shrink : at low temperature, because $Z_v D_v C_v < Z_i D_i C_i$; at high temperature because the evaporation of the loop.

The only chance to find a stable vacancy loop (dR/dt > 0) is to look for it in the vicinity of a free surface, which is a sink for interstitial ; C_i is locally so small that the relation $S_i D_i C_i > Z_v D_v C_v$, valid in bulk material, no longer holds. Edge dislocations are also efficient interstitial umbrella for vacancy dislocation loops : near a dislocation, a depletion in interstitial concentration does occur and explains the observation of rows of vacancy loops parallel to the dislocation line. But of course this dislocation line climbs away from adjacent vacancy loops ; when the dislocation is far enough so as C_i recovers its bulk value, the bias driven shrinkage of vacancy loops begins and goes on until complete disappearance.

The other chance that a vacancy loop has a non ephemera existence occurs if the irradiated material is stressed in compression. Then, a vacancy loop with habit plane not far from perpendicular to the compression axis, would be stabilized [19]. This mechanism certainly exists ; it explains irradiation growth [20].

Fig. 24, 28, 29 respectively collect selected values of activation energies for interstitial migration, vacancy migration binding energies between vacancies and impurities.

Fig. 30 and 31 give point defect formation energies and maximum equilibrium concentrations.

CONCLUSIONS

The radiation induced damage phenomena are fairly well understood. P.K.A. spectra due to each kind of particle are rather well known. E_d measurements are numerous and good : but the damage function is not too well known, because the radiation annealing due to subthreshold events.

Very good simulation models have been developed but unfortunately the in-put data are often quite uncertain : the point defect properties, i.e. the formation energies E_F^V E_F^{2V} E_F^{3V} E_F^{4V} ... E_F^{2i} E_F^{2i}, i.e. binding energies E_b^{12} E_b^{V2} E_b^{V3} ... the binding energies of v and i to impurities, the volume changes induced by these defects that determine the sink strength ..., have been measured only in a few cases.

We certainly need much more information, and among other ways, High Voltage Electron Microscopes can give much.

REFERENCES

1   M.W. THOMPSON – Defects and Radiation damage in Metals, Cam-
              bridge University Press (1969).
     M.T. ROBINSON – in Radiation Damage in Metals (p.1) American
              Society for Metals (1976).

2   O.S. OEN – ORNL Reports : ORNL 3813 (1965), ORNL 4897 (1973).

3   G.B. GIBSON, A.N. GOLAND, M. MILGRAM and G.H. VINEYARD – Phys.
     Rev. 120, 1229 (1960).
     C. ERGINSOY, G.H. VINEYARD and A. ENGLERT – Phys. Rev. 133A,
     595 (1964).

4   VON JAN and A. SEEGER – Phys. Stat. Sol. 3, 465 (1963).

5   A. BOURRET – Rad. Effects 5, 27 (1970).

6   P. VAJDA – Rev. of Modern Physics 49, 481 (1977).

7   G.H. KINCHIN and R.S. PEASE – Rep. Prog. Phys. 18, 1, (1955).

8   D.N. SEIDMAN – In Radiation Damage in Metals (p. 28), American
              Society for Metals (1976).
     C.Y. WEI and D.N. SEIDMAN, Cornell University, Ithaca,New York,
     COO4022 Report (1978). To be published in Applied Physics
     Letters (1979).

9   K.L. MERKLE – in Radiation Damage in Metals (p.58) American
              Society for Metals (1976).
              and J.N.M. 69,70, 78b (1978).

10   J.M. LANORE – CEA report B.1567 (1972).

11   R. SIZMANN – J.M.N. 69–70, 386 (1978).

12   H. WIEDERSICH – in Radiation Damage in Metals (p.157) American
              Society for Metals (1976).

13   M. KIRITANI and H. TAKATA – J.M.N. 69–70, 277 (1978).

14   B.L. EYRE, M.H. LORETTO and R.E. SMALLMAN – Harwell report
     R 8621 (1975).

15   A. BOURRET – Phys. Stat. Sol. (a) 4, 813 (1971).

16   M.K. HOSSAIN and L.M. BROWN – Rad. Eff. 31, 203 (1977).

17   L.D. GLOWINSKI et al. J.N.M. 61, 8, 22, 29, 41 (1976).

18   M.R. HAYNS – Harwell report AERE R 8806 (1977).

19   P.T. HEALD and M.V. SPEIGHT – Phil. Mag. $\underline{29}$, 1075 (1974).
      R. BULLOUGH and J.R. WILLIS – Phil. Mag. $\overline{30}$, 855 (1975).

20   J. LETEURTRE and Y. QUERE – Irradiation effects in fissile
      materials, Series Defects in Crystalline Solids, vol. 6,
      Edited by S. Amelinckx, R. Gevers and J. Nihoul – North
      Holland.

21   M. KIRITANI, H. TAKATA and K. MORIYAMA – Phil. Mag.
      (in the press).

PRECIPITATION PROCESSES IN IMPLANTED MATERIALS [*+]

J. A. Borders

Sandia Laboratories [**]

Albuquerque, NM 87185

ABSTRACT

Ion implantation is a nonequilibrium process. It is possible
to implant materials with impurities to concentration levels which
far exceed the solid solubilities. The return of the system to
thermodynamic equilibrium is often accomplished by precipitation
of the implanted species or a compound involving atoms of both the
host and the implanted species. This process may involve very long
time scales when taking place at room temperature or it may take
place dynamically during the implantation. In order to understand
the metallurgy of implanted systems at high concentrations, an
understanding of the basic phenomena controlling second phase
precipitation is necessary. These processes will be briefly
reviewed and illustrated with examples of data from implanted
systems.

INTRODUCTION

Ion implantation is a method of introducing impurities into
a solid material in a nonequilibrium fashion. Atoms of the desired
impurity are ionized, accelerated to a high energy (generally
10 keV to 1 MeV) and directed onto the surface of the solid to be

* This article is based on a lecture delivered at a N.A.T.O.
  Advanced Study Institute held in Aleria, Corsica between
  September 10th and September 23rd, 1978.
+ This work supported by the United States Department of Energy
  under Contract #AT(29-1)-789.
**A U. S. DOE Facility.

doped. The ions penetrate the material to a depth characteristic
of the ion, the material, and the ion energy. When the ions come
to rest they will reside in a region of composition determined by
the fluence of impurity ions incident upon the solid. The morphology of the implanted layer may be a nonequilibrium structure,
it may be affected by the interaction of the damage created by the
slowing down of the incident ions, or it may be a normal equilibrium structure that could have been predicted from the binary phase
diagram of the system involved. In order to understand the physical
properties of the implanted layer, an understanding of the stoichiometry and morphology of the layer is necessary. Many implanted
materials have been found to contain nonequilibrium phases such as
supersaturated solid solutions.[1] Phases such as these will prefer
to assume an equilibrium structure and the most common phenomenon
by which the equilibrium structure evolves is second phase precipitation. This paper will review the processes which govern the
precipitation of a second phase and briefly illustrate the discussion with examples using implanted materials.

Any discussion of second-phase precipitation must be based on
an understanding of the phase diagram of the system under study.
As commonly used, a phase diagram is a schematic representation of
the equilibrium phases of a system plotted as a function of composition on the horizontal axis and temperature on the vertical axis.
A phase is nothing more than a portion of matter which has homogeneous chemical and physical properties and which is normally
separated from other phases by well-defined boundaries or interfaces.

If we are working at constant temperature and pressure, which
is usually the case, the equilibrium value of any free parameter of
a thermodynamic system is that which minimizes the Gibbs Free
Energy at that temperature and pressure. The Gibbs Free Energy is
given by the expression

$$G = E - TS + PV \qquad\qquad\qquad (1)$$

which when combined with the Euler relation is shown to be similar
to the chemical potential and, for a one-component system, is
identical to the molar chemical potential. Thus an equilibrium
phase is that phase for which, at constant temperature and pressure,
the Gibbs Free Energy is minimized. In order to construct a phase
diagram then, we merely calculate the Gibbs Free Energy of the
various possible phases. This is seldom simple and usually
impractical, but in the discussion that follows we will assume that
the Gibbs Free Energy of a phase can be calculated as a function
of composition.

Figure 1 shows an example of the construction of a phase
diagram for the simple case of a liquid and a solid exhibiting
complete miscibility, but having liquidus and solidus boundaries

which are separated in temperature. In the corners of Figure 1 are
plotted the Gibbs Free Energies as a function of composition at
four different temperatures. At T = T_1, the liquid phase has
a lower free energy at all compositions, so on the phase diagram
in the middle of the figure, at a temperature of T_1, the liquid is
found at all compositions. At T = T_2 the liquid has a lower free
energy at all compositions except for pure A. Thus the phase
diagram indicates that at T_2 the liquid is found for all compositions except pure A, where liquid and solid coexist. At T = T_3,
the solid has a lower free energy from a composition of pure A to
a composition of M and the liquid has a lower free energy from
a composition of M to pure B. Between M_1 and M_2, the lowest free
energy is actually achieved by a mixture of solid composition M_1
and liquid of composition M_2. Thus a two-phase region is found in
the phase diagram at temperature T_3 extending from M_1 to M_2. This
is a very common phenomenon, regions of single phase in the phase
diagram are usually separated by two-phase regions. The situation
depicted at T = T_4 is the same as that at T = T_3 but the two-phase
region is now shifted to compositions between N_1 and N_2. The
example shown in Figure 1 was hypothetical. Actually phase diagrams
are generated by physical or chemical measurements of the phases
present or of phase transformations and not determined by calculating
the Gibbs Free Energy of the system at all compositions and temperatures. There are extensive tabulations of these diagrams. [2-4]

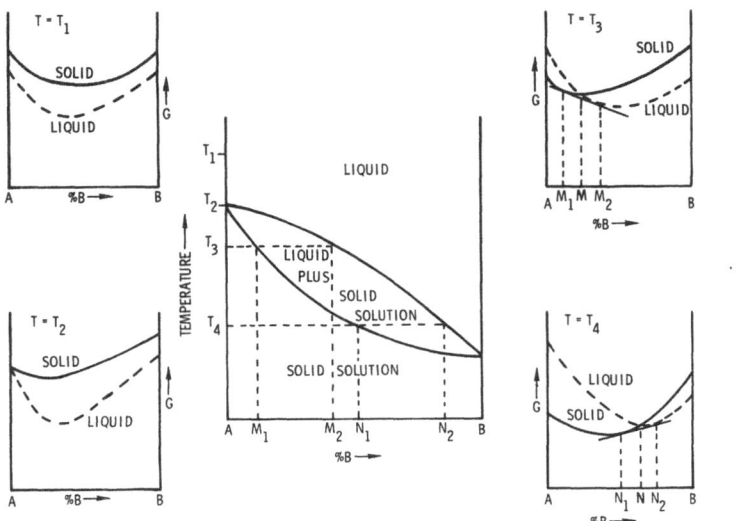

Figure 1. The graphs in the corners show the variation of the Gibbs
Free Energy for a system with only a solid and liquid phase at four
different temperatures: T_1, T_2, T_3, T_4. The center portion of
the figure shows the phase diagram which is constructed from these
data.

Let us imagine now that we can generate a phase so rapidly
that the structure of the material is such that the Gibbs Free
Energy is a local but not an absolute minimum when plotted against
some generalized structure coordinate. If we can make this phase,
it will be stable against small changes in the structure which
would tend to evolve the equilibrium structure and only fairly
large-scale diffusive rearrangement can produce the equilibrium
phase. This is a metastable phase and the existence of such structures has been well documented and experimentally explored using
rapid quenching methods.[5] The Cu-Ag system is the classic example.
In Figure 2 is shown the equilibrium phase diagram of the Cu-Ag
system. According to some fairly successful criteria for the prediction of solid solubility[6] this system should be a complete
series of solid solutions. However only very limited solubility
is found. By preparing melts of various Cu-Ag combinations and
splat-cooling[5] droplets of the liquids, solid solutions of any
composition can be produced. These solutions are metastable and
precipitate into the equilibrium phase, a mixture of the two solid
solutions upon annealing. Estimates of the cooling rates operable
during splat-cooling experiments are in the range 10^5-10^8 K/sec.

Another technique which has also been used to produce metastable alloys such as these is vapor quenching.[7] In this method
a vapor is condensed into a thin film on a cryogenically-cooled
substrate. Estimates of the cooling rates in these experiments
are in the range of 10^{12}-10^{15} K/sec. It has recently been shown
that ion implantation can produce metastable solid solutions of
Cu-Ag.[8] For the case of ion implantation, cooling rate estimates
have been made based on calculations of the lifetime and maximum
temperature of the energy spike created by the collision cascade
of the slowing-down ion. These estimates are about the same as
those for vapor quenching.

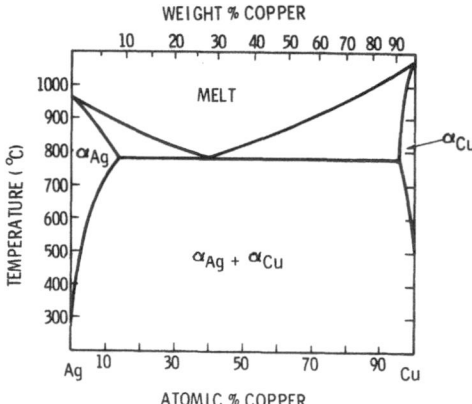

Figure 2.   Phase diagram of the Ag-Cu system.

All of these metastable phases show the same behavior upon
thermal annealing. Precipitates of the equilibrium phases begin to
form and grow. We will briefly review the formation and growth of
precipitates before looking in more detail at an ion implanted
system.

Processes Governing Precipitation

There are basically four processes that interact to determine
if precipitation will take place. First, nucleation; small regions
of the second phase must form. Secondly, diffusion of atoms to the
nuclei to supply additional material so that the second phase can
grow. Thirdly, reaction; the atoms at the interface must rearrange
themselves into the structure of the new phase, either absorbing or
giving off energy in the process. Fourthly, diffusion of excess
atoms from the reaction away from the interface. The second and
fourth steps involve the same process and will be discussed together.
More detailed discussion of these processes can be found in most
books on physical metallurgy; a particularly complete reference has
been edited by Cahn.[9]

Nucleation. The process by which stable aggregates of atoms
in the required arrangement are formed is called nucleation. There
are two types of nucleation: homogeneous and heterogeneous. Homogeneous nucleation takes place when the new phase forms unformly
throughout the volume of the parent phase, in heterogeneous nucleation the new phase forms at inhomogeneaties in the parent phase,
such as impurities, grain boundaries or in the case of a liquid or
gas, sharp points protruding from the container vessel. Most reactions are of the heterogeneous type and all of the reactions that
have been observed in ion-implanted materials are of this type.
The theory of heterogeneous nucleation starts from the consideration
of the free energy of formation of a small spherical precipitate of
radius r. This can be written as

$$\Delta G(r) = \frac{4}{3}\pi r^3 \Delta G_v + 4\pi r^2 \gamma + \Delta G_s \qquad (2)$$

where ΔG_v is the volume free energy change due to formation of the
precipitate, ΔG_s is the strain energy, and γ is the interfacial
energy. Since both ΔG_s and γ will be positive, the volume free
energy of formation ΔG_v must be negative if formation of the precipitate is to lead to a lower system free energy. A plot of
Equation 2 as a function of radius is shown schematically in
Figure 3. At very small r, both ΔG and $\partial \Delta G/\partial r$ are positive so that
nuclei are unstable and any that come into existence because of
entropy considerations will shrink. At some critical radius,
$\partial \Delta G/\partial r$ becomes negative and although nuclei of this radius are still
energetically unstable they will tend to grow as they can reduce

their free energy by becoming larger. Finally at a larger radius,
ΔG is negative and the nuclei will be stable. One might very well
ask, how can those nuclei which are unstable exist since this implies a rise in the total free energy of the nuclei? But the
existence of some of these unstable nuclei will increase the entropy
of the system and thus decrease the total system free energy. Thus
we must have nuclei of some critical size before precipitation will
commence.

    Diffusion. Once suitable stable nuclei have formed, nuclei
growth becomes an important process. Since, in general the new
phase and the parent phase do not have the same stoichiometry,
there must be either a flow of atoms to and/or from the precipitate.
Diffusion provides this flow of atoms. Fick's law defines the diffusion constant D in terms of the atomic flux, J, and the concentration gradient, c.

$$J = -D(\partial c/\partial x) \tag{3}$$

From Equation 3 and the continuity equation which is a statement
that matter is conserved, we can arrive at the one-dimensional
diffusion equation

$$\frac{\partial c}{\partial t} = \frac{\partial}{\partial x}\left[D\frac{\partial c}{\partial x}\right]; \quad \text{if } D \neq D(x) \text{ then } \frac{\partial c}{\partial t} = D\frac{\partial^2 c}{\partial x^2}. \tag{4}$$

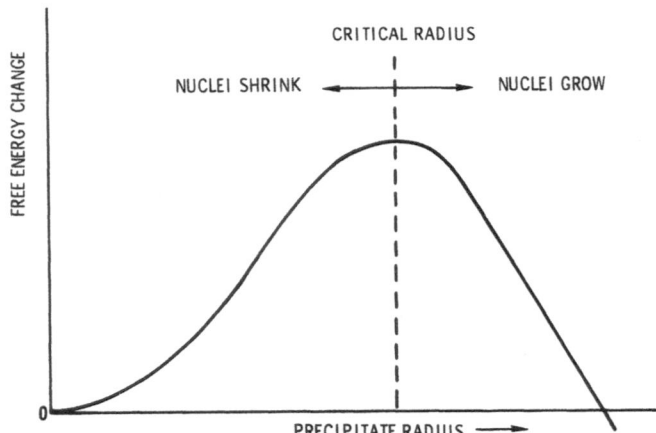

Figure 3. Free energy of a spherical precipitate as a function of
precipitate radius.

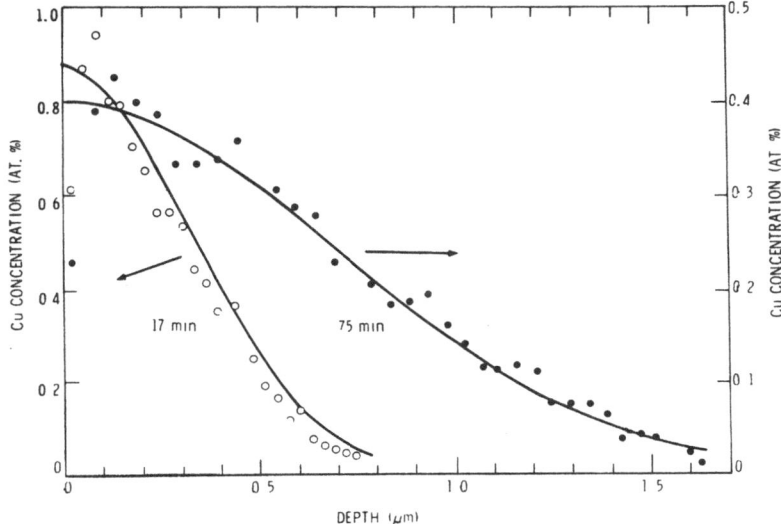

Figure 4. Diffusion profiles of ion implanted Cu in Be. The solid
curves are fits of solutions of the diffusion equation to the experimental data. The data are for 17 minutes at 595C (o) and 75 minutes
at 595C (●).

Many volumes have been written on the solutions to this equation.[10]
Typical data of the concentration gradients of a diffusing species
are shown in Figure 4. These data from the work of Meyers et al[11]
are for the case of Cu diffusing in Be. In these experiments, ion
implantation was used to prepare a bulk sample of Be with a thin
layer containing Cu and Rutherford backscattering analysis[12] was
used to analyze the results.

Experimentally it is found that diffusion coefficients obey
an Arrhenius-type equation

$$D = D_o \exp(-Q/RT) \tag{5}$$

where D_o and Q are constants, R is the gas constant and T is the
absolute temperature. This is easily seen by plotting the log of
the diffusion coefficient against inverse absolute temperature.
The quantity Q is referred to as the activation energy and is
related to the enthalpy change as a diffusing atom moves from one
site to another.

Reaction. Once nuclei are formed and the proper numbers of
atoms are present at the interface between the nuclei and the parent
phase, it is necessary that a reaction take place. This is usually
not a problem, the kinetics of precipitation are generally limited
by either nucleation or diffusion, but it is possible to find
situations where the time development of the precipitates is limited

by the reaction rate at the interface. Reactions are commonly
classified into reaction orders depending on how the rate of reaction varies with the concentration. For example, for a first
order reaction, the negative of the rate of change of the concentration is proportional to the concentration or $-dc/dt = Kc$ where
c is the concentration. This equation can be simply solved to show
the kinetics of the reaction will be such that $\ln(c/c_0) = k(t-t_0)$
where t is the time and the subscripts represent initial conditions.
Thus by looking at the time behavior of a system we may be able to
determine the reaction order or even if the system is reaction-rate
controlled.

This discussion has not meant to be extensive or even complete.
For further information the reader is referred to the list of
references. Now that we have briefly reviewed the processes that
govern precipitation, we will discuss two examples of precipitation
in ion-implanted systems.

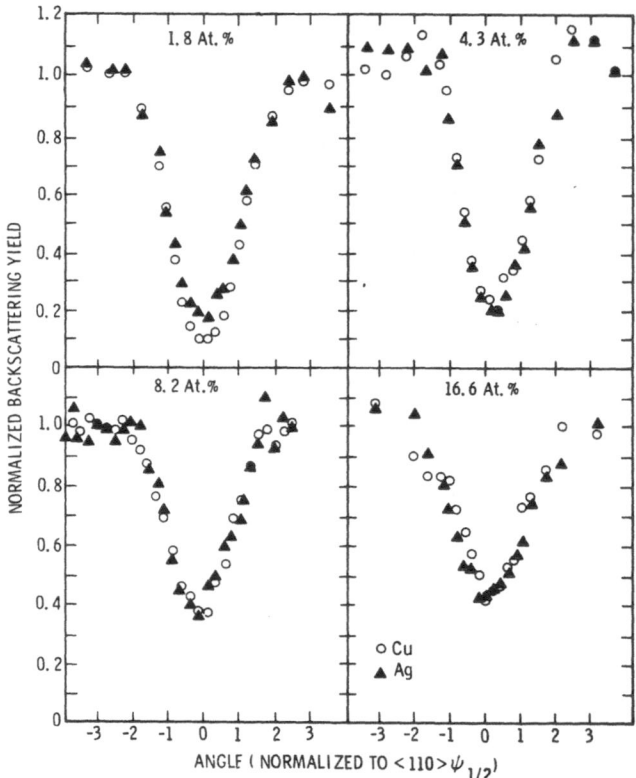

Figure 5. ⟨110⟩ angular distributions of Cu single crystals
implanted with Ag to four different peak concentrations, all
above the room temperature solubility limit.

EXPERIMENTAL EXAMPLES

As mentioned briefly above, the Cu-Ag system is the classic metallurgical example of a binary alloy system that is predicted to show complete solid solubility, but in fact shows very limited solubility. It has been demonstrated[8] that it is possible to implant Ag in Cu up to the sputtering limit and that the result is a 100% solid solution, even at concentrations outside the equilibrium solubility limits. In Figure 5 are shown channeling data for Ag implanted into single crystal Cu at four different concentrations, all above the room temperature solubility limit. The fact that the angular scans for the Ag and Cu coincide indicates that the Ag atoms lie 100% on Cu lattice sites and thus they have formed a solid solution. Thermal annealing of this sample gives rise to precipitation of α-silver in the α-copper matrix. In Figure 6, angular scans are shown as a function of annealing temperature for the sample which had been implanted to a maximum concentration of 8.2% Ag. It is clear that the Ag angular scan is showing a less distinct dip than

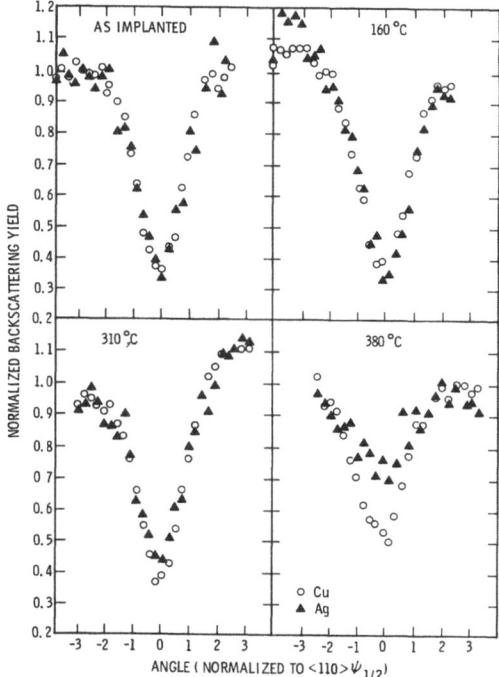

Figure 6. $\langle 110 \rangle$ angular distributions of a Cu single crystal implanted with Ag to a peak concentration of 8.2 at. % and annealed at four temperatures.

the Cu indicating that the Ag is no longer completely on Cu lattice
sites at the highest temperatures. At lower temperatures it is
possible to obtain coherent precipitates that show 100% substitutionality as measured by channeling, but that can be imaged separately
from Cu using electron diffraction in the transmission electron
microscope.

Another example of precipitation of an ion implanted species
is found in the work of Arnold and Borders[13] on Ag-implanted lithia-
alumina-silica glass. These workers were implanting Group I atoms
in order to attempt to form a glass-ceramic in the implanted layer.
In this process the Group I atoms aggregate into colloids and if
the colloids are of the correct dimensions, crystalline phases can
be nucleated on the metal colloids. This is another example of
heterogeneous nucleation. In the case of Au atoms the process
worked well, but with Ag there was no glass ceramic formation seen

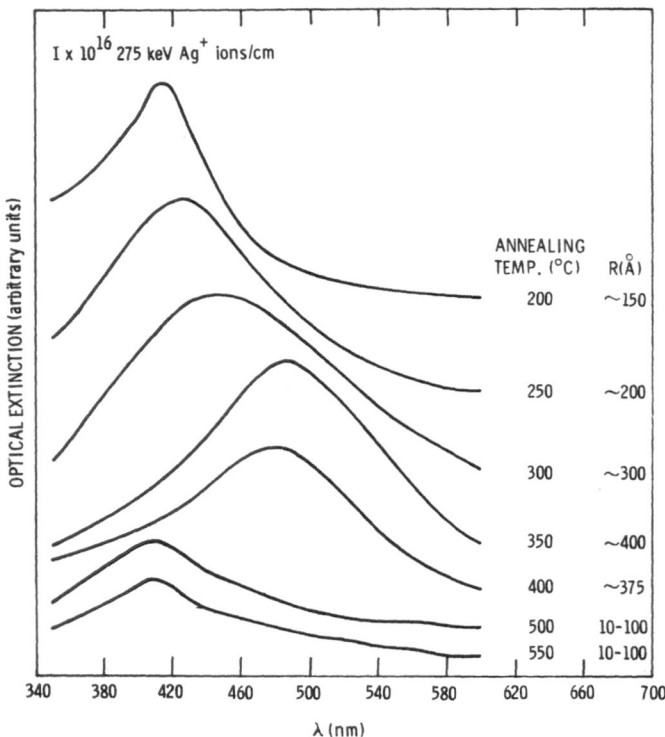

Figure 7. Optical extinction versus wavelength for a lithia-alumina-
silica glass implanted at room temperature with 275 keV 1×10^{16}
Ag/cm^2 and annealed. The optical results at various temperatures
have been arbitrarily offset in the vertical direction for clarity.

using infrared spectroscopy. Using a combination of optical
(visible) spectroscopy and energetic ion backscattering to measure
the depth distribution of the Ag, the reason for the lack of
glass-ceramic formation after Ag-implantation was determined.
Figure 7 shows the optical results as a function of annealing temperature. From the position and the width of the resonance band,
the size of the Ag colloids can be calculated. The optical results
indicate that the colloids grow in size in a temperature regime
between 200 and 350 C and further thermal annealing causes a shrinkage of the colloids. Furthermore the number of colloids present
after annealing at 550 C is much less than that present after room
temperature implantation and 200 C annealing. Either silver is
being lost from the sample or the precipitates are dissolving back
into the glass and the resonance absorption due to the metal colloids
is disappearing.

   The explanation is found in the Rutherford backscattering
results, some of which are shown in Figure 8. As the sample is
annealed, the implanted Ag migrates to the surface of the glass and
most of the silver is on the surface at the temperatures where the
largest colloids are observed. At higher temperatures the Ag is
beginning to diffuse back into the glass, but at very low concentrations and in atomic form. Thus at the temperatures necessary

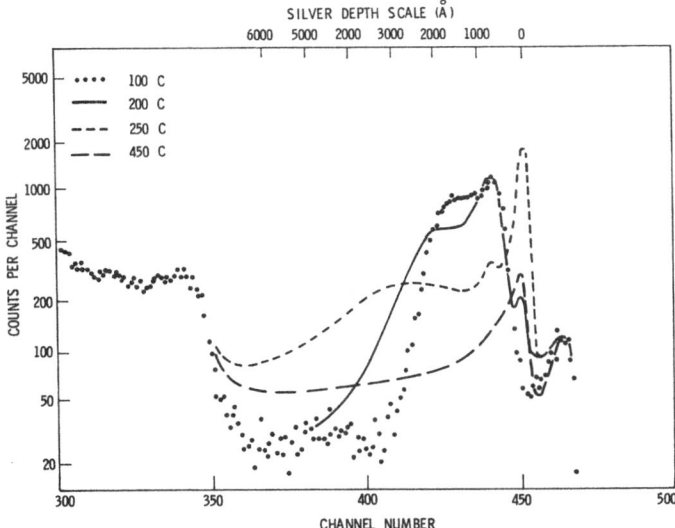

Figure 8. Depth distribution of Ag in the implanted glass from
Figure 7 as determined by Rutherford backscattering after annealing
at various temperatures. The plateau near channel 350 is due to
a potassium component of the glass and the small Ce peak is due to
residual polishing compound on the surface.

for the formation of the glass-ceramic, there are no longer colloids
of a size large enough to nucleate the crystalline phase.

There are many other aspects to the examples cited above and
the discussion of the mechanisms governing precipitation has been,
of necessity, rather brief. This article is intended to acquaint
the reader with the large role played by precipitation processes
in ion-implanted materials and to present a variety of suggested
reading.

REFERENCES

1.   J. A. Borders, Annual Reviews of Materials Science, Vol. 9,
     in the press.

2.   M. Hanson, Constitution of Binary Alloys, McGraw-Hill,
     New York, 1958.

3.   R. P. Elliot, Constitution of Binary Alloys, First Supplement,
     McGraw-Hill, New York, 1965.

4.   F. A. Shunk, Constitution of Binary Alloys, Second Supplement,
     McGraw-Hill, New York, 1969.

5.   P. Duwez, Progress in Solid State Chemistry, Vol. 3, ed. by
     H. Reiss, p. 377, Pergamon Press, New York, 1967.

6.   W. Hume-Rothery, R. E. Smallman and C. W. Haworth, The
     Structure of Metals and Alloys, Institute of Metals, London,
     1969.

7.   S. Mader, A. S. Nowick and H. Widner, Acta metall. 15, 203
     (1967)

8.   J. M. Poate, J. A. Borders, A. G. Cullis and J. K. Hirvonen,
     Appl. Phys. Lett., 30, 365 (1977).

9.   Physical Metallurgy, ed. by R. W. Cahn, North-Holland,
     Amsterdam, 1970.

10. J. Crank, The Mathematics of Diffusion, Oxford Univ. Press,
     London, 1956.

11. S. M. Myers, S. T. Picraux and T. S. Prevender, Phys. Rev. B,
     187, 3953 (1974).

12. J. A. Borders and S. T. Picraux, these proceedings.

13. G. W. Arnold and J. A. Borders, J. Appl Phys. 48, 1488 (1977).

EQUILIBRIUM PHASE FORMATION BY ION IMPLANTATION*

S. T. Picraux

Sandia Laboratories †

Albuquerque, New Mexico 87185

1. INTRODUCTION

This chapter examines the fundamental processes involved in the formation of equilibrium phases in metals by ion implantation. These processes will be described within the framework of conventional solid state behavior and special emphasis will be given to those aspects which are uniquely modified by implantation. It is convenient in this regard to consider three major sequences of events indicated schematically in Fig. 1: (1) the migration of implanted atoms, (2) the formation of precipitates, and (3) the dissolution of the precipitated layer. These processes are dealt with in turn and form the overall outline of the present chapter. The primary objective here is to give a qualitative overview of the processes involved and the degree to which they are understood. For more detailed discussions or surveys of the literature, the reader should consult recent conference proceedings and reviews [1-6].

Temperature is a key parameter in the formation of equilibrium phases by implantation. Implanted atoms are introduced in a statistically dispersed fashion and at sufficiently low temperatures will be immobile. Thus only diffusionless phase transitions would be expected when the appropriate concentrations are reached unless appreciable enhanced migration due to the implanting beam occurs. Upon heating the implanted impurities will

*This work supported by the U. S. Department of Energy (DOE) under Contract DE-AC04-76-DP00789.
†A U. S. Department of Energy Facility.

1. MIGRATION OF IMPLANTED ATOMS

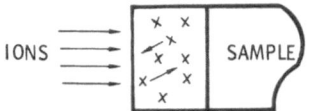

2. FORMATION OF PRECIPITATES

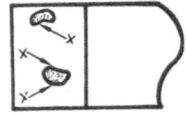

3. DISSOLUTION OF PRECIPITATED LAYER

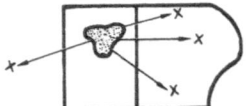

Fig. 1   Schematic representation of the evolution of an
         equilibrium phase formed by ion implantation.

become mobile and may agglomerate to form precipitates of new
phases. Further heating can lead first to growth of the precipitates and finally to their dissolution by migration of the
implanted atoms into the substrate. For implantations at sufficiently high temperature all of these processes may occur
during implantation, however it will be useful here to consider
them in turn to best illustrate the limits of our understanding
of these processes.

     In studies of equilibrium phase formation it is important to
determine directly both the compositional changes and the micro-
structural changes within the implanted layer. Often the technique of ion backscattering is used to determine the composition
vs depth profile and study the kinetics of the profile evolution
with time and temperature, and transmission electron microscopy
(TEM) is used to observe the microstructure and identify the new
phases which are present. This combination of techniques has
proven quite powerful and will be given primary emphasis here. Of
course, other combinations of techniques for the study of
implanted layers such as SIMS measurements for composition vs
depth determinations and glancing angle x-ray analysis for micro-
structural studies may be fruitfully employed.

2.   MIGRATION OF IMPLANTED ATOMS

Since formation of equilibrium phases by implantation usually
requires agglomeration of the implanted atoms, we first consider
the process of implanted atom migration. Of primary concern is
whether and under what conditions normal diffusion-controlled processes will take place. In this regard it is important to consider
atom displacements created by the implantation beam and the close
proximity of the surface. In this section we first present a
case where a conventional diffusion treatment can be shown to
describe the migration process and then give results to illustrate
the importance of complicating factors such as enhanced diffusion
and solute segregation.

2.1   Diffusion

Diffusion from the implanted layer is demonstrated by the
example of Cu implanted into Be [7]. Ion backscattering measurements showed that for peak concentrations ~ 3-4 at.% the Cu depth
profile broadened upon isothermal annealing with no loss of Cu
from the sample or preferential trapping of Cu at the surface. The
observed depth profiles are indicated schematically in Fig. 2a.

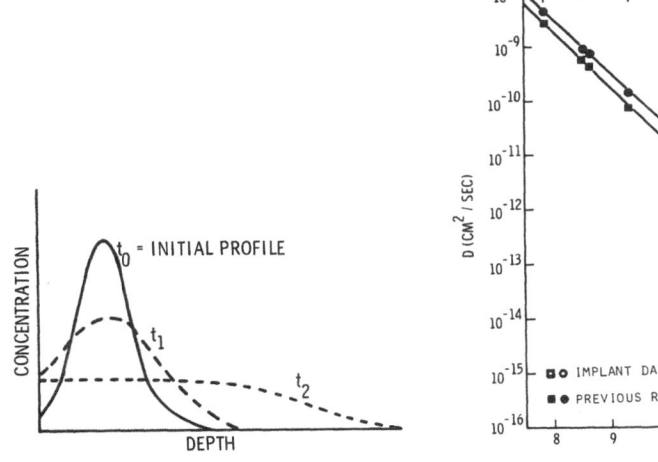

Fig. 2.   The left figure (a) shows a schematic of the solute depth
          profiles after implantation and after subsequent isothermal
          anneals and corresponds to the conditions of Eq. (1). The
          right figure (b) shows the measured diffusivity vs
          reciprocal temperature determined from such profiles for
          Cu implanted in Be. From [7].

Solution of the diffusion equation under a reflecting surface
boundary condition for an initial profile $C_0(x)$ gives

$$C(x,t) = (4\pi Dt)^{-1/2} \int_0^\infty C_0(x') \left[e^{-(x' - x)^2/4Dt} \right. $$

$$\left. + e^{-(x' + x)^2/4Dt} \right] dx' \quad , \tag{1}$$

where t is time and D is diffusivity. The initial condition
$t = t_0$, for an implanted profile can be closely represented by a
Gaussian (or sum of Gaussians) and then the integral is simply a
Gaussian (or sum of Gaussians) $\exp(-x^2/4Dt)$ where t is replaced by
$(t + t_0)$ throughout. The increase in the square of the half width
of the profile of initial width W_0 will go as $(4Dt \ln 2 + W_0)$ and
at long times the profile will look like a half Gaussian emanating
into the bulk sample from the surface. It was shown that once
the Cu had diffused more than twice the implanted range of 900 Å
both the shape of the profile and time dependence of the increase
in width were accurately described by Eq. 1 for a single value of
diffusivity at each temperature. Further, as seen in Fig. 2b the
diffusivities so obtained are in good agreement with previous
measurements using conventional radiotracer and stripping techniques, and allow extension of the data to more than 4 orders of
magnitude lower D values.

Thus the migration of the Cu from the implanted layer is
well described by conventional bulk diffusion. However at very
short times and diffusion distances there were some deviations
from this behavior with initially a slower and then faster spread-
ing of the profile width than predicted. This suggests some trapping and enhanced diffusion effects may occur during early stages
of migration within the implanted layer. Such effects in general
may limit predictive accuracy in modeling new phase formation and
should be checked for experimentally. Trapping effects can be
particularly significant in the case of the motion of light interstitials, since their migration involves lower activation energies
relative to a given trap energy. Also migration occurs at lower
temperatures so that less annealing of implant-induced defects may
have occurred. Enhanced diffusion or mixing becomes most significant at high implant fluxes or fluences as discussed below.

2.2 Enhanced Diffusion

Defects created by the implanted atom collision cascade can
result in an enhancement in the diffusion of implanted atoms. This
may take place during implantation, for example by the creation of

excess vacancies and interstitials. Alternatively, enhancement in
the diffusion within the implanted layer may occur subsequent to
implantation, for example, via vacancies released upon heating or
via pipe diffusion along dislocations created during implantation.
Many studies have demonstrated enhanced migration of previously
implanted atoms by light ion irradiation. However, relatively
little work has been carried out on enhanced diffusion during heavy
ion implantation in metals, this being of primary interest relative
to equilibrium phase formation.

As an example of enhanced diffusion we discuss results from a
detailed study of enhanced Zn diffusion in Al due to Ne bombardment [8]. Displacement rates of the order of one displacement per
atom per second (dpa/sec) resulted in enhancement factors in the
Zn diffusion coefficient of 2 to 4 orders of magnitude. In the Zn
depth profile a tail emerged from the implanted Zn profile to
depths extending ~ 1000 Å beyond the Ne defect generation region.
Results were interpreted in terms of free vacancy and interstitial
induced enhancement in the migration over thermal equilibrium values.
Theoretical calculations [9], as shown in Fig. 3, demonstrated semiquantitative agreement with the observed behavior provided a more

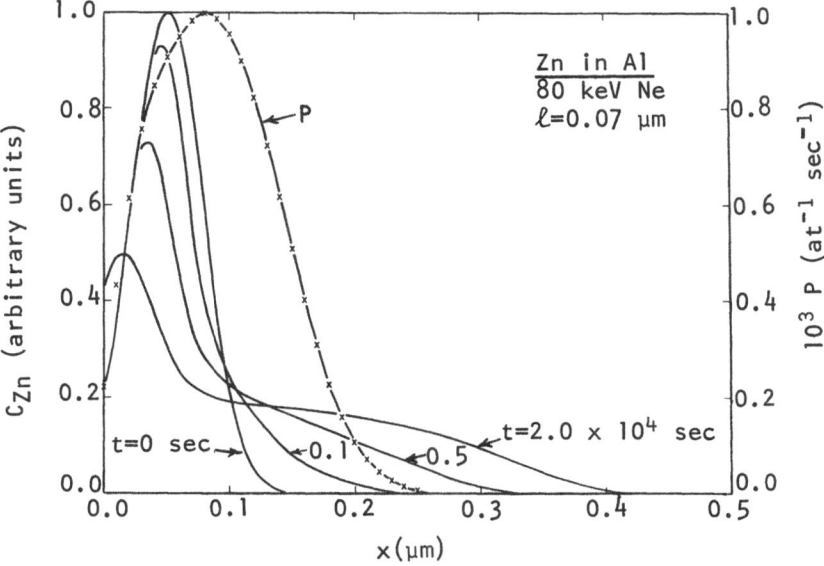

Fig. 3.   Calculated time evolution of implanted Zn depth profile
under Ne—enhanced diffusion. Crosses give Ne damage deposition profile used vacancy—interstitial source. From [8,9].

rapid annihilation rate was used at depths $\leq$ 500 Å. Comparison of experiment and theory indicated that at deeper depths the approximate mean diffusion length for free interstitials and vacancies before annihilation was ~ 1000 Å and that only ~ 1 in 40 vacancies or interstitials escaped annihilation in the primary damage cascade. The observed linear dependence on ion flux indicated that annihilation of mobile defects occurred at a time independent density of fixed sites, as opposed to vacancy-interstitial recombination which would have resulted in a square root dependence on the flux. These results show that such enhanced diffusion processes can be important during implantation and may lead to the formation of precipitated phases at lower temperatures than anticipated from conventional diffusion. However, relatively high displacement rates, (~ 1 dpa/sec) and fluences (~ 10^{17}/cm^2) are required for appreciable effects.

2.3 Solute Segregation and Trapping

There are a number of additional effects which lead to deviations in the implanted atom distribution from that expected for simple implantation and diffusion processes. One case is that which can result when point defects created by the implanting beam are mobile during implantation but interact differently with solute and host atoms. For example, a positive vacancy-impurity binding energy and a net sink for vacancies at the surface of the implanted sample can lead to a net flow of the impurity. This solute segregation results from the flow of point defects in combination with binding of defects to impurities and has been nicely demonstrated for the case of 1 at.% Si in Ni by 3.5 MeV Ni irradiations which resulted in appreciable Si segregation to the Ni surface [10]. Again, relatively high fluences, $\gtrsim$ 1 dpa and elevated temperatures were required to observe the effects.

In contrast to enhancement effects, implanted damage may also reduce migration due to solute trapping and therefore slow the formation of new phases via agglomeration. This can result if defects are immobile and trapping of the solute to the defects occurs. As mentioned previously, this is often more important for the case of interstitial atoms.

The above possible deviations from normal diffusion behavior need to be considered in a detailed microscopic description of the evolution of the implanted layer. However, many of these effects occur only at relatively high temperatures and implantation fluences and often conventional diffusion can give a good first order description of the processes involved.

3.   PRECIPITATE FORMATION

3.1   Nucleation of Equilibrium Phases

The phases are dependent on the composition in the implanted
layer.   Ion implantation allows the near-surface composition of
solids to be adjusted independently of thermodynamic constraints.
This gives great versatility to the implantation technique, not
only in the formation of metastable phases but also with regard
to equilibrium phase formation.   For example, difficulties in reaching equilibrium due to such effects as highly disparate melting
points, persistent intermediate phases, oxide barriers, or strong
tendencies to form competing phases with contaminants such as oxygen
or carbon often can be bypassed by implantation.

The composition one can achieve by ion implantation is
limited by sputtering of the implanted ions.   In the limit, the
maximum implanted atom fraction can be estimated by the recipro-
cal of the sputtering coefficient (typically of the order 3 to 10
for heavy ion implantation) provided transient effects, enhanced
diffusion, or enhanced mixing do not predominate.   Typical concentrations, therefore, are of the order of 10–30 at.% for heavy
metal ion implantation into a monoatomic substrate.   However, this
should not be viewed as an absolute limit, since such processes
as alloying or codeposition to adjust the starting composition, or
enhanced mixing of films by the ion beam may allow other compositions to be achieved [11].

At temperatures where implanted atom mobility is nil a homogeneous mixture of the implanted and substrate atoms is formed.
If the resulting composition lies outside the single phase
region on the phase diagram then precipitation may be expected
upon heating.   For equilibrium phases, consideration of equilibrium phase boundary data can normally be used to predict the
new phases which will form.   However, the occurrence of metastable phases at intermediate temperatures can give rise to
unexpected phases, so that it is preferable to directly confirm
the phases present by transmission electron microscopy or
glancing angle x-ray diffraction.

The intimate mixtures formed by implantation results in
short reaction times.   The reason for short times as well as
the reason that precipitation should occur rather than simply
dissolution of the implanted atoms into the bulk can be seen by
considering the dimensions involved.   In general, an implanted
layer thickness is ~ 1000 Å, whereas the mean distance between
atoms and the distance that atoms must move initially to form a
second phase region are generally $\leq$ 100 Å.   Since the

characteristic time to diffuse a distance $\bar{\ell}$ is the order of $\bar{\ell}^2/D$, then the short distance involved in implanted layers implies very short reaction times. Further, the order of magnitude larger distance atoms must move to leave the implanted layer rather than to form precipitates implies a factor of ~ 100 longer times for dissolution than for precipitate formation.

Available evidence suggests that there is an excess of nucleation sites for new phase formation in implanted systems. This is supported by two points. First, the density of precipitates is observed to be much higher in implanted systems than found in precipitated systems formed by conventional metallurgical treatments. This very high density of small precipitates typically found suggests the presence of many nucleation sites. Secondly, there has been no reported evidence, to this author's knowledge, of the observation of nucleation-limited phenomena in implanted systems regarding equilibrium phase formation. The reason for the high density of nucleation sites is believed to be related to the displacement damage which accompanies implantation. The resulting defects presumably provide ample sites for new phase nucleation.

3.2 Morphology

Relatively little work has been carried out concerning the morphology of implanted equilibrium phases. In general, similar precipitate shapes are expected for implanted systems relative to those formed by other means, since this essentially involves a minimization of the interfacial free energy. Possible deviations from this would be expected primarily when mobilities are sufficient that precipitate formation and evolution can occur during implantation. Under such conditions radiation damage effects on precipitate dissolution and growth could give rise to new morphologies. An example of this was given by high energy Ni irradiation of Ni alloys which had previously been processed to form Ni_3Al precipitates by thermal treatment [12]. The cuboidal Ni_3Al precipitates were observed to form notched corners as they evolved to a new size and density distribution under Ni irradiation at elevated temperatures.

3.3 Precipitate Size Evolution

The high density and small size of precipitates is one of the special features of implant-formed second phase systems and leads to interest in such effects as surface hardening due to the fine dispersion of precipitates. The precipitate sizes for implanted systems can range from atomic dimensions to the thickness of the implanted layer and a number of implantation parameters have been found to strongly influence the size distribution. Three major

parameters are the implantation fluence, flux, and temperature.
It has been found that the mean size of precipitates increases and
the number density decreases with increasing implantation fluence,
increasing implantation temperature, and decreasing implantation
flux [13].

The evolution of the mean precipitate size and size distribution can be understood in terms of microscopic growth and dissolution processes. Modeling of the precipitate size evolution has been
carried out and has demonstrated recently observed experimental
trends [13,14]. However, little systematic or quantitative experimental work has been carried out and a predictive capability for the
the precipitate size distribution does not yet exist. In the remainder
of this section, an overview of such modeling and the processes which
affect precipitate growth and dissolution will be discussed. Then
the AlSb system formed by Sb implantation into single crystal Al will
be used to illustrate the various points of this section on precipitate formation.

We first consider the evolution in size of a single precipitate
of radius R at time t as given approximately by

$$\frac{dR}{dt} = \frac{D}{NR}\left[C - C_o \exp\left(\frac{2\sigma v}{RkT}\right)\right] - \frac{FS}{4n}\left[1 - \frac{R\exp(-\varepsilon/\xi)}{R + \varepsilon}\right]$$

$$+ \frac{F}{4N}\left[1 - \exp(-R/a)\right] \qquad (2)$$

where D is the diffusivity, N the number density of solute atoms in
precipitates, C the atomic concentration, C_o the solid solubility,
σ the precipitate–matrix interfacial energy, T the temperature, F
the incident ion flux, S the number of ejected atoms from the precipitate per incident atom (which may be thought of as an internal
sputtering from the precipitate into the matrix), ε the mean range
of the ejected atoms and ξ the mean distance between precipitates [14]. The first term on the right–hand side of the equation
is essentially a thermal one which accounts for the growth of the
precipitate via collection of atoms out of solution. The second
term accounts for recoil dissolution of precipitates due to the
implanted atoms and the third term approximately accounts for
growth by direct capture of implanted atoms within the precipitate
as they come to rest.

An ensemble of precipitates will not have a single size but rather a
distribution of sizes, the evolution of which can be described by

$$\frac{\partial}{\partial t} N(R,t) = \frac{\partial}{\partial R}\left[\frac{dR}{dt} N(R,t)\right] \qquad (3)$$

where $N(R,t)dR$ is the number of precipitates between radius R and
R + dR at time t. Equation 2 can be used to estimate the mean precipitate size either during or subsequent to implantation, whereas
Eqs. 2 and 3, together with the auxiliary equation for the total
implanted atom concentration, describes the evolution of the total
precipitate size distribution. The latter treatment is also needed
to describe the observed change in number density of precipitates
with time. The above equations do not include the variation in
concentration with depth or loss of solute due to dissolution into
the bulk.

We now consider qualitatively the major features in precipitate evolution [14,15]. In the purely thermal evolution part, as
described by the first term in Eq. 2, large precipitates may grow
at the expense of smaller ones, resulting in a net reduction in
the total number of precipitates as well as an increase in the
mean size. The driving force for this well-known phenomena,
referred to as Ostwalt ripening in conventional metallurgical
studies, is the reduction in energy achieved with decrease in surface to volume ratio due to the interfacial free energy of the precipitate-matrix boundary. Although this phenomena does not require
the presence of implantation, it might be expected to be important
in implanted systems due to the large number of very small precipitates formed and corresponding high energy of the system.

In addition, there are several aspects unique to the implant
process in the evolution of precipitates. First, one may consider
simply the effect of solute addition and collection at the second
phase for the case of precipitate growth during implantation. Even
in the absence of thermal ripening this would cause a given distribution of precipitates to increase in radius with implant time if
loss of the implanted atoms from the implanted layer is not appreciable. The second aspect unique to implantation is the enhancement in
precipitate dissolution by the implanting beam. One possible mechanism for this is the direct scattering of atoms out of the precipitate back into the matrix, some fraction of which will escape via
diffusion as given approximately by the second term of Eq. 2. Some
additional enhancement of precipitate dissolution may result from
disordering within the precipitate and a resulting increased ease
of precipitate dissolution. However, these and other possible
effects due to the implantation are not well understood at present.
More detailed quantitative work on the precipitate size evolution
is needed to develop an increased understanding of the radiation
dissolution processes. Also, systematic measurements of the precipitate size evolution can lead to a direct measure of the interfacial

free energy of the precipitate-matrix boundary and relatively few determinations have been made of this quantity by any means [17].

The discussion of precipitate formation and evolution will be illustrated by Sb implantation studies in single crystal Al [13,14]. In this system the Sb mobility is nil at room temperature and becomes appreciable above ≈ 120°C. The binary phase diagram indicates only the fcc phase Al, the hcp phase Sb, and the diamond lattice line compound AlSb are allowed in equilibrium. The solid solubility for Sb in Al is quite low (<<0.1 at.% at temperatures

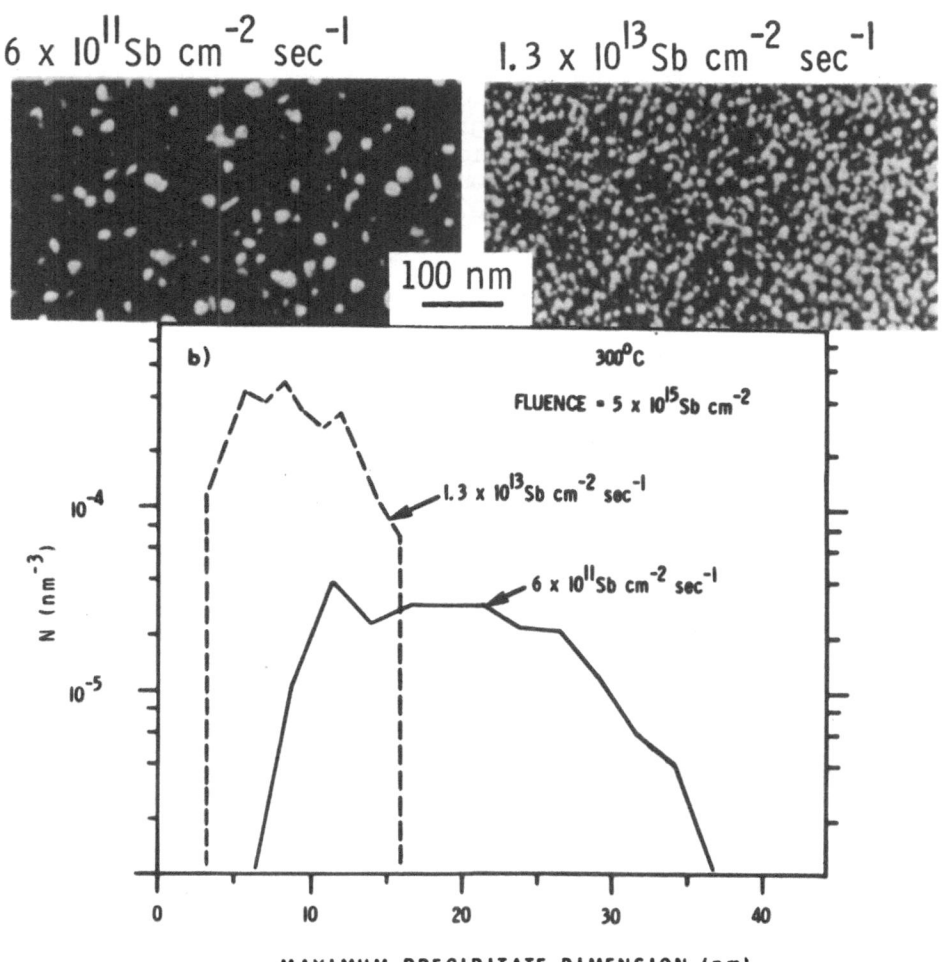

Fig. 4.   Dark-field micrographs and corresponding size distributions of AlSb precipitates as a function of implantation flux for 300°C, 50 keV Sb implants in Al.   From [13].

of interest here). Thus for typical implanted concentrations of
1-20 at.%, one would expect to be in a two-phase region composed
of Al and AlSb precipitates once sufficient mobility of the Sb was
allowed. Consistent with this, triangular AlSb platelets of a
fixed orientation with respect to the Al crystalline matrix were
always observed once the implant or anneal temperature was sufficient to allow the Sb to migrate distances $\gtrsim$ 100 Å. Also there
was some indication that at sufficiently high fluxes the implant
temperature at which the precipitated phase can be formed could be
lowered to near room temperature [16], presumably due to enhanced
diffusion. The precipitate sizes were seen to increase and the
number density decrease with increasing fluence, increasing temperature, and decreasing flux. In Fig. 4 examples of dark-field
micrographs of the AlSb phase and the size distributions determined
from these are shown as a function of flux with all other parameters
held constant. Semiquantitative agreement with the observed behavior was obtained by model calculations using Eqs. 2 and 3. Also it
was shown that AlSb precipitates could be disordered by additional
ion bombardment at room temperature.

4. DISSOLUTION OF PRECIPITATED LAYER

4.1 Dissolution Process

With continued heating at temperatures where implanted atoms
are mobile the implanted layer will dissolve into the sample substrate. In general, the substrate is much thicker than the implanted layer and the final equilibrium state implies a very low concentration of the implanted atoms dispersed throughout the sample.
While this final state is of little interest, the dissolution process often involves a quasi-equilibrium between the implanted multi-
phase region and the substrate. Studies of this process provide
information valuable to metallurgical applications [4,5].

The establishment of quasi-equilibrium between the implanted
layer containing precipitates and the substrate requires that the
flow of implanted species from precipitates into solution within
the layer be greater or equal to the flow from the implanted layer
into the bulk. This latter term is determined by the product of
the diffusivity and the solubility at the temperature of interest.
This process has been described analytically [18] via the diffusion equation with an appropriate source-sink term for the precipitates and the auxiliary equation for the rate of change of
precipitate radius (simplified Eq. 2) as

$$\frac{\partial}{\partial t} R(x,t) = - \frac{D}{NR(x,t)} \left[C_o - C(x,t) \right] \qquad (4)$$

$$\frac{\partial}{\partial t} C(x,t) = D \frac{\partial^2}{\partial x^2} C(x,t)$$

$$+ 4\pi D \ R(x,t) \ N_p(x) \left[C_0 - C(x,t) \right] \qquad (5)$$

where C is the concentration in solution at depth x and time t, D is the diffusivity, R is the precipitate radius, N_p is the number of precipitates per unit volume, C_0 is the solubility, and n is the number density of solute atoms in the precipitates. Under the simplifying assumptions used here of a single radius for all the precipitates and a small radius compared to the average precipitate spacing this model allows quantitative comparison with experiment and has been shown to give a good description of the implanted layer dissolution process for the case of Sb implanted in Fe [18].

We will now illustrate the use of implanted layer dissolution to obtain such fundamental information as diffusivity, solubility, ternary phase boundaries, and solute trapping. In such studies quasi-equilibrium was established as determined by the kinetics of the dissolution process.

4.2   Diffusivity and Phase Boundary Determinations

Consider the example of Cu implanted in Be, which in 2.1 was shown to provide a way of making accurate measurements of the diffusivity at low Cu concentrations. For implants to higher Cu concentrations ($\sim$ 10-20 at.%) the new equilibrium phase Be_3Cu will precipitate within the implanted α-Be layer [19]. Continued heating establishes a quasi-equilibrium between the implanted (two phase) layer and the (one phase) substrate. The Cu concentration at this boundary is determined by the solid solubility of Cu in Be and therefore remains constant with dissolution of the implanted layer for as long as the quasi-equilibrium conditions are held. The diffusion equation can be solved for a constant source giving rise to a complementary error function. As shown in Fig. 5, this analytic treatment gives good agreement with the shape of the Cu profile below the implanted layer and allows extraction of both the solid solubility and diffusivity without treatment of the evolution of the precipitates within the two-phase implanted layer [19]. The results are in good agreement with the diffusion measurements described in 2.1 and also allow solid solubilities to be determined down to quite low temperatures.

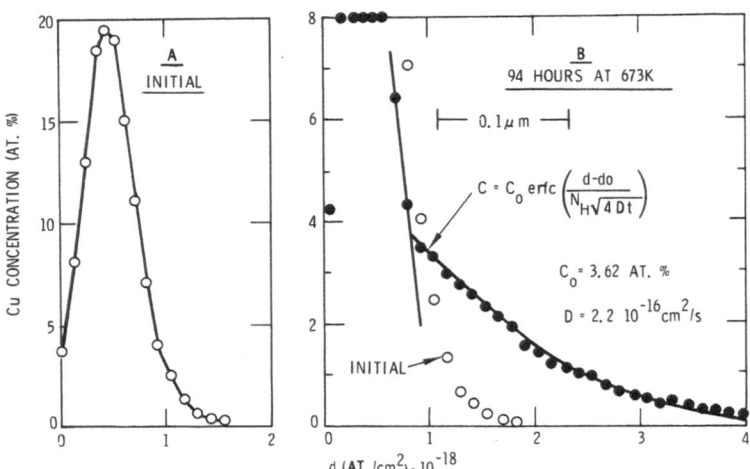

Fig. 5.   Depth profile of Cu implanted into Be to concentrations
           above the solid solubility before and after annealing to
           400°C for 94 hrs. Right hand side of annealed curve
           (solid line) is given by indicated equation. From [20].

        In the Be(Cu) case the surface acted as a reflecting bound-
ary layer and a simple treatment was quite adequate. For more
complex cases such as Sb implanted in Fe where both evaporation of
the Sb from the Fe surface and dissolution into the bulk occurs
upon heating, it is necessary to carry out a more complete treatment of the entire implanted layer such as given by Eqs. 4 and
5 [18]. In Fig. 6 an example is shown of the calculated and measured Sb profile with the shorter dashed line corresponding to Sb
in solution and the peak above the dashed line to the Sb in FeSb
precipitates [18]. Good agreement between theory and experiment
was obtained and the calculated precipitate radius of 91 Å for the
conditions of Fig. 6 was consistent with TEM observations. Such
analytic descriptions are quite valuable and allow the solubility
and diffusivity of the implanted solute to be determined even in
the presence of strong surface evaporation.

        These applications of precipitate evolution theory to quan-
titative studies of dissolution of the implanted layer are particularly valuable for determining metallurgical parameters [4,5].
For example, the low temperature solubility and diffusivity data

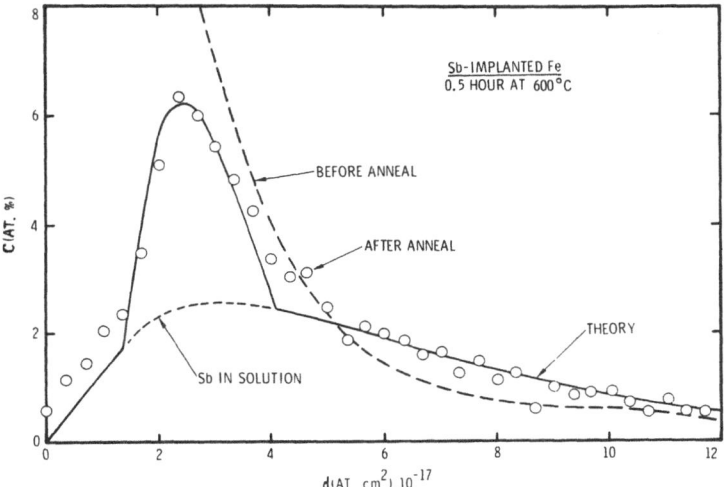

Fig. 6.   Depth profile of Sb implanted into Fe to concentrations
above the solid solubility before and after annealing
to 873 K for 0.5 hrs.   From [18].

for Cu in Be were of interest to alloy processing since Cu is used
to solution strengthen Be alloys.   Also, these techniques by Myers
and coworkers have been extended to ternary systems such as Al and
Fe implanted in Be to determine the Be rich region of the ternary
phase diagram [20].   In this way they have obtained previously
unavailable phase boundary information.   In this case the very
stable ternary precipitate $AlFeBe_4$ is formed which strongly reduces
the solubility of Fe in Be when Al is present.   This is relevant
to the control of unwanted Al impurity accumulation at grain boundaries where it affects the alloy ductility.

4.3 Multiple Layers and Trapping Effects

More elaborate use of the dissolution process in the implanted
layer can be made by the study of implanted multiple layers or of
trapping of implanted species in precipitated layers.   For example,
it has been shown that, by implanting Al and Fe to different depths
in Be, the composition range of the precipitate phase $AlFeBe_4$ can
be obtained [20].   Thus information away from the α–Be (substrate)
phase boundary is obtained by use of multiple layers to establish
a quasi–equilibrium between interior regions of the ternary phase
diagram.

     Trapping of solute atoms can also be studied by first implanting
and thermally processing the implanted layer to form a precipitate
region and then implanting the solute and observing its migration
in the presence of this precipitated region. For example, this
can provide very direct information on the possible use of second
phase formation to reduce the transport of embrittling species in
alloys via trapping at precipitates. A very nice demonstration of
the application of such work has been made for the case of Ti
implanted into Fe to form TiC precipitates which were observed to
provide strong traps for implanted Sb, presumably at the TiC-Fe
matrix interface [21]. Motivation for such studies lies in the fact
that Ti is an alloy addition which reduces temper embrittlement in
steels due to common low concentration impurities Sb, P, As, and
Sn, and the mechanism for this reduction in the embrittlement rate
by the addition of Ti was not previously understood.

                           5.   SUMMARY

     A rather accurate picture exists for the formation of equilib-
rium phases by implantation, their evolution and subsequent dissolution. While quantitative treatment can be given to the overall
evolution of the concentration profile in the formation and dissolution of a two-phase precipitated region, details of the precipitate growth and dissolution process required for a predictive capability for the precipitate size distribution remain to
be understood. In this regard the application of hyperfine techniques may well be a nice complementary tool since hyperfine measurements are microscopically sensitive to new phase formation.

     A particular area of interest area in the formation of equilib-
rium phases by implantation is its application to form microalloys
and study their evolution and physical characteristics. An important use of this has been in the area of implantation metallurgy,
where valuable metallurgical parameters such as diffusivity, solubility, phase boundary, and trapping data can be obtained for complex
systems and at low temperatures. An additional value in the microscopic understanding of implanted new phase formation is its application to the understanding of new physical and chemical properties
obtained by implantation, such as increased hardness, wear resistance, corrosion resistance, or fatigue resistance.

                         REFERENCES

[1]. Proc. Int'l. Conf. on Ion Beam Modification of Materials,
      Budapest, 1978 (to be published by Rad. Effects).
[2]. Ion Implantation in Semiconductors, Ed. by F. Chernow,
      J. A. Borders, and D. K. Brice (Plenum Press, NY, 1977).

[3]. Applications of Ion Beams to Materials, Ed. by G. Carter, J. S. Colligan, and W. A. Grant, Conf. Series No. 28 (Institute of Physics, London, 1976).

[4]. S. M. Myers in Ref. 1 and in Ref. 2, p. 167.

[5]. S. M. Myers, J. Vac. Sci. Technol. 15, 1650 (1978).

[6]. S. T. Picraux, in Ref. 3, p. 183.

[7]. S. M. Myers, S. T. Picraux, and T. S. Prevender, Phys. Rev. B9, 3953 (1974).

[8]. S. M. Myers and S. T. Picraux, J. Appl. Phys. 46, 4774 (1975); S. M. Myers, U. S. Energy Research and Development Administration Report No. CONF-751-006, 1063 (1976).

[9]. S. M. Myers, D. E. Amos, and D. K. Brice, J. Appl. Phys. 47, 1812 (1976).

[10]. L. E. Rehn, P. R. Okamoto, D. I. Potter, and H. Wiedersich, J. Nucl. Materials 74, 242 (1978).

[11]. Z. L. Liau and J. W. Mayer, J. Vac. Sci. Technol. 15, 1629 (1978).

[12]. J. E. Westmoreland, P. R. Malmberg, J. A. Sprague, F. A. Smidt, and L. G. Kirchner in Ref. [2], p. 181.

[13]. R. A. Kant, S. M. Myers, and S. T. Picraux, J. Appl. Phys. 50, 214 (1979).

[14]. R. A. Kant, S. M. Myers, and S. T. Picraux in Ref. [2], p. 191.

[15]. R. S. Nelson, Applications of Ion Beams to Metals, Ed. by S. T. Picraux, E. P. EerNisse and F. L. Vook (Plenum Press, NY, 1974) p. 221.

[16]. P. A. Thackery and R. S. Nelson, Philos. Mag. 19, 169 (1969).

[17]. L. E. Murr, Interfacial Phenomena in Metals and Alloys, (Addison-Wesley, Reading, Mass., 1975) p. 155.

[18]. S. M. Myers and H. J. Rack, J. Appl. Phys. 49, 3246 (1978).

[19]. S. M. Myers and J. E. Smugeresky, Metall. Trans. A8, 609 (1977).

[20]. S. M. Myers and J. E. Smugeresky, Metall. Trans. A7, 795 (1976).

[21]. S. M. Myers, D. M. Follstaedt, and H. J. Rack, Appl. Phys. Lett. 33, 396 (1978).

FORMATION OF NONEQUILIBRIUM SYSTEMS BY ION IMPLANTATION[*]

S. T. Picraux

Sandia Laboratories[†]

Albuquerque, New Mexico 87185

1. INTRODUCTION

This chapter surveys nonequilibrium phase formation by implantation [1-3]. Such crystalline or amorphous metastable systems are
characterized by not being in the lowest energy state and can be
formed by the athermal process of implantation. Crystalline
systems may be characterized by metastable solid solutions or by a
metastable structure which may, for example, be stabilized by the
implanted impurities. Amorphous systems are characterized by no
long range order and their formation and return to equilibrium sometimes involves metastable intermediate crystalline phases. Upon
heating the metastable phases will return to equilibrium either
via diffusion, as in the case of phase separation, or via diffusionless means, as in the case of purely structural transformations.

2. METASTABLE CRYSTALLINE SYSTEMS

2.1 Metastable Solid Solutions

Both substitutional and interstitial solid solutions at compositions not allowed by equilibrium phase diagrams have been formed
by ion implantation [1,2]. The ion channeling technique of determining the lattice location of the implanted species has been
found to be a convenient way to characterize such systems. Figure 1

---
[*]This work supported by the U. S. Department of Energy, DOE, under
 Contract DE-AC04-76-DP00789.
[†]A U. S. Department of Energy Facility.

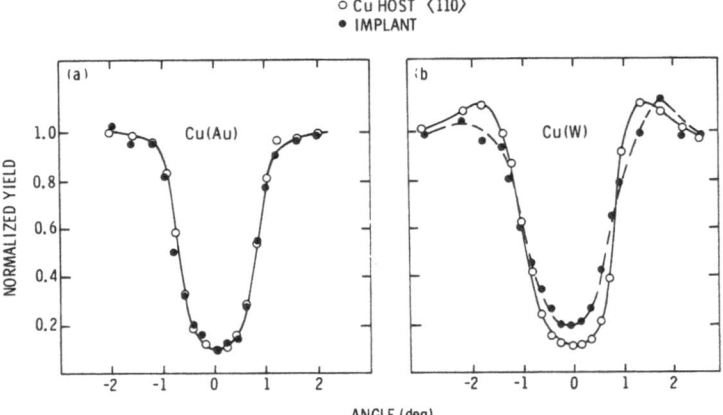

Fig. 1   Channeling angular distributions for Au and W implanted
         in Cu. From [4].

shows the channeling angular scans for both an equilibrium system,
a) Cu(Au) and a metastable substitutional solid solution, b) Cu(W),
where the implanted fluences correspond to ~ 1 at.% [4].   The close
correspondence of the dip for both the Cu host and implanted species
indicates ≈ 100% substitutionality for the equilibrium Cu(Au)
system and ≳ 90% substitutionality for W in Cu.   The Au goes 100%
substitutional even for implants at 15 K, suggesting the final state
is reached via a nonequilibrium process, since no long range migration should occur at these temperatures and the replacement collision probability for Au on Cu is only ~ 80%.   It has been pointed
out [4] that this may imply the energy spike of the collision cascade [5,6] allows the system to relax to its lowest <u>local</u> energy
state which corresponds to Au on Cu lattice sites.   W is insoluble
in Cu according to the binary phase diagram but formation of the
metastable solid solution by implantation suggests that the lowest
local energy configuration appears to be on substitutional sites.
Again the energy spike presumably provides the mechanism to achieve
this.   Heating of the W-implanted Cu system to temperatures where
the W atoms can migrate allows W-W interactions to give rise to
precipitation thereby returning the system to an equilibrium state.

A related study of the Ag-Cu system showed implantation can
form substitutional solid solutions similar to that formed by splat
cooling [1,2].   The Ag-Cu system is a classic eutectic system
with very limited mutual solutilities, even though there is relatively little difference in atomic radii of Cu and Ag.   This
suggested that there might exist a metastable solid solution over
the complete composition range and was directly demonstrated by

splat cooling. Similar results were obtained by implantation for
Ag into Cu for implant concentrations of Ag up to ~ 20 at.%. It may
be pointed out that the local energy spike formed by the collision
cascade may be thought of in terms of a temperature of the order of
10^3 K and with anticipated relaxation times of this temperature into
the lattice of ~ 10^{-11} sec this corresponds to quench rates of
~ 10^{14} K/sec. Thus, in some respects implantation may be considered
as an even faster quenching process than that of splat cooling where
typical cooling rates for the splat cooled droplets are 10^9 K/sec.

Metastable interstitial solid solutions can also be formed by
implantation. The best examples of these are given in the extensive studies of the lattice location of implanted species in Be [7].
It has been shown that substitutional solutions, as well as tetrahedral and octahedral interstitial solutions, can be formed in Be to
concentrations exceeding the equilibrium phase boundaries.

2.2 Metastable Structures

Phase transformations have been observed in thin metallic films
Ni, Fe, Mo, and Ti implanted with reactive species C, O, N, and
with the inert gas species He and Ar [8]. Structural transformations
from bcc to fcc, fcc to hcp, and hcp to fcc were reported and
interpreted as due to the accumulation of radiation defects in the
film. However, it is likely that these phases are in part stabilized
by the implanted or residual impurities present in the film. Chemical reactions to form carbide, oxide or nitride phases may also have
occurred. In general, new compositions and new structures not found
in equilibrium phase conditions may be quite readily formed by
implantation. However much work will be required before a general
view of such systems can be obtained, even for the binary case.

2.3 Binary Alloying Rules

One step in understanding metastable alloys is to first consider the alloying rules for the formation of binary equilibrium
phases. Unfortunately no general theory of binary alloys exists,
although a number of empirical rules have emerged. The Hume-Rothery
rules essentially indicate that complete solid solutions are
expected for a binary system if the elements have the same crystal
structure and only small size and electronegativity differences [9].
For example, in a Darken-Gurry plot a second atom is expected to
form a binary alloy with an element if it lies within a circle or
ellipse of axes ± 15 at.% radius difference and ± 0.4 electronegativity difference. In Fig. 2 an example of a Darken-Gurry plot
for implanted elements in Cu is shown which systematizes channeling results according to whether dilute substitutional solid

solutions are formed [10]. Modified Hume-Rothery rules extending the allowed atom size differences to -15/+40% and electronegativity differences to ± 0.7 (dashed line) gave a better partition of sub-stitutional versus nonsubstitutional character than did the conventional rules (circle). This was then shown to apply reasonably well for available data in Fe and Ni hosts. However for Al the conventional Hume-Rothery rules gave better agreement [10].

Recently a new alloying rule has been given by Miedema and co-workers in which two thermochemical parameters for the elements are used to correctly predict the sign of the heat of formation for almost all of the more than 500 known binary alloy cases [11]. These parameters are ϕ^*, the electronic work function, and n_{ws}, the elec-tron density in a metal at the Wigner-Seitz cell boundary. The change in ϕ^* essentially corresponds to a charge transfer between dissimilar cells, which gives a negative contribution to the heat of formation, and the change in n_{ws} implies a need to provide energy to smooth a discontinuity in the electron density at the cell boundary, which gives a positive contribution to the heat of formation. It is difficult to assess the quality of this empirical approach based on a maximization of a fit of two parameters for 50 elements in a prediction of the sign of the heat of formation for 500 binary com-binations. However, it does appear to be more successful than pre-vious rules and has been applied to a variety of other alloy data

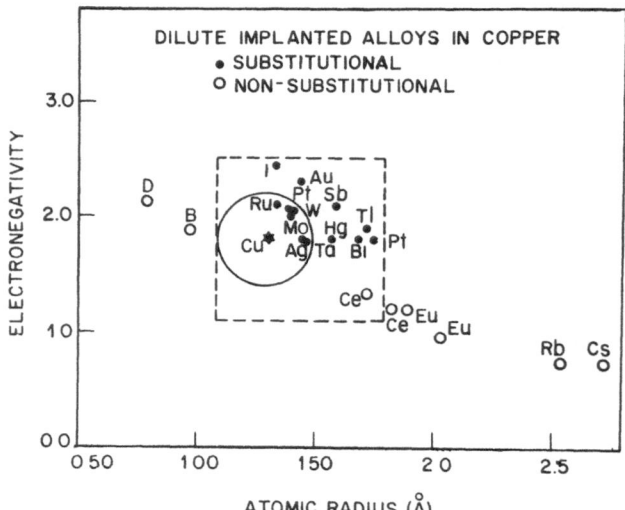

Fig. 2.   Darken-Gurry plot of implanted element location in Cu for
          dilute concentrations. Circle is Hume-Rothery zone and
          dashed rectangle is Sood's modified Hume-Rothery zone.
          From [10].

with some very encouraging results. Recently this has been applied
to provide a consistent description of the substitutional and interstitial solutions which can be obtained by implantation of 30
different elements into Be [7]. A modified Landau–Ginsberg expansion
of the site energy differences was used to determine the boundaries
of the site region and, as seen in Fig. 3, the plot in terms of the
two Miedema thermochemical parameters allows partitioning into the
three regions of substitutional, octahedral interstitial, and
tetrahedral interstitial behavior. This raises a very interesting
question of whether the Miedema parameters might be used not only to
obtain some predictive information about equilibrium phases, but
also to provide some empirical guidelines for predicting the formation of metastable systems.

The above approaches serve to systematize observations but give
little understanding concerning the microscopic processes involved in
metastable phase formation by implantation. In addressing such an
understanding in implanted alloys a number of aspects associated with

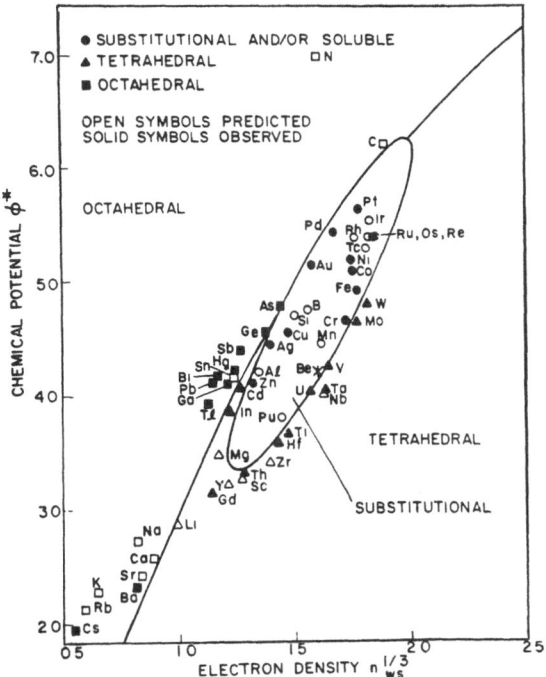

Fig. 3. Plot of Miedema parameters for implanted elements in Be as
function of lattice location. Calculated site boundaries
(solid line) based on site energy differences. [From 7].

the implantation processes need to be considered. For example, it is
important to: 1) determine the relative role of replacement collisions in forming substitutional solutions; 2) assess the importance
of defect association, for example in Al; 3) understand in what sense
one is probing a local equilibrium by implantation; and 4) to more
fully understand the role of the energy spike in this regard.

3. AMORPHOUS SYSTEMS

In this section we first give an overview of amorphous metallic systems, including their characteristics, formation, and methods
of characterization. Then a series of selected examples are given
of amorphous systems which have been formed by ion implantation to
elucidate relevant features. Finally a brief discussion is given.
of the current understanding of such systems.

3.1 General Characteristics

In Table I primary methods of amorphous phase formation are summarized. Amorphous materials would be expected to be in a higher
energy state than their corresponding ordered condition and the
processes for forming such a metastable state have a common property

Table 1. Selected Methods of Amorphous Phase Formation

Method	Process	Form	Quench Rate (K/sec)
Condensation of vapor phase	Co-evaporation or sputtering onto cold surface	film	$\sim 10^{15}$
Quench of liquid phase	Stream of molten metal onto cold rotating drum	ribbon	$\sim 10^6 - 10^9$
Splat cooling	Molten puddle onto cold surface	slug	$\sim 10^9$
Laser glazing	High power heat pulse onto surface	surface layer	$\sim 10^9$
Ion implantation	Athermal atom introduction with "energy spike"	surface layer	$\sim 10^{14}$

ct involving a rapid quench. As a result, the resulting amorphous
regions of material are in a form with limited bulk dimensions. The
laser glazing and ion implantation methods are related in that they
are surface layer treatments and in the future possible combinations
of ion implantation with laser or electron beam pulsed annealing may
further link these approaches.

Amorphous metallic systems can be considered to fall into several
broad classifications including transition metal + metalloid (NiP),
transition metal + rare earth (NiDy), noble metal + metalloid (AuSi),
and pure metals. The transition metal plus metalloid class is one
of the most common and the small metalloid elements such as P and Si
are noted by their glass forming character. Also, there is still
some question as to whether pure metals may form amorphous systems
and, although they have been reported in the amorphous state, there
is the likelihood that they may have been stabilized by impurity
contamination.

A variety of interesting material properties are exhibited by
amorphous alloys. For example, mechanical strength can be quite
high and they also may exhibit hardness and ductility properties of
interest. Often the corrosion resistance is very good, even under
aqueous environments and stress corrosion conditions. Amorphous
alloys can exhibit magnetic ordering even though they are struc-
turally nonordered. They often exhibit uniaxial magnetic anisotropy
and hold applications potential because of such characteristics as
fast switching for magnetic bubble applications and low hysteresis
losses for transformer applications. Also they can be made at low
cost and are quite stable. For example $Fe_{0.8}B_{0.2}$ is estimated to
last 20 years at 200°C before detectable changes in magnetic properties occur.

Various techniques have been used to characterize amorphous
alloys, as summarized in Table 2. The x-ray diffraction technique
has been one of the most important, since by analysis of the radial
distribution intensity function it is possible to obtain information about atom spacings. The result of such studies indicates that
amorphous alloys are characterized by only weak correlations in
atom spacings for ≥ 4 atomic diameters, and that in general nearest
neighbor spacings are typically $\leq 1\%$ larger, but with greater variation, whereas next nearest spacings may be up to 10 or 15% larger
than for the corresponding crystalline phase. Also, the width of
the diffuse halo which characterizes the diffraction pattern of
amorphous systems sets an upper limit to the size of crystallites, if any, which may be present. Typically, these set size
limits of 5 to 10 Å. Electron diffraction is quite convenient to
use for implanted systems, but gives less precision than x-ray techniques. To this author's knowledge only electron diffraction and
ion channeling results have been reported at present in the characterization of implant-formed amorphous systems.

Table 2. Various Methods of Characterizing Amorphous Alloys

Method	Comments
Diffraction - x-rays or electrons	Characteristic diffuse rings. Radial distribution gives atom spacings
Transmission electron microscopy	Observe crystallites or modulation (e.g., spinodal decomposition)
Channeling	Loss of single crystal ordering
Resistivity annealing	Resistivity change at amorphous to crystalline transition
Density	Amorphous metals are closely packed
EXAFS	Local coordination

3.2 Implant-Formed Amorphous Systems

Implantation has only recently been used to form metallic amorphous systems. Therefore we only discuss several examples of
implant-formed systems in this section to illustrate the type of
behavior observed. For detailed experimental discussions or literature surveys the reader is referred to several recent review
articles [1-3].

Copper: We earlier showed how implantation of ~1 at.% W into
Cu results in the formation of a metastable substitutional solid
solution. As the implantation fluences are increased to achieve a
local concentration of 10 at.% a transition into an amorphous alloy
occurs. In Fig. 4 this is suggested by the loss in the channeling
dip for the W signal at the highest fluence corresponding to concentrations above 10 at.% [4]. Similar behavior was observed for Ta
implantation into Cu and the presence of amorphous phase was indicated in both cases by the observation of diffuse rings in the electron diffraction pattern [12]. In the case of the Ta implant the
width of the strongest diffuse peak indicated a maximum crystallite
size ≤ 6 Å and this amorphous phase was stable up to $\approx 600°C$. At
present, there is no evidence that the amorphous phase can be
formed in a pure metallic system by self ion irradiation, experiments
having been reported for Cu, Ni, and Fe. For example, implantation
at 40 K to fluences of $10^{17} Cu/cm^2$ into Cu gave rise to only a small
increase in the channeling signal [4].

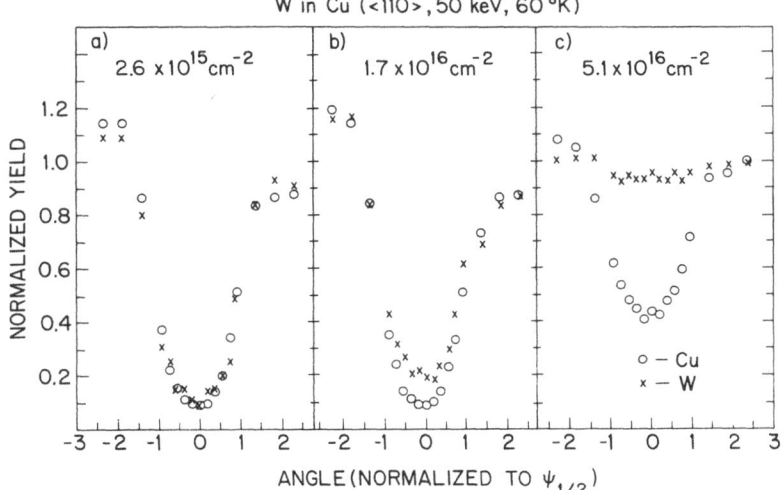

Fig. 4.   Channeling <110> angular distributions for a) 2.6 x 10^15,
          b) 1.7 x 10^16, and c) 5.1 x 10^16 W/cm^2 implanted into Cu
          corresponding to transition from metastable substitutional
          solution in a) (dips coincide) to amorphous alloy in c)
          (no W dip). From [4].

     Nickel: Phosphorus implants into polycrystalline Ni resulted
in structural transformations, first from fcc to a combination of
fcc and hcp Ni, and finally at higher fluences an amorphous Ni(P)
phase [3]. The metastable hcp phase is oriented such that the
closed packed directions for the fcc and hcp phases are parallel.
In Fig. 5 the diffraction intensity distribution as a function of P
implant fluence is shown. This illustrates the transition from
sharp diffraction lines for polycrystalline material to broad diffuse lines for amorphous material. The position of the peaks in
the scattering angle parameter S after 10^{17}P/cm^2 corresponds
closely with that of amorphous material of the same composition
(~ Ni$_{.75}$P$_{.25}$) formed by electrodeposition (not discussed in
Table 1). The position of the diffraction peaks may be used to compare various structural models for amorphous alloys. In Table 4
several of the structural models are summarized. In the case of the
Ni(P) system, it has been shown that the ratio of the first two diffuse peak positions for both ion implanted and electrodeposited
material agreed and were more consistent with the dense random packing model than with the microcrystalline model. In general, the
dense random packing model appears to be currently favored for
describing amorphous metallic alloys.

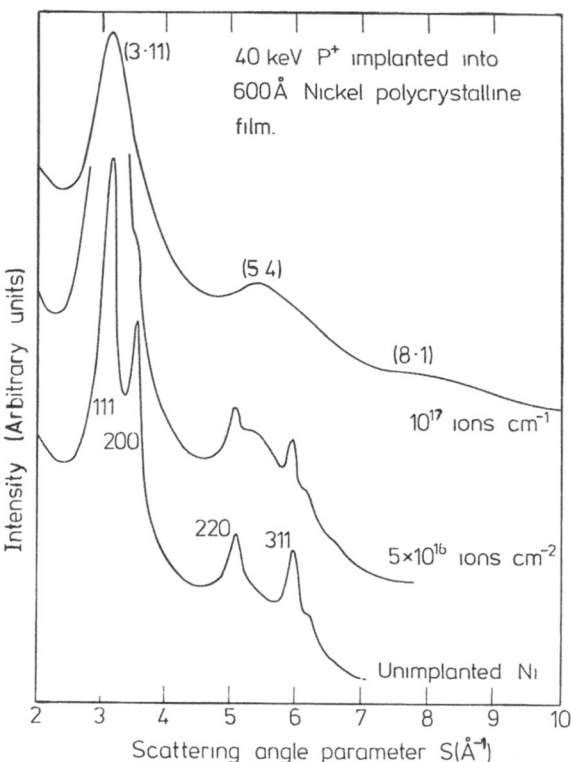

Fig. 5.   Electron diffraction intensity vs scattering angle param-
          eter S = 4πsinθ/λ for Ni as a function of P implant fluence
          showing transition from polycrystalline (unimplanted) to
          amorphous (10^{17}P/cm^2) scattering behavior. From [3].

          Table 4.   Structural Models for Amorphous Systems

Microcrystalline model: high density of tiny microcrystallites

Dense random packing model: alloyed metals; tetrahedron with
        variable bond length − 12 nn, 10 nnn

Continuous random network model: oxides, elemental semiconductors,
        chalcogenides; directional (covalent) bonds − 4 nn, 12 nnn

        Platinum: Gold implanted into Pt to ≈ 15 at.% results in a
modulated microstructure corresponding to spinodal decomposition [13].
Comparisons to splat quenched and aged Au-Pt alloys indicated the
implant-formed amorphous system had appreciably finer modulation.

Stainless Steel: Phosphorus implantation into the more complex fcc system, stainless steel 316, has been shown to result in similar behavior to that for the model fcc system Ni. For SS 316 the structural transition was from fcc to bcc and then followed by the formation of amorphous layer at higher fluences [14].

In general, amorphous phase formation by implantation appears to be indicated for systems in which amorphous alloys may be formed by other methods. Also the resulting amorphous phase appears to be similar in nature to that formed by other methods, although relatively few detailed comparisons have been made as yet. However, one of the key advantages of implantation is that very controlled composition changes can be achieved without addition of unwanted impurities. This provides the possibility of a wide survey of which systems can form amorphous phases without some of the uncertainties which may accompany more conventional means of amorphous phase formation. Also the resulting well-defined layer on a crystalline substrate should allow highly controlled studies of diffusion within amorphous phases and the crystallization of amorphous phases.

3.3 Recrystallization of Amorphous Systems

Amorphous metallic phases are often characterized by a fairly sharp transition in temperature to crystalline phases upon heating. This often seems to be a microscopically homogeneous process and if the equilibrium condition is a two phase system the transition often goes via intermediate (metastable) one phase crystalline structures. The transition temperature indicates the degree of stability of the amorphous phase. Various studies of amorphous systems formed by means other than implantation have indicated that the maximum transition temperatures are of the order of 1/3 the average melting point and that the transition temperature does not correlate with atomic size difference, but does show some correlation with the degree of immiscibility in the crystalline state [15].

In the P-implanted Ni system [3], crystallization was observed by electron diffraction to start in a temperature region of 350-400°C and to be complete at $\approx$ 460°C. At 550°C fairly large crystallites ($\sim$ 500 Å) of Ni and Ni_3P could be observed. While crystallization has been noted in this and a number of other implanted systems, relatively little exploitation has been made of implant-formed amorphous systems to study diffusion or recrystallization processes.

3.4 Theoretical and Empirical Guidelines

There has been relatively little theoretical work on amorphous alloys. In an electronic treatment the tendency to form noncrystalline alloys has been related to an increase in the effective valence of alloys [16]. Also, this theory predicted an increase in stability of noncrystalline forms when the Fermi level lies at a minimum in the density of states. In structural treatments [17], there have been a number of considerations which have been related to the tendency to form noncrystalline alloys, including: (1) large size differences, (2) directional bonding characteristics, (3) systems with many intermediate phases in the equilibrium phase diagram, (4) many polymeric forms to the constituent elements, (5) deep eutectic points nearby which may aid in preventing nucleation of crystalline phases during a rapid quench. Also, it has been observed that the composition range for amorphous alloys appears to increase with differences in the atomic size for binary systems.

More work is needed to obtain a detailed understanding of the amorphous phase formation process during ion implantation. A variety of views of how the amorphous formation process proceeds during implantation may be considered, all of which may be important:

(1) Disorder Stabilization – The displacements created by the implanted atoms result in disorder which builds up. The implanted species may help to stabilize this disorder.

(2) Energy Spike – The high density of energy deposited within a collision cascade provides for local rearrangements and perhaps may assist in barrier-limited diffusionless-type transitions. This seems to be relevant to current understanding of metastable substitutional solutions, but it is less clear as to its significance for the formation of amorphous systems. For example, B-implanted Ni also forms an amorphous phase, whereas for this light specie the energy density is too low to invoke the concept of a true energy spike.

(3) Phase Transformation – When an appropriate atomic concentration is reached the amorphous phase may be the preferred structure in the absence of sufficient temperature to allow phase separation and thereby result in a diffusionless transformation. The randomness required in an amorphous system could be provided simply by the implantation mixing.

Finally, it must be mentioned that extremely high stress levels are present in implanted layers and there is a need to consider whether this may act as a driving force in the formation of amorphous phases by implantation.

4. SUMMARY

Implantation in metals has been shown to form both metastable crystalline and amorphous phases. Typical implanted atom concentrations to form amorphous phases are ~ 10-30 at.%, whereas lower concentrations can be used to form metastable solid solutions. Implant-formed amorphous phases appear to be similar to those formed by other methods while many of the crystalline metastable systems have not previously been observed. As of yet amorphous phases have not been observed in self ion irradiated pure metallic systems. One of the key features of implantation is the high degree of control over composition and the associated purity in achieving a given composition. Thus implantation should provide an ideal method to clarify whether pure monatomic metallic systems may exist in the amorphous phase, as well as to systematically study the formation of binary amorphous and crystalline metastable systems.

REFERENCES

[1] J. A. Borders, Annal Reviews Materials Science, Vol. 9 (to be published).

[2] J. M. Poate, J. Vac. Sci. Technol. 15, 1636 (1978).

[3] W. A. Grant, J. Vac. Sci. Technol. 15, 1644 (1978).

[4] J. A. Borders and J. M. Poate, Phys. Rev. 13, 969 (1976).

[5] R. Kelly, Rad. Effects 32, 91 (1977).

[6] G. Carter, D. G. Armour, S. E. Donnelly, and R. Webb, Rad. Effects 36, 1 (1978).

[7] E. N. Kaufmann, R. Vianden, J. R. Chelikowsky, and J. C. Phillips, Phys. Rev. Letters 39, 1671 (1977).

[8] V. N. Bykov, V. A. Troyan, G. G. Zdoratvseva, and V. S. Kharmovich, Phys. Stat. Sol. (a) 32, 53 (1975); I. M. Belii, F. F. Kamarov, V. S. Tishkov, and V. M. Yankovskii, Phys. Stat. Sol. (a) 45, 343 (1978).

[9] W. Hume-Rothery, R. E. Smallman, and C. W. Haworth, The Structure of Metals and Alloys (Institute of Metals, London, 1969).

[10] D. K. Sood, Phys. Letters 68A, 469 (1978).

[11] A. R. Miedema, Philips Technical Review 36, 217 (1976); A. R. Miedema, F. R. de Boer, and R. Boom, Calphad 1, 341 (1977).

[12] A. G. Cullis, J. A. Borders, J. K. Hirvonen, and J. M. Poate, Phil. Mag. 37, 615 (1978).

[13] S. P. Singhal, H. Herman, and J. K. Hirvonen, Appl. Phys. Letters 32, 25 (1978).

[14] J. L. Whitton, W. A. Grant, and J. S. Williams, Proc. Int'l. Conf. on Ion Beam Modification of Materials, Budapest, 1978 (to be published by Rad. Effects).

[15] M. M. Collver and R. H. Hammond, J. Appl. Physics 49, 2420 (1978).

[16] S. R. Nagel and J. Tauc, Phys. Rev. Letters 35, 380 (1975).

[17] See, for example, C. H. Bennett, D. E. Polk, and D. Turnbull, Acta Metallurg. 19, 1295 (1971).

PERTURBATIONS OF THE SPUTTERING YIELD

P. D. Townsend [1]

LETI/MEP

C.E.N.G. 85X   38041 GRENOBLE CEDEX   FRANCE

ABSTRACT

For low doses of implanted ions there is excellent agreement between the theoretical and measured parameters of ion range and sputtering yield. In the case of high doses the substrate is modified by the implant and one must use revised estimates of the range, energy loss and sputtering efficiency. This review will indicate that important variables include the flatness of the original surface ; radiation enhanced diffusion ; bubble and void formation ; aggregation of new crystalline phases ; and preferential sputtering. Preferential sputtering may occur because one is producing a new material during the implantation ; similarly, bombardments of compound materials frequently produce preferential sputtering. In insulators and semiconductors secondary processes of preferential sputtering result from electronic excitation in the lattice.

(1) On leave of absence from the School of Mathematical and Physical Sciences, University of Sussex, Brighton, BN1 9QH, England.

INTRODUCTION

Energetic ion beams can be used to implant surface layers with impurity ions or to sputter material from the surface. Both effects are always present during ion bombardment but in general the sputtering aspect is important for beams of low energy and heavy ions whereas the penetration into the lattice is dominant for high energy implantations. The theory for both implantation and sputtering is well established and one can readily predict ion ranges, the implant depth profile, the rate of sputtering and reflection of primary ions from the surface in the early stages of an implantation (e. g. 1-7). Historically this has proved to be sufficient for many of the semiconductor applications where the impurity doping levels for electrical changes need only be of the order of 10^{12} to 10^{15} ions cm^{-2}. However with the diversification of ion implantation to fields such as alloy formation in metals or optical waveguides in insulators doses of up to 10^{17} ions cm^{-2} are used. Similarly in work to sputter machine surfaces for devices or depth analysis the density of the implant layer is not trivial. In both the sputtering and implantation a whole new range of secondary effects begin to assume importance. This paper will indicate some of the possible problems and we shall see that most, but not all, of the changes are readily predictable.

BASIC ION RANGE AND SPUTTERING THEORY

The incident ion beam loses energy by two major types of interaction with the solid. At high energy the major component of the energy loss is by electronic excitation (dE/dx). The deposited energy appears in the form of ionisation and excitons. In relation to defect formation or sputtering this term is relatively unimportant in metals as the energy of the excited state is dissipated by the conduction electrons and so excitation does not allow the lattice to relax. However, in insulators and semiconductors the excitation can directly lead to displacements and sputtering (8, 9). At the end of the ion track the ion velocity is reduced and here the dominant energy loss is by

direct nuclear collisions, (dE/dx). This produces a very high rate
of energy transfer by displacement of the lattice atoms and the subsequent collision cascade controls the sputtering and defect formation.
The typical dimensions of such a cascade are some 10 nm. Over a large
part of the ion range both $(dE/dx)_e$ and $(dE/dx)_n$ terms occur together.
This can be important for the insulators and semiconductors as the
ionisation modifies the lattice bonds, allowing relaxation of the
lattice, hence one can produce regions of reduced displacement energy,
more efficient diffusion and reduced surface binding energy.

The details of the calculations of the ion range, damage and the
extent of the collision cascade are influenced by the choice of interatomic potentials and hence by the primary and target masses M_1 and M_2
as well as the energy E_1. For our purposes the most instructive presentation of the results are the contour maps, obtained by computer simulation of the damage, ion range, energy loss, sputtering and primary
beam reflection calculated by following the interactions of a collision
cascade in an amorphous solid [10-13]. In figures 1 and 2 "sputtering"
corresponds to atomic movement across the region defined as the surface in a continuous medium. For a primary ion of mass M_1 and a simple
target of atoms M_2 the range of the collision cascades is a function
of the ratio M_1/M_2 so one alters the range and sputtering yields as
shown by figure 2. It is important to note that in the irradiation of
a compound material $(A_x B_y)$ the sputtering yield and the number of displacements of the atoms A and B are not sensitive to the mass ratios
M_1/M_A and M_1/M_B as within the cascade there is an equipartition of
energy which obscures the effect of the primary ion. Nevertheless
there is still preferential ejection of the lighter ions but, as we
shall see, this is normally explained by a different process [14].

DOSE DEPENDENCE OF THE IMPLANT PROFILE

Although one may commence with a simple elemental material the
implantation produces an "alloy". The alloy has its own characteristic

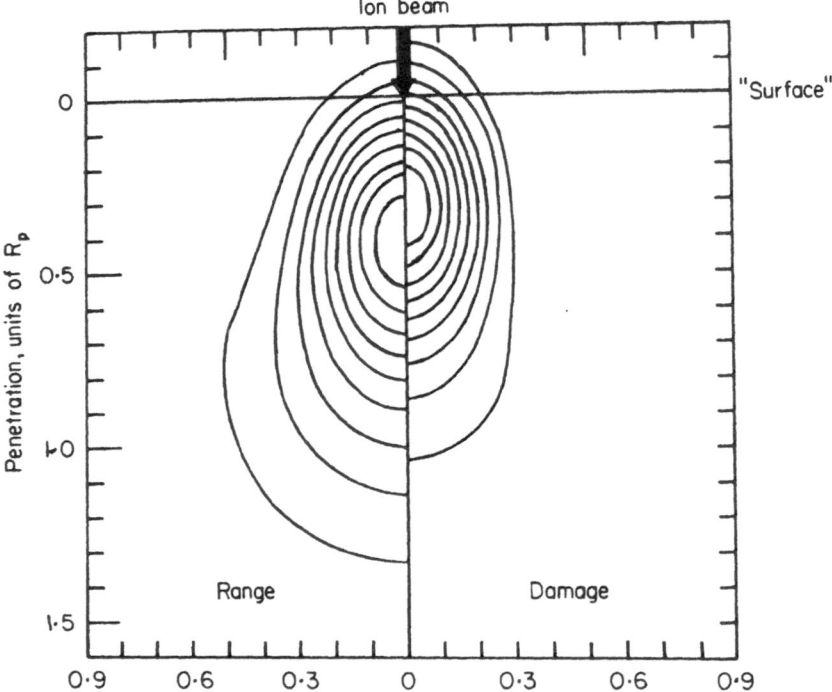

Fig 1. Contours of equal density and equal range calculated for an ion entering a solid ($M_1 = M_2$). Scales are in units of the projected range.

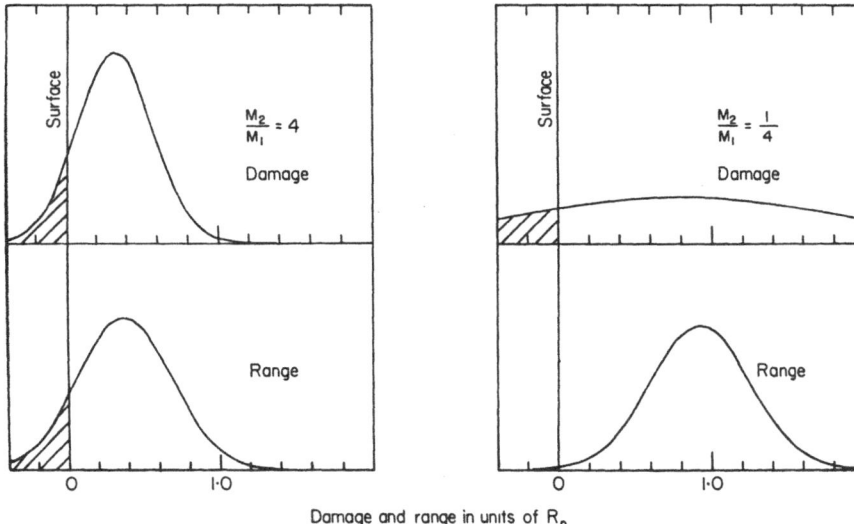

Damage and range in units of R_p

Fig 2. Examples of computed range and damage profiles for target to
ion mass ratios of 4 and 1/4. The vertical line represents the
"surface" in the computer model so the shaded regions may be interpreted as ion reflection or sputtering.

potentials, density and energy loss terms so during the build up of
the alloy the range profiles are progressively modified [15, 16]. For
the majority of the ion target combinations the implant profile extends
to the surface, figure 1, so the accretion of the implant also changes
the sputtering yield. The direction of the change may be either an
increase or a decrease depending on the system. For example the bombardment of copper [17] with 45 KeV ions of Bi^+ increases the sputtering
yield (from 22 to 34) whereas the use of 45 KeV V^+ ions decreases the
yield (from 10 to <1). Both the original values are in accord with the
low dose theory [10]. The extreme example of this alloy sputtering was
shown in the early data of Almén and Bruce [18] where they noted oscillations in the equilibrium (i.e. high dose) sputtering yields as a
function of M_1.

The details of the depth profiles may be revealed by many experimental techniques and one of the most common is the use of Rutherford
Back Scattering (RBS). In the measurements with He^+ ions of lead implanted silicon, Williams[19] determined depth profiles which varied with
ion dose as shown in figure 3. At the lowest doses, $5x10^{14}$ ions cm^{-2},
the lead has a Gaussian depth distribution centred at the predicted
projected range R_p, however at the higher doses the peak concentration
is closer to the resultant surface as the surface level adavances into
the crystal because of the sputtering. Further, at the highest doses
of $5x10^{16}$ ions cm^{-2} there is a further effect of lead accumulation at
the surface as the rates of Pb and Si sputtering are different. One
may even expect that there are other distortions in this measured RBS
profile as the change in material from Si to Pb doped with Si alters
the energy loss term for the He^+ ion probe and so distorts the energy
depth relation of the results.

Rather than measure the detailed depth profile it may be simpler
to record the total concentration of the implant which is retained
within the solid. In the analysis by Carter et al[20] they discuss the
changes in surface position, and implant concentration and note after
a layer of thickness of some $(R_p + 2\Delta R_p)$ has been sputtered then the
sample will reach a maximum concentration for the implant (assuming
equal sputtering probabilities for the two types of atom M_1 and M_2).
The accumulation follows a saturation curve which is energy (i.e. R_p)
dependent[20,21,22]. Therefore at low energies the saturation level
of the implant is less than with high energy implantations.

UNEQUAL SPUTTERING RATES

In compounds, or pseudo compounds of implant plus target atoms,
there are many reasons why the sputtering rate can be different for
the component atoms. A list of the reasons may include the following.
(i) Different probabilities of escape from the cascade.
(ii) Radiation or stress enhanced diffusion to the surface of one
     species.

(iii) Formation of gaseous ions which diffuse and escape.

(iv) Evaporation of one surface species.

(v) Electronic sputtering processes which only operate for one component.

The recent work of Liau et al[23] on the sputtering of PtSi provides good examples of the first three cases. They found that their
surfaces became enriched with the heavier element, Pt and reached an
equilibrium concentration of Pt_2Si. The enhancement is not a function
of the energy of the bombarding ion (Ar) which therefore suggests the
enrichment is not directly related to the primary collision but instead
is controlled by the collision cascade. If there is equipartition of
energy within the cascade the sputtering will be controlled by the
loss from the cascades which intersect the surface. The lighter ions
of silicon will have greater velocity so have a greater probability
of escape, even if they start from sites one or two layers below the
surface. The result is preferential sputtering. In these experiments
the layer of enrichment extends down some $750°A$ below the surface when
using 80 KeV Ar^{++} ions. To maintain such a thick layer of modified
stoichiometry there must be a secondary process which transports the
light ions up to the surface for sputtering. As the dimensions of the
cascades are much less than the depth of the depleted layer the cascade is not the controlling factor. It thus seems reasonable to assume
than an additional stage of radiation enhanced diffusion might operate.
Indeed the level of non-stoichiometry follows the damage and implant
concentration and in all cases the depth of the excess Pt was the same
as the argon range. Both observations are consistent with radiation
enhanced diffusion. In more general terms one might also consider the
possibility of stress enhanced diffusion. An important consequence of
this type of preferential enhancement is that the saturation is approached very rapidly at low energy. Therefore in a situation of ion beam
machining, or sputter cleaning, the surface may be rapidly converted
into a new material even though the total layer removed is only a few
nanometres.

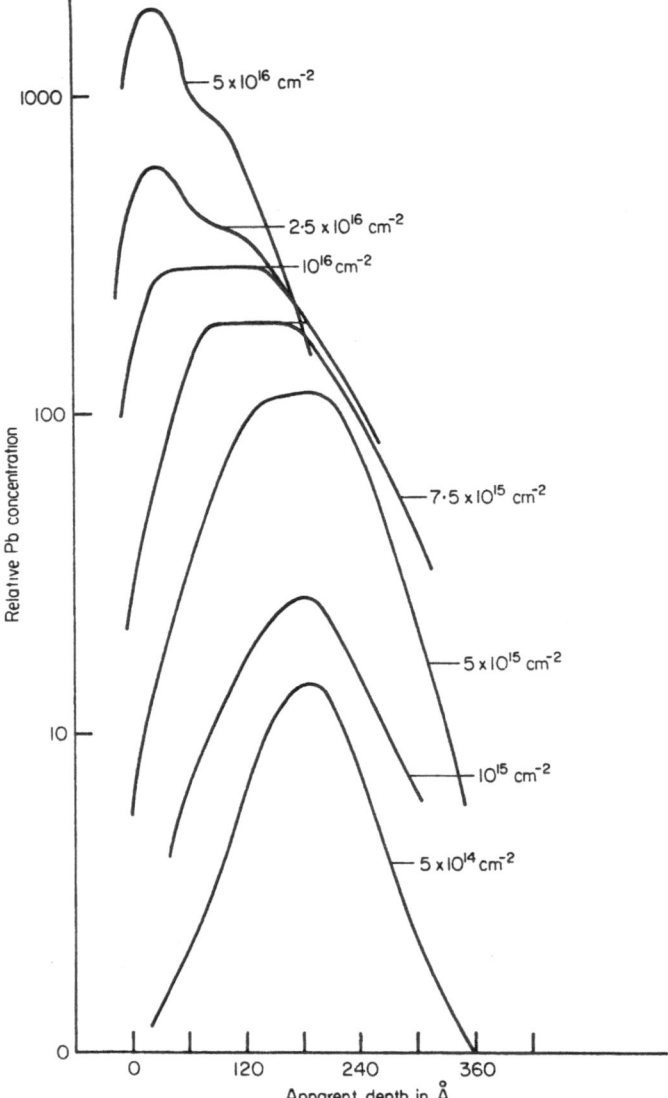

Fig 3. The range distribution of 20 KeV Pb$^+$ implanted at room temperature into Si as a function of dose, from 5×10^{14} to 5×10^{16}. Above a
dose of 5×10^{15} the original Gaussian profile flattens and then develops
a double peak structure, believed to be associated with precipitation
of lead near the surface.

In the same work[23] there is also an example of diffusion of a
gaseous species, the argon, which nucleates into bubbles. The gas
within the bubbles is released when the sputtering reduces the walls
so that rupture takes place. The argon gas release thus follows
a cycle of gas build up in bubbles followed by the simultaneous escape
of many gas atoms. Such bubble rupture may also cause small scale surface roughness.

To continue with the possibilities (iv) and (v) for preferential
sputtering we may consider data obtained with insulators and semiconductors[e.g. 8, 9]. In the alkali halides the mechanism of defect
formation and sputtering by exciton production is well understood.
Briefly, one imagines the exciton is trapped by a pair of adjacent
halogen ions (e.g. $Cl^- + Cl^- + e^- + h^+$), these relax along a $< 110 >$
direction to a new excited state in the form of $(Cl_2^- + e^-)$ and in
doing so extract momentum from the lattice.

If there has only been a small movement of the ions from their
normal lattice sites then the decay of the bound electron from the
upper state produces a luminescence. The luminescence energy differs
from the original exciton energy as both the ground and excited states
are a function of the separation of the ion pair. A more interesting
decay path occurs if the adjacent halogen ions have moved into close
proximity. In this case the new ground and upper states of the ion
pair (Cl_2^-) are essentially at the same energy and the production of
two electrons in the ground state (Cl_2^{--}) causes an explosive separation of the ions along their axis. This results in the formation of
a vacancy with a trapped electron, and by a sequence of replacement
collision sequences along the $< 110 >$ direction, a crowdion interstitial. Alternatively if the replacement collision sequence intersects
the surface the mechanism provides a directional sputtering of the
halogen ions[24].

Less complex models in other materials, for example AgBr, merely
use the exciton energy to change the excitation state of lattice ions
so that they are no longer bonded into the lattice. The resultant

neutral atoms are then assumed to diffuse through the lattice. In the case of AgBr the Ag ions cluster into stable metal precipitates whereas the bromine escapes from the crystals. Whilst such models were originally proposed to explain effects from purely ionising radiation it is important to realise that the electronic energy losses from ion implantation achieve exactly the same result but whereas in photon or electron excitation the effect is often inhibited after a few monolayers of material have been released (for example in Auger electron spectroscopy of ZnO or SiO_2) the ion implantation can continually expose new material and radiation enhanced diffusion can transport the sensitive species up to the surface layer.

In a similar situation, the preferential loss of oxygen from Nb_2O_5 during implantation suggested to Murti and Kelly[25] that not only did oxygen diffuse and escape but also the residual material formed into a new stable phase of NbO. Precipitates in the form of new phases have been observed in other materials but the subject is complicated by the facts that the phase diagrams of a mixture may be quite temperature sensitive so variations in the dose rate, and hence the heating of the layer, may sample different parts of the diagram. The fact that the implantation profile with a single energy implant is not a step function also produces a variation of phase along the ion range. The problems can be formidable and a particularly complex phase diagram is considered for the Cu-Al system[26].

Flux effects may also be apparent if there is radiation enhanced diffusion es the equilibrium between different types of point defects is frequently a function of the rate of formation of the defects[27]. It is possible to vary the rate of defect formation without changing the power dissipation in the sample by making comparisons between results obtained with ion and molecular ion beams (i.e. a 2 μA beam of 1 MeV N^+ ions is thermally equivalent to a 1 μA beam of 2 MeV N_2^+ ions). Whilst evidence exists for the changes in the equilibrium defect concentrations obtained under these two extremes of flux[28] one should also be aware that such changes alter the power in the

collision cascades and hence change the sputtering yields.
For implantations with heavy ions and molecules the sputtering yield from the molecular bombardments is appreciably
greater[29].

SURFACE ROUGHNESS

In the preceding comments we have assumed that the
ion bombarded surface was flat and the beam profile was
uniform. For these ideal conditions the sputtered surface
advances as a flat plane with only minor roughening from
random collision cascades. Even these small, less than 10nm
features may be partially smoothed by surface diffusion.
However if the surface is not planar then one must consider
the effect of the angular dependance of the sputtering yield.
At nearly normal incidence this is approximately a sec Θ
function but shows a maximum at high angles of incidence[7]
The result for an amorphous material is that the surface
topography is time dependent and there can be development
and subsequent disappearance of cones, ridges and valleys
on the surface. Even the directions of equal sputtering
rate do not instantly produce a permanent surface structure
as the equilibrium planes inclined to the direction to the
ion beam can travel sideways across the surface. With amorphous materials one normally reaches a stable condition
consisting of planes which sputter at the same rate.

Equally if one commences with a planar surface decorated with contaminants of different material, or inclusions,
and therefore different sputtering rates, these can cast
a shadow which generate cones and other surface features
(e.g. 30-33). Having once developed the features one may
sputter the original contaminant but there is no guarantee
that the planar surface will reappear. In depth analysis
this is particularly unfortunate as it means one is sampling

different depths of the sample at different regions of the
surface.

     With crystalline targets there is a further complica-
tion that the surface binding energy is a function of the
crystal face so for a polycrystalline sample the sputtering
reveals the crystal facets.

                          DEPTH RESOLUTION

     Particularly clear examples of the effect of surface
roughness on depth resolution are provided by the work of
Hofmann et al$^{(34-36)}$. In their experiments they formed a
multilayer sandwich structure by depositing alternate
layers of Ni and Cr. They then analysed the composition of
the surface layer by Auger electron spectroscopy (AES).
As they exposed different parts of the sandwich structure
by ion beam sputtering with 1KeV Argon ions. Their results
are not perturbed by interdiffusion of the two elements
and as both metals sputter at approximately the same rate
the depth profile is approximately proportional to the
sputtering time. However we see in figures 4a and 4b that
the AES measurement does not record a step profile, in part
because the depth of escape for the electrons is finite.
More important than the rounding of the step function of
the layers is the progressive change in width and magnitude
of the signals from successive layers. This is a direct
result of the surface roughness which is time dependent
with the sputtering and even for the approximately flat
sample, figure 4a, the development of pits, cones and
valleys allows the electrons to sample more than one layer
at the same time from different regions of the surface. In
the case of the sample shown by figure 4b the original surface was extremely rough, relative to the layer thickness,
and one sees that after only four or five layers the AES data does not
give useful information as to the depth composition.

In other experiments with single sandwich layers Hofmann[35] has also noted the effects of assymetric broadening of the thin layer by both knock-on implantation and, for rough samples, by redeposition.

CONCLUSION

The theory for ion range and sputtering yield during ion implantation is well understood and for low dose implantations the theoretical predictions are in accord with the observations. At high levels of implantation there are a

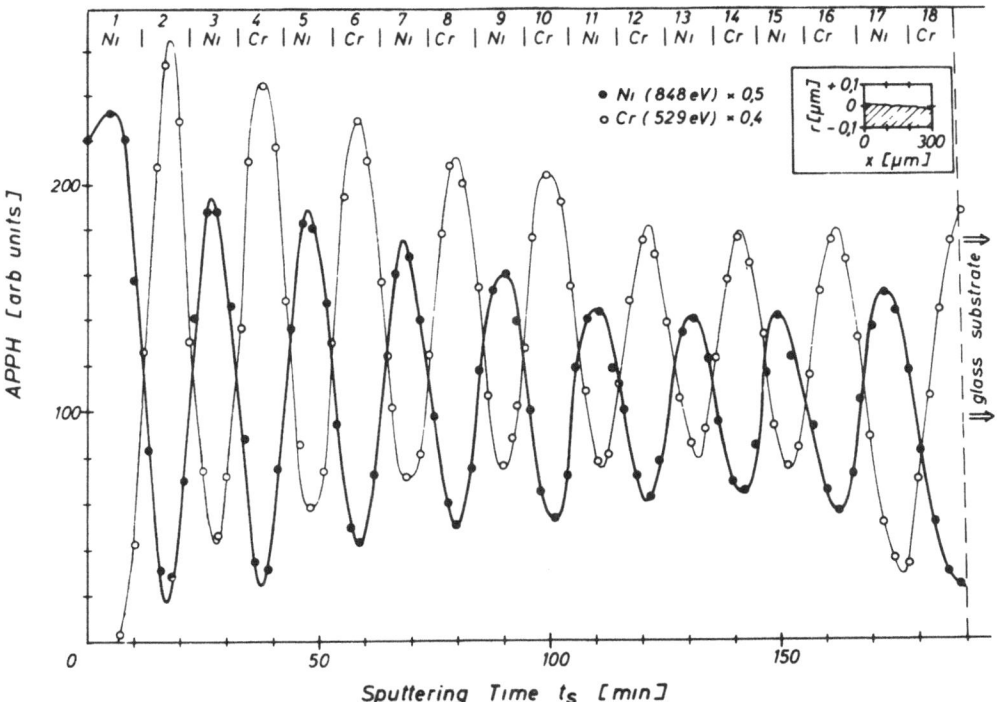

Fig 4A. The AES depth profile of a Ni (115 Å)/Cr (115 Å) multilayer sandwich stucture for specimen B_s : surface roughness not greater than 100 Å ; total thickness of twenty layers is 2350 Å ; ●, Ni (848 eV)x0.5; 0, Cr(529 eV)x0.4. Note the effect of the increasing oxygen content of the layers beyond the thirteenth layer.

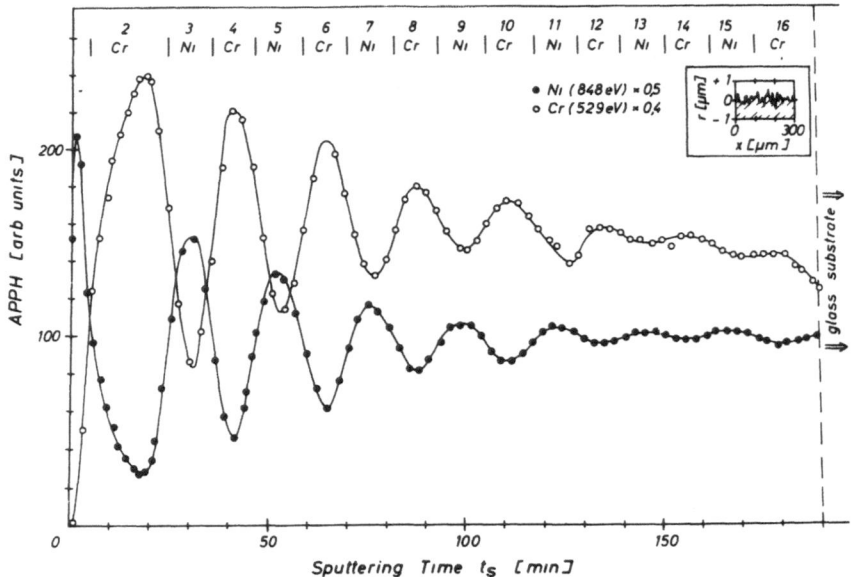

Fig 4B. The AES depth profile of a Ni (115 Å)/Cr(115 Å) multilayer
sandwich for specimen B_r : surface roughness is about 1μm ; total thickness of twenty layers is 2350 Å ; ●, Ni(848 eV)x0.5 ; O, Cr(529eV)x0.4.
The profiles of the first two layers have been disturbed by alterations
of the argon pressure during sputtering ; therefore the time scale is
not linearly related to depth until t_s = 20 min.

range of secondary effects which must be considered that
change the ion range, energy loss, sputtering rate and composition of the implanted layer. Points to consider include
solubility saturation of the implant, radiation enhanced
diffusion, alloy or new phase formation, preferential sputtering of one type of ion, gas bubble growth and release,
and additional specialised effects such as sputtering of
one constituent by electronic processes. In applications
where the sputtering is used as a means of machining the
surface for devices or for depth analysis there is an extremely important problem of surface roughness.
        This review is indicating possible problems and one
should realise that with a careful choice of parameters

one may reduce them. For example in depth analysis by ion
sputtering the use of low energy ion beams at non-normal
incidence reduces the depth of the perturbed layer and one
can quickly establish equilibrium sputtering rates which
are representative of the target composition. However for
analysis of an inhomogeneous material it is wiser to compare
the results obtained from a range of techniques of surface
analysis. Although the list of successful examples of the
applications of ion implantation and sputtering is continually increasing we should remain aware of the secondary
and high dose problems, particularly as the techniques are
being more generally applied in a routine fashion.

REFERENCES

1. Carter,G. and Colligon,J.S. "Ion Bombardment of Solids",
   Heinemann, 1968.
2. Johnson,W.S., Gibbons,J.F. and Mylroie,S.W. "Projected
   Range Statistics", Halstead Press, 1975.
3. Mayer,J.W., Eriksson,L. and Davies,J.A., "Ion Implantation in
   Semiconductors", Academic Press, 1970.
4. Gibbons,J.F., Proc IEEE, 60,1062,1972.
5. Dearnaley,G., Freeman,J.H., Nelson,R.S. and Stephen,J. "Ion Implantation", North Holland, 1973.
6. Wilson,R.C. and Brewer, G.R., "Ion Beams", Wiley, 1973.
7. Townsend, P.D., Kelly,J.C. and Hartley,N.E.W., "Ion Implantation,
   Sputtering and their Applications", Academic Press, 1976.
8. Saidoh,M. and Townsend,P.D., Rad. Effects, 27,1,1975.
9. Townsend,P.D. "Sputtering by electrons and photons", in
   "Sputtering by ion bombardment" Ed,R. Behrisch, Springer-Verlag.
10. Sigmund,P., Phys. Rev 184,383,1969.
11. Sigmund,P., Rev. Roum. Phys, 17, 823,969,1079,1972.
12. Sigmund,P., J.Mat. Sci.8,1545,1973.
13. Brice,D.K., Rad. Effects 6,77,1970 ; 11,227,1971.

14. Brice,D.K., to be published.

15. Blank,P. and Wittmaack,K., Phys Lett, 54A,33,1975.

16. Krautle,H., Nucl. Inst. and Meth., 134,167,1976.

17. Andersen,H.H., Rad. Effects 19,257,1973.

18. Almén,O. and Bruce,G., Nucl. Inst. and Methods,11,279, 1961.

19. Williams,J.S., Phys Letts 51A,85,1975.

20. Carter,G., Baruah,J.N. and Grant,W.A., Rad. Effects, 16,107,1972.

21. Whitton,J.L., Carter,G., Baruah,J.N. and Grant, W.A., Rad. Effects, 16,101,1972.

22. Schulz,F. and Wittmaack,K., Rad. Effects, 29,31,1976.

23. Liau,Z.L., Mayer,J.W., Brown,W.L. and Poate,J.M., J. Appl. Phys., , ,1978.

24. Townsend, P.D., Browning,R., Garlant,D.J., Kelly,J.C., Mahjoobi,A., Michael,A.J. and Saidoh,M., Rad. Effects, 30,55,1976.

25. Murti,D.K. and Kelly,R., Thin Solid Films,33,149,1976.

26. Arminen,E., Fontell,A. and Lindroos,V.K., Phys. Stat. Sol (a) 4,663,1971.

27. Marwick,A.D. and Piller,R.C., Rad. Effects,33,245,1977.

28. Saidoh,M. and Townsend,P.D., J. Phys C 10,1541,1977.

29. Andersen,H.H. and Bay,H.L., Rad. Effects, 19,139,1973.

30. Ducommun,J.P., Cantagrel,M. and Marchal,M., J.Mat.Sci. 9,725,1974.

31. Carter, G. Colligon,J.S. and Nobes,M.J., Rad. Effects, 31,65,1977.

32. Stewart,A.D.G. and Thompson,M.W., J.Mat. Sci. 4,56,1969.

33. Wilson,I.H., Rad. Effects, 18,95,1973.

34. Hofmann,S., Erlewein,J. and Zalar,A., Thin Solid Films 43,275,1977.

35. Hofmann,S., Proc 7th Int Vac. Cong., Vienna 1977, p2613.

36. Hofmann,S., Appl. Phys, 9,59,1976 ; 13,205,1977.

Part IV

Applicability of Solid
State and Nuclear Methods

INTRODUCTION TO SOME SOLID STATE METHODS FOR THE STUDY OF
ISOLATED IMPLANTED ATOM SITES IN METALS

L.M. STALS *

Materials Physics Group, Limburgs Universitair Centrum

B-3610 DIEPENBEEK (Belgium)

1. Introduction

A number of methods exists in the field of solid state physics for
the study of implanted atom sites in metals. This review will be
concerned with two methods which are especially suited for this
purpose and which furthermore can be combined with nuclear physics
methods. These methods are : resistivity measurements and thermal
desorption spectrometry. For other methods, see the contributions
of other lectures to this Summerschool. In chapter 4 some techniques are described for the analysis of experimental annealing
data . In agreement with instructions given to the lectures,
namely "that the principal goal of the Summerschool is that nuclear
physicists and solid state physicists teach their methods to each
other", only the principles of these methods will be explained
without going into much detail.

2. Resistance Measurements

2.1. General Remarks

The electrical resistivity of a metal is inversely proportional to
the mean free path of the conduction electrons [1,2] . Since the
mean free path of electrons is equal to the mean distance between
scattering centres, and since defects are to be considered as
scattering centres, the electrical resistivity of a metallic specimen increases as defects are introduced. If the various kinds
of scattering centres contribute independently to the resistivity,
one has

$$\rho = \sum_{i=1}^{n} \rho_i \qquad (2.1)$$

* Also at S.C.K./C.E.N Mol (Belgium)

where ρ = total resistivity, ρ_i = resistivity due the ith scattering centre. This is <u>Matthiessen's rule</u> [3,4]. A large number
of deviations from this rule have been published [5-13], however.
Equation (2.1) can be written in the following form :

$$\rho(T) = \rho_i + \rho_s + \rho_f \qquad (2.2)$$

where ρ_i = resistivity of impurities (extrinsic defects), ρ_s =
resistivity of structural defects (intrinsic defects) and ρ_f =
resistivity due to the electron-phonon interactions. If
Matthiessen's rule is valid, then ρ_i and ρ_s are independent of
temperature and, as a consequence, the total temperature dependence of ρ is reflected in ρ_f. For normal alloys - these are the
alloys which do not show a resistivity minimum at low temperatures
[1] - one has $\rho_f \simeq 0$ at 4.2 K. For very pure metals ($\rho_i \simeq 0$), the
resistivity at 4.2 K is therefore nearly entirely determined by
the structural defects. When studying defects by means of electrical resistivity measurements, one, therefore, follows the variation of ρ at 4.2 K. In the case that the metal becomes supercon-
ducting, one uses either a higher reference temperature or one
applies a magnetic field to suppress superconductivity. In implantation experiments the resistivity increase at 4.2 K is determined
both by the implanted atoms and the structural defects created during implantations. The sum $(\rho_i+\rho_s)$ at 4.2 K is called the residual resistivity ρ_r. Some data exist about the contribution of impurities and structural defects to ρ_r. For normal alloys <u>the rule
of Norbury</u> [14] states that the addition of B atoms to an A matrix
results in a residual resistivity increases $\Delta\rho_r$ proportional to Z^2,
where Z is the difference in valence between A and B atoms. Addition of 1 at % of B-atoms usually yields an increase $\Delta\rho_r$ of a few
10^{-8} Ωm for copper [14]. Evans and Eyre [17] have determined the
resistivity of nitrogen in molybdenum. They found :
$\Delta\rho_r = 2.4 \ 10^{-8}$ Ωm/at%. The contribution of noble gases to the resistivity of metals is not known, but it will be at the maximum
that of intrinsic point defects [15]. As far as intrinsic point
defects are concerned, theoretical and experimental data point to
a contribution $\Delta\rho_r = \rho$ (273 K) per at % Frenkel pairs [16,18-23].
This is the so-called 'rule of thumb of Lucasson-Walker' [22].
One may thus conclude that in first approximation the residual resistivity increase due to 1 at % extrinsic or intrinsic defects is
equal to the resistivity of the metal at 273 K. This rule may be
applied for the determination of the purity of, say, a single crys-
tal. In order to do so, one determines the ratio $\Gamma \equiv \rho_{273K}/\rho_{4.2K}$.
This ratio can be larger than 10^5, which corresponds to a total
impurity concentration less than 10^{-5} at %. It should be mentioned,
however, that the method is only a qualitative one, because the
contributions of the various kinds of impurities are measured simultaneously, and because the above mentioned rule is only an
approximation. Resistance measurements can very simply be performed and their accuracy and resolution is high. This is one of the
major reasons why such measurements have been used in the past and

are still used today so intensively in the study of defects in
metals.   The resolution of the method is about 5.10^{-14} Ωm, which
for molybdenum would allow the detection of a change of Frenkel
pair concentration of 10^{-2} at ppm.   In order to obtain this resolution in radiation damage experiments and for reasons of homo-
geneity during irradiation one has to work with small specimens.
This means that for wires and sheets or thin films the condition
$s/l \lesssim 5.10^{-7}$m has to be fullfilled, where s is the cross section
and l the length of the specimen[1].   For very pure specimens
($\Gamma_{bulk} > 10^4$ to 10^5) and at low temperatures (4.2 K) the mean free
path of the electrons may become comparable to or larger than the
smallest dimension of the specimen.   For pure copper ($\Gamma_{bulk} > 10^5$),
for instance, one has : mean free path at 300 K $\simeq 4.10^{-8}$m and at
4 K $\simeq 3.10^{-3}$m [24].   As a consequence, the mean free path in thin
foils, wires or films at liquid helium temperature is mainly reduced by the surface and therefore the resistivity is increased
with respect to its bulk value.   This phenomenon is known as the
size effect, and is theoretically described by Sondheimer [26].
The theory has been worked out in the form of tables for correc-
ting the size effect of the electrical resistivity for foils and
wires by Dworschak et al. [24,25].   They have shown [25] that for
pure metals (Γ_{bulk} : 10^3-10^4) the resistivity increase per unit
defect concentration can easily be a factor two larger than the
bulk value for thin foils ($\lesssim 10^{-5}$m thick) or even a factor ten for
thin films ($\lesssim 5.10^{-7}$m thick).   The size effect, which introduces
a strong non-linear dependence of the resistivity increase on the
defect concentration, may be of importance in low energy implantation experiments on thin films.
Resistance measurements have been used in the past in a large variety of experiments.   They have been used for example for the
determination of :
1. Frenkel pair concentrations in low temperature irradiation experiments [28] and threshold energies for damage production [27].
2. Equilibrium vacancy concentrations at high temperatures [29].
3. Vacancy formation energies from quenching experiments [30].
4. Diffusion parameters in metals [31].
5. Characteristics of gas-metal reactions [39].
6. Order-disorder parameters [32-33].
7. Recovery parameters of defects in metals [34-38].
In the framework of this Summer school, the discussion will be concentrated on recovery studies by means of resistance measurements.
This technique seems to be most suitable to be combined with
nuclear physics methods for elucidating the annealing processes in
the various recovery stages of pure metals [40].

---

1) From Pouillet's law $R = \rho \frac{l}{s}$ it follows $\Delta\rho = \Delta R \frac{s}{l} = \frac{\Delta E}{I} \frac{s}{l}$;
   for $\Delta E = 10^{-8}$V and I = 10^{-1} A it follows that $\frac{s}{l} \lesssim 5.10^{-7}$m for
   $\Delta\rho \lesssim 5.10^{-14}$ Ωm.   If l=10^{-2}m, then s = 5×10^{-9} m^2

2.2. <u>Short review of recovery experiments</u>

A large amount of papers have been published during the last two
decades on the use of resistance measurements for the study of the
recovery of defects in metals (see the general references [34-38],
[41-44]). Quite generally one can say that if a metal is irradiated, say with electrons of a few MeV or with thermal neutrons
at 4.2 K, and the recovery of the irradiation damage is followed
upon subsequent annealing by means of resistance measurements,
one observes a decrease of the resistivity due to the annealing
of point defects (vacancies and interstitials) created during
irradiation. This decrease is observed to be enhanced in certain
temperature ranges, which gives rise to the appearance of so-called
recovery stages. One can plot ρ or $\frac{d\rho}{dT}$ as function of temperature.
In general the recovery stages are more clearly visible in the
plot $(\frac{d\rho}{dT},T)$, which is called the 'recovery spectrum'. A typical
example is shown in fig. 2.1..

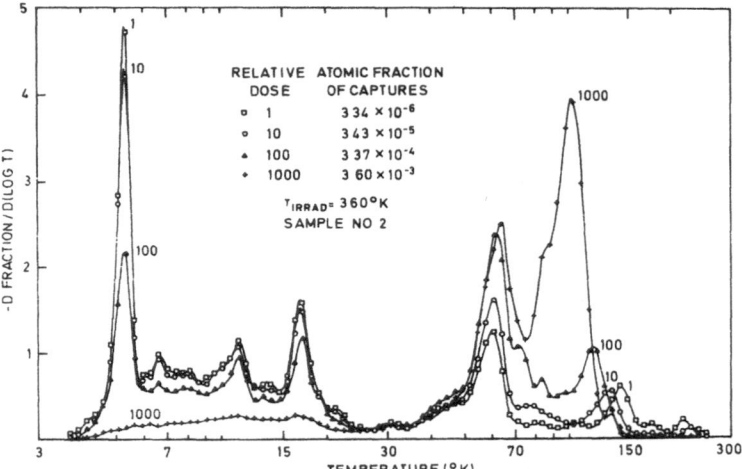

Fig. 2.1. Recovery spectrum of thermal neutron irradiated Cadmium
[34]

In the analysis of recovery spectra, one looks for the free migration peaks[2]. These peaks have a characteristic concentration
dependence, in that the peak temperature shifts to lower temperature with increasing amount of recovery. This can, for instance,
be observed in fig. 2.1. for the peak between 90 K and 150 K,
which shifts from ~ 140 K to ~ 95 K as the thermal neutron dose
increases by a factor 10^3. All other peaks in the spectrum remain
at their position as the thermal neutron dose is increased. The
characteristic concentration dependence of free migration peaks
can be understood as follows. Let us assume that the specimen con-

2) With 'free migration' is meant "uncorrelated" free migration of
   point defects.

tains only vacancies and interstitials with the initial concentration : $c_{i,o}$ for interstitials $c_{v,o}$ for vacancies. We assume further $c_{i,o} = c_{v,o} = c_o$, where c_o stands for the initial concentration
of Frenkel pairs, and that during annealing the following reaction
takes place :

$$i + v \rightarrow \text{annihilation}$$

If $E_i^m < E_v^m$, where $E_{i,v}^m$ is the migration energy of respectively
interstitials and vacancies, this reaction can be described by
means of the isothermal equation

$$\frac{dc_r}{dt} = - K_o \exp(- E_i^m/kT) \, f \, (c_r) \qquad\qquad (2.3.)$$

where $c_r = \dfrac{c}{c_o}$ is the relative Frenkel pair concentration, and

$$K_o = 4\pi D_{o,i} \, r_o c_o^{\beta-1}, \quad D_{o,i} = \zeta n \, \nu_o \lambda^2$$

with $D_{o,i}$ = diffusion constant for interstitial diffusion
   n = number of saddle points across which the interstitial
        can jump to an equivalent position
   ν_o = attempt frequency
   λ = jump distance
   ζ = factor determined by the configuration of the inter-
        stitial
       = 1/6 for three-dimensional migration of $< 100 >$-dumbbell
        (in f.c.c. metals)
       = 1/2 for one-dimensional migration of $< 110 >$ -crowdion
        (in f.c.c. metals)
   r_o = capture radius for i-v annihilation
   β = highest power of c in the function f(c).

The function $f(c_r)$ can take various analytical forms. If we should
number the interstitials from 1 to N and if the probability of annihilation of the j-th interstitial would not be influenced by the
annihilation of one of the N-1 remaining interstitials, then
$f(c_r) \sim c_r$. This is for instance the case for close pair annihilation or correlated recovery annihilation, where each interstitial
annihilates with its own vacancy :

$$i_k + v_k \rightarrow 0 \qquad k : 1...N$$

If, however, the annihilation probability for the j-th interstitial
depends on the annihilation of the other interstitials, as it is
the case for random three-dimensional migration of interstitials,
(uncorrelated free migration), then for the case of low capture
radius r_o and/or low concentration c_o : $f(c_r) \simeq c_r^2$ [45]. The function $f(c_r)$ can be calculated from diffusion theory [46,47,48]. For
instance, for the case under consideration it has been shown that
[45] :

$$f(c_r) = \frac{1}{[1+\frac{1}{4}r_o^{-3}c_o^{-1}(\frac{1}{c_r} - 1)^{1/2}]-1} \; c_r^2$$

The function $f(c_r)$ can be equated to a single power of c_r [45]:

$$f(c_r) = \alpha c_r^{\gamma}$$

where α and γ depend on c_r. Equation (2.3.) then becomes

$$\frac{dc_r}{dt} = -K_o \alpha \exp(-E_i^m/kT) \, c_r^{\gamma} \qquad (2.4)$$

In analogy to the chemical reactions, γ is called the order of reaction. $\gamma=1$ means first-order reaction; $\gamma=2$ means second-order reaction. Equation (2.4) can be transformed for linear heating into

$$\frac{dc_r}{dT} = -\frac{K_o}{\mu} \alpha \exp(-E_i^m/kT) \, c_r^{\gamma} \qquad (2.5)$$

where $\mu \equiv \dfrac{dT}{dt}$ is the heating rate. The peak temperature T_p can be

found by putting $(\dfrac{d^2 c_r}{dT^2})_{T=T_p} = 0$. This yields the relation :

$$\frac{dT_p}{dc_o} \simeq (\frac{E}{kT_p^2} + \frac{2}{T_p})^{-1} (\frac{1-\gamma}{c_o}) \qquad (2.6)$$

From equation (2.6) it follows :

$$\frac{dT_p}{dc_o} = 0 \text{ for } \gamma = 1$$

$$\frac{dT_p}{dc_o} < 0 \text{ for } \gamma > 1$$

For free uncorrelated migration of point defects one has $\gamma > 1$ at the peak temperature [45,49]. From equation (2.6) it follows that the peak shift per unit concentration is larger for lower initial concentrations, as observed experimentally. It should be mentioned that a peak shift to lower temperatures can also occur if only one defect species is present, e.g. vacancies. A frequently occurring annihilation reaction for vacancies is the disappearance at the surface or at internal sinks, e.g. dislocation loops. If together with a larger vacancy concentration a larger sink density is build up, then the vacancy migration stage will show the typical negative peak shift of a free migration stage. This is also the case if the reaction $v_1 + v_1 \rightarrow v_2$ predominates and if furthermore divacancies (v_2) are much more mobile than monovacancies (v_1).

2.2.1. F.C.C. Metals
For the labelling of the recovery stages in f.c.c. metals van Bueren's nomenclature [50,51] has been accepted. A similar labelling is also used for b.c.c. and h.c.p. metals. According to this labelling, there are five recovery stages I to V in f.c.c. metals. Corbett et al [52,53] have shown that following electron irradiation stage I in copper is subdivided in five substages, labelled I_A to I_E. Fig. 2.2 shows stage I recovery spectra for some of the most intensively investigated f.c.c. metals following electron irradiation. The substage I_E is that substage of stage I which shifts to lower temperatures with increasing defect concentration. It can in

general only be observed at low defect concentration ($<10^{-5}$), because it merges very rapidly into stage I_D (see fig. 2.2.b). All
spectra of fig. 2.2. show the five typical substages, except gold
(see fig. 2.2.g). These substages are interpreted as being due to
close pair recombination (A,B,C) and interstitial migration (D,E).
The substages I_D and I_E are observed to have the same activation
energy (e.g. 0.12 eV for Cu [52]). Table 2.1 gives as an example
the characteristics of the other recovery stages in copper. For
the other f.c.c. metals the reader is referred to the original
literature [34-38, 41-44].

The first high temperature stage above $T \simeq 0.1\ T_m$ which shows the
typical dose dependence of free migration of a point defect is called stage III. The stages III and I_E are, according to their kinetics, ascribed to the migration of simple point defects. There are
only two types of simple point defects : interstitials and vacancies. The interpretation could therefore be very simple and
straightforward : interstitials migrate in stage I_E and vacancies

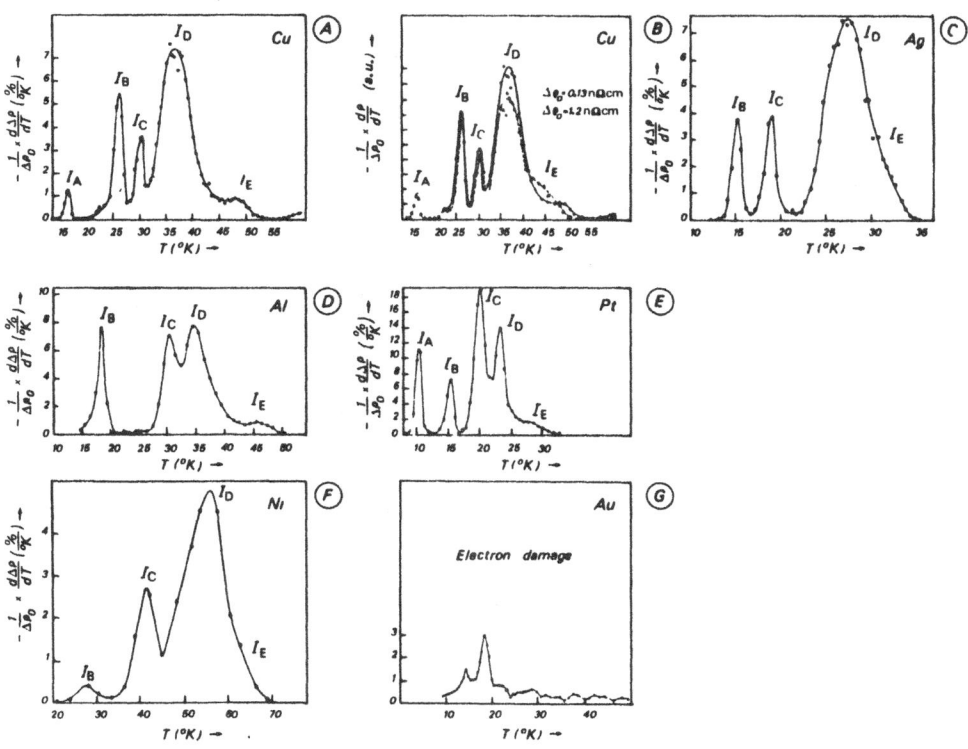

Fig. 2.2. Stage I recovery in f.c.c. metals [34]

in stage III.   There are, however, some metals, e.g. Cu, which
show a third dose dependent stage, namely stage IV.   For these
metals one needs, therefore, at least three simple point defects.
Without going into the details, it should be mentioned, that there
exists a strong controversy about the configuration of the interstitial which migrates in stage I_D and I_E and about the mechanism for
stage III.

Table 2.1 Characteristics of recovery stages in copper

Stage	Temp. range (K)	Activation Energy
I	< 60	0.1
II	60 - 220	0.1 - 0.4
III	220 - 330	0.6 - 0.8
IV	330 - 450	1.1
V	> 450	2.1

Following Seeger and Frank [54] stages I_D and I_E are due to the migration of the metastable < 110 > -crowdion : on-line   crowdions
(I_D) and off-line crowdions (I_E), in those metals for which the
thermal conversion energy E_C^C of the crowdion, i.e. the energy necessary for converting the < 110 > -crowdion into the more stable
< 100 > -dumbbell, is larger than the migration energy E_C^m of the
crowdion.   For gold E_C^C < E_C^m and therefore golf behaves differently
in that only close pair peaks are observed (no free migration) in
stage I.   Following Schilling [34,37,38,74] it is the stable < 100>
dumbbell interstitial which migrates in I_D (correlated) and in I_E
(uncorrelated).   No satisfactory explanation of the "anomalous" behaviour of gold can be given within Schilling's model [55,56,77,79].
This controversy has of course its influence on the interpretation
of the high temperature stages, especially of stage III.   Following
Seeger and Frank stage III has to be attributed to the migration of
stable dumbbell interstitials, whereas stage IV is due to the migration of vacancies[3].   Following Schilling and coworkers, vacancies
migrate in stage III.   Since in the interpretation of Schilling
only one interstitial is needed, the corresponding model, which has
in fact originally been proposed by Corbett, Smith and Walker
[52,53], is called the one-interstitial model.   The Seeger-Frank
model is called the conversion-two-interstitial model, because they
additionally assume that the stage III interstitials (dumbbells)
are converted crowdions.   At the present day, after two conferences one
held at Gatlinburg (1975) [37] and the other at Argonne (1976) [38],

3) Au and Al do not show stage IV recovery after electron-irradia-
   tion ; Cu on the other hand has a well-defined stage IV whereas
   some controversy exists about stage III and IV in Pt [57].   In
   the Seeger-Frank model vacancies and interstitials (dumbbells)
   have the same or nearly the same activation energy in Au and Al.
   Therefore, stages III and IV are not separated in these models.

it is the general feeling among the majority of the scientists involved in this research, that the presently available experimental
information is largely in favour of the one-interstitial model [74].
The following remarks, however, have to be made in this respect.
In spite of extensive research, people were unable to detect a free
migration stage in gold at low temperatures [79] [4]. Until now, no
satisfactory explanation for this particular behaviour of gold has
been given, except by Seeger and Frank within the framework of
their model. The identification of stage III in platinum is still
a matter of controversy [57]. In platinum there are two distinct
recovery stages : from 300 K to 350 K and from 420 K to 650 K. The
300-350 K range is called stage III by Seeger and Frank, whereas
the broad range 420 K-650 K is called stage III by Sonnenberger et
al. [58]. In both models the higher temperature part of the
420 K-650 K range is attributed to monovacancy migration. The
lower temperature part is attributed by Sonnenberger et al. to di-
vacancy migration, whereas Seeger and Frank have strong arguments to
attribute this part of the 420 K-650 K range to the decay of small
interstitial clusters. The temperature range 300 K- 350 K is ascribed by Seeger and Frank to the migration of dumbbells. The
particular behaviour of stage III (300-350 K) in platinum, i.e. the
fact that its amplitude decreases with increasing irradiation dose,
is explained on the basis of a larger tendency of interstitials to
cluster in platinum than in the other f.c.c. metals. This controversy on the labelling of recovery stages in platinum has its effect
on the interpretation of electron irradiation results. Depending
on the choice of the temperature range of stage III, recovery is
complete following electron irradiation, as it is also the case for
Au and Al, or incomplete, as it is the case for Cu.

2.2.2. B.C.C. Metals

Recovery experiments in b.c.c. metals have started somewhat later
than on f.c.c. metals. Initially the recovery spectra for these
metals showed the same general behaviour as for f.c.c. metals.
Therefore people are using the same labelling and a similar interpretation as for f.c.c. metals. For review papers we refer to the
references [34,36,44,59]. Fig. 2.3 shows a selection of recovery
spectra for b.c.c. metals at low temperatures [70,75,78]. As for
f.c.c. metals, one observes in the low temperature region a number
of substages. The Fe-spectrum approaches best the spectra for
f.c.c. metals. Iron and tungsten are in fact the only b.c.c.

---

4) Interstitial clustering is, however, observed in Au even at 0.3 K
during electron irradiation [77]. This might be an indication of
long range migration of interstitials at that temperature.
Radiation induced clustering can, however, not be excluded as a
possible mechanism [79]. The formation of vacancy clusters far
below stage III has, for instance, been observed in Cu [80] and
Pb [81] during electron irradiation. The formation of these
clusters cannot be a result of long range migration of vacancies.

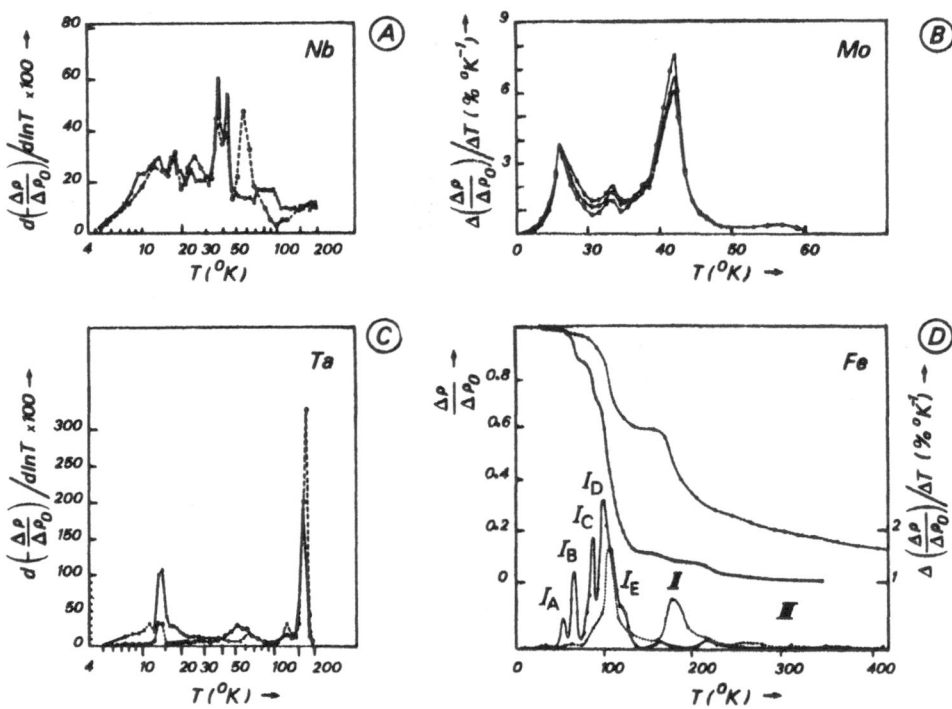

Fig. 2.3. Recovery spectra for b.c.c. metals ([78] for A,C; [70]
          for D and [75] for B)

metals for which a dose dependent substage (stage I_E) in stage I
could be observed [60,61,83]. This would imply in the one-interstitial model that no uncorrelated migration of interstitials (for
b.c.c. metals most probably < 110 > -dumbbells) similar to that
for copper takes place in nearly all b.c.c. metals [5].
The high temperature recovery is very similar to that of f.c.c.
metals, at least for the b.c.c. metals of group VI a (chromium
group). As an example, fig. 2.4.a shows the recovery spectrum
above 300 K for neutron-irradiated molybdenum [62]. Four major
recovery stages can be observed : stage III (450-500 K), stage IV
(550-600 K), stage V (900-1000 K), stage VI (1000-1100 K). Similar
recovery stages are observed following coldwork [63]. The corresponding activation energies, as determined by means of the change
of slope technique, are shown in fig.2.4.b . For b.c.c. metals

---

5) Recently Dausinger et al [78] have shown, however, by means of
   damage rate measurements that free migration of interstitials
   might take place in Nb at T ≤ 33 K and in Ta at T ≤ 17 K. Similar observations have been made for Mo [82].

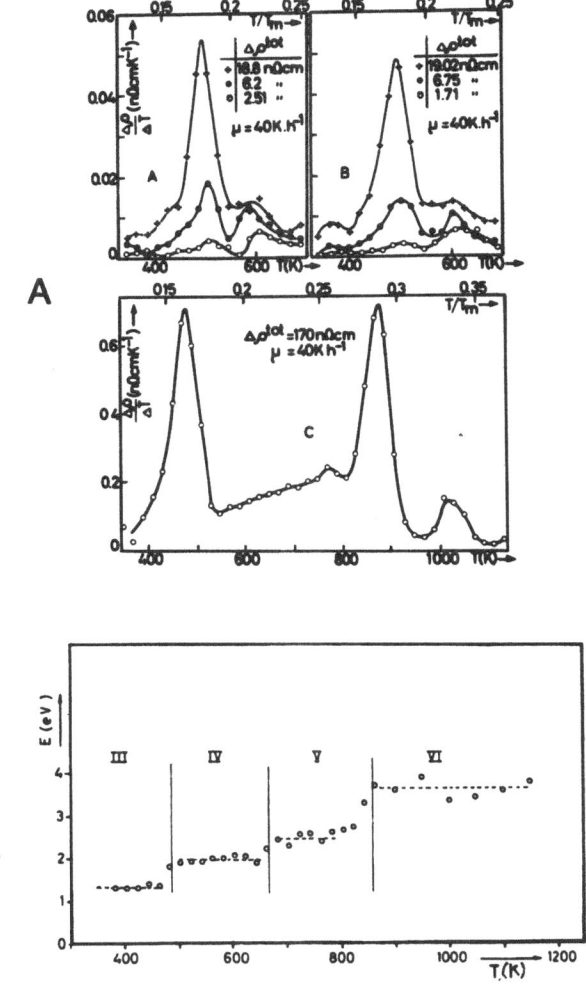

Fig. 2.4. Recovery spectrum of fast-neutron-irradiated and cold-
         worked molybdenum above 300 K (A) [62]. Activation ener-
         gy for cold-worked molybdenum in the recovery stages III
         to VI (B) [63].

there exist two interpretations, very similar to those for f.c.c.
metals. It is shown by Johnson [64] that for b.c.c. metals the
most stable interstitial configuration is the $< 110 >$ -dumbbell.
In the one-interstitial model this dumbbell migrates at low temperatures. In the two-interstitial model the $< 110 >$ -dumbbell migrates at higher temperatures (stage III), whereas the $< 111 >$
crowdion could migrate freely at low temperatures. In the one-
interstitial model vacancies migrate in stage III, which for W and
Mo is situated around 0.16 T_m [36,63,65,70] and for Fe around
0.12 T_m [61,70]. There is, some experimental evidence that in Mo
vacancies migrate in stage III [66,67][6]. The situation is, however, less clear for Fe and for the group Va b.c.c. metals V, Nb
Ta (vanadium group).

There is experimental evidence that vacancies in Fe do not migrate
below 400 K [68,69], their activation energy for migration being
~ 1.3 eV [68,69,75] whereas stage III is Fe is situated around 200 K
([70,76]; E^{III}=0.6 eV). The situation for Ta is similar : the vacancy migration energy is 2.2 eV [72] whereas stage III is situated
at 260 K-300 K ([71]; E^{III}=0.7 eV). Table 2.2. summarizes the
present situation. It is clear that a lot of work remains to be
done in this field before any definite conclusion can be drawn
about the recovery processes which take place in the various
recovery stages in b.c.c. metals.

2.2.3. H.C.P. Metals

In fig. 2.1 and 2.5 some representive recovery spectra for h.c.p.
metals are reproduced. Again the spectra look very similar to
those of f.c.c. metals. For gadolinium and cobalt one observes
the "normal" I_D-I_E behaviour, whereas Cd, Zn and Re behave similar
to gold and to the b.c.c. metals (with the exception of Fe and
tungsten), in that no dose dependent substage (I_E) is observed.
In the case of cadmium the stage III recovery range (80-100 K) is
shown in fig. 2.1. A careful analysis in the light of the Gösele-
Seeger theory for diffusion in anisotropic media, demonstrates that
stage III in cadmium and zinc can be explained in terms of preferential one-dimensional diffusion of c-dumbbells [73]. Vacancy

---

6) It is observed that the τ_2-component of the positron lifetime
   increases during stage III annealing in Mo. This observation is
   used as an argument for vacancy clustering, and as a consequence
   for vacancy migration in stage III. However, the lifetime of
   positrons increases also during annealing around 200 K in Fe
   where it is believed that stage III has to be situated for that
   metal [68]. Vacancies migrate in Fe at much higher temperatures.
   Therefore, no direct conclusion about vacancy migration can be
   drawn from the observation of increased positron lifetimes.

Table 2.2 Data on the activation energy for stage III recovery
(E^{III}) and for vacancy migration (E^m_V) for b.c.c. metals.
E^{SD} is self-diffusion energy

Metal	E^{SD}	E^f_V	$E^m_V = E^{SD} - E^f_V$	E^{III}
Mo	4.74+0.05 [91]	3.0+0.2 [88]	1.74+0.25	1.3+0.1 [63]
W	6.08 [90]	4.0+0.3 [88]	2.0 +0.3	1.8+0.1 [70]
Ta	4.1 [84]	2.7+0.7 [88]	1.4 +0.7	0.7 [71]
		1.9+0.35 [72]	2.2 +0.35	
Nb	3.62 [84]	2.5+0.5 [88]	1.12+0.5	0.54 [89]
α-Fe	2.99 [86]	1.60+0.15 [87]	1.29+0.15	0.53 [76]
		1.4 +0.1 [85]	1.59+0.1	

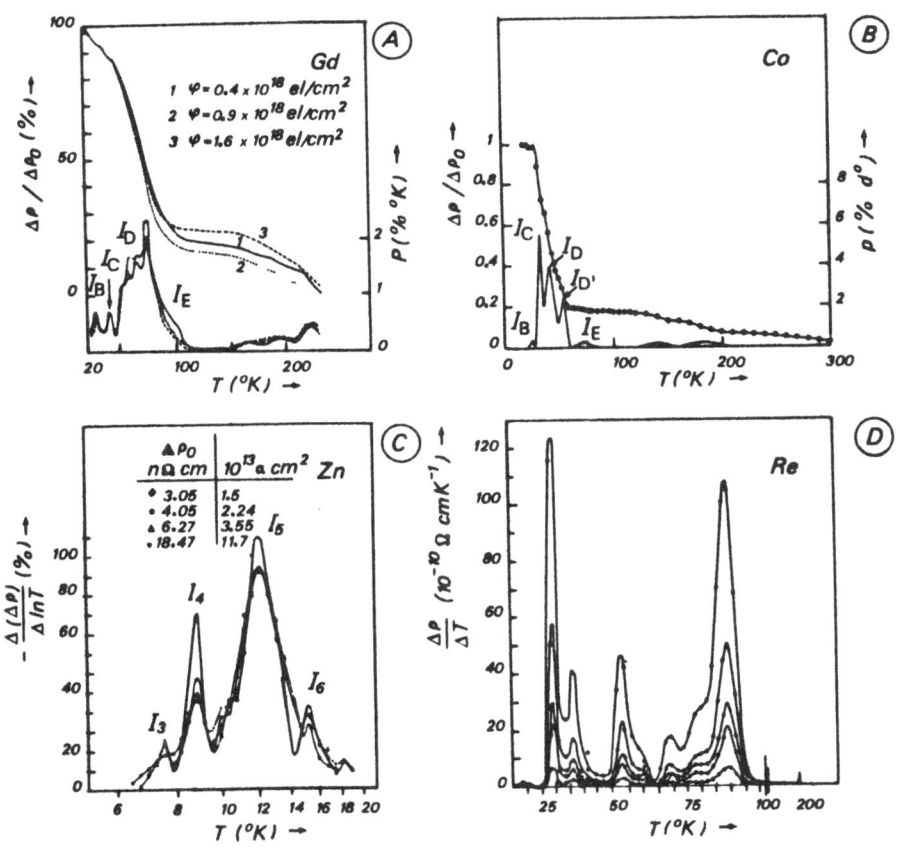

Fig. 2.5. Recovery spectra for h.c.p. metals [34,73,74]

migration at the end of stage III can, however, not be excluded. If a similar interpretation would be valid for the other h.c.p. metals, gadolinium and cobalt, then at least two types of interstitials are needed : a low temperature type, most probably the $<11\bar{2}0>$-crowdion, and a high temperature type, the c-dumbbell, for instance. Anyhow, people try to interpret their measurements either in a one-interstitial model or a two-interstitial model for h.c.p. metals which are very analogous to those for f.c.c. metals. As for b.c.c. metals a lot of work remains to be done in the field of h.c.p. metals, before any definite conclusion can be drawn about the recovery processes in h.c.p. metals.

References

[1] C. Kittel, Introduction to solid state physics, 4th edition, (1971) (Wiley & Sons).

[2] A. Sommerfeld, H. Bethe, Handbuch der Physik bd. 24, 2 Springer Verlag (1933) p. 333.

[3] A. Matthiessen, Ann.Phys.Chem. 110, 190 (1860).

[4] A. Matthiessen, C. Voigt, Ann.Phys.Chem. 122, 19 (1864)

[5] J. Bass, Advances in Physics 21, nr. 91, 431-604 (1972)

[6] B. Lengeler, Jül - 593 - FN (1969).

[7] G. Revel, Mém.Sci.Rev.Met. LXV, nr.2, 181-184, (1968).

[8] K. Mišek, Cryst.Latt.Def. 1, 223-227 (1970).

[9] O.R. Alldredge, J.W. De Ford, A. Sosin, Phys.Rev. 11, 2860-2870, (1975).

[10] J.C. Jousset, Phys.Stat.Sol., 29, K 127-131,(1968).

[11] F.V. Burckbuchler, C.A. Reynolds, Phys.Rev. 175, 550-555,(1968).

[12] M.C. Kazamergin, C.A. Reynolds, F.P. Lipschutz, P.G. Klemens. Phys.Rev. 6, 3624-3632, (1972).

[13] R.S. Seth, S.B. Woods, Phys.Rev. 2, 2961-2972, (1970).

[14] A.J. Dekker, Ned. T. Natkde 28, 329 (1962).

[15] J. Hillairet, Rapport CENG - C.R. Fin d'Etude 'Métallurgie' nr. 74.7.1085 (Grenoble 1976). J.R. Cost and D.L. Johnson, J. Nucl.Mat. 36, 230 (1970).

[16] A. Seeger, J. Phys.Rad. 23, 616 (1962).

[17] J.H. Evans, B.L. Eyre, Acta Met. 17, 1109, (1969).

[18] C.J. Meechan, A.L. Sosin, Phys.Rev. 113, 424 (1959). T.G. Nilan, A.V. Granato, Phys.Rev. Lett. 6, 171 (1961).

[19] H. Wenzl in 'Vacancies and Interstitials in Metals' Proc. Jülich Conf. 23-28 sept. 1968 (North-Holland (1970))p.364-423.

[20] A. Seeger, H. Stehle, Z.Phys. 146, 242 (1956).

[21] J. Friedel in 'The interaction of Radiation with Solids'. Proc. Summerschool Mol 1963, p. 114 (North-Holland).

[22] P.G. Lucasson, R.M. Walker, Phys.Rev. 127, 485 and 1130 (1962).

[23] J. Roggen, J. Nihoul, J. Cornelis, L. Stals, J.Nucl.Mat. 69-70 (1978).

[24] F. Dworschak, W. Sassin, J. Wick, J. Wurm, Jül. - 575-FN (1969).

[25] F. Dworschak, H. Schuster, W. Wollenberger, J. Wurm, Phys.Stat. Sol. 21, 741 (1976).

[26] E.H. Sondheimer, Advances in Physics 1, 1-42 (1952).
[27] P. Vajda, Reviews of Modern Physics 49, nr.3, 481-521, (1977).
[28] See for instance : the general references [34-38], [41-44]
[29] R.O. Simmons and R.W. Balluffi, Phys.Rev. 125, 862 (1962).
[30] See for instance : M. Suezawa and K. Kimura, Scripta Met 5, 121, (1971).
[31] S. Ceresara, T. Federighi, F. Pieragostini, Phys.Stat.Sol. 16, 439 (1966).
[32] E. Torfs, J. Van Landuyt, L. Stals, S. Amelinckx, Phys.Stat. Sol.(a) 31, 633 (1975).
[33] E. Torfs, L. Stals, J. Van Landuyt, P. Delavignette, S. Amelinckx, Phys.Stat.Sol.(a) 22, 45 (1974).
[34] Vacancies and Interstitials in Metals, North-Holland (1970)
[35] Journal of Physics F : Metal Physics 3, nr. 2, (1973).
[36] Defects in Refractory Metals, Mol (1972).
[37] Proc.Int.Conf. on Fundamental Aspects of Radiation Damage in Metals, Gatlinburg (1975).
[38] Properties of Atomic Defects in Metals, J.Nucl.Mat. 69-70, (1978).
[39] See for instance E. Gebhardt, H.D. Seghezzi, Z. Metallkunde 48, 430 (1957).
[40] W. Mansl and G. Vogl, J. Phys.F. : Metal Phys.7, nr.2, 253-271, (1971).
     H. Rinneberg, W. Semmler, G. Antesberger, Phys.Lett.66A, nr.1, 57, (1968).
[41] Y. Quéré, Défauts ponctuels dans les Métaux, Masson et Cie, Paris (1967).
[42] A.C. Damask and G.J. Dienes, Point Defects in Metals, Gordon and Breach (1963).
[43] J.W. Corbett, Electron Radiation Damage in Semiconductors and Metals (Acad.Press 1966).
[44] M. Doyama and S. Yoshida, Progress in the study of Point Defects, University of Tokyo Press (1977).
[45] J. Nihoul and L. Stals, Phys.Stat.Sol. 17, 295 (1966).
[46] T.R. Waite, Phys.Rev. 107, 463 (1957).
[47] T.R. Waite, Phys.Rev. 107, 471 (1957).
[48] W. Frank in : Defects in Refractory Metals, eds. R. De Batist, J. Nihoul, L. Stals, Mol (1972) p. 199.
[49] L. Stals, Ibidem p. 171.
[50] H.G. Van Bueren, Z. für Metallkde 46, 272 (1955).
[51] H.G. Van Bueren, Imperfections in Crystals, North-Holland (1960).
[52] J.W. Corbett, R.B. Smith, R.M. Walker, Phys.Rev. 114, 1460 (1959).
[53] R. Schindler, Rad. Eff., 35, 17 (1978).
[54] R. Schindler, W. Frank, M. Rühle, A. Seeger, M. Wilkens, J. Nucl.Mat. 69-70, 331, (1978).
[55] ref. 34 , p. 225.
[56] ref. 38 , p. 489.
[57] W. Frank and A. Seeger, J. Nucl.Mat. 69-70, 708, (1978).

[58] K. Sonnenberger, W. Schilling, K. Mika, K. Dettmann,
Rad.Eff. 16, 65 (1972).

[59] Defects and Defect clusters in b.c.c. Metals and their Alloys,
Nucl.Mat. 18, (1973).

[60] V. Hivert, P. Groh, P. Moser, W. Frank, Phys.Stat.Sol.(a) 42,
511 (1977).

[61] W. Mensch, J. Diehl, Phys.Stat.Sol.(a), 43, K 175 (1977).

[62] L. Stals, G. Goedeme, J. Nihoul in 'Defects in refractory
metals' p. 24, Mol (1972).

[63] L. Stals and J. Nihoul, Physica 42, 165, (1969).

[64] R.A. Johnson, Phys.Rev. 134, A 1329, (1964).

[65] J. Nihoul in : Radiation Damage in Reactor Materials, Vienna
(1969), vol.I, p.3.

[66] J.H. Evans, M. Eldrup, Nature 254, 685, 1975.

[67] K.D. Rasch, R.W. Siegel, H. Schultz, J. Nucl.Mat. 69-70, 622,
(1978)
H. Schultz, Scripta Met. 8, 721 (1974).

[68] M. Weller, J. Diehl, W. Trifshaüser, Solid Stat.Comm. 17, 1223,
(1975).
J. Diehl, U. Merbold, M. Weller, Scripta Met. 11, 811, (1977).

[69] S.R. Reintsema, Ph.D. Thesis Groningen (Netherlands) (1976).

[70] J. Nihoul, in 'Vacancies and Interstitials in Metals' North-
Holland (1970), p. 846.

[71] K. Faber, J. Schweikhardt, H. Schultz, Scripta Met. 8, 713
(1974).

[72] K. Maier, H. Metz, D. Herlach, H.E. Schaefer, A. Seeger,
Phys.Rev.Lett. 39, 484 (1977).

[73] J. Roggen, J. Nihoul, J. Cornelis, L. Stals, J.Nucl.Mat.
69-70, 700 (1978).

[74] H. Vandenborre, L. Stals, J. Cornelis, J. Nihoul, Rad.Eff. 21,
137 (1974).

[75] H. Wagenblast, A.C. Damask, J.Phys.Chem.Solids 23, 221 (1962)

[76] L.J. Cuddy, Acta Met. 16, 23 (1968).

[77] W. Schilling, J. Nucl.Mat.72, 1-4, (1978).

[78] F. Dausinger, J. Fuss, J. Schweikhardt, H. Schultz, J. Nucl.
Mat. 69-70, 689, (1978).

[79] P.S. Gwodz and J.S. Koehler, Phys.Rev. B 8, 3616 (1973).

[80] K. Urban and W. Jäger, Phys.Stat.Sol.(b), 68, K1 (1975).

[81] K. Urban and A. Seeger, Phil.Mag. 30, 1395 (1974).

[82] J.N. Lomer, R.J. Taylor, Phil.Mag. 19, nr.159, 437-448 (1969).

[83] F. Dausinger and H. Schultz in ref. [37] p.438.

[84] R.E. Einziger, J.N. Mundy, H.A. Hoff, Phys.Rev.17, 440 (1978).

[85] S.M. Kim and W.J.L. Buyers, J.Phys.F. : Metal Phys. 8, L103
(1978).

[86] G. Hettich, H. Mehrer, K. Maier, Scripta Met. 11, 795, (1977).

[87] H.E. Schaefer, K. Maier, M. Weller, D. Herlach, A. Seeger,
J. Diehl, Scripta Met. 11, 803 (1977).

[88] K. Maier, M. Peo, B. Saile, H.E. Schaefer, A. Seeger,
Phil. Mag., to be published.

[89] K. Faber, H. Schultz, Rad. Eff. 31, 157, (1977).

[90] N.L. Peterson in ref [38] p.3.
[91] K. Maier et al. to be published.

3. Thermal Desorption Spectrometry

3.1. Introduction

In the framework of the study of ionic pumps for the production of
ultra-high vacuum, much attention has been paid in the past to the
interaction of low energy noble gas ions (Ar, Xe, Kr, Ne, He ;
E $\lesssim$ 5 keV) with metals (see ref. 1-11, and references there in).
People were interested in data about sticking probability, spontaneous and induced reemission and saturation. Clean metallic surfaces were bombarded with ions and the reemission or desorption
was measured after bombardment as function of the target temperature. In these experiments the derivative of the partial pressure
(dP/dT) of the desorbed gas is measured by means of a mass spectrometer during a linear rise of the target temperature. It is always
observed that the desorption does not take place in a continuous
way, but that the function $\frac{dP}{dT}$(T) displays several more or less pro-
nounced peaks, which constitute the "desorption spectrum". Typical
desorption spectra are shown in fig. 3.1 for tungsten after bombardment with Ne$^+$, A$^+$, Kr$^+$ and Xe$^+$-ions [1]. In all these experiments the heating rate was 66 K/s. It can be seen that the desorption spectra all contain a number of distinct peaks which show pro-
nounced variation with increasing energy. There are a number of
invidual peaks in each spectrum. The position of the lower temperature peaks is independent of the ion energy, which points to desorptive release whereas the peak temperature of the last peak in
the spectrum depends on ion energy which indicates diffusive
release (see §3.2). It was furthermore observed that the activation energy for desorption, although different for the various
peaks, was independent of the implanted atoms. This observation
suggests that the release is governed by properties of the target
and not by properties of the implanted atoms. The highest desorption peak is ascribed to diffusion of the implanted atoms.
Thermal desorption = spectrometry (TDS) has yielded a large number
of data on the behaviour of inert gases (Kr, Xe, Ne, A, He) in
metals. The desorption spectra for the heavy ions show in general
two types of peaks : at the lower temperature side of the spectrum
first order desorption peaks due to the release of gas atoms trapped within a few layers from the surface, and at the higher temperature side diffusive desorption peaks due to the release of more
deeply implanted gas atoms. The He-desorption spectra show in geng-
ral only first order desorption peaks at low doses ($\leqslant 10^{13}$ ion/cm^2)
[22] whereas the spectrum becomes more complicated at higher doses
($\geqslant 10^{15}$ ions/cm^2) [13] (see §3.3). In thermal desorption spectrometry two types of experiments can be performed. In the first type
of experiment one studies the desorption of the implanted gas atoms
(Xe, Kr, Ne, A, He). In this case the experiment consists of two
steps (see fig. 3.2.a) : (1) implantation of gas atoms by means of

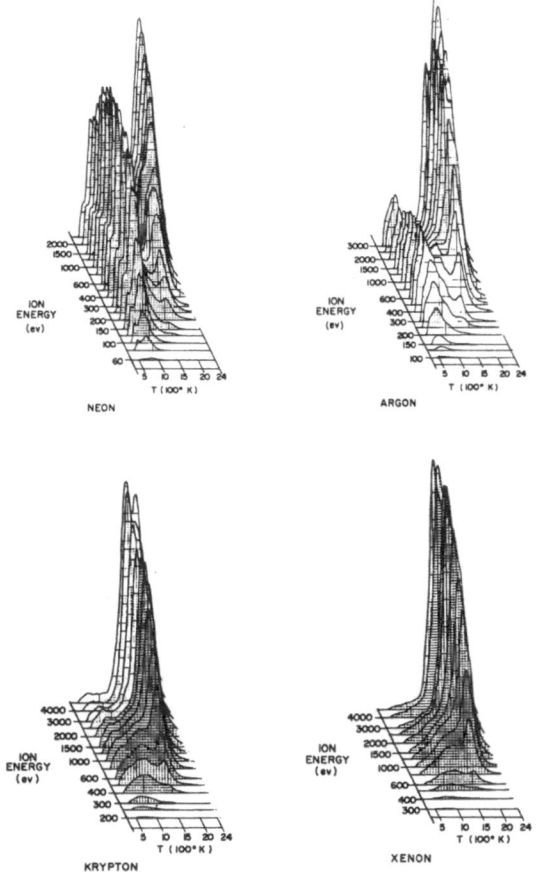

NEON

ARGON

KRYPTON

XENON

Fig. 3.1. Thermal
desorption spectrum
for tungsten for Ne,
A, Kr, Xe as func-
tion of incident ion
energy [1].

low energy ion irradiation (energy < 10 to 20 keV) ; (2) study of
subsequent release upon heating. From a theoretical point of view
it is easier to treat the data if linear heating is applied,
although other heating programme may be used as well [23]. This
type of experiment we call TDS : Thermal Desorption Spectrometry.

In the second type of experiment, a low dose (usually $\leqslant 10^{12}$
ions/cm^2) of heavy ions (Kr, Xe, A, Ne) is implanted in order to
create lattice damage. This implantation is followed by a subthreshold helium irradiation to doses in excess of 10^{12} ions/cm^2 at
a temperature where interstitial helium is mobile (see fig. 3.2.b).
As a consequence of this helium implantation, the vacancy-type defects created during the first irradiation are filled-up. During
subsequent heating of the specimen the helium is released. This
type of experiment was first proposed by Kornelsen in 1972 [12]
and may be considered as a rather ingenious method for studying

defects especially vacancy type defects, in metals. We call this
technique "He-probing-technique" (HPT).

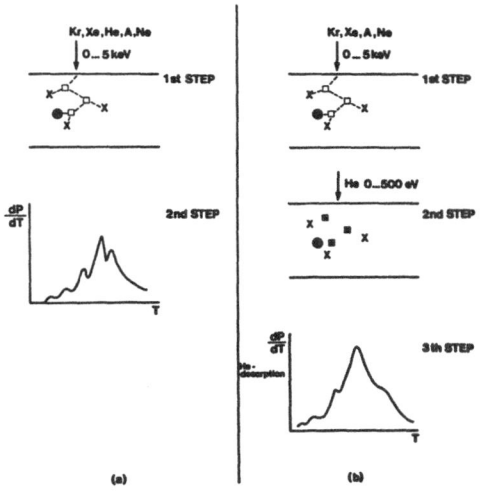

Fig. 3.2 :
Schematic representation of
(a) Thermal desorption spec-
    trometry experiment
(b) He-probing technique

☐ vacancy
X interstitial
● gas atom
▣ helium filled vacancy

3.2. Physical basis

3.2.1. Implantation process

When a clean metallic surface is irradiated with low energy noble
gas ions (Kr^+, Xe^+, Ne^+, Ar^+, He^+) the ions become generally neutralised [14-18] before entering the metal. The gas atom penetrates the metal and loses its kinetic energy in elastic collisions. The energy transferred to the metal atom during a head-on
collision is given by the well-known equation

$$T = \frac{4m_1 m_2}{(m_1 + m_2)^2} E \qquad (3.1)$$

where T = transferred energy
      E = energy of incident ion
     m_1= mass of incident ion
     m_2= mass of metal ion
Some typical values for T are given in table 3.1 for E = 5 keV

m_2 \ m_1	He^4	Ne^{20}	Ar^{40}	Kr^{84}	Xe^{132}
Al (26,98)	2249	4890	4811	3615	2818
Ni (58,71)	1194	3791	4820	4842	3925
W (183,85)	417	1770	2935	4006	4865

Table 3.1. : Values for transferred energy in eV for head-on
              collisions of gas ions of 5 keV with metal ions.

If $T > E_d$, where E_d is the displacement energy, which is typically
between 10 and 110 eV for most metals and most crystallographic
directions [19], the metal atom is displaced from its normal lattice position into an interstitial position, thus leaving behind
a vacant lattice site. If the metal atom has gained enough kinetic
energy, it will in its turn displace other atoms thus creating
lattice damage over several lattice distances and sputtering of
surface. The implanted atom will continue its way creating possibly other primary knocked-on atoms. At the end of its path it will
be at some interstitial position and depending on the target temperature it will migrate interstitially[1]. Davies and Jespersgård
[20] have shown that Xe^{125} migrates interstitially in W with an activation energy between 0.3 and 1.0 eV. On the other hand, it has
been calculated by Bauer and Wison [21] that the interstitial migration energy of He in metals, with the exception of Pd, is below
1 eV. These data point to interstitial migration of inert gas
atoms in metals at 300 K[2]. During this migration the gas atoms
will be trapped in a substitutional position or they will escape
from the target (desorption). The escape probability decreases
with increasing incident ion energy (see for instance E.V.
Kornelsen [1]). This can be understood as follows. Let us suppose
that the gas atom ends its path at a depth z planes below the surface. The number of times the plane at depth z will be visited by
the gas atom during its random walk is $\sim z$. If there are n_a atomic
sites in a plane and n_t trapping sites, then the probability of
being trapped in plane z is $\sim z \frac{n_t}{n_a}$. Because z increases with E,
the trapping probability increases also with E. In the Helium-
Probing-Technique the energy of the helium ions must be such that
no additional damage is created. From equation (3.1) one can
easily derive that the critical energy E_{cr}, below which no additional damage is created, is given by

$$E_{cr} = \frac{(1+m_2/4)^2}{m_2} E_d$$

for $^4He^+$-ions. Some typical values of E_{cr} are given in table 3.2
(E_d-values are taken from ref. [19]).

---

(1) If the gas atom ends its path in the vicinity of a vacancy it
    will be captured by that vacancy. Other interactions between
    gas atoms and defects are also possible. This depends on the
    ratio of atomic radii of gas atom and metal atom.

(2) On the basis of atomic calculations, de Hosson [43] comes to
    the conclusion that in the system Ar-Cu, the argon atoms are
    always in substitutional position even if therefore copper
    atoms have to be pushed in interstitial position.

E_d \ m_2	Ta (180,91)	Mo (95,94)	α-Fe (55,84)	Ni (58,71)
< 100 > : 33	390	-	-	-
< 100 > : 35	-	228	-	-
< 100 > : 17	-	-	68	-
< 100 > : 38	-	-	-	159

Table 3.2 : Values for E_{cr}. All energy-values are in eV.
Values of E_d are taken from ref [19].

3.2.2. Release processes
Following implantation the target is heated, say linearly, with a
heating rate $\mu = \frac{dT}{dt}$ (K/s). The implanted atoms are released. Two
types of release processes can take place : desorptive release and
diffusive release.

Desorptive release
The release rate is only determined by the binding energy of the
atom to the trapping site. This is the case for atoms trapped at
sites near the surface (within $\lesssim$ 10 Å). Due to the proximity of
the surface no random walk takes place. Another very important
case of desorptive release is the helium release in the He-probing-
technique. During release of helium, the helium atoms are pushed
into interstitial position. The migration energy for interstitial
helium is low ($\sim$ 0.3 eV for most metals) [21] compared to the activation energy for desorption ($\sim$ 2.0 eV [12,24]), so that the
rate controlling step is the release from the trapping centre.
Desorptive release is described by a first-order rate equation :

$$\frac{dn}{dt} = - K_o \, n \, e^{-E/kT} \qquad (3.2)$$

where n = number of trapped atoms
    K_o = rate constant ($\sim 10^{13} s^{-1}$)
    E = desorption energy (eV)
    k = Boltzmann constant (8.617 X 10^{-5} eV/K)
    T = absolute temperature (K)
The isothermal rate equation (3.2) can be transformed into the
rate equation for linear heating by substituting

$$dt = \frac{dT}{\mu} .$$

This gives[3] : $\frac{dn}{dT} = - \frac{K_o}{\mu} \, n \, e^{-E/kT}$ \qquad (3.3)

or in relative concentration $n_r = n/n_o$ (n_o is the initial number of
trapped atoms, n is the number of trapped atoms remaining at temperature T) :

$$\frac{dn_r}{dT} = \frac{K_o}{\mu} \, n_r \, e^{-E/kT} \qquad (3.4)$$

---

(3) For an arbitrary heating programme given by t = g(T) one has of
course $\frac{dn}{dT} = -K_o g'(T) \, n \, e^{-E/kT}$, where $g'(T) = \frac{dg}{dT}$.

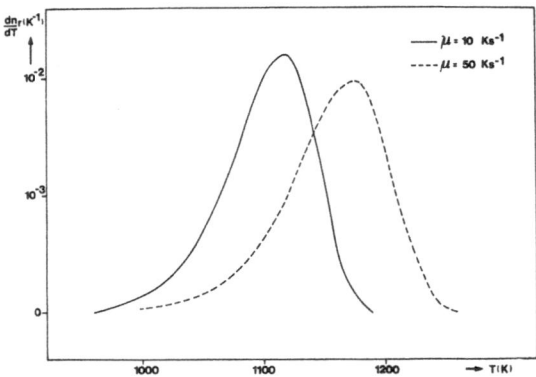

Fig. 3.3 : First order
desorption peak for a
linear heating rate of
μ_1=10 K/s and μ_2=50 K/s.
E = 3.0 eV and
K_o=$10^{13}s^{-1}$; n_r = n/n_o.

The activation energy for a first order desorption peak can be determined from the position of its peak temperature. The peak temperature can be found by differentiation of equation (3.3).
At the peak temperature one has $\frac{d^2n}{dT^2}$ = 0. From the condition
$\frac{d^2n}{dT^2}$ = 0 follows for the peak temperature T_p :

$$\frac{K_o}{\mu}\, e^{-E/kT_p} = \frac{E}{kT_p^2} \qquad (3.5)$$

If K_o and μ are known, the activation energy for desorption can be
determined by means of equation (3.5) by measuring T_p. If K_o is
not known, at least two desorption runs are needed for the determination of E : one run at a heating rate μ_1 and the other run at a
heating rate μ_2. It then follows from equation (3.5)[4] :

$$E = -k\,\frac{T_{p1}T_{p2}}{T_{p1}-T_{p2}}\, \ln \frac{\mu_1 T_{p2}^2}{\mu_2 T_{p_1}^2} \qquad (3.6)$$

where T_{p1} and T_{p2} are the peak temperatures for the heating rates
μ_1 and μ_2. The effect of the heating rate on T_p is demonstrated
in fig. 3.3, which shows first-order-desorption peaks for
E = 3.0 eV, K_o = $10^{13}s^{-1}$ and μ_1 = 50 K/s and μ_2 = 10 K/s.
How can one identify a fist order desorption peak in a desorption
spectrum ? The most important criteria are : (i) the theoretical
width of the peak is 0,08 T_p and its shape is asymmetrical in that
the slope is **steeper** at the higher temperature side (ii) the peak
position is independent of the amount of gas atoms trapped in so
far that no interaction between trapped atoms can occur (see equation 3.4) (iii) the peak temperature is independent of the implan-
tantion energy. Let us take an example : $Ne^+ \rightarrow$ (100)-W [7].
Fig. 3.7.A shows the desorption spectra for Ne^+ irradiation on
(100)-W for various ion energies in the range 40 eV to 5000 eV.
The position of all peaks below 1650 K (4 peaks) is independent of
the implantation energy. Furthermore, Kornelsen and Sinha [7]

---

4) Equation (3.6) holds in first approximation for each pair of
     temperatures (T_1,T_2) at the same value of n_r.

mention that the position of these peaks is independent of dose.
Fig. 3.4 shows the comparison of the 800 K peak with a first order
desorption peak [7]. The width of the theoretical curve at half
maximum is 62 K, whereas that of the experimental curves is 94 K
and 104 K for heating rates of 15.3 K s^{-1} and 40.4 K s^{-1} respectively. Although the experimental width is too large probably as a
consequence of retrapping, the shape of the peaks shows the typical
asymmetry for a first order desorption peak i.e. a steeper slope
at the higher temperature side. The four peaks below 1650 K are
ascribed to desorption from traps in the surface layer ($\lesssim$ 10 Å).

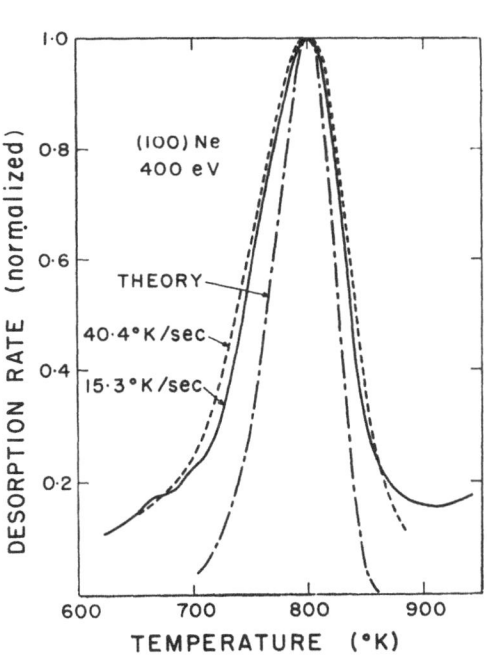

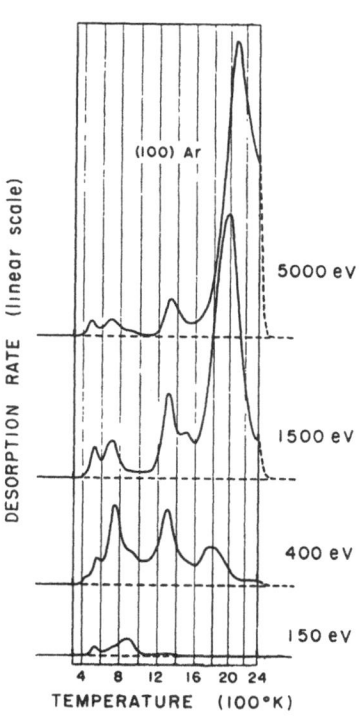

Fig. 3.4 : Comparison of a theoretical first order desorption peak
with the 800 K-desorption peak for
Ne$^+$ (400 eV) → (100) W [7]

Fig. 3.5 : Desorption spectra
for Ar$^+$ → W(100) [7]

Diffusive release
The rate controlling process is the random diffusion of the gas
atom to the surface. The position of the peaks corresponding to
diffusive release from the bulk depends on implantation energy in
that the peak temperature increases with increasing implantation
energy. This is illustrated for the case of Ar$^+$ on W(100) [7]
(fig. 3.5). The peak due to diffusive release, this is the peak at
the highest temperature, appears in the spectrum already at ~ 1800 K

at 400 eV and shifts to $\sim$ 2100 K at 5000 eV. Following Kornelsen
and Sinha [7] this can be qualitatively explained as follows. The
number of jumps executed per unit time by a particular atom is
equal to $K_0 \exp^{-E/kT}$ where E is the energy for diffusive motion of
the atom. If an atom has to perform on the average N jumps before
desorption then the mean desorption time is

$$\tau = \frac{N}{K_0 \exp^{-E/kT}}$$

since $\frac{dn}{dt} = - \frac{n}{\tau}$ it follows

$$\frac{dn}{dt} = - \frac{K_0}{N} n \, e^{-E/kT} \qquad\qquad (3.7)$$

Equation (3.7) replaces equation (3.2) in the case of diffusive
release. It is furthermore known [25] that during random walk the
mean displacement is proportional to N. Thus if the atom starts
its random walk p jump distances below the surface, it follows
p^2 = N, so that equation (3.7) becomes :

$$\frac{dn}{dt} = \frac{K_0 n}{p^2} e^{-E/kT} \qquad\qquad (3.8)$$

For a linear heating rate equation (3.8) can be transformed into :

$$\frac{dn}{dT} = - \frac{K_0 n}{p^2} e^{-E/kT}$$

from which follows for the peak temperature :

$$\frac{K_0}{p^2} e^{-E/kT_p} = \frac{E}{kT_p^2}$$

Fig. 3.6 a,b shows T_p as function of p and $\frac{dn}{dT}$ (T) for different
values of $p(K_0 = 10^{13}$; E=3.0 eV). It can clearly be seen that T_p
increases with p, in agreement with experiments.

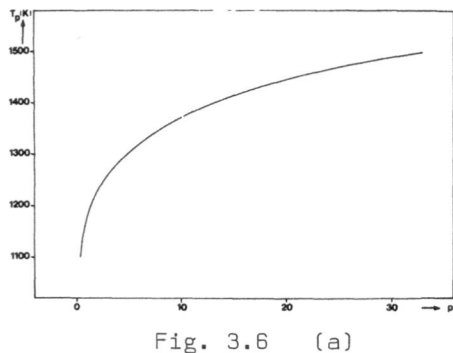

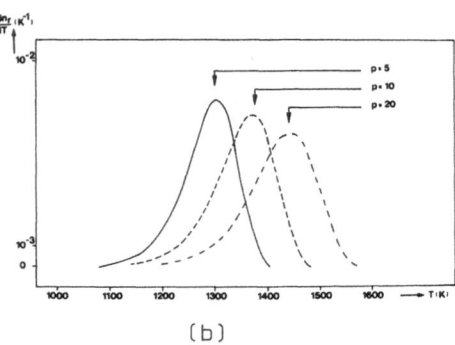

Fig. 3.6   (a)   (b)

3.3. Results obtained by means of TDS and HPT

3.3.1. TDS

The injection of energetic ions into metals followed by a measurement of the release rates of the injected gas is a valuable tool
for studying the mechanism of implantation and migration of the gas

atoms.   This we shall illustrate with a few examples.   We restrict
the discussion to the noble gases (Xe, Kr, Ne, A, He) because they
are chemically inert and do not form chemical bondings as for instance H, O, ... which would complicate the spectra.   The most important data for these gases, which we need for further discussion
are given in table 3.3.

Table 3.3

Element	Mean mass	Most abundant isotope	Natural Abundance (%)
He	4	He^4	99.9999
Ne	20,2	Ne^{20}	90.92
Ar	40,0	Ar^{40}	99.600
Kr	83,8	Kr^{84}	56.90
Xe	131,3	Xe^{129}	26.44
		Xe^{132}	26.89

Xe, Kr, Ne, A
Gas release studies for noble gases from metals have been performed
by a number of authors [1,3,7,8,27,28,29].   A rather complete study
has been performed by Kornelsen [1] and Kornelsen and Sinha [7,8].
They studied the release of inert gases from polycrystalline [7]
and monocrystalline tungsten [7,8] following implantation at 300 K
of the inert gases Xe, Kr, Ne and A in the energy range 40 eV to
5 keV. In the case of polycrystalline material 6 different release
peaks were originally observed (see fig. 3.1).   The activation ener
gy for the various peaks, as derived from first order kinetics, are
tabulated in table 3.4.

Table 3.4   Activation Energy for Desorption of noble gases from
            polycrystalline W (eV) [7]

Gas	Peaks					
	1	2	3	4	5	6
	α-group				β-group	
Ne	1.50	2.03	2.30	2.90	4.1	4.6
A	1.58	1.87	2.27	2.80	3.7	4.4
Kr	1.44	1.78	2.24	2.70	3.7	4.3
Xe	1.45	1.81	2.20	2.64	4.0	4.6

Kornelsen divides the peaks into two groups : from 1 to 4 the α-
group, and 5 and 6 the β-group.   It is believed that the α-group
peaks have to be associated with trapping in the surface layer
($\lesssim$ 10 Å) because the four peaks appear simultaneously in the spectrum at the sticking threshold energy (this is the minimum energy
for absorption, which varies from 31 eV to 170 eV in going from Ne
to Xe [1]).   The β-group peaks have to be associated with the release of more deeply (> 10 Å) implanted atoms.   The activation

energy of all peaks seems to be independent of the gas species,
whereas the activation energy of peak 6 approaches the self-diffusion energy in tungsten (6.08 eV, [26]). It is therefore believed
that the highest temperature peak in the spectrum is due to impurity diffusion. The interpretation of Kornelsen of peak 6 has been
confirmed by Corkhill and Carter [27] . No explanation was given
at that time for peak 5. Kornelsen and Sinha have continued their
experiments on single-crystals [7,8]. Some of their spectra are
shown in fig. 3.7 and fig. 3.8. From the spectra on fig. 3.7 one
can conclude : (i) that below 1650 K all peaks, these are the peaks
1 to 4 or 5 in fig. 3.1 (4 for Ne, and not 5 as thought originally,
5 for Ar, Kr, Xe), are first order desorption peak (see § 3.2)
which are due to release from surface trapping sites ($\lesssim 10$ Å) ;
(ii) that the single peak (peak 6) above 1700 K is due to random
migration of the gas atoms which are more deeply implanted ($\gtrsim 10$ Å)
(activation energy : ~ 4.5 eV). This latter result is in agreement with the data shown on fig. 3.8, where one can observe that
the peak temperature of the impurity diffusion peak at constant implantation energy increases with decreasing mass of the implanted
atom (see table 3.3), in agreement with § 3.2.2. and fig. 3.6.
(iii) the desorption spectra are very sensitive to the crystal face
through which the implantation has occurred.

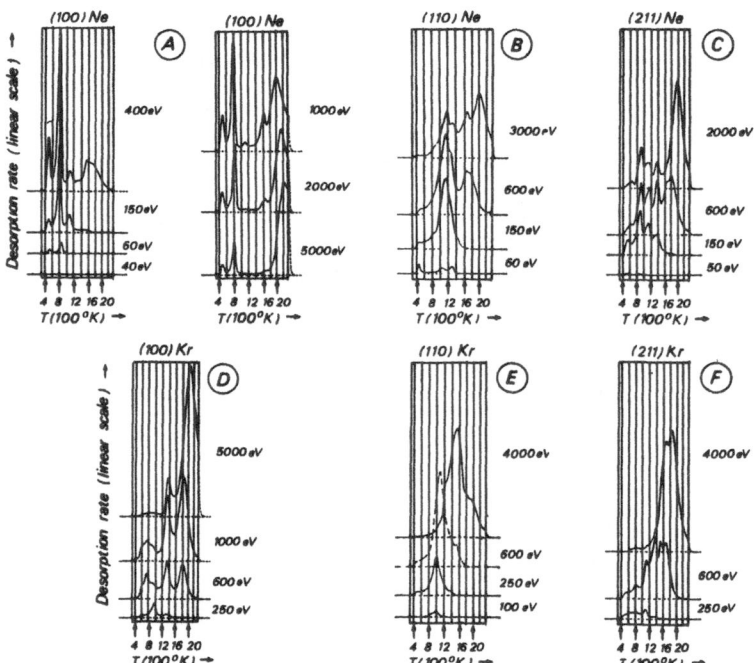

Fig. 3.7 : Desorption spectra for monocrystalline tung-
           sten [7,8]

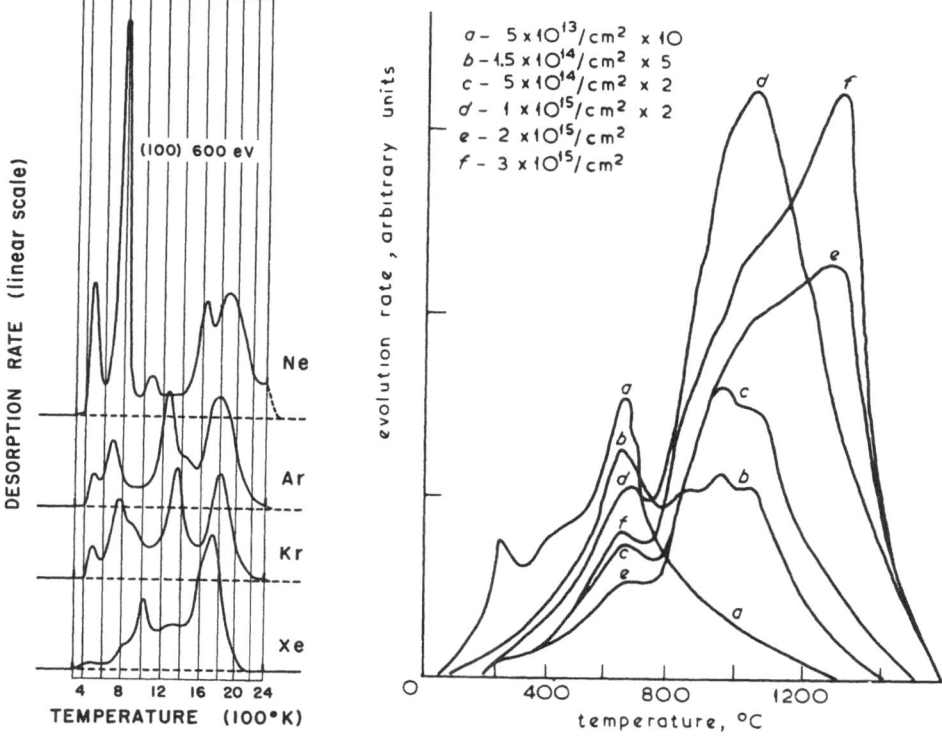

Fig. 3.8. Desorption spectra for
tungsten, which show the effect
of the mass of the implanted atom
on the position of the diffusion
peak [7].

Fig. 3.10. He-release spectra
in the high dose range for
nickel [10]

He
Kornelsen [9] has bombarded at room temperature (100) and (110)
tungsten surfaces with helium ions in the energy range 5 eV to
2000 eV. The thermal desorption spectrum was subsequently measured
during linear heating (μ = 40 Ks^{-1}) up to 2400 K. For energies below 400 eV trapping occurs in the surface layer (see § 3.2.1),
whereas for ion energies > 400 eV trapping occurs predominantly at
bulk sites, created by the incident ions (see fig. 3.9). All peaks
in fig. 3.9 are first order peaks. No peaks for diffusive release
are observed. A number of characteristic binding states are observed with binding energies varying from 2.65 eV to 5.40 eV corresponding to peak temperatures for desorption varying from 1030 K to
2060 K (for a heating rate of 40 K s^{-1}). Further investigations on
the system He → W brought Kornelsen [12] to the conclusion that the
bulk binding sites were vacancy-type defects, and that the method

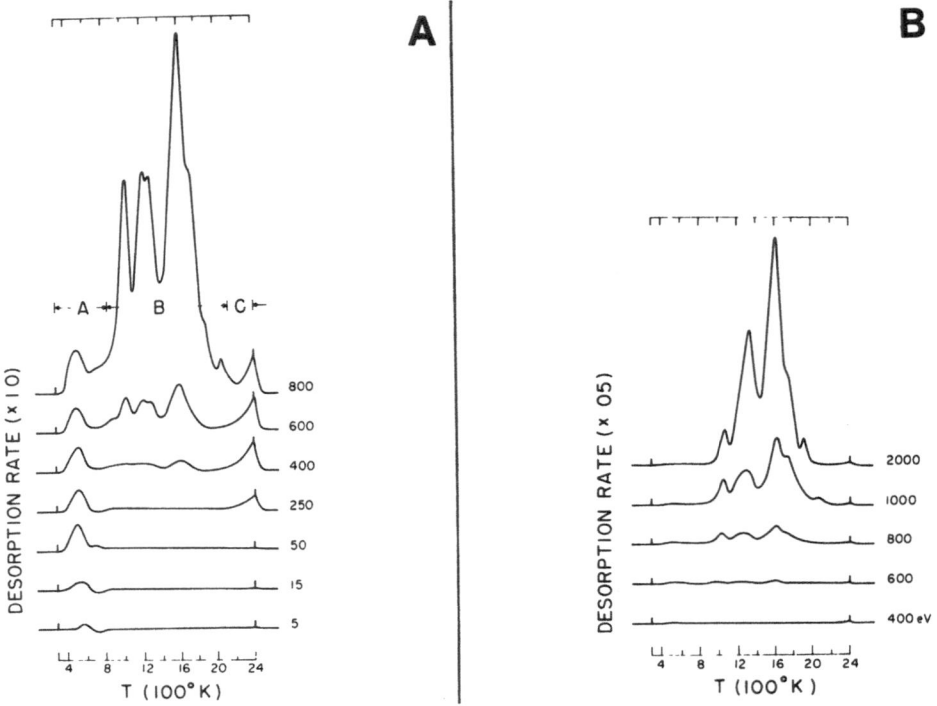

Fig. 3.9. : He-desorption from monocrystalline tungsten [9]

of He-TDS could be used for the study of defects created during implantation of heavy ions (see § 3.2.2).   Edwards and Kornelsen[22]
have studied the release of helium implanted at room temperature
into a nickel (100) crystal.   At a dose of 5.6 X 10^{12} ions/cm^2 and
an energy of 400 eV ($E_{threshold}$ = 160 eV see § 3.2.1) they observe
a first order peak with a maximum at (677 ± 4) K for a heating
rate of 18.4 K s^{-1}.   The activation energy of the peak is
1.70 ± 0.14 eV.   A second apparent first order peak is observed at
(835 ± 4) K with an activation energy of (2.00 ± 0.1) eV.
The position of both peaks is independent of implantation energy
(100 eV to 5 keV) and injected dose (3 X 10^{11} to 8 X 10^{13} ions/cm^2),
which fact suggests, together with the typical asymmetry of the
peaks for first order kinetics, that the underlying process in both
peaks is single step release i.e. desorptive release instead of diffusive release.   Careful analysis of the high temperature peak has
however revealed that this peak cannot be described by the single
activation energy of (2.00 ± 0.10 eV) but that instead a better description may be obtained by assuming K_0 or E or both to be function
of temperature.   It is indeed observed that E decreases with increasing temperature.   The release of helium at higher doses
(> 10^{14} He$^+$/cm^2) from monocrystalline as well as from polycrystalline nickel has been investigated by Reed, Armour, Carter [10]

and by Reed, Harris, Armour, Carter [13]. A typical release spectrum is shown in fig. 3.10 after implantation at 1 keV and at 300 K.
Below doses of 5 X 10^{13} He$^+$ ions/cm^2 release is complete at ~ 800°C,
whereas at' 5 X 10^{14} He$^+$ ions/cm^2 release proceeds until ~ 1200°C.
At the higher doses release moves towards temperatures where considerable evaporation of nickel atoms occurs [13]. The highest
peak in spectrum f is attributed to this evaporation. The authors
consider as the basic absorption reaction, the formation of
complexes He$_n$ V and He$_n$ V$_m$ and attribute the various release peaks
to the dissociation reactions

$$He_n V \rightarrow He_{n-1} V + He \qquad below\ 650°C$$

$$He_n V_m \rightarrow He_{n-1} V_m + He \quad above\ 650°C$$

3.3.2. HPT
The helium probe technique derives from TDS and has first been proposed by Kornelsen [12] in 1972 as an alternative method for studying defects in metals, especially vacancy-type defects. Since
then a number of papers have been published dealing with the study
of defects by means of HPT [11,30-41]. The method consists basically in creating damage in a specimen by means of heavy ion irradiation (Kr, Xe, Ne, A), with energies say 5 keV, and subsequently
irradiating the specimen with He$^+$-ions with energy below the threshold energy for damage (see § 3.2.1, table 3.2, see also fig. 3.2).
The helium ions will be neutralised by an Auger transition [14-18]
before entering the target. They will penetrate into the lattice
a few lattice distances as interstitial atoms, and start threedi-
mensional interstitial diffusion if the temperature of the target
is high enough in order that such diffusion can take place. If the
helium atom encounters a trapping site, e.g. a vacancy, it will be
captured. Wilson and Johnson [42] have, indeed, shown that the formation energy of a helium atom in a vacancy is much smaller than the
helium interstitial formation energy. As a consequence there is a
positive energy balance for the reaction He + V → He V. After the
helium implantation, the vacancy type defects are "colored" with
helium. Upon heating the target, one gets a spectrum which is a
finger print of this "coloration". Fig. 3.11 shows the original
results of Kornelsen [12] obtained for a (100) tungsten crystal.
Fig. 3.11 (A) shows the results for 250 eV-He$^+$ irradiation to a
dose of 2.4 X 10^{13} ions/cm^2 without prior bombardment (a) and with
a prior damaging bombardment of 5 keV Kr$^+$ ions to a dose of
2.4 X 10^{11} ions/cm^2 (b). The helium desorption spectrum is drastically changed by the previous damaging bombardment. A spectrum de-
convolution based on six discrete binding states for helium and
bulk desorptive release is shown in fig. 3.11 B. The low temperature peak which is present in fig. 3.11 (A) (a) and (b) is due to
surface desorptive release. The six peaks in the spectrum are
labelled E, F, G, H', I and are ascribed by Kornelsen to the reactions given in table 3.5. The alternative reactions as calculated
by Caspers, Van Dam, Van Veen [24] by means of their programme

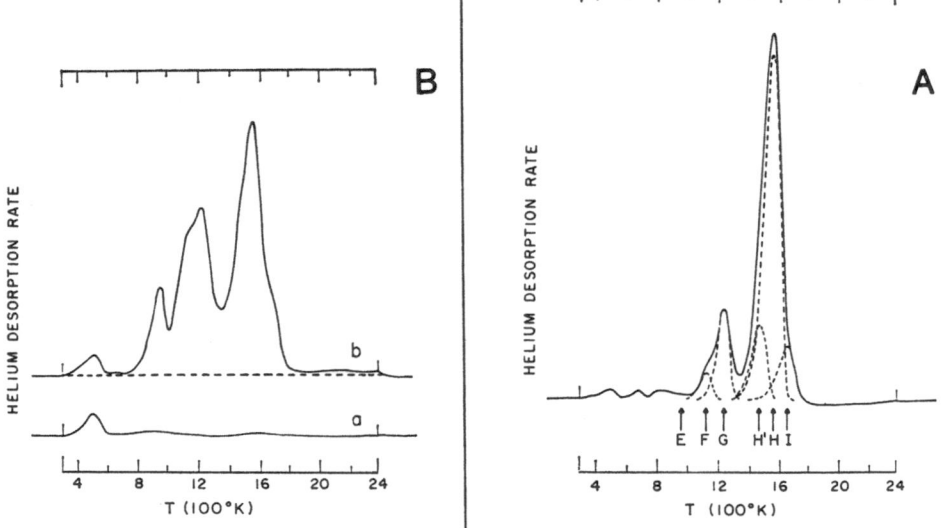

Fig. 3.11: Helium-probing-technique applied to tungsten [12]

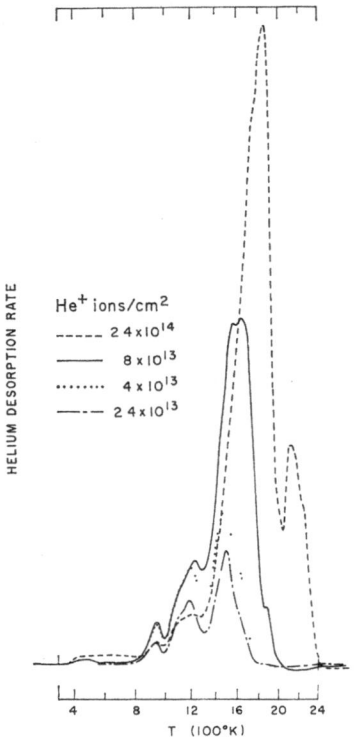

Fig. 3.12

Table 3.5

Peak	$T(\mu=40\ Ks^{-1})$	Activation Energy (eV)		Desorption reaction	
		Kornelsen [12]	Relax [24]	Kornelsen [12]	Relax [24]
E	950	2.52	2.94	$He_3^{II}V \to He+He_2V$ or $He_4V \to He+He_3V$	$He_4V \to He+He_3V$
F	1120	3.0	3.02	$He_3V \to He+He_2V$	$He_3^{I}V \to He+He_2V$
G	1220	3.25	3.43	$He_2V \to He+HeV$	$He_2^{I}V \to He+HeV$
H	1560	4.15	5.07	$HeV \to He+V$	$HeV \to He+V$
I	1675	4.3	5.27	$He_2V_2^{II} \to 2He+2V$	$He_2V_2^{I} \to He+HeV_2$ $(HeV_2 \to He+V_2)$
J	~ 1850	4.8	-	$He_2V_2^{I} \to 2He+2V$	$He_xV_{y\geqslant4} \to He^+$ $He_{x-1}V_{y\geqslant4}$
K	~ 2200	5.8	-	$He_xV_3^2 \to He+He_{x-1}V_3$	

II_{V_2} : second neighbour divacancy

He_3^{II} and $He_3^{I}V$: two configurations of three He-atoms in one vacanc

RELAX are also shown. The peaks J and K appear at high helium doses (see fig. 3.12).

Since peak H_1' which occurs at 1480 K, depends on the mass of the damaging ion, Kornelsen ascribes this peak to the reaction :

$$He(VKr)^{II} \to He + (VKr)^{I}$$

$(VKr)^{II}$: vacancy - Krypton pair in second neighbour position
$(VKr)^{I}$: vacancy - Krypton pair in first neighbour position.

Peaks J and K are again attributed to multiple occupation of trap- ping sites, but in view of the higher activation energy for desorpti the traps must undergo a change in nature. A strong argument in favour of this, is that at doses in the range 10^{13} -10^{15} the trap- ped helium fraction increases by about 50 % [12,13]. Kornelsen suggests as mechanism for changing the trapping site the possibili- ty that an arriving helium atom pushes a lattice atom in intersti- tial position, thus giving rise to bubble growth. Kornelsen has indeed observed that the growth of peaks J and K is prohibited du- ring irradiation at 1100 K. This is the temperature of peak F. If a third helium atom is prevented to be attached to a vacancy, then this mechanism of bubble growth cannot start. Recently Caspers et al. [39] have performed high dose experiments on <110>-molybdenum. If they add more helium to a specimen annealed below peak H they observe the appearance of new high temperature peaks : a triple

I peak, (I_1 I_2 I_3) and a J-peak, in agreement with Kornelsen's data on tungsten [12]. They have also shown by means of atomistic cal-culations that the mutation reaction $He + He_n V \rightarrow He_{n+1} V_2^I + I$ is more like-ly for $n \geqslant 4$ or 5. This means that trap mutation of the complex $He_n V$ occurs only at high doses of helium in agreement with the experi-ments. Van Veen et al [38] have studied the trapping of helium to krypton and krypton-vacancy complexes in tungsten. From their study it follows that, contrary to what has been supposed previous-ly by Kornelsen [12], the annealed state of the krypton-vacancy complex is substitutional krypton. If helium atoms are injected they are trapped by the vacancy-krypton complex, the trapping posi-tion being a neughbouring octahedral interstitial site of the b.c.c. lattice (see fig. 3.13 B). If one helium atom is trapped it desorbs at peak A (see fig. 3.13 A). If two helium atoms are trap-ped, they occupy adjacent octahedral sites and desorb in peak B. Additional He-atoms, from third to sixth, fill the other octahedral sites as indicated in fig. 3.13 B and desorb all in peak C.

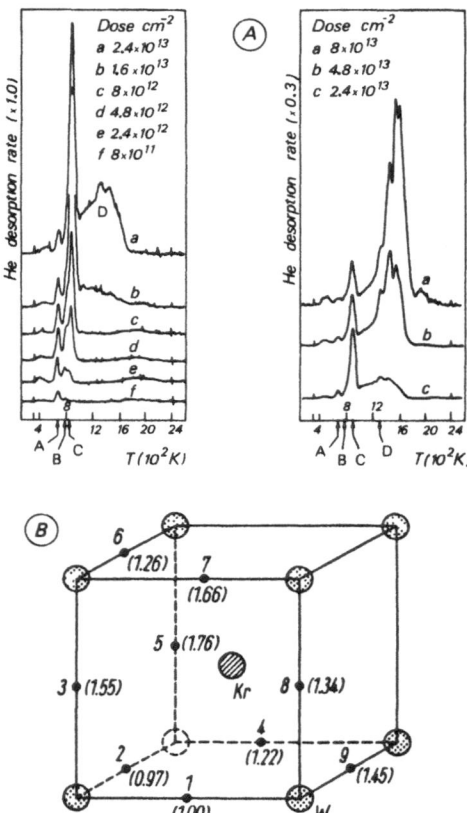

Fig. 3.13 :
$He_n KrV$-complex for-mation in W [38]

Upon further addition of helium trap mutation occurs following the reactions :

$$He_6Kr\ V + He \rightarrow He_7Kr\ V_2^I + I$$

The desorption of helium from that mutated trap occurs in peak D.

References

[1] E.V. Kornelsen, Can.Journ. of Physics 42, 364-381,(1964).
[2] P.A. Redhead, Vacuum 12, 203, (1962).
[3] K. Erents, R.P.W. Layson, G. Carter, Journal of Vacuum Science and Technology, 4, nr. 5, 252-256, (1966).
[4] K. Erents and G. Carter, Vacuum 16, nr. 10, 523-527, (1966).
[5] K. Erents and G. Carter, Vacuum 17, nr. 4, 215-218, (1967).
[6] K. Erents, G. Farrel and G. Carter, Proc. 4th Intern. Vacuum Congress (Institute of Physics and Physical Society, London) (1968).
[7] E.V. Kornelsen and M.K. Sinha, Journ. Applied Physics 39 nr. 10, 4546-4555, (1968).
[8] E.V. Kornelsen and M.K. Sinha, Journ. Applied Physics 40 nr. 7, 2888-2894, (1969).
[9] E.V. Kornelsen, Can.J. Applied Physics 48, 2812-23, (1970).
[10] D.J. Reed, D.G. Armour, G. Carter, Vacuum 24, nr. 10, 455-461, (1974).
[11] G. Carter, Vacuum 26, nr. 8, 329-332, (1976).
[12] E.V. Kornelsen, Rad.Eff. 13, 227-236, (1972).
[13] D.J. Reed, F.T. Harris, D.G. Armour, G. Carter, Vacuum 24, nr.4, 179-186, (1974)
[14] Hagstrum H.D., Phys. Rev. 96, 336-365, (1954).
[15] Hagstrum H.D., Phys. Rev. 96, 325-335, (1954).
[16] Hagstrum H.D., Phys. Rev. 104, 309-316, (1956).
[17] Hagstrum H.D., Phys. Rev. 104, 317-318, (1956).
[18] Hagstrum H.D., Phys. Rev. 123, 758   , (1961).
[19] P. Vajda, Reviews of Modern Physics 49, nr. 3, 481-521, (1977).
[20] J.A. Davies and J. Jespersgård, Canad.J. of Physics 44, 1631-1638, (1966).
[21] W. Bauer and W.D. Wilson, Proc. Int. Conf. on Radiation induced voids in Metals, Albany(1971)p. 230.
[22] D. Edwards and E.V. Kornelsen, Surface Science 44, 1-10, (1974)
[23] G. Carter and D.G. Armour, Vacuum 19, nr. 10, 459-460, (1969).
[24] L.M. Caspers, H. Van Dam, A. Van Veen, Delft, Progress report on Chemistry and Physics, Chem. Eng. 1, 39-44, (1974).
[25] See for instance P.G. Shewmon, Diffusion in Metals, Mc Graw Hill (1963), p. 47.
[26] N.L. Peterson, J. Nucl. Mat. 69-70, 3-37, (1978)
[27] D.P. Corkhill and G. Carter, Phys. Lett. 18, nr.3, 264, (1965).
[28] R. Kelly and F. Brown, Acta Met. 13, 169 (1965).
[29] E.V. Kornelsen and M.K. Sinha, App. Phys.Lett. 9, 122 (1966).
[30] E.V. Kornelsen, A. Van Veen, L.M. Caspers, H. Bakker, Phys.Stat. Sol.(a) 39, K 143-K 146, (1977).
[31] D. Edwards, E.V. Kornelsen, Rad. Eff. 26, 155-160, (1975).

[32] E.V. Kornelsen and D.E. Edwards, Appl, Ion beams to Metals,
     Ed. S.T. Picraux   E.P.Eermisse, F.L. Vook, Plenum Press,
     (1974) p. 521-529.
[33] C. Roodbergen, A. Van Veen, L.M. Caspers, Delft Progress
     Report, Chemistry and Physics, Chem. Eng. 1, 107-114, (1975).
[34] A. Van Veen and L.M. Caspers, Proc. 7th Int.Vac.Congress and
     3rd Int. Conf. on Solid Surfaces, Sept. 12-16, 1977, Vienna
     (2637-2640).
[35] R.H.J. Fastenau, L.M. Caspers, A. Van Veen, Phys.Stat.Sol. (a)
     34, 277-289,(1976).
[36] A. Van Veen, A.A. Van Gorkum, L.M. Caspers, J. Nihoul,
     L. Stals, J. Cornelis, Phys.Stat.Sol. (a) 32, K 123-K 126,
     (1975).
[37] L.M. Caspers, A. Van Veen, A.P. Van Gorkum, A. Van den Beukel,
     C.M. Van Baal, Phys.stat.Sol. (a) 37, 371-383, (1976).
[38] A. Van Veen, L.M. Caspers, E.V. Kornelsen, R. Fastenau,
     A. Van Gorkum, A. Warnaar, Phys.Stat.Sol.(a) 40, 235-246,
     (1977).
[39] L.M. Caspers, R.M.J. Fastenau, A. Van Veen, W.F.W.M. Van
     Heugten, Phys.Stat.Sol.(a) 46, 541-546, (1978).
[40] W.F.W.M. Van Heugten, V. Van de Berg, L.M. Caspers, A. Van Veen,
     Delft Progress Report, 3, 97-106, (1978).
[41] F. Van de Berg, W. Van Heugten, L.M. Caspers, A. Van Veen,
     to be published in Solid State Chemistry (1978).
[42] W.D. Wilson and R.A. Johnson, in 'Interatomic Potentials and
     Simulation of Lattice Defects' Eds. P.C. Gehlen et al.
     Plenum Press (1972), p. 375.
[43] J. Th. M. de Hosson Phys. Stat. Sol. (a) 40, 293 (1977).

4. Analysis of experimental annealing data

4.1. Annealing Programme

The diffusional properties of structural defects can be studied by
means of annealing experiments. In such experiments one measures
the variation of the defect sensitive physical property of the bulk
material as a function of time at a constant annealing temperature,
(isothermal annealing) or, as a function of temperature at a varying
annealing temperature. In some experiments e.g. stored-energy
measurements, combined length change-lattice parameter experiments
or thermal desorption spectrometry it is suitable to measure the
variation of the physical property at the annealing temperature itself. In other experiments e.g. electrical resistivity, lattice
parameter, length-changes or Hyperfine Interaction experiments, it
is more suitable to measure the variation of the physical property
at a constant reference temperature. The latter has to be chosen
low enough so that at that temperature and within laboratory times,
no changes of the physical properties involved occur. If measurements are carried out at a reference temperature, one has to inter-
rupt the annealing treatment for each measurement. Isothermal annealing treatments are then performed stepwise i.e. the sample is

heated at a well-defined temperature T_1 for a certain length of time Δt_1, next cooled down very rapidly to the reference temperature for measurement, after which the sample is heated again to T_1 for a length of time Δt_2 ($\leqslant \Delta t_1$). One obtains a heating programme as shown in fig.(4.1.):

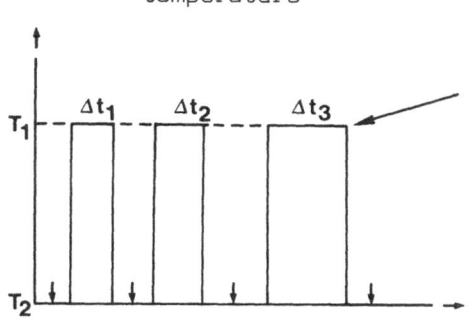

temperature

total annealing time at
the temperature
$T_1 = \Delta t_1 + \Delta t_2 + \Delta t_3$

time

Fig. 4.1.   Isothermal pulse annealing programme. The arrows indicate the moments at which measurement takes place at
the reference temperature T_r.

In addition to isothermal annealing programmes other temperature-
time programmes are used in which the temperature varies with time.
Two programmes are mainly used : <u>continuous heating</u> and <u>pulse
heating</u>. In the first case the temperature T is a continuous function of time t :

$$t = \varphi(T) \qquad (4.1)$$

A special case is the linear continuous heating programme, for
which the temperature T is a linear function of time t :

$$t = \frac{1}{\mu} (T - T_o) \qquad (4.2)$$

where μ represents the (constant) heating rate and T_o the temperature at the moment $t = t_o = 0$ i.e. the initial temperature. Such programme, can easily be realized if the measurements take place at
the annealing temperature itself. These measurements can be performed continously or periodically. One obtains a programme as shown
on fig. 4.2. :

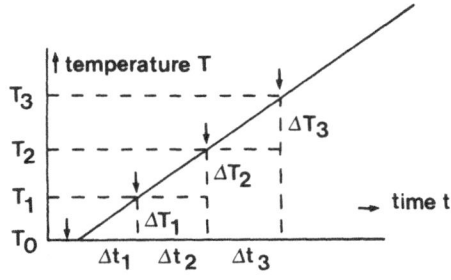

Fig. 4.2.   Linear continous
annealing programme. The
arrows indicate the moments at
which a measurement takes place
at the corresponding tempera-
ture $T_1 (= T_0 + \Delta T_1)$, $T_2 (= T_1 + \Delta T_2)$,
etc. In general one has

$\Delta T_1 \neq \Delta T_2 \neq \Delta T_3$ and $\Delta t_1 \neq \Delta t_2 \neq \Delta t_3$

If, however, the physical property is measured at a reference temperature, the programme has to be interrupted periodically. This
is shown on fig. 4.3.

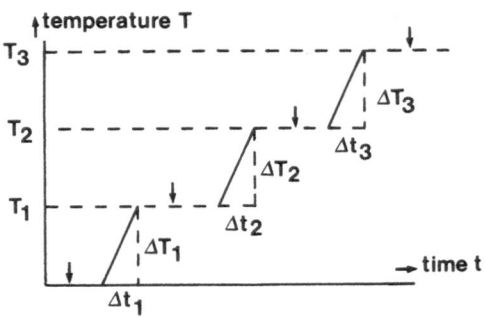

Fig.4.3. Linear continuous
annealing programme with periodical interruptions. The
arrows indicate the moments at
which measurement at the reference temperature T_r takes
place. In general one has :
$\Delta T_1 \neq \Delta T_2 = \Delta T_3$ and $\Delta t_1 \neq \Delta t_2 \neq \Delta t_3$

It is clear that in practice a programme as shown on fig.4.3. is
difficult to realize. One uses, instead, a pulse heating programme.
The most general form of such a programme is shown on fig. 4.
The sample is heated at a temperature T_1 for a certain length of
time Δt_1, next cooled down to the reference temperature T_r for
measurement, after which the sample is heated again to a temperature
$T_2 > T_1$ for a length of time $\Delta t_2 \neq \Delta t_1$ or $= \Delta t_1$.

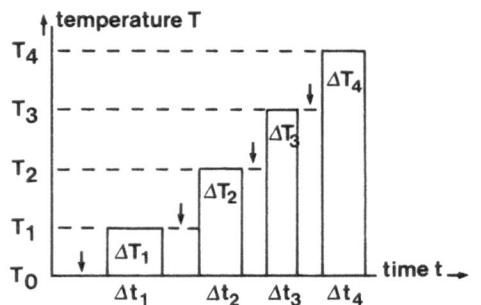

Fig. 4.4.   The most general
pulse heating programme. The
arrows indicate the moments at
which measurement at the reference temperature T_r takes
place. $\Delta T_1 \neq \Delta T_2 \neq \Delta T_3$

$\Delta t_1 \neq \Delta t_2 \neq \Delta t_3$

From fig. 4.4. one can deduce that actually an infinite number of
(ΔT, Δt) combinations are possible, each of which gives rise to a
pulse heating programme. However, only two programmes are of
practical interest, namely <u>isochronal pulse heating</u> and <u>linear
isochronal pulse heating</u>. The first programme is defined by :

$$\Delta t_1 = \Delta t_2 = \Delta t_3 = \ldots$$
$$\Delta T_1 \neq \Delta T_2 \neq \Delta T_3 \quad \ldots \qquad (4.3.)$$

The linear isochronal pulse heating programme is defined by :

$$\Delta t_1 = \Delta t_2 = \Delta t_3 = \ldots$$
$$\Delta T_1 = \Delta T_2 = \Delta T_3 = \ldots \qquad (4.4.)$$

The heating rate, defined by

$$\Delta t = \frac{1}{\mu} \Delta T \qquad (4.4.)$$

is constant in the latter case only.
In practice square pulses can only be realized partially : a good
approximation is obtained by using temperature-controlled bathes.
This technique, however, can only be used for moderate temperatures
(between 20°C and 200°C) and for metals which do not react with
the bath liquid. For higher or lower temperatures one usually uses
temperature controlled electrical furnaces. In that case it is
very difficult and sometimes impossible to obtain a square pulse
programme. This, however, is not an inseparable difficulty. One
can, indeed, correct for this effect. Suppose, therefore, that the
pulse is as shown on fig. 4.5.. The measured pulse time is t_p,
which is apparently to short.

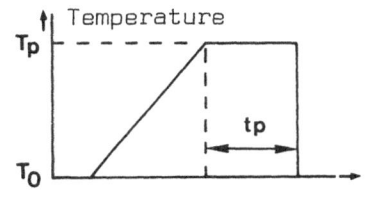

Fig. 4.5.   Idealized pulse
during heating of a sample
in a temperature controlled
electrical furnace

The sample is in fact during a longer time $(t_p + \Delta t)$ at the temperature T_p. The value of Δt can be calculated as follows.
Suppose that the temperature increases linearly with time from a
value T_o to T_p. Following principle of equivalent times one can
write :

$$\Delta t \exp(-\frac{E}{kT_p}) = \frac{1}{\mu} \int_{T_o}^{T_p} \exp(-\frac{E}{k\vartheta}) d\vartheta \qquad (4.$$

where E represents the activation energy of the (singly-activated)
process, k is Boltzmann's constant and $\mu = \frac{d\vartheta}{dt}$, is the heating rate
for the interval (T_o, T_p). Eq. (4.5.) results in :

$$\Delta t = \frac{k}{\mu E} [T_p^2(1- \frac{2kT_p}{E})-T_o^2 e^{\frac{E}{k}(\frac{1}{T_p} - \frac{1}{T_o})} (1- \frac{2kT_o}{E})] \qquad (4.6.)$$

A real pulse does not look like as shown on fig. 4.5.,   but rather
like as shown on fig. 4.6..

temperature

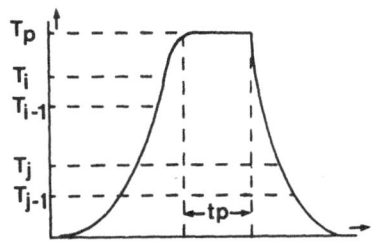

Fig. 4.6.   Real heating
pulse during heating of a
sample in a temperature controlled electrical furnace.

In that case one can still apply eq. (4.6.) by dividing the
heating and cooling curve into several parts, which can be approximated by straigth lines. For each straight part one can write :

for the heating curve

$$\Delta t_i = \frac{1}{\mu_i E} [T_i^2 (1 - \frac{2kT_i}{E}) - T_{i-1}^2 \; e^{\frac{E}{K}(\frac{1}{T_i} - \frac{1}{T_{i-1}})} (1 - \frac{2kT_{i-1}}{E})] \qquad (4.7.a)$$

and for the cooling curve :

$$\Delta t_j = \frac{1}{\mu_j E} [T_j^2 (1 - \frac{2kT_j}{E}) - T_{j-1}^2 \; e^{\frac{E}{k}(\frac{1}{T_j} - \frac{1}{T_{j-1}})} (1 - \frac{2kT_{j-1}}{E}] \qquad (4.7.b)$$

μ_i and μ_j are the heating rates and cooling rates for the intervals being considered. For the effective pulse length one obtains :

$$(t_p)_{eff} = t_p + \sum_i \Delta t_i + \sum_j \Delta t_j \qquad (4.8.)$$

Most of the theoretical equations which describe the annealing process are only applicable to isothermal and linear continuous heating data. They are, however, frequently used for pulse heating data. It should be mentioned, however, that these equations can only be applied for linear isochronal pulse heating data and then still only if the programme meets special requirements and further- more, if the experimental data are corrected in the right way. The corrections which have to be made are derived in detail in [1] and are summarized in the next paragraph (§.4.2.) (Some aspects of these corrections are also treated in [2,3]).

4.2. Normalization to the same heating rate

It frequently occurs that measurements of various authors have to be normalized to the same heating rate. In order to do so one has to perform two operations : (i) if the data are obtained for a linear isochronal pulse programme, then a first correction has to be per- formed by adding $(1-y)\Delta T$ to the temperatures of the step annealing programme. This correction changes the step annealing programme into a continuous annealing programme (ii) the second operation con- sists of a normalization of two continuous annealing programmes to the same heating rate. As far as the first correction is concerned, it is clear that the concentration of defects after pulse annealing at a temperature T_r is not equal to the concentration after continuous annealing up to the temperature T_r but up to the tempe- rature $T_r + (1-y)\Delta T$, where y depends on the annealing temperature and on the temperature step ΔT of the pulse programme. The value of y can be computed as follows. If the change of the physical parameter being studied has to be the same after the pulse at temperature T_r as after continuous annealing up to $T_r + (1-y)\Delta T$, one has

$$e^{-E/kT_r} \Delta T = \int_{T_r - y\Delta T}^{T_r + (1-y)\Delta T} e^{-E/kT} \; dT$$

or

$$e^{-E/kT_r} \Delta T = [\frac{1}{\Delta T} \frac{kT^2}{E} e^{-\frac{E}{kT}} (1 - \frac{2kT}{E} + \frac{6k^2 T^2}{E^2})]_{T_r - y\Delta T}^{T_r + (1-y)\Delta T}$$

$$(4.9.)$$

Equation (4.9.) gives y as function of T_r and ΔT. Fig. 4.7.

shows y as function of T_r with T as parameter. It is clear, that, if ΔT is small the correction is also small

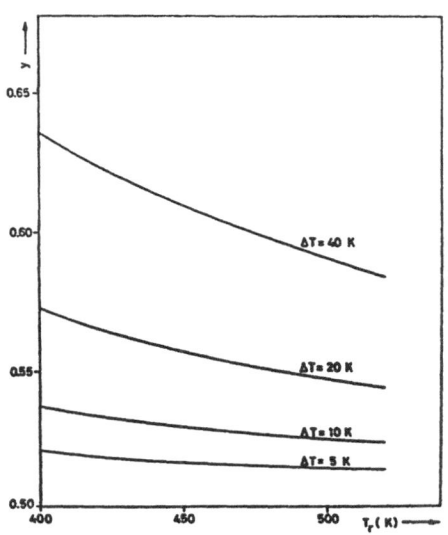

Fig. 4.7.   Dependence of the correction fac- tor y on the pulse annealing temperature T_r and temperature step ΔT [4].

The second correction can be performed as follows. Suppose that the equation which describes a singly-activated annealing process can be written as follows :

$$- \frac{dc}{dT} = \frac{1}{\mu} f(c)\ e^{-E/kT} \qquad (4.10.)$$

For two different heating rates μ_1 and μ_2, one will arrive at the same amount of recovery at two different temperatures T_1 and T_2. One has

$$\int_{co}^{c} \frac{dc}{f(c)} = \frac{1}{\mu} \int_{To}^{T} e^{-E/kT}\ dT \qquad (4.11.)$$

from which it follows

$$\frac{1}{\mu_1} \int_{T_0}^{T_1} e^{-E/kT}\ dT = \frac{1}{\mu_2} \int_{T_0}^{T_2} e^{-E/kT}\ dT \qquad (4.12.)$$

To a very good approximation, eq (4.12.) can be replaced by :

$$\frac{1}{\mu_1} \frac{kT_1^2}{E} e^{-E/kT_1} (1-2\ \frac{kT_1}{E}) = \frac{1}{\mu_2} \frac{kT_2^2}{E} e^{-E/kT_2}(1-2\ \frac{kT_2}{E}) +$$

$$(\frac{1}{\mu_1} - \frac{1}{\mu_2}) \frac{kT_0^2}{E} e^{-E/kT_0}(1-2\ \frac{kT_0}{E}) \qquad (4.13.)$$

If μ_1, μ_2, T_2 and E are known one can calculate the value of T_1 by means of equation 4.13.
To conclude : if annealing data are available for a step annealing

programme, they can be transformed to a continuous linear annealing
programme by means of a two-fold correction : the new temperature T
can be computed from the experimental temperature T_{exp} by means of
$T = T_{exp} + (1-y)\Delta T - \Delta T_{progr}$, where $\Delta T_{progr} = |T_1 - T_2|$
(see eq. 4.13.).   Fig. 4.8.   shows an example of this normalisation
procedure [5].   The figure shows electrical resistivity data of
four different laboratoria [6-9] on stage III recovery in molyb-
denum.   The figure represents $\Delta \rho_i^{III}$ as function of $1/T_i^{III}$ where
$\Delta \rho_i^{III}$ is the fraction of $\Delta \rho^{III}$ left at the inflection point T_i^{III}.
Fig. 4.8.a   shows the data without any correction ; fig. 4.8.b
shows the data after correction for step annealing ; finally, fig.
4.8.c   shows the data after correction for the heating rate.   It
can clearly be seen that after the two corrections the points of
the four laboratories come fairly well on one and the same continuous
curve.   Without correction the peak temperatures scatter by about
20 K, after correction the scattering is about 5K. It should be
remarked, that in resistivity experiments or hyperfine interaction
measurements the heating rates are generally moderate ($\sim$ 10 Kh^{-1}
to 50 Kh^{-1}).   In thermal desorption experiments the heating rates
are typically 30 Ks^{-1}.   Therefore large corrections have to be
applied if one wants to compare TDS-data with say electrical resistivity data or HFI-measurements.

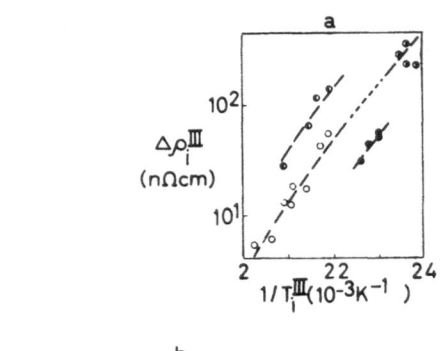

Fig. 4.8.   Normalization of
stage III recovery data of
molybdenum to the same heating
rate (see text).

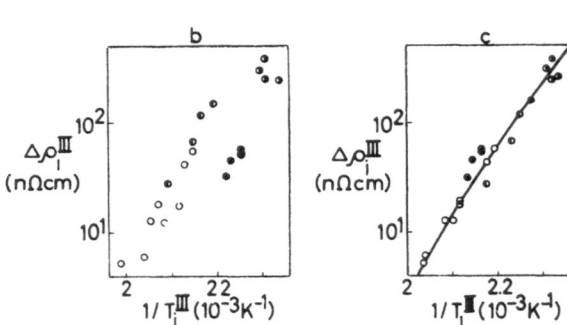

References

[1] R. Gevers, J. Nihoul, L. Stals, Phys.Stat.Sol.15, 701-716 (1966).
[2] M. Balarin, R. Rattke, A. Zetzsche, Phys.Stat.Sol.22, 123 (1967) ; Zfk-123 (1966).
[3] F. Bell and R. Sizmann, Phys.Stat.Sol. 15, 369 (1966)
[4] G. Goedemé, doctor Thesis K.U. Leuven (1971)
[5] J. Nihoul in "Defects in Refractory Metals" Mol (1972) p. 385
[6] L.K. Keys, J.P. Smith, J. Moteff, Phys.Rev.Lett. 22, 57 (1968)
[7] G.H. Kinchin, M.W. Thompson, J. Nucl.Energy 6, 275 (1958).
[8] D.E. Peacock, A.A. Johnson, Phil.Mag. 8, 563 (1963).
[9] L. Stals, G. Goedeme, J. Nihoul in "Defects in Refractory Metals", Mol (1972) p. 23.

APPENDIX : Reaction Order of Recovery Processes

The isothermal recovery of point defects in solids can quite generally be described by the equation

$$- \frac{dc_r}{dt} = K \, f(c_r) \tag{A.1}$$

where K is a rate constant and $c_r = c/c_0$ is the relative concentration of the defects being considered, with c in the concentration at time t and c_0 at t=0. It is often assumed that the function $f(c_r)$ varies as a power of the concentration so that equation (A.1) may be written as :

$$- \frac{dc_r}{dt} = K \, \alpha \, c_r^\gamma \tag{A.2}$$

γ is called the order of reaction : $\gamma=1$ means first order reaction $\gamma=2$ means second order reaction, etc...
α is a proportionality factor which for constant order reactions is equal to 1. There exist a number of methods for determining the reaction order from experimental annealing data [A1-A5] These methods are all based on the assumption that γ is constant over the whole recovery process. This is, however, in most cases an invalid assumption [A6-A8]. If the function $f(c_r)$ is known, then γ and its concentration dependence can be calculated by means of the expression [A6] .

$$\gamma(c_r) = c_r \frac{f'(c_r)}{f(c_r)} \tag{A.3}$$

where $f'(c_r) = \frac{df(c_r)}{dc_r}$. One observes in general that γ varies with concentration [A.6-A.9]. If experimental isothermal annealing data are available, then γ can be determined by plotting $\ln(-dc/dt)$ as function of $\ln c$. From equation (A.1) and (A.3) it can indeed be shown that

$$\gamma = \frac{d \, \ln(-dc/dt)}{d \, \ln c} \tag{A.4}$$

The value of γ and its variation as recovery proceeds is often used to distinguish between various recovery processes. One should however, be vary careful in doing so, since different processes can

give rise to a similar variation of γ as function of concentration"
[A.6,A.7,A.9]. Kinetics arguments may only be used in the opposite
way : if some recovery mechanism is put forward, then the observed
kinetics should at least be consistent with the model being proposed.

References

[A1] C.J. Meechan and J.A. Brinkman, Phys.Rev. 103, 1193 (1956).
[A2] J.H. Evans, Phys.Letters 17, 105 (1965).
[A3] F.E. Fujita, A.C. Damask, Acta Met. 12, 331 (1964).
[A4] M. Balarin, A. Zetzsche, Phys.Stat.Sol. 2, 1670 (1962) ; 3,
       K 387 (1963).
[A5] H. Schultz, Acta Met. 12, 331 (1964).
[A6] J. Nihoul and L. Stals, Phys.Stat.Sol. 17, 295 (1966).
[A7] L. Stals in "Defects in Refractory Metals" Mol (1971).
[A8] A. Seeger and U. Gösele, Rad. Eff. 24, 123, (1975).
[A9] G. Goedemé, Ph.D.Thesis Leuven (1971).

Acknowledgement

The many helpful discussions with dr. J. Nihoul, dr. J. Cornelis,
dr. H. Janssen and Miss H. Van Swijgenhoven are gratefully
acknowledged.
Many thanks are due to Mrs. Magda Ieven and Miss Ingrid Dewispelaere
for photographic work and drawings. The careful typing of the final
text by Mrs. Lea Verboven is highly appreciated.

AN INTRODUCTION TO SEVERAL SOLID STATE TECHNIQUES FOR THE STUDY
OF ION IMPLANTED MATERIALS[*+]

J. A. Borders

Sandia Laboratories[**]

Albuquerque, NM 87185

ABSTRACT

The study of ion-implanted materials requires methods which
are sensitive to the local structure and chemistry of the implanted
atoms. Optical spectroscopy and transmission electron microscopy
have proven to be among the most useful solid state methods. For
the study of materials implanted to very high fluences, the use of
surface analysis methods may also provide some unique information.
The characteristics of these methods will be reviewed and examples
presented which show how the techniques can be used to analyze implanted materials.

INTRODUCTION

Ion implantation is a method of introducing impurities into
a solid in a nonequilibrium fashion. The method was developed for
producing very thin doped layers in semiconductors and it is now
a successful commercial process. For semiconductors, the number
of ions implanted into a surface is relatively low because it is
the electrical properties of the implanted material which are of
interest and large changes in the electrical characteristics of
a semiconductor can be produced with a small number of dopant atoms.

*This article is based on a lecture delivered at a N.A.T.O.
 Advanced Study Institute held in Aleria, Corsica between
 September 10th and September 23rd, 1978.
 +This work supported by the United States Department of Energy
 (DOE) under Contract #AT(29-1)-789.
**A U. S. DOE facility.

Typical implantation doses for semiconductors range from 10^{11}cm^{-2} to 10^{14}cm^{-2}. In metals and insulators, however, interest centers on changing the optical or mechanical or electrochemical properties and much higher doses are required, in the range 10^{15}cm^{-2} to 10^{18}cm^{-2}. Doses in this range can produce maximum atomic concentrations of implanted atoms of up to 10-30 atomic percent. At these high doses, interactions between dopant atoms are commonly seen, for example the production of new phases and high levels of damage in the host material can be produced.

During the development of ion implantation as a commercial semiconductor process, the implanted layers were first characterized by measuring their electrical properties such as sheet resistivity and Hall mobility. But electrical effects integrate over the effects of all the implanted atoms and their associated damage. A basic understanding of the damage processes and the electrical activation of the implanted atoms was achieved only after techniques were applied which were capable of measuring defect inventories as a function of annealing temperature: techniques such as optical spectroscopy, electron paramagnetic resonance, and transmission electron microscopy (TEM).

In order to study the state of the implanted atoms and the damage in the implanted layer in metals and insulators, we must also apply experimental methods which are sensitive to the microscopic structure of damage such as point defects and dislocations and to the more macroscopic structures such as regions of new phases.

This paper will briefly discuss optical spectroscopy and TEM. Additionally we will briefly describe the electron spectroscopy methods of Auger electron spectroscopy (AES) and x-ray photoelectron spectroscopy (XPS) and their possible applications to the study of materials implanted to very high fluences. We have included general references to optical methods[1], TEM[2] and the electron spectroscopy methods[3] which the reader may use for further study concerning these techniques.

OPTICAL SPECTROSCOPY

Optical spectroscopy methods rely on the absorption, or reflection or scattering of light from a sample. Simply stated, a beam of monochromatic light is made incident on the surface of a sample. By measuring the absorption or reflection or scattering of the incident light as a function of the incident frequency or wavelength of the light, the interaction of the light with the material of the sample can be examined. This interaction is described by solving Maxwell's equations along with the equations which describe the response of the material. This is generally quite complicated. Some general ideas of how light interacts with materials can be gained by considering the simple classical Lorentz model which views the electrons of a solid as masses bound to their respective nuclei by springs of natural frequency ω_0 and a damping factor Γ. By solving the equation of motion of the system, this model can express the complex dielectric function in terms of the natural frequency by

$$\epsilon = 1 + \frac{4\pi N e^2}{m} \left\{ \frac{1}{(\omega_0{}^2 - \omega^2) - i\Gamma\omega} \right\} \tag{1}$$

where N is the number of atoms per unit volume assuming one transition per atom, ω is the angular frequency of the light and e and m are the charge and mass of the electron. This function is divided into real and imaginary parts and we can find that

$$\epsilon = \epsilon_1 + i\epsilon_2 \tag{2}$$

$$\left.
\begin{aligned}
\epsilon_1 &= n^2 - k^2 = 1 + \frac{4\pi N e^2}{m} \left[\frac{(\omega_0{}^2 - \omega^2)}{(\omega_0{}^2 - \omega^2)^2 + \Gamma^2\omega^2} \right] \\[2em]
\epsilon_2 &= 2nk = \frac{4\pi N e^2}{m} \left[\frac{\Gamma\omega}{(\omega_0{}^2 - \omega^2)^2 + \Gamma^2\omega^2} \right]
\end{aligned}
\right\} \tag{3}$$

where n is the refractive index and k is the extinction coefficient. The real and imaginary parts of ϵ are related by the Kramers-Kronig dispersion relations as indicated in Equations (4) below.

$$\left.
\begin{aligned}
\epsilon_1(\omega) - 1 &= \frac{2}{\pi} \, \mathbf{P} \int_0^\infty \frac{\omega'\epsilon_2(\omega')d\omega'}{(\omega')^2 - \omega^2} \\[2em]
\epsilon_2(\omega) &= \frac{-2\omega}{\pi} \, \mathbf{P} \int_0^\infty \frac{\left[\epsilon_1(\omega') - 1\right]}{(\omega')^2 - \omega^2} d\omega'
\end{aligned}
\right\} \tag{4}$$

Similar relations can be shown to exist between the index of refraction and the extinction coefficient or between the amplitude and the phase shift of the complex reflectivity. What these relations mean physically is that in the frequency region near an absorption there is a region of anomalous dispersion. It also means that the dispersive part of the complex dielectric function can be calculated from measurements of the absorptive part of the dielectric function.

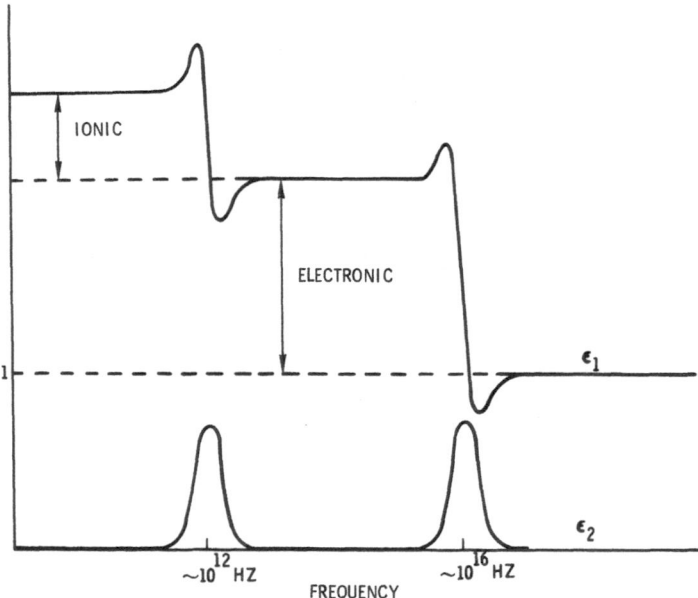

Figure 1.   Schematic plot of the real (ϵ_1) and imaginary (ϵ_2) parts of the dielectric function on the vertical axis versus frequency on the horizontal axis for an idealized ionic crystal.

For the simple case of an ionic crystal, such as an alkali halide, a schematic plot of the real and imaginary parts of the dielectric function is shown in Figure 1. What this illustrates is that at low frequencies, the ions and the electrons are capable of following the electric field of the incident electromagnetic wave and both contribute to the polarizibility and thus to the dielectric function. In the region around 10^{12} Hz, the ions lose their ability to respond to the rapidly varying electric field, which results in a change in the dispersive part of the dielectric function and a peak in the absorptive part. This absorption due to the lattice modes of the crystal is generally in the infrared and, in ionic crystals where this absorption is so strong that no light is transmitted, is called the "Restrahlen." A similar feature is seen in the dielectric function at higher frequencies where the electrons can no longer follow the applied field.

Much of the optical absorption work on implanted materials has been directed at electronic transitions. In an insulator, with a completely filled valence band and an empty conduction band, electrons can be excited across the band gap to investigate valence and conduction band states. States within the gap due to defects or impurities can be investigated by observing transitions from these states to the valence or conduction bands.

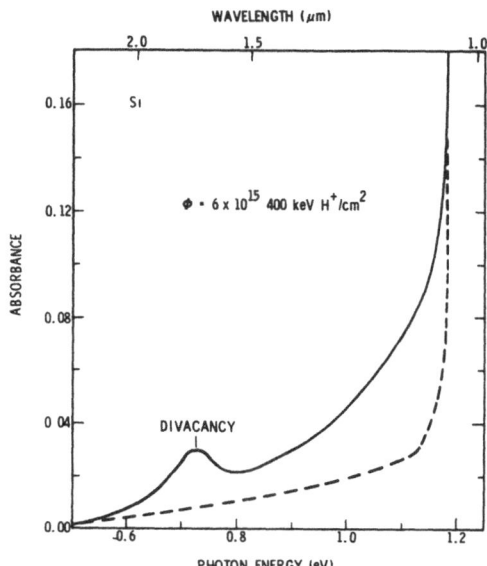

Figure 2.   Optical absorption spectrum of hydrogen implanted Si
showing the divacancy absorption band near 1.8 μm.

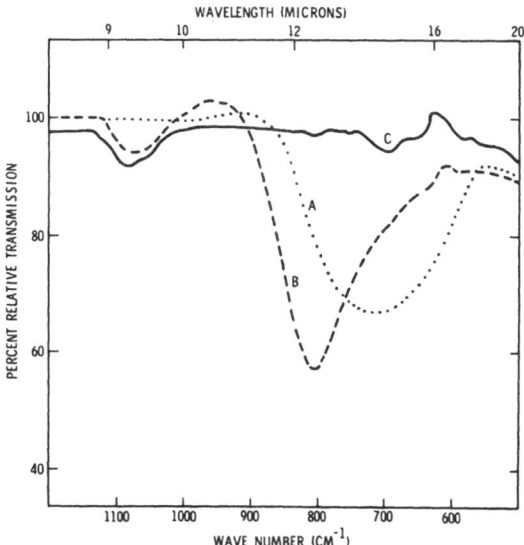

Figure 3.   Optical transmission spectrum of A.   carbon implanted
silicon after implantation B.   same sample after annealing at
1000°C and C.   unimplanted silicon.

Because implantation is intimately concerned with the final state
of impurity atoms and with damage due to the implantation process,
optical absorption is a very powerful tool for the study of implanted insulators. The same is true for semiconductors because
the band picture is essentially the same except for a smaller
energy gap between the valence and conduction bands. For metals
where the highest occupied band is either not empty or overlaps
an empty band, conventional optical spectroscopy is not very useful
and special techniques must be applied.

There are numerous examples of the use of optical spectroscopy
to study implanted semiconductors and insulators, but lack of space
dictates that these cannot be discussed in detail. Examples of
damage studied by optical means would include the identification
of the divacancy as a specific defect produced by implantation in
silicon[4]. This defect was the first defect in Si to be identified
as being produced by ion implantation. There is a characteristic
absorption band at 1.8 μm and data [5] from a typical implanted
sample is shown in Figure 2. By measuring the absorption of various
defects as a function of implanted ion, fluence and annealing temperature, it is possible to obtain a better understanding of the
implantation damage process. This understanding is very important
for the semiconductor device industry.

The F-center and the associate F-aggregate centers can be
observed in ion implanted alkali halides[6]. These centers are identified by a characteristic broad absorption band due to a 1s to 2p
transition of an electron bound to a halogen vacancy. Because these
centers can be produced by electronic as well as lattice energy,
they can be used to investigate the partition of the energy of the
incident ion into atomic and electronic processes.

In silicon implanted with high doses of carbon, it is possible
to observe the change from isolated Si-C bonds to SiC by monitoring
the infrared absorption due to the lattice modes. Figure 3 shows
the development of the absorption band due to SiC formed by implantation or carbon into Si. Using methods like this, it is possible
to produce buried layers of compounds such as SiC, SiO_2 or Si_3N_4.[7]

In recent years, the development of fast Fourier transform
algorithms have enabled Fourier transform infrared spectroscopy to
become a laboratory tool.[8] Although this technique has not been
specifically applied to implanted materials yet, the advantages of
greater light throughput and time efficiency could prove very useful.
A schematic diagram of a Fourier transform spectrometer and the
equations describing the measured intensity (I) as a function of
the movable mirror displacement (δ) in terms of the intensity (B)
as a function of wave number ($\bar{\nu}$) are shown in Figure 4.

Optical spectroscopy methods have proved to be valuable tools
for the study of ion implanted materials, particularly insulators

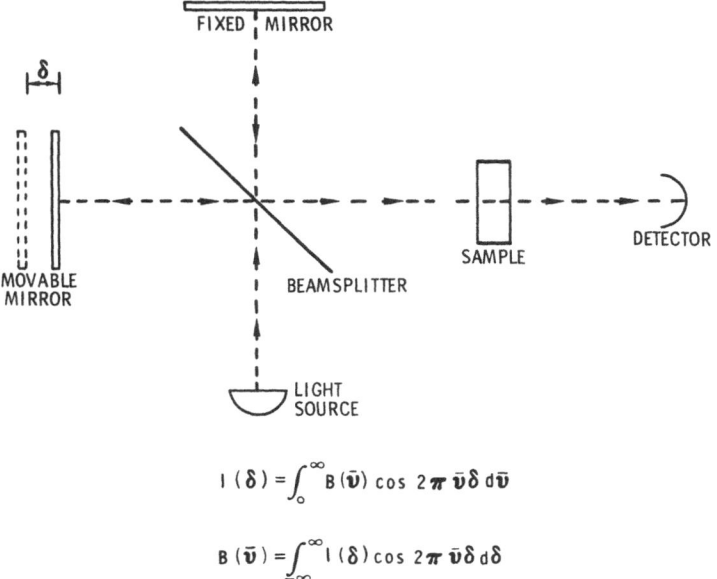

$$I(\delta) = \int_{0}^{\infty} B(\bar{v}) \cos 2\pi \bar{v}\delta \, d\bar{v}$$

$$B(\bar{v}) = \int_{-\infty}^{\infty} I(\delta) \cos 2\pi \bar{v}\delta \, d\delta$$

Figure 4. Schematic diagram of a Fourier transform spectrometer. The detector measures the intensity of light (I) as a function of mirror displacement (δ) and a computer is used to transform these data into the intensity function of wave number $B(\bar{v})$.

and metals. A brief description of an example of the uses of optical techniques to study precipitation can be found in another article from these proceedings.[9]

TRANSMISSION ELECTRON MICROSCOPY

Transmission electron microscopy is probably the most widely used "solid state" tool for the characterization of ion implanted materials. The use of electron rather than light waves in a microscope has the advantage that the electrons have a much shorter wavelength than visible light and thus much smaller structures are resolvable. The wavelength of an electron is given by the de Broglie relation which can be written

$$\lambda(\text{\AA}) = 12.26/\sqrt{E(V)} \tag{5}$$

where E is the electron energy in volts. A 100 keV electron has a wavelength of 0.039 Å.

Electrons being charged particles, interact with matter much more strongly in general than do light waves. Because of this strong

interaction, specimens in an electron microscope must be very thin
(~1000Å) or the electrons will be completely absorbed or scattered.
Figure 5 shows a schematic diagram of a transmission electron
microscope.   Electrons are emitted by a source and a condenser lens
system is used to form the electrons into a beam which is incident
on the sample.   In the sample, electrons will be diffracted,
scattered and/or absorbed, and the electrons which penetrate the
film are magnified and imaged by a series of lens systems until
they form an image on some sort of recording medium, generally
a fluorescent screen or photographic film.   For a magnification of
20,000X, the objective lens will typically contribute 25X; the
intermediate lens, 8X; and the projector lens 100X.

    Transmission electron microscopy has another very useful fea-
ture of being able to determine the crystalline structure of microscopic features of the sample.   Figure 6 shows schematic diagrams
of the operation of the electron microscope in two different modes.
The upper part of the figure illustrates the use of the microscope

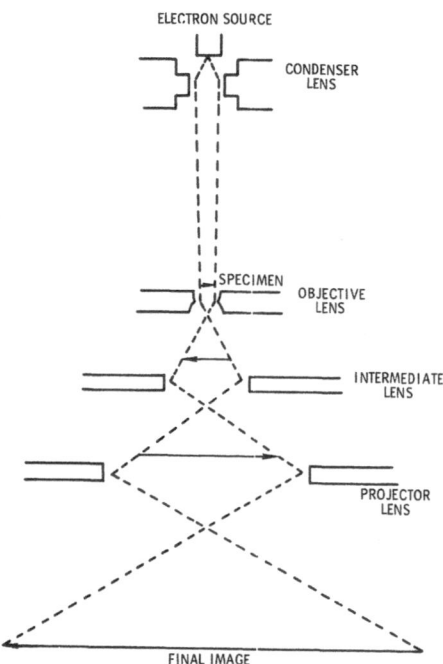

Figure 5.   Schematic illustration of a transmission electron
microscope   The final image is usually recorded on photographic
film.

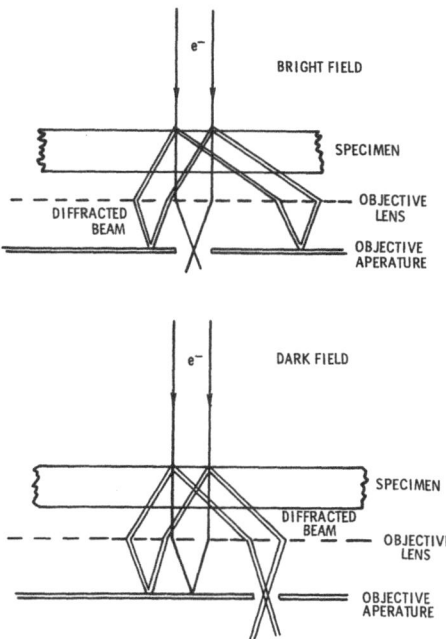

Figure 6.   Illustration of imaging in the bright field mode (top)
and dark field mode (bottom).   In dark field only one specific
reflection is imaged.

in the bright field mode.   This is similar to an optical microscope
in that the image is formed by the electrons which penetrate the
sample without scattering or absorption.   Thus contrast is caused
by features which do scatter or absorb electrons appearing darker
in the final image.   Crystalline features within the imaged area
can Bragg-reflect the electrons in well-defined patterns that are
characteristic of the crystal structure.   If we move the objective
aperature of the microscope such that only electrons that have
been Bragg-reflected at a specific angle are transmitted to the
magnifying lenses and the image forming system, then only features
of a given crystal structure will appear bright in the final image.
This is known as dark-field microscopy.   Actually modern microscopes deflect the incident beam and keep the objective aperature
fixed but the result is the same.   Dark-field microscopy is particularly useful for looking at the distribution of second-phase
precipitates within a sample.   Examples of bright and dark-field
microscopy are shown in Figures 7 and 8.   Figure 7 shows a bright-
field image of a thin multi-grain gold film.   The contrast is due
to the different amounts of Bragg-reflection of the electrons which
pass through gold grains of differing orientations.   This film was

Figure 7.   Bright field photomicrograph of a gold film.

ion-milled to perform the final thinning of the sample. The dark "speckled" features appearing on the brightest of the grains are evidence of damage due to the ion bombardment.[10] In Figure 8 is the dark-field micrograph of a sample of Be metal which has been implanted with Al and Fe and annealed at 775°C. This image is formed by Bragg-reflected electrons from AlFeBe$_4$ and the bright feature near the middle of the figure is evidence of a twinned precipitate of the AlFeBe$_4$ phase.[11]

Transmission electron microscopy is an exceedingly useful technique for the examination of implanted materials because it can measure morphology, crystal structure and damage on a microscopic scale. The technique is generally useful for all solid materials. When it is combined with analysis of the x-rays emitted because of excitation of atoms by the high energy electrons, it can be a valuable tool for the measurement of chemistry on a microscopic scale. The reader is referred to an excellent book on microscopy for further information.[2]

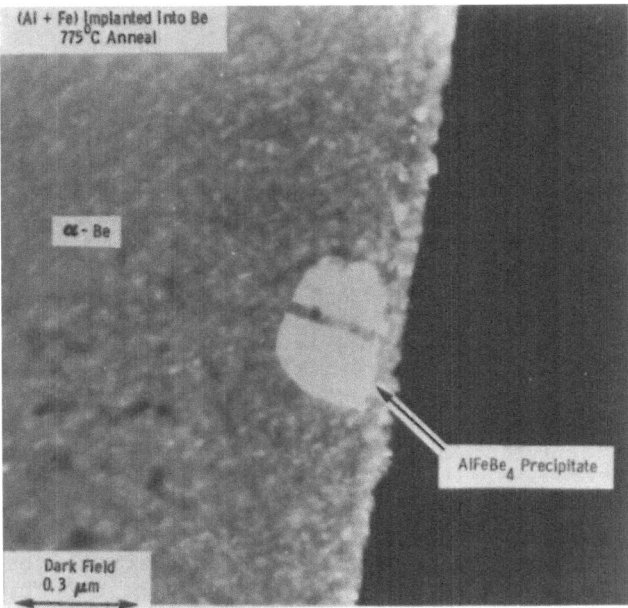

Figure 8.   Dark field photomicrograph of Be metal implanted with
Fe and Al and annealed at 775°C. The image is being formed by
a reflection from $AlFeBe_4$.

ELECTRON SPECTROSCOPY TECHNIQUES

The last decade has seen the development of powerful surface
sensitive techniques for the examination of the first few monolayers of a surface.[3] In general these methods consist of the
excitation of electronic levels of surface or near-surface atoms
and the measurement of the energy distribution of the electrons
emitted from the solid as a consequence of the excitation.   Two
methods are very widely used today, Auger Electron Spectroscopy
(AES) and X-Ray Photoelectron Spectroscopy (XPS), and although
these methods have not been widely employed in the study of implanted materials, their sensitivity to the surface region suggests
that they may be of valuable use in the future. These methods are
only applicable to concentrations of impurities above those normally present in ion-implanted semiconductors,[12] but because of the
requirements for higher doses in metals and insulators, they may be
ideally suited for studies of these materials when ion implanted.
Insulators can present some unique problems by charging up as the
sample is irradiated with electrons.

The basic processes governing these methods are indicated in
Figure 9. In AES, an inner shell electron vacancy is created,

usually by an external electron beam. This excited atom can now
decay by a higher level electron occupying the inner shell state
with the excess energy and momentum being carried off by another
electron from the same shell as the electron which "fell" into the
inner shell vacancy. The energy of the Auger electron is a function
only of the type of atom from which it originated. For example if
the inner shell vacancy is created in the K-shell and an L-shell
electron falls into the K-vacancy, the measured energy carried off
by the KLL Auger electron is given by

$$E_{KLL} = E_K - E_L - E_L - \phi \tag{6}$$

where E_x is the energy of an z-shell electron and ϕ is the work
function of the Auger spectrometer. Similar expressions can be
written for other transitions, but it is obvious that since the
electron energy levels for a given atom are well-defined, the Auger
electron energies will also be well-defined.

    XPS involves measuring the energy spectrum of electrons emitted
after irradiation of the sample with a monochromatic beam of x-rays.
Since the x-ray must give up all its energy to the electron, the
energy of the electron will be simply given by

$$E = h\upsilon - E_z - \phi \tag{7}$$

where $h\upsilon$ is the x-ray energy and Z is the atomic shell from which

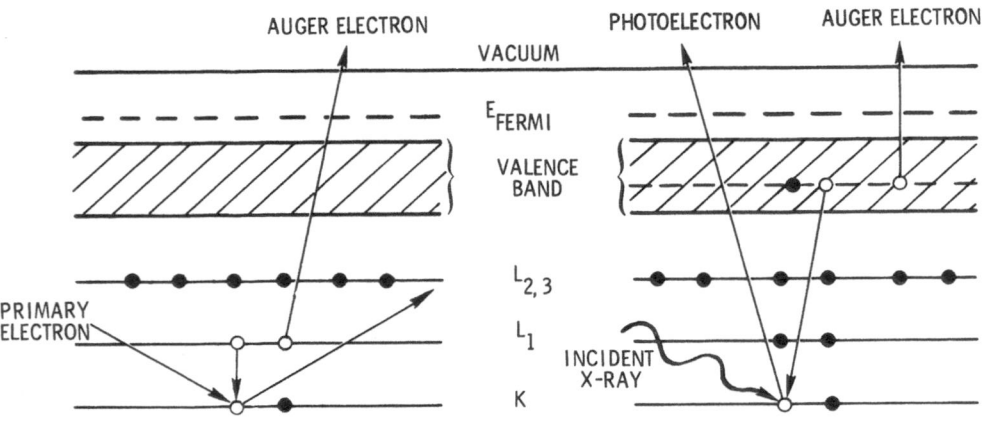

Figure 9. Energy level diagrams for Auger Electron Spectroscopy
(left) and X-ray Photoelectron Spectroscopy (right).

the photoelectron was excited. Since photoelectrons also create
inner shell vacancies, there will also be Auger electrons present
in the measured spectra, but the Auger and photoelectrons are very
easy to distinguish. The energies of the photoelectrons will vary
directly with the x-ray energy, but the Auger energies will remain
constant, since they are a function only of the atom from which they
were emitted.

The Auger and photoelectrons excited in the surface region of
a solid interact strongly with the atoms of the solid. For this
reason, these methods are only useful for the analysis of atoms
within 20-100 Å of the surface, depending on the particular material
since photoelectrons from deeper in the material cannot escape without scattering. Electrons escaping from atoms deeper within the
solid are scattered and absorbed. These techniques can be used to
obtain depth information when used in conjunction with ion sputtering. Measurements are made as the ion beam is sputtering through
the surface layer of a sample and from the behavior of a signal as
a function of sputtering time the depth distributions of particular
species in a material can be determined. Measurements such as
these have been of valuable assistance in analyzing contamination
on thin film semiconductor devices.

There has been very little work in the application of these
electron spectroscopy methods to ion implanted materials. For the
cases of metals and insulators particularly, as the work in these
materials increases there may be many valuable contributions that
can be made. These techniques are particularly sensitive to the
first few tens of Angstroms of a solid where the effects of ion
implantation in changing the surface properties of the solid will
be evident. In addition there is the possibility of obtaining
chemical information from Auger and photoelectron spectroscopies
about the state of the implanted ions and the state of the host
atoms as they have been modified by the implantation process.
This may be particularly important in understanding the role of
metastable phases which are produced in many systems which are
implanted to high fluences.[13]

In conclusion, we have tried to briefly describe two of the
most powerful solid state methods, optical spectroscopy and transmission electron microscopy for studying ion-implanted metals and
insulators. In addition we have discussed a third class of electron spectroscopy techniques that may be very useful for studying
the surface properties of implanted materials, particularly metals.
The discussion has been brief and far from complete, but it is
hoped that the material we have presented will enable the students
to progress further on their own.

REFERENCES

1. M. Born and E. Wolf, Principles of Optics, Pergamon Press,
    London, 1959.

2. P. B. Hirsch, A. Howie, R. B. Nicholson, D. W. Parkley and
    M. J. Whelan, Electron Microscopy of Thin Crystals,
    Butterworths, London, 1970.

3. T. A. Carlson, Photoelectron and Auger Spectroscopy, Plenum
    Press, New York, 1975.

4. H. J. Stein, F. L. Vook and J. A. Borders, Appl. Phys. Lett.
    14, 328 (1969).

5. H. J. Stein, unpublished data.

6. P. E. Thompson and R. B. Murray, Rad. Effects 25, 127 (1975).

7. J. A. Borders and W. Beezhold, Proc. II Int'l Conf. on Ion
    Implantation in Semiconductors, p. 241, Springer-Verlog,
    Berlin, 1971.

8. P. R. Griffiths, Chemical Infrared Fourier Transform
    Spectroscopy, John Wiley & Sons, New York, 1975.

9. J. A. Borders, these proceedings.

10. T. J. Headley, unpublished data.

11. G. J. Thomas, unpublished data.

12. Ion Implantation in Semiconductors - 1976, ed. by F. Chernow,
    J. A. Borders and D. K. Brice, Plenum Press, New York, 1977.

13. J. A. Borders, Annual Reviews of Materials Science, Vol. 9,
    in the press.

NUCLEAR METHODS FOR STUDYING AGGREGATION PROBLEMS

H. de Waard

Laboratorium voor Algemene Natuurkunde, University of

Groningen, The Netherlands

1. INTRODUCTION

In studies of the aggregation of impurities or defects in
metals by nuclear hyperfine interaction techniques it is useful to
distinguish between the following effects:
A. formation of small monatomic impurity clusters (dimers, trimers,
    or, in general, *oligomers*),
B. formation of larger, monatomic, impurity-clusters. These are
    called *precipitates*, or, if the impurity atoms normally form
    a gas, *bubbles*. If the impurity atoms are situated in regular
    positions with respect to the atoms of the host lattice the
    precipitate is called *coherent*, if not, it is called *incoherent*.
C. formation of vacancy clusters. These are called *voids*.
D. formation of polyatomic aggregates by chemical reactions between
    host and impurity atoms. Here we may discern between *compound
    formation*, formation of an *ordered alloy* or formation of an
    amorphous phase. The first two again may be coherent or inco-
    herent, the last is incoherent by definition.

Hyperfine interaction techniques often make it possible to dif-
ferentiate between these various types of aggregates and, moreover,
they sometimes allow a determination of the size of the aggregates,
especially for small clusters containing fewer than, say, 10 atoms.
Growth and phase transitions of aggregates can be followed in detail,
as long as these processes are not too fast. Experiments of the following types have been carried out:
- Annealing of samples implanted with insoluble impurities,
- Annealing of samples implanted above the solubility limit of
   impurities,
- Quenching, cold working and annealing of oversaturated solutions

413

produced at a high temperature.
By far the most used method in these experiments is Mössbauer
spectroscopy, mostly of the 14.4 keV transitions in ^{57}Fe. The spectra
give fingerprints of the particular state of the iron in the aggregates. In this chapter we shall only discuss a few examples of such
experiments.

2. VOIDS AND BUBBLES

As pointed out in Chapter 6, section 3.4, the recoilless
fraction f of gamma emission decreases sharply when the space around
the gamma emitting nucleus increases. Reintsema [1] was able to
determine what fraction a_z of gamma emitting nuclei in implanted
sources of $^{133}XeFe$ has a value of f too small to be observed, as a
function of annealing temperature. His results and an interpretation
are shown in Fig. 1. In this interpretation, the terms "void" and
"bubble" are defined as clusters containing at least 4 vacancies or
xenon atoms, respectively. Electron microscopy has shown that much
larger clusters, defined as voids and bubbles in the more usual
way, are also formed under similar implantation and annealing conditions. The interpretation of bubble formation in the XeFe samples
requires that Xe-atoms can migrate above about 450°C. This is confirmed by the observation that ^{133}Xe activity begins to be desorbed
from the sample around 500°C.

3. MÖSSBAUER STUDIES OF THE PRECIPITATION OF IRON IN METALS

A large amount of work has been done on the precipitation of
iron atomically distributed in copper and a fair amount also on

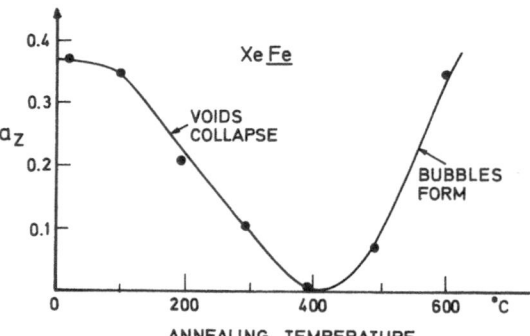

Fig. 1. *Fraction a_z of implanted Xe atoms in iron in sites with very
small f-value as a function of isochronous annealing temperature (duration of each step 15 m). Deduced by Reintsema [1]
from Mössbauer spectra of 81 keV transition in ^{133}Cs.*

iron in aluminium, as a function of annealing, cold working and
ageing treatments. In both host metals the iron has a very small
solubility at room temperature. It is initially incorporated in
atomic distribution either by fast quenching from the melt or by
implantation.

TABLE I

*Mössbauer investigations of precipitation of iron in metals. Q is
quenched, I is implanted.*

sample	prep.	conc. or dose	treatment	ref.
CuFe	Q		anneal	Gonser et al. [2]
"	Q	1%	anneal, roll	Window [3]
"	Q	0.5-4.6%	anneal, dose	Campbell et al. [4,5]
"	Q	0.9,2.7%	age, anneal	Clark et al. [6]
"	I	10^{15}-$5 \cdot 10^{16}$/cm^2	anneal, dose	Longworth & Jain [7]
AgFe	I	10^{15}-$5 \cdot 10^{16}$/cm^2	anneal, dose	Longworth & Jain [8]
AlFe	I	10^{14}-$2 \cdot 10^{17}$/cm^2	anneal, dose	Sawicki et al. [9]
"	Q	0.5-5%	dose	Nasu et al. [10]

In Table I a survey is given of investigated cases, with some
particulars, referring only to a small number of papers of interest
for our discussion. We shall mainly focus our attention on the CuFe
case and subsequently discuss what can be learnt about the nature
of precipitate formation from measurements at room temperature and
at low temperatures without and with an externally applied magnetic
field.

3.1. Analysis of room temperature $Cu^{57}Fe$ Mössbauer spectra

Typical spectra for CuFe samples of moderate Fe concentration
are shown in Figures 3 and 4. We see an unresolved pattern and one
may wonder at first how unambiguous conclusions can be drawn if
such spectra must be decomposed into more than two components.
The answer is that (i) consistent assignments of the components
can be achieved for a large number of spectra obtained after
different annealing treatments and at different implanted doses,
(ii) measurements at low temperatures yield additional information
(section 3.2) and (iii) from measurements in an external magnetic
field further conclusions can be drawn about the structure of the
spectrum (section 3.3).

Bearing this in mind, there is now considerable agreement about the assignment of the following spectral components that **may** be found in spectra taken at room temperature (isomer shifts S are relative to α-iron at 300 K, Δ is the splitting of the outermost lines of a particular component):

1. Single line with S = 0.22 mm/s: isolated Fe-atoms.
2. Sum of doublets with S ∿ 0.2 mm/s, Δ ∿ 0.7 mm/s: oligomers of Fe.
3. Single line with S = 0.09 mm/s: gamma-iron (f.c.c.).
4. Six line spectrum, S = 0, Δ = 10.56 mm/s: alpha-iron (b.c.c.).
5. Broad line with S ∿ 1 mm/s: oxide or sulfide.

Results for unannealed samples. Both Window [3] and Campbell and Clark [4,5] mainly observe components 1 and 2 in their quenched (Q) alloys; Window [3] ascribes the asymmetric doublet structure (see e.g. Fig. 4) to a range of oligomers (with n = 2, 3, 4 Fe-atoms) with slightly different values of S and Δ such that the left hand lines remain close together for each n while the right hand lines spread. Only at the highest concentration used by Campbell and Clark [5], namely 4.6% Fe, a magnetically split component 4, due to α-iron, is observed. In view of the fact that the high temperature (1094°C) solubility of Fe in Cu is only 4%, this is not surprising.

Longworth and Jain [7] performed conversion electron Mössbauer spectroscopy (see Ch. 6, section 3.4) on implanted (I) $Cu^{57}Fe$ samples. Their implanted doses correspond to maxima in the concentration profile ranging from 0.2-10%. They have also observed components 1 and 2, but in addition a component 3, due to γ-iron, is present at all but the lowest implanted dose. The f.c.c. γ phase of iron is unstable at room temperature in bulk, but it is stabilized in copper because its lattice constant is very close to that of the (f.c.c.) copper. For this reason it is present as a coherent precipitate. Typical spectra are shown in Fig. 3.

There are interesting differences between Q and I alloys:
a. At low concentrations, the iron in the I samples is less randomly distributed than in the Q samples. This is shown in Fig. 2, where the solid dots give the fraction of isolated iron atoms for the I alloys and the open circles for the Q alloys. Both lie well below the full curve which represents the isolated fraction expected for a random distribution of iron atoms (binomial law). This behaviour is explained by assuming that the iron atoms have an attractive interaction. In the Q samples this causes them to cluster somewhat during cooling but in the I sample the stronger process of *vacancy enhanced diffusion* occurs. The implanted iron atoms produce vacancy rich zones which can trap other iron atoms. This process enhances the probability of forming oligomers and thus reduces the isolated fraction.

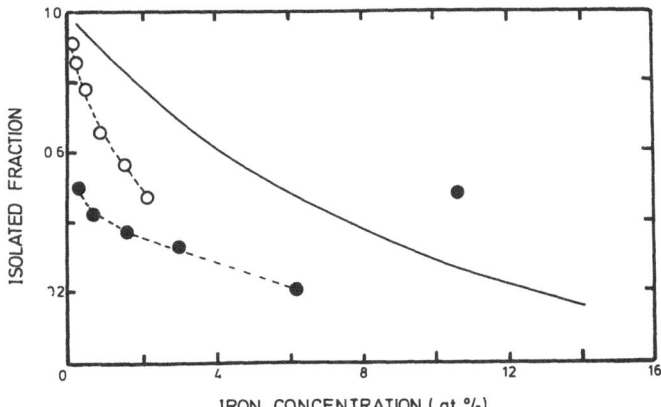

ISOLATED FRACTION

IRON CONCENTRATION (at.%)

*Fig. 2. Variation of the fraction of isolated iron atoms with iron
concentration in copper for implanted alloys (•), solution
quenched alloys (o) and on the basis of a random iron
distribution (full curve). Taken from Longworth and Jain [7].*

b.   At high implanted doses a γ-iron precipitate is found in the I
     samples, but not in Q samples fast quenched from 1000°C. There,
     it can only be produced by annealing more slowly or by an inter-
     mediate annealing step at 500°C. On the other hand, no α-iron
     is found in the I-samples even at the highest dose and after
     annealing but, as we have already seen, it is present in the
     high concentration quenched samples.

     Annealing and ageing effects. The change that occurs first of
all when heating Q and I samples of CuFe alloys is a decrease of
the isolated fraction along with an increase of small oligomeric
clusters. There is, however, a marked difference between Q and I
samples: the temperature where clustering starts is about 100°C
lower in I than in Q samples. In Fig. 3 an example is given   for
an I sample. Clearly, the single line component at 0.22 mm/s, which
corresponds to isolated Fe-atoms, starts to decrease already around
175°C. Another feature of the conversion electron Mössbauer spectra
shown in Fig. 3 is the reduction of the spectral area as a result
of annealing. This is caused by diffusion of iron atoms away from
the surface. Since the low energy (8 keV) conversion electrons of
the 14.4 keV transition have a very small range (about 100 nm),
the C.E.M.S. spectral area is a very sensitive indicator of such
diffusion. Both the low temperature clustering and diffusion are
interpreted as caused by vacancies that have been induced by the
implantation.

     In Q samples, the reduction of the isolated atom component only
starts at annealing temperatures close to 300°C. An example is given

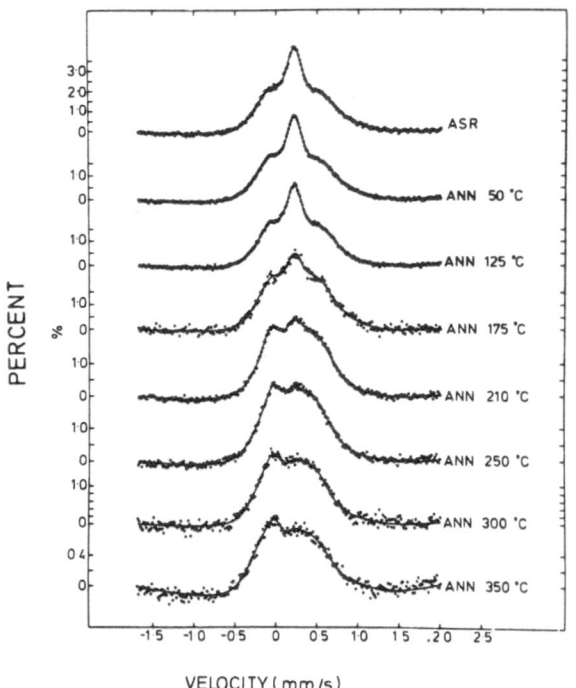

Fig. 3. *Conversion electron Mössbauer spectra of iron implanted in copper to a dose of 5×10^{15} ^{57}Fe atom/cm^2; as a function of annealing temperature. Full curves represent least squares fits. Isomer shifts relative to α-iron. Taken from Longworth and Jain [7].*

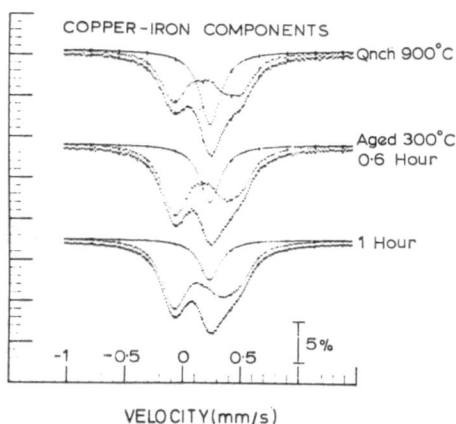

Fig. 4. *Mössbauer spectra of Cu 1% Fe quenched sample in early annealing stage. Spectra have been decomposed in a single line (isolated atoms) and an asymmetric doublet (oligomers).*

in Fig. 4. The γ-iron precipitate in the higher dose I samples, which is not found in the Q samples, increases a little at annealing temperatures up to 300°C, but somewhere above this temperature it starts to dissolve. It is most persistent in the highest dose (5 × 10^{16}/cm^2) sample, where it remains up to 500°C. As long as it stays, it blocks the diffusion of the iron deeper into the sample. At high temperatures, the γ-phase finally dissolves, leaving small iron clusters.

Recently, Clark et al. [6] have studied ageing effects in a Cu 2.7% Fe quenched alloy. In one of their old samples, the isolated atom fraction was followed for 10^8 s at room temperature. It was seen to decrease slowly, at a rate corresponding to an activation energy $\phi_m \sim 1$ eV for migration and a hopping rate p $\sim$ 4-10 × 10^{-10}/s for isolated iron atoms at room temperature.

They also studied ageing at 400°C. Keeping a 0.9% quenched sample at that temperature they carefully analyzed spectra taken in subsequent time intervals. In their analysis they used components 1 (isolated atoms) and 3 (γ-iron) as before, but the mixed doublet component 2 (oligomers) was further decomposed into three separate doublets: L (dimers, n = 2), M (n = 3-5) and N (n = 6-8). The results are shown in Fig. 5. A striking phenomenon is observed: γ-iron precipitates start to be formed only after all single Fe-atoms have clustered in about 200 minutes. As long as there are single Fe atoms, the N clusters keep growing at the expense of single atoms

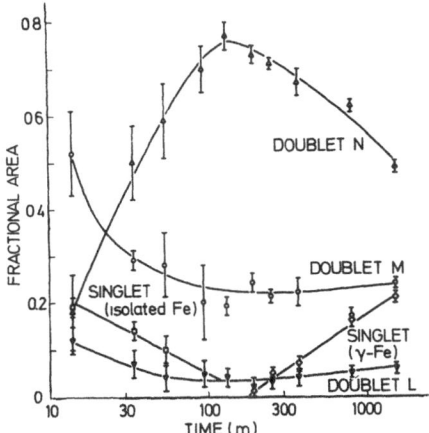

Fig. 5. Ageing of a quenched Cu 0.9% Fe sample at 400°C. Fractional areas are given of components of Mössbauer spectra corresponding to isolated Fe atoms, dimers (doublet L), n = 3-5 clusters (doublet M), n = 6-8 clusters (doublet N) and γ-iron. Figure taken from Clark et al. [6].

and of L and M clusters, but after that their number decreases while
the γ-iron precipitates grow. It must be concluded that N clusters
consisting of 6-8 iron atoms move as a whole to form γ-iron precipitates.

3.2. Low temperature $\underline{Cu}^{57}Fe$ Mössbauer spectra.

One of the weak points in the analysis of RT spectra, discussed
in the previous section, is the identification of γ-iron. It yields
a single line with a velocity shift of -0.9 mm/s, very close to that
of one of the lines of the quadrupole doublet corresponding to small
iron clusters (see e.g. Fig. 3, spectra after annealing above 250°C).
Below T_N = 67 K, γ-iron orders antiferromagnetically, and the magnetic
hyperfine field of the iron atoms in the γ precipitate is quite small,
B_{HF} = 2.7T at liquid helium temperature. On the other hand, much
larger hyperfine fields are experienced at that temperature by single
iron atoms (B_{HF} = 8T) and by iron atoms in small clusters (B_{HF} =
14-25 T), provided the iron is magnetically ordered at that tempe-

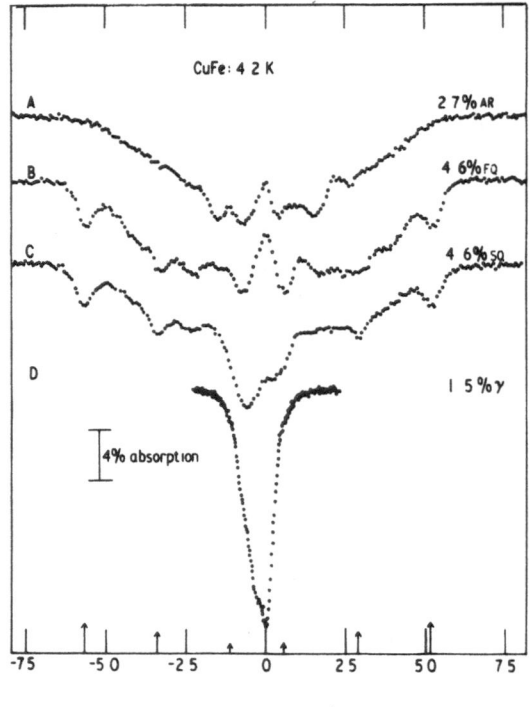

Fig. 6. Spectra of $\underline{Cu}Fe$ alloys at 4.2K. A. 2.7% as rolled; B. 4.6%
        fast quench; C. 4.6% slow quench; D. 1.5% γ-iron. Arrows in
        velocity scale give line positions expected for α-iron. Taken
        from ref. 5.

rature. This is the case at Fe concentrations above 0.5% [5]. The
characteristic differences between the Mössbauer spectra of small
clusters, α-iron and γ-iron are shown in Fig. 6, taken from the
work of Campbell and Clark [5]. In the spectra indicated by A and B
there is evidence of isolated Fe atoms and small Fe-clusters, giving
rise to a distribution of fields and, therefore, to a poorly resolved structure. In B there is, moreover, a contribution of α-iron
(sharp lines at higher velocities), because this sample has an iron
concentration above the high temperature solubility limit. Spectrum
C has the same γ-iron contribution (same iron concentration) but a
component with small hyperfine field appears near the center. The
sample from which spectrum C was obtained was quenched about 10 times
slower than that from which spectrum B resulted. Apparently the
difference in quench rate leads to the production of some γ-iron.
This is confirmed by comparing spectrum C with spectrum D, obtained
from a sample that contains mainly γ-iron as a result of a special
annealing treatment. The γ-iron component can be observed to grow
in samples like the one that produced spectrum C by annealing at
about 500°C.

3.3. Use of external fields

Suppose that we observe a ^{57}Fe spectrum consisting of 2 lines
of about the same intensity. Is this (i) a quadrupole doublet or
(ii) do these two lines correspond to two different iron sites each
giving a single line, but with different isomer shifts? This is a
common problem in ^{57}Fe Mössbauer spectroscopy which may be resolved
by applying an external magnetic field large enough to produce a
Zeeman splitting in excess of the linewidth of the individual components. Such a field will produce different hyperfine pattern in
case (i) and case (ii). This can easily be explained by referring
to Ch. 6, section 2, where the influence of magnetic dipole and
electric quadrupole interactions on the Mössbauer spectrum was
discussed, using the 14.4 keV transition of ^{57}Fe as an example.
(i) For a quadrupole doublet, the influence of a magnetic field is
     to split the $\pm 3/2 \rightarrow \pm 1/2$ transition into 4 components and the
     $\pm 1/2 \rightarrow \pm 1/2$ transition into 2 components with a smaller spacing
     than that of the other components. Clearly, this yields an
     asymmetric 6 line pattern around the center of gravity of the
     original spectrum. In practice, the situation is somewhat com-
     plicated by the fact that the field gradient axes are randomly
     oriented in space with respect to the magnetic field axis, but
     it is not difficult to calculate the resulting pattern also
     in this case.
(ii) Two independent single lines on the other hand are each split
     into 6 components by a magnetic field, resulting in a symmetric
     12 line pattern around the center of gravity.
The patterns obtained for cases (i) and (ii) are obviously quite
different.

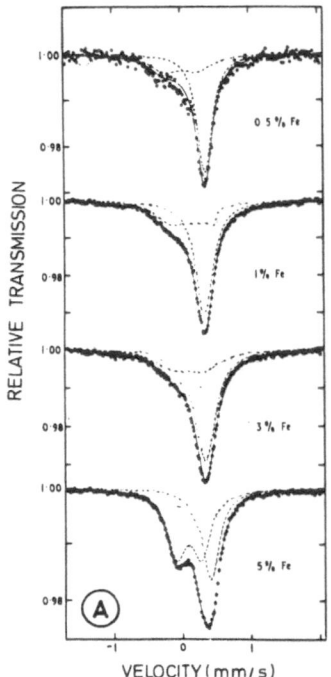

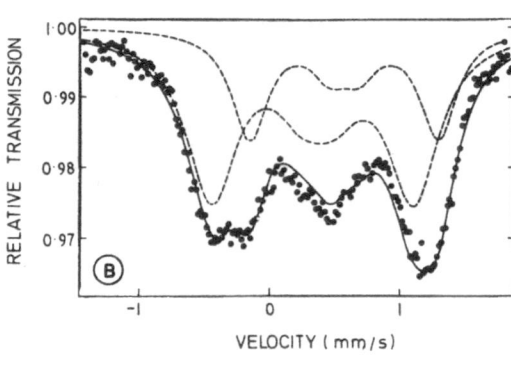

Fig. 7. A. ^{57}Fe Mössbauer spectra obtained at room temperature for AlFe alloys with Fe concentrations from 0.5% (top) to 5% (bottom). B. Spectrum obtained for 6% alloy in 5T longitudinal field at 43K. Fitted as discussed in text. Taken from Nasu et al. [10].

More complicated cases, for instance one single line and one doublet, can also be tackled. An example of this is given in Fig. 7. In Fig. 7A some spectra obtained by Nasu et al. [10] for AlFe alloys without external field are shown. These can be fitted just as well with two single lines as with one single line and one doublet. In a longitudinal field of 5T the spectrum shown in Fig. 7B results for a 6% alloy. This can only be fitted with the two components shown as broken lines in that figure. The one on the right hand side is a symmetric 4 line pattern resulting from a pure magnetic interaction (lines 2 and 5 are suppressed because the field is along the direction of the gamma rays) and the one on the left hand side is an asymmetric pattern resulting from a combined quadrupole and magnetic interaction.

Another application of the use of external fields is to study the paramagnetic behaviour of clusters at a temperature low enough to approach paramagnetic saturation but not so low as to produce a magnetically ordered system. If we assume the magnetic moments μ_1

in an iron cluster containing n atoms to be parallel, the whole cluster has a giant moment $\mu = n\mu_1$ and the effective hyperfine field has a Brillouin function dependence

$$H_{eff} = -H_{sat}^n B_J(\mu H_o/kT) \qquad (1)$$

on the externally applied field H_o. The saturation field H_{sat}^n of a cluster of n iron atoms can be obtained from a measurement below the magnetic ordering temperature. Then, the fit of a plot of H_{eff} vs. H_o to Brillouin curves with different J values yields J as well as the moment μ. A knowledge of the moment μ_1 of an isolated iron atom then suffices to derive the cluster size n. Window [3] used this method in his measurements on CuFe alloys to determine Fe cluster sizes.

3.4. Comparison of clustering of iron implanted in f.c.c. and b.c.c. metals

We have already seen that the isolated fraction of Fe atoms implanted in f.c.c. copper at a low dose is well below the limit of random distribution. The same is true for iron in aluminium [10,11], silver [8,11] and gold [11]. The results for these f.c.c. metals are given in Fig. 8B. The solid lines in this figure give the expectation for the isolated atom fraction S_S/S in the case of random distribution in a cubic lattice (upper curve) and in a lattice of densely packed spheres (lower curve). A different picture results for b.c.c. metal hosts (Fig. 8A), where the isolated fractions lie close to the random expectation except for tungsten. The systematic difference between the f.c.c. and b.c.c. hosts is not clearly understood. On

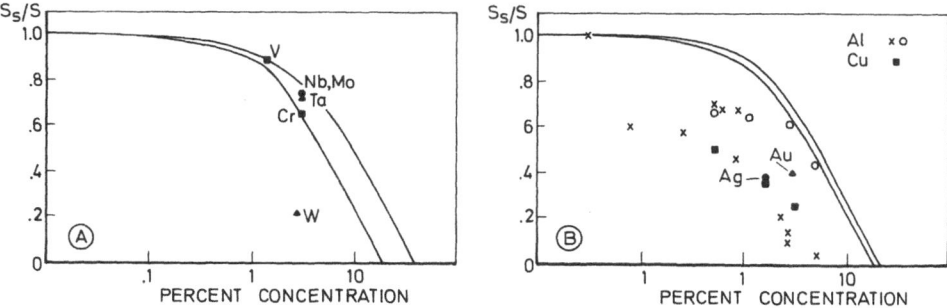

Fig. 8. A. Isolated fraction of implanted Fe-atoms S_S/S in b.c.c lattices vs Fe concentration. Solid lines: expectations based on random distribution of iron atoms in cubic lattice (upper curve) and lattice of densely packed spheres (lower curve). B. Same as A, but for f.c.c. lattices. Taken from the work of Sawicka and Sawicki [11].

the one hand, vacancy enhanced diffusion may be more effective in
f.c.c. than in b.c.c. metals, but it may also be that the attractive
Fe-Fe potential is larger in the f.c.c. metals studied than in the
b.c.c. (transition) metals.

4. CONCLUSIONS

The examples given in the previous section serve to show that
hyperfine interaction techniques sometimes provide a sensitive tool
for studying aggregate formation at its very beginning, when the
cluster size is still far too small for observation in the electron
microscope. In particular, the measurements immediately yield the
extent to which clustering has occurred in excess of the expectation
based on a random impurity distribution. The spectra obtained sometimes depend very strongly on the physical nature of the clusters
formed, for instance on their magnetic behaviour. The 14.4 keV
transition in ^{57}Fe has, by its very suitable parameters for Mössbauer spectroscopy, made the investigation of dilute iron alloys an
ideal case. The relevant parameters of most other isotopes suitable
for Mössbauer spectroscopy are much less ideal. This restricts the
application of this technique. Little has been done so far to study
impurity aggregation by other hyperfine interaction techniques. There
may be interesting applications of the perturbed angular correlation
method here.

REFERENCES

1. S.R. Reintsema, S.A. Drentje and H. de Waard, Hyperfine Interactions 5 (1978) 167.
2. U. Gonser, R.W. Grant, A.H. Muir and H. Wiedersich, Acta Metal-
   lurgica 14 (1966) 259.
3. B. Window, Phil. Mag. 26 (1972) 681.
4. S.J. Campbell, P.E. Clark and P.L. Liddell, J. Phys. F2 (1972)
   L 114.
5. S.J. Campbell and O.E. Clark, J. Phys. F4 (1974) 1073.
6. P.E. Clark et al., J. Phys. F, to be published.
7. G. Longworth and R. Jain, J. Phys. F8 (1978) 351; R. Jain and
   G. Longworth, ibid. F8 (1978) 363.
8. G. Longworth and R. Jain, J. Phys. F8 (1978) 993.
9. B.D. Sawicka, J.A. Sawicki, J. Stanek and M. Drwiega, Hyperfine
   Interactions 5 (1978) 147.
10. S. Nasu, U. Gonser, P.H. Shingu and Y. Murakami, J. Phys. F4 (1974)
    L 24.
11. B.D. Sawicka and J.A. Sawicki, 1978 Kyoto conference on appli-
    cations of the Mössbauer effect, to be published in J. de Physique.

COMPARISON BETWEEN NUCLEAR PHYSICS METHODS AND SOLID STATE PHYSICS METHODS FOR THE STUDY OF IMPLANTED ATOM SITES IN METALS: SOLID STATE METHODS

L.M. STALS[a] and S. REINTSEMA[b]

Materials Physics Group, Limburgs Universitair Centrum, Diepenbeek (Belgium) (a)

Instituut voor Kern- en Stralingsfysika, Departement Natuurkunde, Katholieke Universiteit Leuven (Belgium) (b)

In the study of implanted atoms and associated radiation damage in metals solid state methods can yield information on the defect pattern created during implantation as well as on the site of the implanted atoms.

Isolated implanted atoms can, for instance, be studied by means of the helium probing technique [1], whereas clustering of implanted atoms can be studied by means of transmission electron microscopy [2]. In general, the results obtained by means of these techniques have to be compared with the results of atomistic calculations [1,2]. For example, Van Veen et al. [1] have studied the desorption of helium from krypton-implanted and annealed tungsten. After annealing at 1400 K, the krypton is in a substituting position. The He-desorption spectra from that annealed Kr-implanted W crystal show three peaks at low temperatures, labelled A,B,C, with single-step desorption energies of 1.7 eV, 2.0 eV and 2.2 eV. These peaks could be assigned by means of computer simulations to desorption of helium from octahedral interstitial sites in the neighbourhood of the KrV-complex. De Hosson [2] has studied the formation of argon-vacancy clusters in copper by means of transmission electron microscopy and computer simulation. On the basis of atomistic calculations he could show that even at low temperatures where thermal vacancies are only present in very low concentrations argon atoms dissolve substitutionally in the copper lattice, even if a lattice atom has to be pushed into an interstitial position.

Although in general the results, obtained in the field of charac-terization of implanted atom sites by means of solid state methods,

have to be supported by computer simulations, as it is shown by the
two examples, the great advantage of these methods is that they
are much less time consuming than the majority of the nuclear
methods. Furthermore, there is no need for specific radioactive
isotopes with the correct decay scheme, so that "metal-implanted
atom" systems, which cannot be studied by means of nuclear techniques, can still be studied by means of solid state methods. An
exception is, however, the Rutherford Backscattering Technique
which can be used equally well for non-radioactive isotopes.
The defect pattern created during implantation can easily be
studied by means of electrical resistance measurements and transmission electron microscopy. Especially, high-voltage electron
microscopy, in which the damage is created by the 1 MeV electrons,
is very well suited for the observation of the initial stages of
defect agglomeration and of the further development of the defect
clusters. Not only reactions among lattice defects in the
immediate surroundings of the implanted atoms and between the implanted atoms and lattice defects can be studied, but also reactions between lattice defects themselves at large distances from
the implanted atoms can be followed, say, during annealing. In this
case electrical resistance measurements are very useful. Furthermore, in order to study the possible influence of the implanted
atoms on the annealing of defects, so-called dummy experiments can
be performed, in which the same defect pattern is present in the
specimen but without the implanted atoms. This can for instance
be achieved by means of electron irradiation of non-implanted and
of implanted and annealed specimens.
Solid state methods, especially electrical resistivity measurements
and the helium probing technique (HPT), are very well suited for a
study of the kinetics of the various processes which take place.
The kinetics of cluster formation can be studied in the high-
voltage electron microscope, as already has been said.
In order to derive values for the important parameters such as :
  - activation energy for migration of elementary defects, such
    as interstitials and vacancies ;
  - activation energy for migration of implanted atoms (inter-
    stitially or substitutionally) ;
  - binding energies between lattice defects and implanted atoms ;
  - number of jumps of migrating defects ;
  - reaction order of a specific recovery process ;
one is obliged to perform a large number of experiments and also
to anneal the specimen either stepwise or continuously in a large
temperature range. Furthermore, in order to reveal the structure
of the annealing curve thus obtained one needs a rather large number
of measurements of the physical property under study.
Let us take as a first example the determination of the binding
energy of helium to vacancies. By variation of the heating rate,
which is typically of the order of 50 Ks^{-1}, one obtains in a HPT-
experiment (see [3]) desorption spectra which are shifted with
respect to each other. As explained in [3], one can deduce from

this peak shift the binding energy of the He_nV_m-complex.   Because
each desorption experiment takes about 1h, an appreciable number of
experiments can be performed in one day.   The resolution of HPT-
experiments is typically 10^5 to 10^6 He-atoms in a desorption volume
of 1 liter, so that experiments at low defect concentration can
still be performed.   An example of the technique is shown in
figure 1.

Let us take as another example the determination of the activation
energy of a recovery process in pure metals.   The electrical resistivity method has been shown to be very useful for this purpose.
In order to determine the activation energy of a recovery process

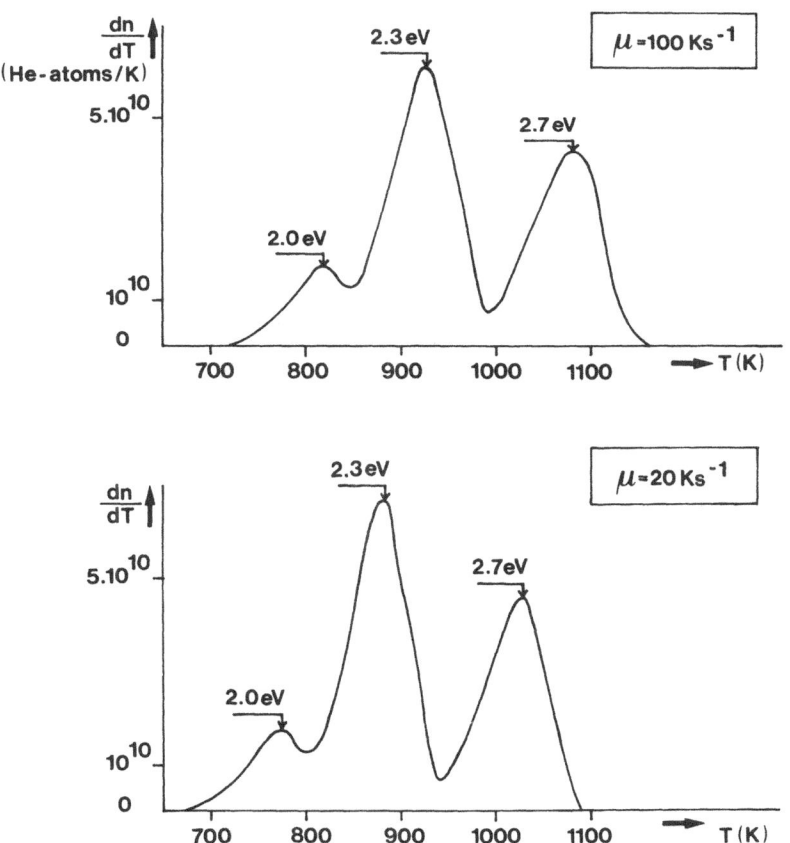

Fig. 1. Determination of the binding energy of helium to vacancy
        type defects in metals (hypothetical case) from the peak
        shift of the various peaks with heating rate.

we need for instance (i) either <u>at least three isothermal curves</u>
for the application of the cross-cut method [9](ii) or <u>one linear</u>
isochronal and one isothermal curve for the application of the
Meechan-Brinkman method [10] (iii) or <u>two-linear isochronal</u> curves
measured at different heating rates for the application of the peak
shift method [11,12] (iv) or a <u>series</u> of <u>isothermal curves</u> at
successive higher temperatures (Overhauser method) [13]. In all
cases one needs a large number of measured points. A typical application of the Overhauser method is shown in figure 2.

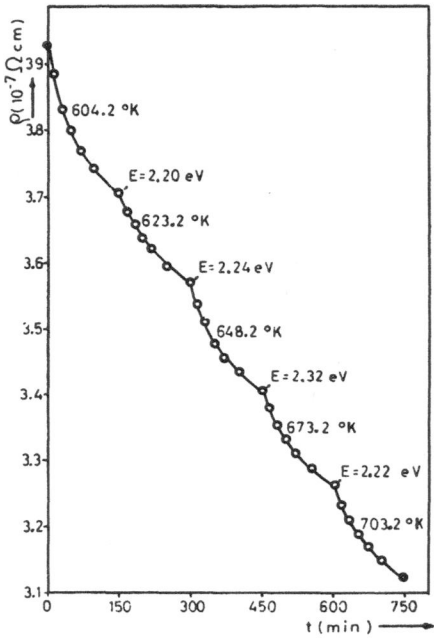

Fig. 2. : Application of the
Overhauser method for the determination of the activation energy
of recovery processes in cold-
worked rhenium [14].

The recovery process can be described by the equations :

$$-(\frac{dc}{dt})_{T_1} = f_1(c)\, e^{-E/k\, T_1} \quad \text{at temperature } T_1$$

$$-(\frac{dc}{dt})_{T_2} = f_2(c)\, e^{-E/k\, T_2} \quad \text{at temperature } T_2$$

At the concentration $c = c*$ where the temperature jump from T_1 to
T_2 occurs, one has $f_1(c*) = f_2(c*)$, if the temperature change does
not influence the defect structure. One obtains finally :

$$E = \frac{k\, T_1 T_2}{T_2 - T_1} \ln \frac{(dc/dt)_{T_2}}{(dc/dt)_{T_1}}$$

where E is the activation energy of the process.
These are only a few examples of so many to show how solid state
methods can be used to study the behaviour of defects and implanted

atoms in metals. It should, however, be mentioned that solid-state methods are in general less suited for a direct study of the interaction between implanted atoms and lattice defects, as this can be done by means of HFI methods (see for instance ref. 4-8).

Solid state methods and nuclear physics methods have their own specific advantages and disadvantages. When they are used in the study of the same problem they can yield very useful complementary information. This can for instance be illustrated for stage III in molybdenum. This recovery stage occurs between 400 K and 550 K and is singly-activated. Its kinetics, as studied by means of electrical resistivity measurements, points to the migration of a simple point defect [3,15]. Data obtained by means of the helium probing technique [16] indicate that single vacancies disappear during stage III annealing. One can, however, not conclude directly on the basis of HPT-data that vacancies migrate freely in stage III, because they can equally well be annihilated by migrating interstitials. Positron annihilation data [17] seem to support vacancy migration, and Mössbauer data on stage III can also be interpreted in a consistent way if vacancies are assumed to migrate to the Mössbauer impurity [18] during stage III annealing. These Mössbauer data seem to be confirmed by TDPAC experiments [19].

References

[1] A. Van Veen, L.M. Caspers, E.V. Kornelsen, R. Fastenau,
     A. Van Gorkum, A. Warnaar, Phys.Stat.Sol.(a), 40, 235 (1977).
[2] J.Th.M. De Hosson, Phys.Stat.Sol.(a), 40, 293 (1977).
[3] L.M. Stals, "Introduction to some solid state methods for the
     study of isolated implanted atom sites in metals", p.
     this book.
[4] A.C. Damask and G.J. Dienes, Point Defects in Metals, Gordon
     and Breach (1963).
[5] C. Dimitrov and O. Dimitrov, Phys.Stat.Sol., 34, 545 (1969).
[6] C. Dimitrov, F. Moreau, O. Dimitrov, J. Phys.F : Metal Phys. 5,
     385 (1975).
[7] R. Rizk, Y. Loreaux, P. Vajda, F. Maury, A. Lucasson, P. Lucasson,
     C. Dimitrov, O. Dimitrov, J.Appl.Phys., 47, 809 (1976).
[8] O. Dimitrov, C. Dimitrov, P. Rosner, K. Böning, Rad.Effects, 30,
     135, (1976).
[9] W.E. Parkins, G.J. Dienes, F.W. Brown, J.Appl.Phys. 22, 1012
     (1951).
[10] C.J. Meechan and J.A. Brinkman, Phys.Rev. 103, 1193 (1956).
[11] A.C. Damask and G.J. Dienes, Point Defects in Metals, Gordon
     and Breach (1963).
[12] F. Bell and R. Sizmann, Phys.Stat.Sol. 15, 369 (1966).
[13] A.W. Overhauser, Phys.Rev. 90, 393 (1953).
[14] H. Vandenborre, L. Stals, J. Nihoul, Jül-Conf 2-1968.
[15] Defects in Refractory Metals, Mol (1972) Eds. R. De Batist,
     J. Nihoul, L. Stals.

[16] Caspers L.M., Van Veen A., Van Gorkum A.A., Van den Berkel A.,
       Van Baal C.M., Phys.Stat.Sol.(a) 37, 371-383 (1976).
[17] Eldrup M., Mogesen O.E., Evans J.H., J.Phys.F : Metal Physics
       499-521 (1976).
[18] S.R. Reintsema, E. Verbiest, J. Odeurs, H. Pattyn, J.Phys. F :
       Metal Physics, to be published.
[19] G. Schatz, E. Recknagel, communicated at the poster session
       of the Summer School.

COMPARISON OF NUCLEAR METHODS WITH SOLID STATE METHODS: NUCLEAR METHODS

R. Coussement

University of Leuven
Instituut voor Kern- en Stralingsfysika
Celestijnenlaan 200 D, B-3030 Heverlee, Belgium

Most of the methods using nuclear radiation measures the hfi of the radioactive atom. Let us start the discussion of the information we will get from this hfi, magnetic interaction, quadrupole interaction, isomer-shift and recoilless fraction.

The effective magnetic field sensored by a radioactive impurity is proportional to the ns electron spin densities[1]. These electron spin densities are due to the interaction of electrons of the impurity with the polarized electrons of the host material. The problem is complicated and up to now no calculations of the hyperfine magnetic field can be done with satisfactory success. However changes of the magnetic behaviour of the host material will show up directly on the hyperfine magnetic interaction.
In that way the magnetic behaviour of a material can be followed on implanted impurity atoms which are well adapted to some specific nuclear methods. In that way Cd^{111}, Hf^{181} etc, is used in TDPAC; Sn^{119} and Fe^{57} in ME. Some anomalous magnetic behaviour has been found and no satisfactory explanation exists. But we will not discuss this point as implantation is only involved to introduce the probe in the material. We will concentrate on the changes of the hyperfine magnetic interaction introduced by the defects. The concentration of the defects is that low that the bulk magnetic behaviour of the host material is not affected. It means that the mean electron polarization will not change and thus no effect will be shown on the impurity atom as well. However, the electron wave function will be changed on the nearest neighbourhood of the defect. As a consequence only when the impurity probe has defects at the nearest neighbours we will see the effect on the magnetic hyper-hyperfine field. The fact that oversized (or undersized) impurities trap vacancies (or interstitials) results in well defined

431

stable association of the impurities with defects, and to each association (called "site") there exists a different magnetic hyperfine field. As an example, when Xe is implanted in $Fe^{2,3)}$, we will
have different sites having different well defined magnetic hyperfine fields.
We will have

$$XeV_1 \qquad H(V_1) = 1540 \pm 19 \text{ kG}$$

$$XeV_2 \qquad H(V_2) = 1245 \pm 25 \text{ kG}$$

$$XeV_3 \qquad H(V_3) \leqslant 350 \text{ kG}$$

The different sites can easily be distinguished by the magnetic
hyperfine field. We must note that when a defect is trapped, the
cubic symmetry is broken and quadrupole interaction is present.
Then, one has to analyse hfi spectra of combined dipole plus electric
quadrupole interaction. The analysis of experimental data becomes
complicated. In that case, energy resolution is very important.
Neither Me nor TDPAC have sufficient resolution and NMR/ON reveals
to be a promising tool.

    The fieldgradient associated to a defect site contains a lot
of information as in fact we deal with a tensor, with five inde-
pendent parameters : V_{zz}, η and the Euler angles α, β, γ, (giving
the direction of the principal axis system). These five parameters
determine the defect configuration trapped by the implanted atom.
Calculations of the fieldgradients are not reliable enough because
the relaxation of the lattice around the impurity is badly known.
However, the sign and the asymmetry parameter η are very important
to reveal the nature of the defect configuration. One can easily
perform simple calculations of lattice sums to evaluate the infor-
mation from $|V_{zz}|$, η, and the sign of V_{zz}.

$$V_{zz} = \sum_i \frac{e_i (3z_i^2 - r_i^2)}{r_i^5}$$

We show the results of such a calculation in Fig. 1 and Fig. 2 in
function of the relaxation of the impurity in direction of the associated vacancy(ies).
The fieldgradient components are given in arbitrary units as they
strongly depend on the lattice distances, a, and the relaxation of
the whole surrounding lattice. We assume, however, that the sign
is conserved and that the relative variations on the figure reflect
reality.
For monovacancy site (Fig. 1), the Z-axis of the PAS is directed towards the vacancy. Thus for bcc in <111>, for fcc in a <110> direction. Thus for bcc we always have threefold symmetry, and thus $\eta = 0$,
while for fcc only twofold symmetry exists and thus $\eta \neq 0$ if the impurity relaxes. The $(V_{zz})_{latt.}$ has always a negative value.

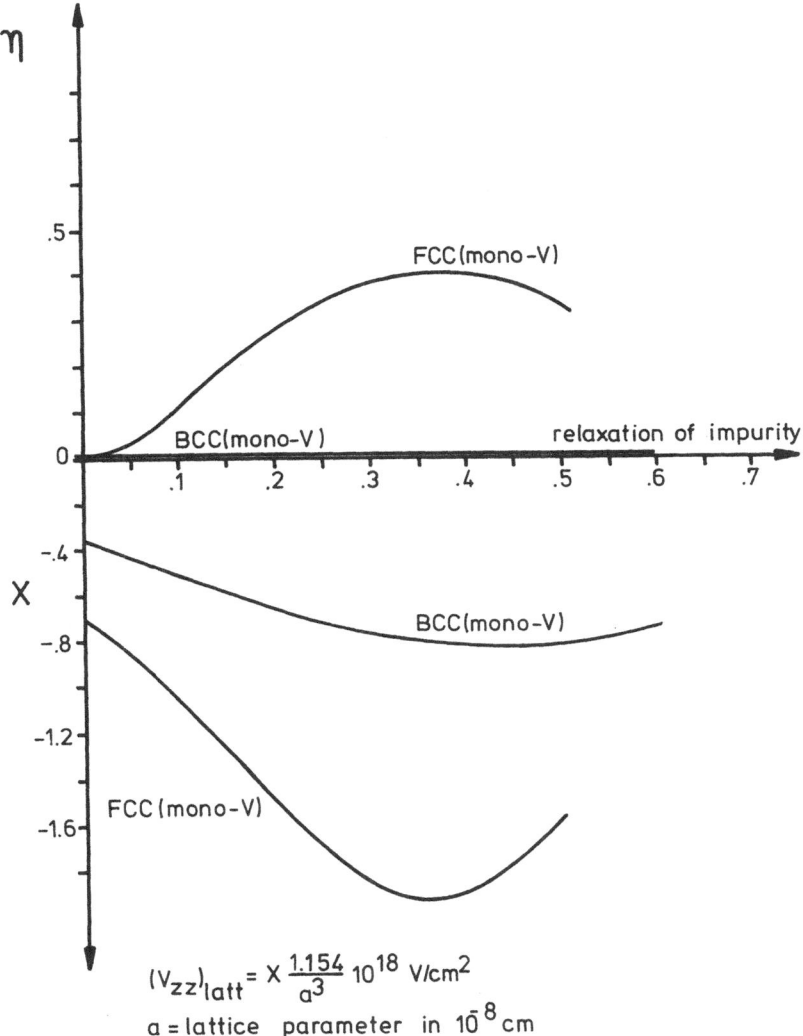

Figure 1

For divacancies (Fig. 2) we only display the fcc case in order to avoid to complicate the figure. We have quoted η, V_{xx}, V_{yy} and V_{zz} in function of the relaxation of the impurity in the direction of the gravity center of the two vacancies as indicated on the figure. When the impurity is not relaxed then it is clear that the Z-axis will be directed in the plane of impurity-vacancies, on the bis-

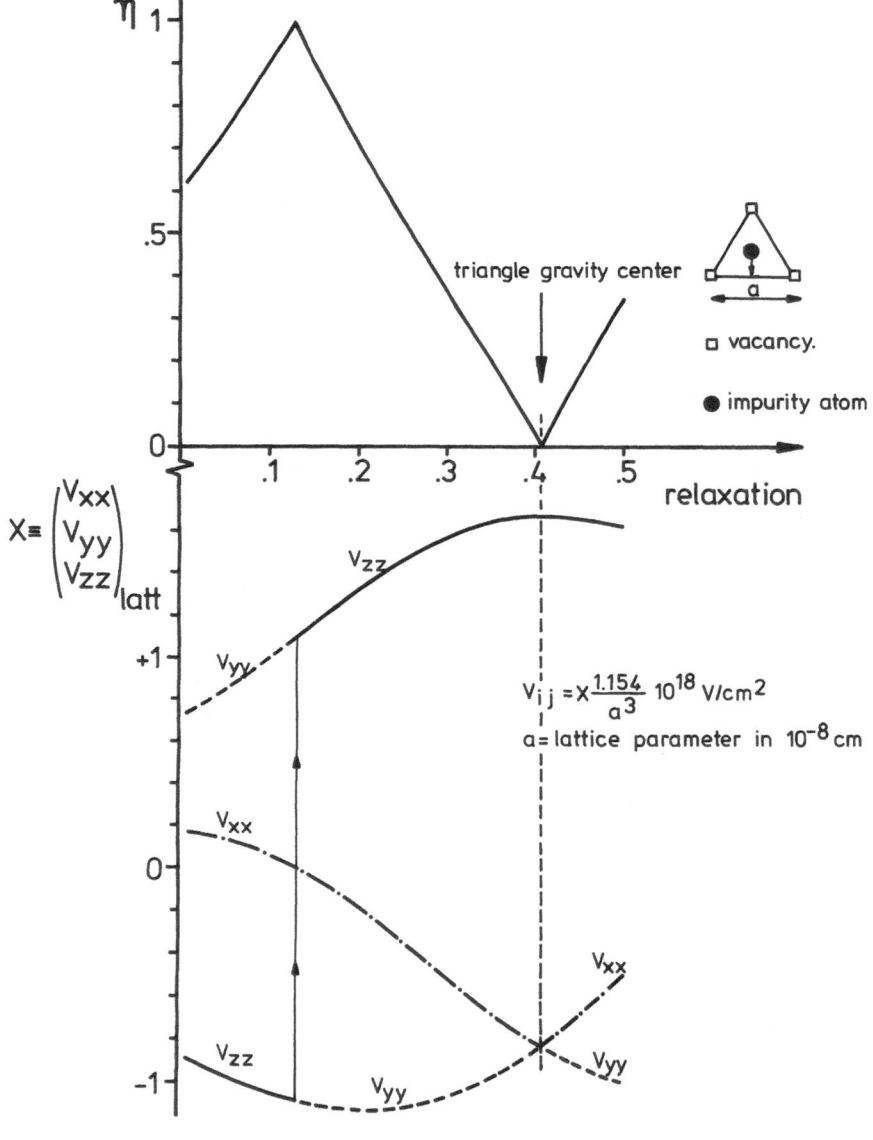

Figure 2

sectrice of the triangle. We have only twofold geometry and $\eta \neq 0$.
If we gradually relax the impurity , we reach a point where $V_{xx}=0$
and $|V_{yy}| = |V_{zz}|$. Due to the condition

$$|V_{xx}| \leq |V_{yy}| \leq |V_{zz}|$$ (1)

we must change the axis system. Then the Z-axis becomes perpendicular to the triangle plane.

At that point $\eta = \dfrac{V_{xx} - V_{yy}}{V_{zz}} = 1$. As we have changed the axis sys-

tem, η again decreases, and we jump from negative V_{zz} to positive
V_{zz}. Reaching the gravity center of the triangle $\eta = 0$ and we have
to interchange the x- and y-axis to further fulfill condition (1).
This change of axis is reflected again as a discontinuity in η.
We expect that in reality the impurity relaxes near to the gravity
center where $\eta = 0$ or small. We thus expect $(V_{zz})_{latt.}$ positive and
perpendicular to the plane of the triangle. Thus by determining η,
the sign of V_{zz} and the direction of the Z-axis of PAS to the lattice directions, one can reasonably distinguish between mono- and
divacancy-sites.
For more defect-sites, symmetry increases and we expect η to be
small and also V_{zz} to be smaller than for mono- or divacancy-sites.
One of the Euler angles, called usually β gives the angle between
the Z-axis of PAS and the laboratory system determined by the direction of the radiation detectors to the sample. For single-crys-
tal samples the defects will introduce fieldgradients with a small
number of angle β, with each angle β having a defined multiplicity.
With multiplicity we mean, the relative occurrence of such a vacan-
cies-impurity situation with angle β. Thus for each site, we have
different β-sites, depending on the crystal direction to the detector system. If we choose the symmetry axis of the crystal to
coincide with the detector axis, the number of β sites mostly reduces
to only two.
For polycrystalline samples all β angles are present and fitting
procedures sometimes become very long due to integration processes
on β. However, in AD experiments integration on all angles makes
the radiation isotropic and all information is lost.
We have to remember that it is most difficult to get all these parameters with one single experiment. Therefore, the conclusions about
the defect configuration are not always unique but sometimes rather
the confirmation of a proposed model. The limitation originates from
the fact that some experiments are rather insensitive to parts of
the parameters and secondly because the resolution is too bad.

    Isomer shifts are very easily obtained by ME and are indeed very
useful to resolve different implantation sites. The first question
before planning an experiment here, is to know if the isomer shifts
are large enough to be resolved in a Mössbauer spectrum. And therefore the number of useful Mössbauer isotopes is limited to the few
with large lifetimes of the Mössbauer level, and thus with small
natural linewidth.
From isomer shifts of chemical compounds, we can have an idea of
the expected isomer shifts. In most cases the resolution seems
too poor to resolve the different shifted lines. However, it turns
out that the situation is more favourable. Due to the enormous
compression of the lattice at an oversized impurity, the isomer

shift is much larger and in fact enormously sensitive to the association of a defect. We can easily understand that a trapped vacancy releases the large compression around an oversized impurity,
and therefore the isomer shift of such a vacancy associated system
is very different from the pure substitutional one and the different
sites can be resolved. It seems essential to us to use in such
studies an oversized or undersized implant, not only to stabilize
the association of a defect, but also to increase the isomer shift
between different sites. Such experiments on which the different
sites are characterized by different quadrupole or magnetic interactions have been used for systematic [4, 5, 6] annealing studies
and were compared to resistivity measurements in the host
material.

I like to make some remarks about the recoilless fraction.
We can easily understand that the phonon spectrum of the impurity
will change if defects are associated or not. We can thus expect
that each site will have his own recoilless fraction. Such recoilless fractions can be obtained from measuring the absorption
depths in function of temperature[6]. It would be worthwhile to
study theoretically this change. In this summer school De Waard
reported on a possible correlation between isomer shift and recoilless fraction. We can expect that the association of a vacancy will
decrease the recoilless fraction and therefore the measurement of
that parameter can help to identify the defect configuration.
Furthermore in order to convert the spectroscopic intensities of
each site to populations, we need to know the recoilless fraction
too.

We will end with an example.
Results are shown of implantation and annealing of $\overrightarrow{Xe}^{133}Mo$ (Fig. 3).
Xe^{133} decays to an excited state of Cs^{133} which decays by emission
of a gamma ray of 81 keV to its groundstate. The lifetime is 6 ns.
Therefore this transition is a good Mössbaeur effect transition.
The different defect sites are separated by the isomer shift and the
non-substitutional sites show quadrupole splitting. These facts
cannot be seen clearly in a single spectrum. But the analysis is
performed on a series of measurements on which different sites
(Fig. 3) are predominant. In that way a decorrelation of the hfi
parameters belonging to different sites is achieved. The implantation was done at RT and then annealed for 1 hour, at different
higher temperatures. After each annealing step the Mössbauer
spectrum was taken. One can easily see the change in the spectra
by increasing the annealing temperature.

Detailed analysis provided the population of each site after
each annealing. The result shows (Fig. 4) that the substitutional
site is predominantly populated by RT implantation and that dramatic
changes occur around 400 °K. This coincides with the stage III of
resistivity measurements. However, we have to be careful in com-

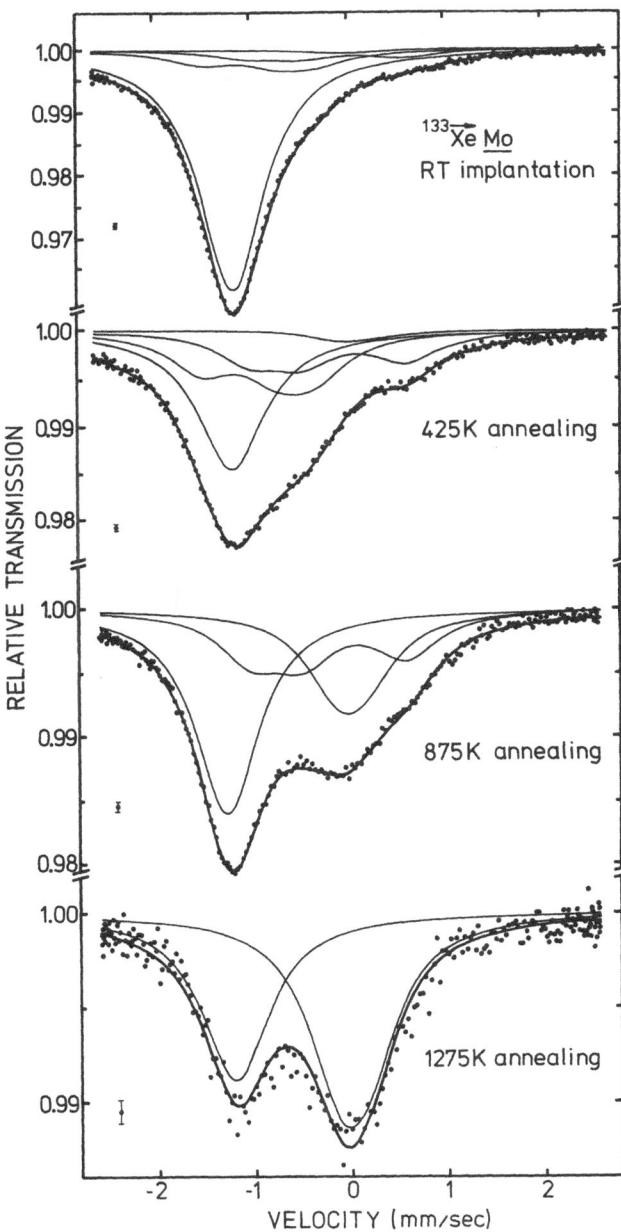

Figure 3

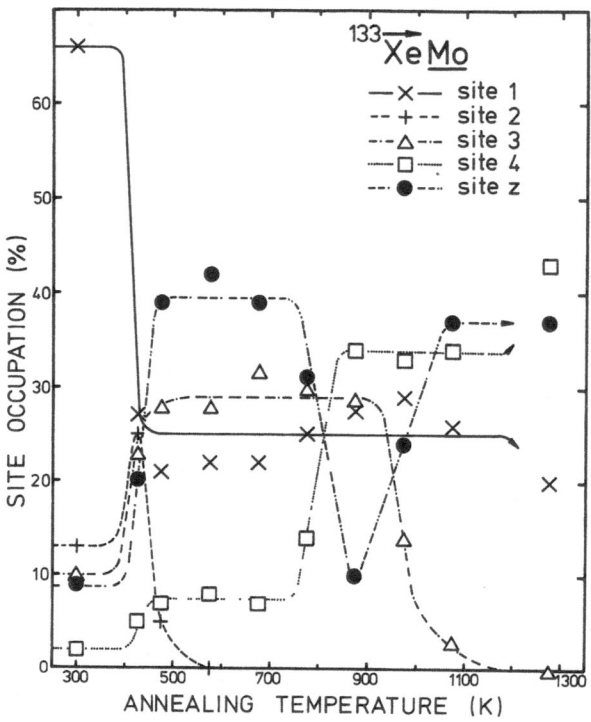

Figure 4

paring the temperature as the annealing cycles are very different.
We assume that at that temperature vacancies become mobile and are
trapped by the oversized substituted Xe^{133} atoms.

    We can conclude that from the hfi parameters, we can identify
the site configuration. This identification is one of the most important advantages of the radiation methods. Combined with the
change of the relative populations in the annealing stages, we can
conclude on the nature of the defect (vacancy or interstitial)
released or mobile at that particular annealing temperature. In
that way one can appreciately help to identify particular annealing
stages. However, the experiments are long, expensive and laborious.
Up to now most experiments were more or less exploratory, but in
future well directed experiments have to be designed.

REFERENCES

1. Freeman & Watson in Hyperfine Interactions, editors : A.J.
    Freeman and R.B. Frankel, New York, Academic Press 1967.
2.a) E. Schoeters, R. Coussement, R. Geerts, J. Odeurs, H. Pattyn,
    R.E. Silverans and L. Vanneste, Phys. Rev. Lett. 37, 302, (1976).

b) E. Schoeters, R. Coussement, R. Geerts, C. Nuytten, R.E. Silverans, and L. Vanneste, to be published.

c) M. Van Rossum, G. Langouche, H. Pattyn, G. Dumont, J. Odeurs, A. Meykens, P. Boolchand, and R. Coussement, J. Phys. Colloq. 35, C6-301 (1974).

3) H. de Waard in Mössbauer Spectroscopy and its applications (I.A.E.A. Vienna, 1972), p.132, and ref. 6.

4) S.R. Reintsema, E. Verbiest, J. Odeurs, H. Pattyn, to be published.

5) C. Hohenemser, A.R. Arends, H. de Waard, H.G. Devare, F. Pleiter, and S.A. Drentje, Hyperfine Interactions 3, 297 (1977).

6) S.R. Reintsema, Ph.D. thesis, Rijksuniversiteit Groningen 1976, and ref. 3.

Part V

New Fields of Applications

SURFACE PROPERTY MODIFICATION IN ION-IMPLANTED METALS

G. Dearnaley

AERE Harwell, England

INTRODUCTION

It has for many years been evident that ion bombardment will
alter the composition and physical state of a wide variety of
materials, but it is only during the present decade that deliberate
attempts have been made to alter and study the surface behaviour
of metals and alloys in this way. In many respects this field
differs from that of inducing electrical conductivity changes in
semiconductors. In that case the ion doses are relatively low and
the ion species are simply those which had previously been
introduced by diffusion. Defect production and aggregation is
totally different in metals and covalently-bonded semiconductors
such as silicon.[1] It is important to restore the semiconductor
lattice by thermal annealing after implantation, but this is
usually unnecessary in the case of metals.

In relation to the topic of the present meeting, it is note-
worthy that the effects of ion implanted species to be discussed
will all involve the interaction between the implanted atoms and
some aggregated defects e.g. vacancies, vacancy clusters, dislocations and grain boundaries. The impurities rarely exert their
influence as single atoms, substitutionally located within the host
metal matrix, as would be analogous to the semiconductor case.

A number of very important material properties (figure 1) are
strongly affected by the composition and structure of the surface
layers of a material, let us say within 1μm of the surface. All
these have been shown to be altered by ion implantation, but I
shall concentrate attention upon two areas of considerable
industrial importance: these are respectively wear and high

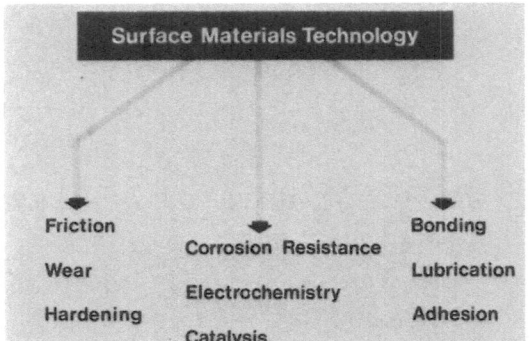

Fig. 1

temperature oxidation. There is a very great deal of scope for
research into all these fields, and the topic of electrochemical
changes and aqueous corrosion is being investigated systematically
by Dr. Ashworth and his colleagues at the University of Manchester
Institute of Science and Technology.

There are two ways in which ion implantation may benefit
these areas of technology. Firstly it can be used to prepare
a well-controlled surface layer which may be exposed to wear or
corrosion. The effects observed may be correlated with composition, and the re-distribution of implanted atoms may be followed
e.g. by SIMS or ion beam analysis. This research, coupled with an
understanding of the behaviour of implanted atoms and defects,
can throw much new light on the mechanisms taking place. I believe
that this research role of ion implantation has already been well
demonstrated.

The second means of application of ion implantation is for
the actual treatment of metallic components in order to improve
their performance. As a metal finishing technique ion implantation
is in competition with a wide variety of coating methods, thermochemical treatments and other processes already commercially
available. It must therefore offer technical or economic
advantages in order to become adopted.

It is important to recognize that ion implantation will not
produce a coating. This is because of the process of sputtering,
which is the kinetic ejection of surface atoms as a result of
collisional energy transfers. As a rule more atoms are sputtered
than are injected as ions and so a dynamic equilibrium is eventually
established in which the proportion of foreign atoms which can be
incorporated into the matrix is usually below 50 per cent. At
the same time there is almost no change in the dimensions of the
workpiece, and for this reason the treatment may be carried out at
the final stage of surface preparation. There is no plane of
weakness which could be subject to decohesion or interfacial
corrosion as is so often the case in coated materials.

The implanted atoms must be chosen so as to exert the required influence within the matrix of the original material and they may often be of gaseous species. In this respect the process resembles that of diffusion, but an important distinction lies in the fact that there is no necessity for the ions chosen to diffuse or even to dissolve in the material. Ion implantation is a non-equilibrium process which produces metastable systems which have, nevertheless, remarkable degrees of stability.

Diffusion, moreover, is frequently impeded by the barrier presented by surface oxides or other contamination films, and its rate is affected by the presence of dislocations or strain. Ion implantation, being carried out at particle energies about a million times greater than typical thermal energies, is free from this uncertainty. In addition it can be monitored continuously by measurement of the ion beam current. It is for just these reasons that the method has become so extensively used[1] in the manufacture of semiconductor devices of all kinds, and the increased controllability has led to improvements in device yield by around a factor of ten.

Ion implantation clearly does not demand a high temperature, as in the case of various thermochemical treatments for diffusing mobile species such as nitrogen or boron. It differs from plasma nitriding or ion nitriding in this respect. There is therefore no distortion which can in these other cases arise due to a relaxation of machining stresses.

An optimisation of the ion implantation process does, however, require considerable understanding of the ways in which the implanted atoms can inhibit wear or corrosion to depths which are many times (perhaps a thousand times) greater than the implantation depth.

MECHANICAL PROPERTY CHANGES

The first work in this field was carried out at Harwell by Hartley[2], mostly in steels such as the chrome-carbon bearing steel (En352). Soft metals (Pb^+, Sn^+, Ag^+ and In^+), rare gases (Kr^+, Ar^+), sulphur, Mo^+ and S^+ and Mo^+ in combination (with the aim of forming MoS_2) were implanted generally to doses of around 3.10^{16} ions/cm^2. The coefficient of friction was then measured using equipment in which a small tungsten carbide ball is loaded in contact with the specimen, and traversed horizontally.

All ions except the rare gases and sulphur produced signifi-cant changes in friction under non-lubricated conditions, the largest effect being seen in the case of tin (fig. 2). It was interesting to find that $Mo^+ + 2S^+$ dual implantation gave an

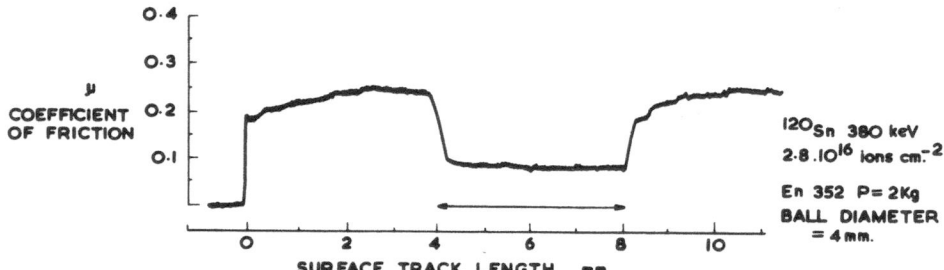

Fig. 2 The effect of ion implantation of Sn+ on the coefficient
of friction of steel.

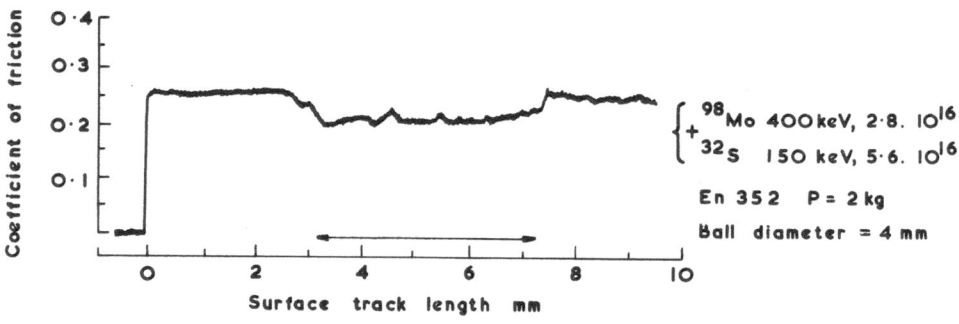

Fig. 3 The effect of implantation of
Mo+ and S+ on the coefficient
of friction of steel.

appreciably lower friction coefficient than either of the constituents alone (fig. 3), although there was no direct evidence that
MoS_2 was formed. Under lubricated conditions only this combination of ion species gave any marked reduction in friction, and
sulphur alone seemed ineffective under all conditions.

Wear is more important, and more complex than friction. Hard
and tough materials such as steels and cemented carbides have been
developed to provide wear resistant surfaces and improved performance is often achieved either by diffusion (e.g. in nitriding
or boronising) or by coatings (e.g. of TiC or TiN).

The principle in use here is the introduction of a high
concentration of material which will form very strong three-
dimensional interatomic bonds with the other constituents. It is
not surprising that ion implanted nitrogen, carbon or boron should
bring about similar effects. Nitrides and carbides are indeed

observed in implanted ferrous materials[3] but we now know that there are important consequences of the non-equilibrium nature of ion implantation and one of these is the relative mobility of ion implanted atoms.[3]

Fig. 4   An Avery-Denison T62 wear tester of the pin and disc type.

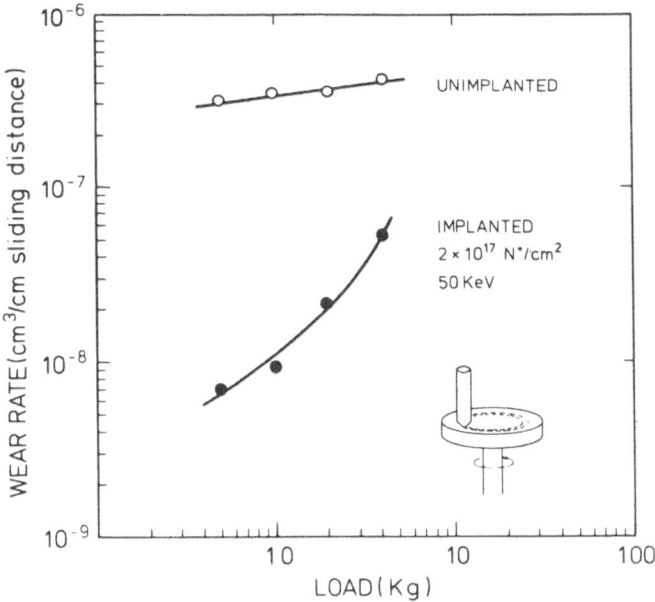

Fig. 5   The effect of N^+ implantation on the wear rate of steel

Measurements of the improvement in wear resistance of ion implanted metal surfaces have been reported by Hartley[4], Dearnaley and Hartley[5] and Hartley and Watkins[6] using a modified Avery-Denison T62 wear tester of the pin-and-disc design. A loaded pin wears against an ion-implanted disc and the wear rate of the couple is assessed mainly from the loss of material from the pin, together with examination of the total wear debris. Reduction in the rate of production of wear debris retards the transition from a mild to severe wear regime and as shown in figure 4 this can reduce the volumetric wear rate by one or two orders of magnitude. A reduction in wear rate by a factor of at least 200 was recorded for a silver steel pin wearing against a mild steel disc, as a result of N^+ implantation.[6]

There are many parameters to vary in such tests: the composition of pin and disc, the load, the sliding speed, the lubricant and its temperature and flow rate. A minimal lubrication system was adopted in this work, with a kerosene flow to carry away wear debris and provide cooling. The implantation of N^+, C^+ or B^+ was found to be effective (fig. 5) in all steel alloys tested, including mild steel, carbon steel, nitriding alloy (En40B), stainless steel (En58A) and a hardenable tool steel. A comparison with a conventionally gas-nitrided disc of En40B showed that the wear resistance of the nitrogen-implanted specimen was comparable, and initially superior. Other tests showed that N^+ implantation can improve the wear resistance of already nitrided steel, a result which suggests that the nitrogen occupies a more favourable site in the material than in the thermal equilibrium case.

A remarkable early discovery was the fact[5] that ion implantation provides a long-lasting wear resistance despite the very shallow initial penetration. Wear tests in which the track formed in the disc reached a depth of 12 μm showed no sign of a breakdown in wear resistance, and nitrogen was detectable in the base of the track at a level of 30-40 per cent of the initial concentration.

There are several ways in which this may come about. The temperature within the work-hardened zone below each microscopic load-bearing asperity is likely to rise[7] during friction and wear by as much as several hundred °C. As a result, mobile implanted species such as N, C or B atoms will migrate interstitially to decorate sub-surface dislocations generated as a consequence of local Hertzian and shear stresses. Such a process has been proposed[8] to account for strain-ageing in steel, and the decorated dislocation network has sometimes been called a 'Cottrell atmosphere'. The movement of dislocations is impeded by the impurities and the metal is thereby hardened: we have direct evidence of this from fatigue data (see below). As wear progresses, dislocations

are driven deeper and deeper and this serves to carry the
implanted atoms forward so as to re-create a hard surface of just
the thickness which can be most effective.

The Mössbauer conversion electron spectroscopy, which is
sensitive to conditions within about 0.1µm of the surface, has
proved a most powerful means of investigating these phenomena.   It
has shown[3], for example, that the lattice site occupancy in
nitrogen-implanted iron changes at temperatures as low as 275°C.
This is quite unlike the behaviour of nitrided steel, which is
of course stable up to the nitriding temperature, usually around
550°C.   The explanation may lie in the work of Wuttig, Stanley
and Birnbaum[9] on neutron irradiated iron containing dissolved
nitrogen or carbon.   These workers found, by magnetic permeability
measurements, that nitrogen and carbon would come out of solution
at relatively low temperatures, only to redissolve at about 300°C.
Quenched but unirradiated iron containing carbon gave rise to
Fe_3C which was stable up to 600 C.   Wuttig et al. proposed that
the interstitial impurities tend to decorate interstitial
prismatic dislocation loops induced by irradiation, and that these
sites are less stable than the vacancies present in quenched iron.

The same behaviour would account for the relative instability
of nitrides and carbides formed by ion implantation in iron and
steels.   It favours considerably the production, during wear, of
a mobile interstitial fraction which can migrate to, and decorate,
the sub-surface dislocation network which is propagated by
contact stresses.

Longworth and Hartley [3] found evidence for nitrogen martensite
in ion implanted iron which had been thermally annealed.   This
distorted form of the α-iron lattice is extremely hard, at the
nitrogen concentrations which are present, and it is hoped to
carry out electron diffraction measurements of the c/a ratio of
the tetragonal martensite lattice.

Some of the clearest evidence for surface hardening and the
pinning of dislocations in nitrogen-implanted steel may be found
in the improvements in fatigue life observed by Hartley[4] and by
Hu et al.[10] in rotating-bend tests.   The latter authors found that
the most striking effects occurred only after a period of ageing
at room temperature, or by a brief heat treatment.   This is
closely analogous to strain ageing and may be explained by the
diffusion of implanted interstitial nitrogen (or carbon) to
decorate dislocations in a manner which will impede their subsequent movement.

These laboratory results led to the testing of many ion-
implanted tools and components under industrial conditions, in a
programme funded by the U.K. Department of Industry.   Striking

improvements in the working life of dies, press tools, punches, cutting knives, taps, bearings, injection moulding equipment etc. have now been reported. Factors of two to twelve increase in life are typical. The advantages of the process are such that there is no directly competing method of improving the wear resistance. Figure 6 shows a steel press tool and a (sectioned) tungsten carbide wire-drawing die, to typify the types of widely-used items which benefit from ion implantation.

Equipment of the design described in the lecture on 'Ion Implantation Procedure' is now coming on to the commercial market, for application to engineering components. As far as can be judged at this stage the economic advantages are significant for a diverse and growing number of purposes. The next step is to encourage the adoption of the process and to develop it so as to optimize its value to the user.

Figs. 6, 7 A steel press tool and a (sectioned) tungsten carbide wire drawing die: two examples of tools which have benefited from ion implantation.

THERMAL OXIDATION

Relatively small amounts of certain impurities are known to have a strong effect upon thermal oxidation in metals. Often it is not known how this comes about, although it is desirable to do so in order to optimise the behaviour.

Ion implantation is attractive in this field from two different points of view. Materials may be conserved if the beneficial species is introduced only into the surface where it can be effective. In many cases alloying throughout may degrade some bulk property such as ductility, electrical conductivity, nuclear properties, etc. Secondly, the ion implanted layer can be controlled in composition and, since it is well-defined, the subsequent migration of these atoms can be followed during corrosion so that transport processes may be determined. This is proving to be a powerful research tool in corrosion science.

Once again the cases of most interest are those in which the effects are prolonged. This may come about either because the implanted species is transported forward as oxidation proceeds, or alternatively it may establish favourable conditions of protection during the initial stages of nucleation and growth of the oxide film.

Benjamin and Dearnaley[11], and Dearnaley et al.[12] were able to reduce the oxidation of titanium at 600°C in O_2 by the implantation of alkaline earth and certain divalent rare earth ions characterised by the ability to combine vigorously with oxygen and to form stable perovskite structures (fig. 8) such as $BaTiO_3$ or $EuTiO_3$. Protection is thought to result from the blocking action of such compounds which precipitate along fast diffusion paths such as dislocations in the oxide and in the metal. Both the implanted species and oxygen ions can penetrate rapidly along these defects, as was shown by the ability to influence oxidation on both sides of a 100 μm foil.[11] Lucke et al.[13] have since been able to identify the presence of $BaTiO_3$ in Ba^+-implanted Ti after oxidation, and have shown that it precipitates preferentially on to the cubic suboxide phase TiO. A more detailed discussion of the role of perovskites and spinels in high temperature oxidation is given elsewhere by Dearnaley[14].

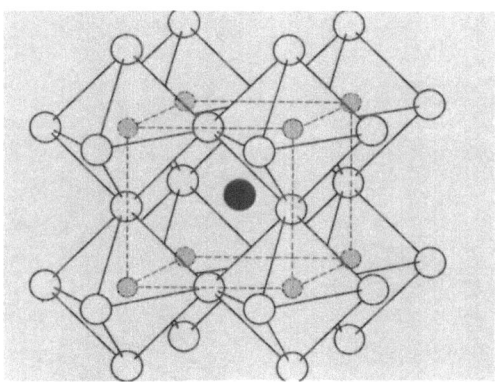

Fig. 8 The cubic perovskite lattice. The larger cation occupies the central site. The open circles indicate oxygen atoms.

In zirconium, the effects of ion implantation are quite different, as a result of the contrasting crystallographic properties of ZrO_2 and TiO_2. Dearnaley et al.[12] and Berti et al.[15] have shown that beneficial species all possess an ionic size lying between 80% and 100% of that of the Zr^{4+} ion. They have argued that the incorporation of smaller cations, such as Ni^{2+}, into the oxide lattice will reduce the energy for propagation of dislocations, and as a result the oxide plasticity will increase. The oxide is then better able to protect the underlying metal without fracturing due to mechanical growth stresses which arise because ZrO_2 has a much larger specific volume than that of the metal.

Yttrium and rare earth elements have long been known to bring about useful reductions in the high temperature oxidation of chromium-alloyed steels. However, they degrade the ductility and tensile strength of the alloy by grain boundary segregation of phases such as Y Fe_9. Ion implanted into the surface of the steel yttrium brought about a long-lasting protection without impairing bulk properties. Figure 9 shows the results of tests by Antill et al.[16], over 8 months in CO_2 at $800°C$. The yttrium was found to be carried forward, remaining close to the metal-oxide interface where it may act by formation of the perovskite $YCrO_3$ which is likely to serve as an impermeable barrier to chromium out-diffusion. Bennett et al.[17] have shown recently that implanted cerium is equally effective.

Alloys of iron, chromium and aluminium owe their oxidation resistance to the formation of a protective film of Al_2O_3. Often, however, it may be difficult to tolerate throughout the alloy a content of Al sufficient to form pure Al_2O_3 (i.e. about 4.5% by weight). For this reason Bernabai et al.[18] chose to implant additional Al^+ ions into the surface of Fe-Cr-Al alloys containing yttrium and observed at striking improvement of the oxidation

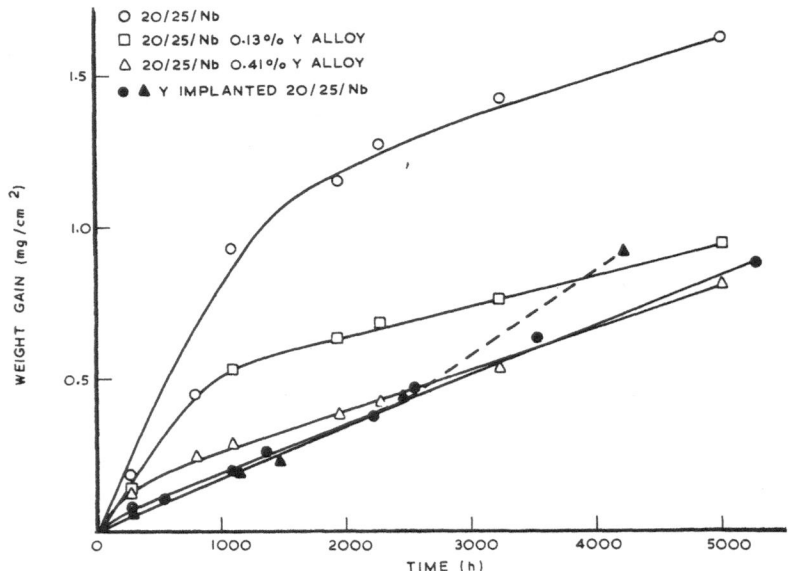

Fig. 9 The effect of yttrium, either as an alloyed addition or implanted to a dose of 3.5×10^{15} ions/cm^2, upon the oxidation of 20%Cr/25%Ni/1%Nb stainless steel in CO_2 at $800°C$.

resistance. At 1100°C a dose of 10^{17} Al$^+$ ions/cm^2 reduced the
weight gain in oxygen by a factor of twelve. In related experiments Bennett[19] et al. introduced yttrium into Fe-Cr-Al alloys
(with a fairly high Al content) and obtained the benefits of
reduced oxidation and spalling normally achieved by adding yttrium
as an alloying constituent. This work suggests that yttrium
(or rare earths) and aluminium can have a joint action in the
protection of high-temperature alloys. It seems likely that in
these cases the presence of Al close to the surface favours the
rapid formation of a protective film, and the subsequent out
diffusion of chromium is resisted. Further aluminium from the
bulk alloy is subsequently able to migrate to the surface to
reinforce and repair this layer.

The degree to which oxide spalls off during thermal cycling
is often important. Bennett et al.[17] showed that 10^{16} Ce$^+$
ions/cm^2 will completely eliminate spalling in the case of 20% Cr
25% Ni stainless steel oxidised in CO_2 at 825°C. There is then no
risk of losing the protective implanted layer. It is not yet
clear whether this effect is brought about by improved plasticity
of the oxide or by a mechanism which eliminates the voids caused
by condensation of vacancies associated with Cr out-diffusion.
Stoneham[20] has pointed out that a reduced out-diffusion rate will
lessen the likelihood of vacancies aggregating to form voids.
Muhl[21] found that rare earth ions implanted into chromium would
reduce the oxidation rate by about an order of magnitude at 750°C,
and the resulting oxide was smooth, fine-grained and extremely
adherent to the metal.

The effects of yttrium and hafnium implanted into a series of
chromium and nickel containing alloys has been investigated recently
by Pivin.[22] Hafnium appears to dissolve in the film of Cr_2O_3 formed,
and accelerates its growth. This may be due to the fact that
substitution of quadrivalent Hf for trivalent Cr will lead to a
compensating excess of Cr vacancies, which will facilitate
cationic diffusion. Yttrium may act by improving the plasticity
of the Cr_2O_3, by trapping Cr vacancies near the metal-oxide interface and by providing Y_2O_3 nucleation sites for the growth of
Cr_2O_3.

AQUEOUS CORROSION

The presence of a chemically active electrolyte and the need
to determine the potential of the metal with respect to this
liquid tend to make studies of aqueous corrosion somewhat more
complicated than in the above examples.

Ashworth, Procter[23] and their colleagues in Manchester have
embarked on a study of the effects of implanting both species which
are used commonly as alloying additions (e.g. Cr) and those which

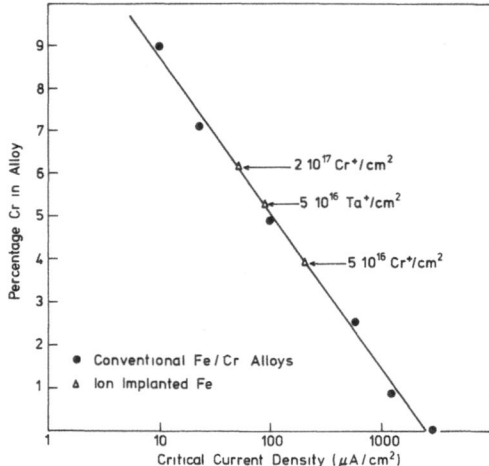

Fig. 10   *The passivating effect of ion implanted Cr^+ and Ta^+ on the aqueous corrosion of iron, compared with the behaviour of various Fe-Cr alloys (from ref. 23).*

would normally be insoluble (e.g. Ta). Potentiokinetic measurements show that the implanted metal usually behaves in a very similar manner to that of an alloy with the same surface composition (figure 10). The behaviour of the insoluble species is also relatively predictable. One common effect is that ion bombardment leads to an increase in the thickness of the air-formed oxide, prior to aqueous exposure.

Sartwell et al.[24], at the US Bureau of Mines, have carried out similar experiments on iron, and showed that implanted Cr^+ or Ni^+ would greatly improve the resistance to attack.

Hubler and McCafferty[25] have observed a remarkable improvement in the resistance of titanium to boiling H_2SO_4 as a result of the ion implantation of palladium. By Rutherford backscattering analysis it was shown that the palladium remains within the surface layers of the metal, becoming progressively concentrated within a shallower depth as the titanium is preferentially leached.

These results encourage the idea that metal components might be protected by ion implantation, but it remains to be demonstrated that the implanted layer will maintain its effectiveness during exposure to abrasion and erosion. Removal of part of the surface may lead to the formation of a couple, and intensified local attack.

Pitting corrosion is a common form of localised dissolution, the causes of which are not well understood. Ashworth et al.[26]

have found that Mo^+ ions implanted into a stainless steel free of molybdenum will impart a resistance to pitting corrosion, just as does alloyed Mo. The same effect was observed also in aluminium, and in this case Mo cannot be alloyed because it is insoluble. This suggests that the molybdenum ions play a direct part in the process, rather than by interacting with some other consituent and modifying the heterogeneity of the alloy.

Corrosion fatigue is another important topic which involves both mechanical and chemical factors. Heavy doses of nitrogen improve the performance of steel in this respect, as was shown by de Anna et al.[27] This is another way in which the performance of tools such as cutting knives, exposed to aqueous conditions, may be improved by ion implantation, since their edges are exposed to a combination of corrosion and repeated mechanical stresses. Attempts to improve resistance to stress corrosion cracking in steels by ion implantation have so far, however, proved unsuccessful[28] due to the rate of removal of the implanted layer by general corrosion.

CONCLUSIONS

Ion implantation is now established as a very versatile and controllable technique in materials science, particularly in the important fields of corrosion and wear. As a practicable means of treating metallic components, the implantation of nitrogen ions has been shown to be a feasible and economic method of improving wear resistance in steels and cemented carbides. Equipment for carrying out this process is expected to be on the market by 1979. Other applications are expected to follow subsequently, as the appropriate ion source technology develops.

The understanding and optimization of these valuable processes depends upon an awareness of the ways in which the implanted atoms interact with the host material and particularly with defects in its structure. Mössbauer and other hyperfine interaction techniques are proving to be powerful methods of examination.

REFERENCES

1. G. Dearnaley, J.H. Freeman, R.S. Nelson and J. Stephen, "Ion Implantation" (North-Holland, Amsterdam, 1973).

2. N.E.W. Hartley, Wear 34, 427 (1975).

3. G. Longworth and N.E.W. Hartley, Thin Solid Films 48, 95 (1977).

4. N.E.W. Hartley, Inst. of Phys. Conf. Series 28, 210 (1976).

5. G. Dearnaley and N.E.W. Hartley, Proc. 4th Conf. on Sci. & Industr. App. of Small Accelerators, Denton (IEEE, N.Y., USA, 1976) p.20.

6. N.E.W. Hartley and R.E.J. Watkins (to be published).

7. F.P. Bowden and K.E.W. Ridler, Proc. Roy. Soc. A 154, 640 (1936).

8. A.H. Cottrell, "Introduction to Metallurgy" (Arnold, London, 1973).

9. M. Wuttig, J.T. Stanley and H.K. Birnbaum, Phys. Stat. Sol. 27, 701 (1968).

10. W.W. Hu et al., Scripta Met. 12, 697 (1978).

11. J.D. Benjamin and G. Dearnaley, Inst. of Phys. Conf. Series 28, (1976).

12. G. Dearnaley, J.D. Benjamin, W.S. Miller and L. Weidman, Corros. Sci. 16 (1976).

13. W.H. Lucke, J.A. Sprague and J.K. Hirvonen, Corros. Sci. (to be published).

14. G. Dearnaley, "Ion Implantation in Materials Science & Technology", J.K. Hirvonen (Ed.) (Academic Press, N.Y., 1979).

15. M. Berti et al., Corros. Sci. (to be published).

16. J.E. Antill et al., Corros. Sci. 16, 729 (1976).

17. M.J. Bennett et al., Corros. Sci. (to be published).

18. U. Bernabai, M. Cavallini, G. Bombara, G. Dearnaley and M.A. Wilkins, Corros. Sci. 19 (to be published).

19. M.J. Bennett, G. Dearnaley and M.R. Houlton, Corros. Sci. 19 (to be published).

20. A.M. Stoneham (AERE) priv. comm. (1976).

21. S. Muhl, Ph.D. Thesis, Univ. of Lancaster (1978).

22. J.C. Pivin, Ph.D. Thesis, Univ. of Paris-Sud (1978).

23. V. Ashworth, D. Baxter, R.P.M. Procter et al., Proc. Conf. on Ion Implantation in Semiconductor & Other Materials, Osaka, S. Namba (Ed.) (Plenum Press, N.Y. '1975) p.367.

24. B.D. Sartwell, Proc. Conf. on Metallurgical Coatings, San Francisco (to be published in Thin Solid Films, 1979).

25. G.K. Hubler and E. McCafferty, Corros. Sci. 19 (to be published).

26. V. Ashworth and A. Mohammed, (to be published).

27. P. de Anna et al., Corros. Sci. (to be published).

28. I.S. Craig and R. Parkins, Univ. of Newcastle-upon-Tyne (Unpublished report 1977).

IMPLANTATION IN OPTICAL MATERIALS

P.D. Townsend [1]

LETI/MEP
CEN/G.
85 X   38041   GRENOBLE CEDEX (France)

The review will emphasise the uses of ion implantation for control
of optical properties of materials. In most respects the addition
of ions by implantation produces similar results in both insulators
and semiconductors because of new levels in the valence-conduction
band gap. Consequently one is able to control such properties as
luminescence, optical absorption, ESR and refractive index. Historically the techniques involving ion beams have received little
attention for applications to property changes in insulators but
the range of examples will indicate the possibilities are worth
exploiting. The studies mentioned include searches for trace impurities in luminescence, determinations of defect structures, and
production of optical waveguides and information storage systems.

INTRODUCTION

Applications of ion implantation for semiconductor physics has
been very successful and it is inevitable that the major part of
the literature is concerned with electrical property changes (1-5).
Nevertheless the techniques are equally applicable to wider energy
gap materials and it is possible to influence or control the optical properties such as luminescence or optical absorption. Several
recent publications have summarised existing applications of ion
implantation to both academic studies and device production in
insulators (5-11). This review will typify the problems which can
be approached by implantation methods but will not discuss the
examples in detail as in most cases the general ideas are more
important than the specific studies. One will note some conflict

(1) On leave of absence from the School of Mathematical and Physi-
    cal Sciences, University of Sussex, Brighton BN1 9QH, England.

between some of the results that have been obtained but this is
understandable as the field is still in its infancy.

    We shall commence with examples related to luminescence and
range over experiments concerned with analysis of trace impurities to phosphor production by implantation. Studies to determine
defect structures also benefit from the simplicity and cleanliness
of the ion beam doping and the techniques of electron spin resonance (ESR and ENDOR) are clarified by the use of single isotopes.
Absorption results will demonstrate the production of very high
defect concentrations in thin transparent layers and we shall see
that flux effects can be clearly emphasised by comparisons of ion
and molecular beam bombardments. Finally the use of ion beams for
control of the refractive index will indicate that ion implantation
may be as valuable a technique in the developing field of integrated optics as it is for semiconductor devices.

IMPURITY CONTROLLED LUMINESCENCE

    Ion implantation is particulary well suited for changing the
dopant levels in insulators in problems concerned with a mixture
of impurities as one can selectively add one impurity at a time in
known concentrations. There is also the possibility of making the
implant at low temperatures so that one may characterise the mate-
riel before and after the addition without secondary problems of
others trace impurities in or out diffusing from the system, as
may occur during high temperature diffusion doping. Although the
implant produces radiation damage this may anneal at much lower
temperatures than are needed for diffusion. The advantages of
implantation are apparent when one is studying the influence of a
gaseous impurity such as oxygen. This is thought to play a role in
the radiation dosimeter material LiF doped with Mg and Ti. In the
LiF the role of the impurity ions is relatively well understood
(e.g. 12-14) and by suitable heat treatment one produces complexes
of (Mg^{++} + (Li$^+$) vacancies) which act as electron traps. Read out
of the dosimeter is made by thermally releasing the electrons.
Some of these are trapped and destroy a second defect which libe-
rates a hole. The hole produces luminescence as it, in turn, is
trapped and converts a Ti^{3+} to a Ti^{4+} ion. The luminescence intensity is a measure of the concentration of trapped electrons and
hence the total ionising radiation received by the dosimeter. A
problem exists as the Ti^{3+} and Ti^{4+} occupy a Li^{1+} site and there is
need for a charge compensator. One possibility is O$^=$ on three adjacent F$^-$ sites and it is probable that the O$^=$ enters the lattice
during the growth stage. Implantation of oxygen showed that this
model is reasonable as it produced more efficient light emission
in samples with a high Ti content which could not be compensated
with the "naturally" occuring oxygen [13, 15]. An anneal at 400°C
was used to remove the radiation damage and also to reproduce the
Mg ions in the same state of dispersion. The results were more
conclusive than in or out diffusion of oxygen as the total change
in impurity level was only a few ppm.

A similar situation occured in CaO which although intentionally doped with Bi ions showed sharp line luminescence characteristic of Gd [16]. Most probably the Gd had entered the CaO during the high temperature diffusion of the Bi but to further refine the starting materials and crucible etc was difficult and the simpler solution was to separately implant Bi or Gd into pure CaO which did not show the line spectra initially. This confirmed the spectra resulted from Gd impurities.

DAMAGE AND IMPURITY RELATED LUMINESCENCE

Radiation damage does not always alter the luminescence response and it was found that in silica, alumina and some orthosilicates[17] that even high dose implants, up to 1% of the lattice sites, with rare earth ions did not degrade the phosphors. Each system continued to show sharp line spectra of the rare earth implanted. However in other systems such as ZnS/Ag [18-19], ZnSiO$_4$/Mn [18], ZnO/Zn [11] or Ca$_2$Al$_2$SiO$_9$/Ce [17] the phosphor efficiency decreased with the defect concentration and it was necessary to anneal the samples to obtain the luminescence effects of the implants.

Implantation produces radiation damage but equally one notes that high temperature diffusion of the impurities is accompanied by changes in the stoichiometry of the lattice and in many cases the impurity ion will only enter the lattice if there is charge compensation from a secondary impurity. Takagi and al [20] made such a comparison between implantation and diffusion doping of Mn in ZnS. In both cases the electroluminescence response was similar but for the diffusion it was necessary to introduce a second impurity for charge compensation.

EFFECTS OF DEFECT MOBILITY

If the luminescence site involves a complex of impurity ions and lattice defects then the resultant site may be a function of the defects which are produced during the implantation. Important parameters to control are the temperature of the implant and the rate of energy deposition into the lattice as both these parameters control the equilibrium concentrations of intrinsic defects. There are far too many parameters that influence the end result for any type of generalisation and one can only list examples from the literature. For example GaAs implanted with Se [21] between 20 and 400°C shows a main photoluminescence at 1.51 eV and a range of features from 1.1 to 1.4 eV which were sensitive to the implant temperature, even after the material had received an anneal at 800°C. Light emitting diodes [22-23] are formed in GaP or GaAsP with Zn implants at 20°C but with N or Bi ions it is necessary to use hot implants. Other work with Zn in GaAsP indicated that implants at 400°C induced defects which annealed at 900°C and these changed the efficiency of the diodes [21]. Similarly, annealing is needed for cathodo and photoluminescence to be observed from ZnTe

implanted with B or Al. The Al implants have the interesting result
that they change the wavelength of the emission [24].

MEASUREMENTS WITH LINE SPECTRA

Many luminescence bands from insulators are relatively broad
and therefore are insensitive to small differences between similar
defect sites. Exceptions are the sharp line spectra which result
from luminescence transitions of ions with incomplete inner shells
(e.g. the rare earths). For thse ions small changes in the interaction with the crystal field produces measurable energy shifts of
the transitions. This probe of the lattice structure may be
achieved by implantation of the rare earth ion or may arise from
ions which already exist in the original structure. In either case
they can be used for defect studies. As an example of implantation
of rare earth ions we may cite the work [25] of Yb^+ in CdTe, which
produces a cathodoluminescence material whose spectra has distinct
changes as a result of annealing. The effects were interpreted in
terms of a range of substitutional sites for the Yb ions which are
linked to intrinsic defects in the CdTe.

SURFACE TEMPERATURE MEASUREMENTS DURING IMPLANTATION

A further use of the changes in the emission energies of
sharp luminescence lines has been made by Chandler [26] to estimate the surface temperature of insulators during implantation. In
his work with H^+ implanted ruby (Al_2O_3/Cr) he looked for the shifts
in the red emission lines which result from beam heating of the
implanted layer. For a calibration one can optically excite the
same transitions to measure their temperature dependence. During
the implant one both heats and excites the same layer so the luminescence provides a direct measure of the average temperature. This
is particulary useful for materials of poor thermal conductivity or
in systems which have weak thermal bonding to a heat sink. The work
also revealed some secondary effects in the degree of polarisation
of the lines during the irradiation which indicated a link between
some of the protons and the chronium ions. This dynamic effect was
not apparent in subsequent photoluminescence.

ISOTOPIC INDENTIFICATION OF DEFECTS

The implantation of ions introduces impurities at a range of
defect sites in the lattice and these change as a function of annealing or interactions with intrinsic defects. For identification of
the structure of the defect sites in insulators the most powerful
techniques are ESR and ENDOR. The strength of these electron resonance methods follows from the detailed structure of the hyperfine
splittings induced by the interaction with the nuclear spins. To
obtain "clean" and easily interpretable spectra it is convenient to
use doping with ions of a single isotope. Fortunately the mass
analysis inherent in most accelerators can be used to provide such
clarity in the ESR and ENDOR spectra as one may select a single
isotopic mass for the implant.

Isotopic effects in optical absorption are rarely apparent except as changes in vibrational resonance frequencies involving H and D bonds. In silica the direct growth of the OH vibrational absorption band in the infra-red has been used to follow the bonding of H^+ into the SiO_2 glass network [27]. The variations in growth rate as a function of dose and dose rate assist in the studies of ion induced compaction of the glass.

More complex vibrational effects are revealed in the low temperature luminescence spectra of SiC implanted with H^+ or D^+ ions [28]. The characteristic CH and CD vibrations give evidence that the hydrogen is substituted on to a silicon vacancy and is linked to a neighbouring carbon atom.

OPTICAL ABSORPTION

Most experiments involving ion implantation and optical absorption measurements have been concerned with colour centres which had been previously detected and identified (e.g. 29). One advantage of colour centre formation by ion beams is that one can produce a saturation concentration of the defects but because it is possible to choose a short ion range the total optical density is still measurable. For studies of growth kinetics and interactions of point defects in high concentrations this is useful and an early experiment by Pooley [30] followed the growth of F centres (the F centre is a halogen vacancy with a trapped electron). Equally fundamental defects of interstitials and di-interstitials were produced in later KBr experiments [31].

One may also follow the growth of impurity bands formed by the implant. In the alkali halides proton irradiations cause the formation of U bands as the hydrogen is bonded into different substitional and interstitial sites [32].

Excess concentrations of alkali metal ions lead to the formation of colloidal metal precipitates. In LiF such aggregates have been formed by implants with Li, Na and K ions [33]. The band positions are a function of the state of aggregation and the alkali metal, the results are in general agreement, but not identical [34], with colloidal bands formed by other methods. However for the preparation of non-stoichiometric crystals the implants have the advantage that the excess metal is not only introduced at relatively low temperature but also one avoids the possible problem of ion exchange between the crystal and the source of the metal. The preparation of LiF/Li is particulary important as the system is not formed by exposure of LiF to Li vapour [35]. A further advantage of the low temperature implants, in this case [33] 77K, is that because the precipitation process is inhibited by the low ion mobility one can develop other alkali-defect complexes that would not be observable by diffusion doping.

Similar types of experiment have been made to form intrinsic defects and colloids in $LiNbO_3$/Ag [36] and MgO/Na, MgO/K [37]. Sodium and potassium colloid centres in MgO evolve like those in alkali halides but one notes both F and F^+ centres (as more charge

states exist in this divalent structure than in alkali halides).
The equilibrium between different charge states is a function of
the rate of ionisation of the lattice, the temperature and the
defect concentration. A measure of the F/F^+ ratio reveals the difference in the type of energy loss of the primary ion. For example
the predominantly nuclear collision loss from heavy ions, argon,
produced mostly F centres whereas the electronic losses from proton implants formed more F^+ centres [38]. Interconversion of
defect types is also possible if one alters the type of bombardment
(e.g. from A^+ to H^+). This changes not only the type of colour
centre induced by the radiation but also produces ionisation assis-
ted annealing of other defects. Ionisation annealing is known to
occur in several oxides MgO [38], Al_2O_3 [39] and SiO_2 [40], for the
damage induced dimensions changes of the lattice. However the
lattice compaction and the colour centres are not directly related
and in the case of Al_2O_3 the ionisation controls the colour centre
production at 6.1 eV, but the nuclear energy loss governs the
compaction. We may also note that the **defect model** of the 6.1 eV
band has been tested by self ion implants of O^+ and Al^+ [41]. The
band appears with the Al^+ implants in accord with earlier models
for the colour centres as being as either an $Al_{vacancy}$ or an
$O_{interstitial}$.

EFFECTS OF RADIATION FLUX

Many insulators form defects when exposed to ionising radiation and as mentioned before the defect equilibrium is a function
of the rate of energy transfer to the solid. Such effects are
apparent if one alters the type of ion but it is much simpler to
vary the energy flux by comparing damage rates during ion and molecular ion beam irradiation. The molecules decompose at the surface
so one produces separate ion beam cascades which overlap in both
space and time. This corresponds to a high flux condition which is
unobtainable with even the highest of currents of single ions. One
also avoids the problems of sample heating by suitably adjusting
the ion and molecule beam energies and currents. The method has
been used with MgO to show that ionisation leads to intrinsic
defects, the ionising source was a beam of H^+ or H_2^+. The latter
produces more colour centres [42].

Defect equilibria between vacancy and di-vacancy centres, and
interstitial and di-interstitial centres in KBr are affected by the
flux from H^+ and H_2^+ ions and the same technique confirmed the di-
interstitial model of the V_4 optical absorption band [43]. An identical approach [26, 44] with H^+, H_2^+ and N^+, N_2^+ ions has been used
to refute an earlier model of a bi-exciton luminescence transition
in NaCl. In this experiment one assumes the higher flux will induce
a higher exciton density in the same region of the lattice which
should favour bi-exciton production and luminescence. The observations showed the emission band tentatively ascribed to such a transition was not changed and instead the earlier workers had been

mislead by reductions in luminescence efficiency caused by the
buildup of radiation damage.

INTEGRATED OPTICS

   With the development of high quality optical fibres, communication systems which operate at optical frequencies have become
viable. This has lead to research to make the devices which will
modulate, mix, switch, filter and detect the light beams [45-46-47].
Because the dimensions of the guides which confine the light are of
the order of a wavelength there is a close analogy between the
optical circuit which control the light and the more familiar
integrated electrical circuits. Optical waveguides require that the
light is confined by a region of higher refractive index than the
surrounding media and they can be formed by several methods in pas-
sive materials (e.g. silica glass [48-49]), electro-optic insulators
(e.g. $LiNbO_3$ [50-53]) or semiconductors (e.g. GaAs [54], ZnTe [55]).
Ion implantation has the same advantages here as are useful for the
semiconductors and one may list advantages of masking, control of
the depth profile of the implant, dose control, temperature of
implant and freedom of choice of the ion used [5,10,11].
   There is of course a major interest in the semiconductor materials as one may wish to combine optical and electrical circuits
and use waveguides coupled to light emitting diodes and lasers [56].
In the semiconductors the refractive index control can be made by
changes in the free electron density or by changes in the absorption which modify both the real and the imaginary part of the
index. In insulators one can increase the refractive index by implantation of polarisable ions (e.g. Boron to form high index
borate glasses [57] ; N or C ions in SiO_2 to form SiON or Si_3N_4
type components [49] ; Ti in $LiNbO_3$ [36]). Equally, changes in
density modify the index as do phase changes. For example, silica
glass compacts with ion implantation and readily forms optical
waveguides [48-49] but because the glass is merely transformed to
a new and stable phase there are no problems of optical absorption
and the change in index is related to the energy deposited by
nuclear collisions.
   More interesting materials are $LiNbO_3$ and $LiTaO_3$ as these show
electro-optic properties and can be used for modulators. However
there are problems with the materials as they decompose above some
900°C which is typical of the temperature needed for metal ion diffusion doping. Even more serious is that the temperature at which
$LiTaO_3$ is poled is a mere 600°C and with such a low Curie temperature the diffusion is destructive for the electro-optic properties.
Ion implantation can avoid these problems by injecting the active
ions at low temperatures. In $LiNbO_3$ the radiation damage reduces the
refractive index but by defining the boundary regions one maintains
high index poled regions of the $LiNbO_3$. To achieve this, Destéfanis
and al [53] used high energy He^+ ions of some 2 MeV so in the first
2 microns of the range the loss of energy is predominantly elec-

tronic and does not damage the lattice. Radiation damage and index changes are thus confined to the end of the ion track where the nuclear energy losses occur. There is a minor side effect of optical absorption but this anneals by 200°C whereas the index change persists up to 400°C.

For integrated optics ion implantation appears to be of real value and will undoubtedly be used more as one optimises the devices and moves away from the constraints set by the selection of ions which can be diffused into the lattices.

INFORMATION STORAGE

To demonstrate the uses of ion implantation in other fields we may record the production of information storage systems by implant induced colour centres (58-59). The injection of Li, B, C, O ions in NaF forms both the vacancy and di-vacancy centres (58). Electron transfer between these centres is achieved by illumination at different wavelengths and is the basis of the writing, reading and erasure stages of the store. The defects formed by the implants are particularly stable and it is estimated the storage sites can be recycled as many as 10^{10} times.

CONCLUSION

This review has sketched a few of the many possibilities for the application of ion implantation to control optical properties of insulators. Some of the techniques are useful for academic determinations of defect structures but it is also apparent that in areas such as integrated optics ion implantation may play a key role in the production methods of the new types of device.

REFERENCES

1. Mayer J.W., Eriksson L. and Davies J.A.
   "Ion Implantation in Semiconductors", Academic Press, 1970.

2. Dearnaley G., Freeman J.H., Nelson R.S. and Stephen J.
   "Ion Implantation", North Holland, 1973.

3. Carter G. and Grant W.A.
   "Ion Implantation in Semiconductors", Arnold, 1976.

4. Agajanian A.H.
   Rad. Effects, 23, 73, 1974.

5. Townsend P.D., Kelly J.C. and Hartley N.E.W.
   "Ion Implantation, Sputtering and their Applications"
   Academic Press, 1976.

6. Crowder B.L.
   (Ed) "Ion Implantation in Semiconductors and Other Materials
   "Yorktown Heights 1972)", Plenum Press, 1973.

7. Crowder B.L.
   (Ed) "Ion Implantation in Semiconductors and Other Materials
   (Osaka 1974)", Plenum Press, 1975.

8. Chernow F.
   (Ed) "Ion Implantation in Semiconductors and Other Materials
   (Boulder 1976)", Plenum Press, 1977.

9. "Applications of Ion Beams to Materials"
   (Warwick 1975), Inst. of Physics Conf. Ser. 28, 1976.

10. Townsend P.D.
    J. Phys. E, 10, 197, 1977.

11. Townsend P.D. and Valette S.
    "Optical Effects of Ion Implantation" in "Ion Implantation in
    Materials Science and Technology"
    (Ed J.K. Hirvoven), Academic Press, 1979.

12. Mayhugh M.R.
    J. Appl. Phys., 41, 4776, 1970.

13. Wintersgill M.C., Townsend P.D. and Cusso-Perez F.
    J. de Phys. C7, 123, 1977.

14. Townsend P.D., Taylor G.C. and Wintersgill M.C.
    to be published.

15. Wintersgill M.C. and Townsend P.D.
    Phys. Stat. Sol. (a), 47, K67, 1978.

16. Hughes A.E. and Pells G.P.
    J. Phys. C., 7, 3997, 1974.

17. Dearnaley G., Goode P.D. and Turner J.F.
    1971, cited p. 714 of ref. 2.

18. Hanle W. and Rau K.H.
    Z. Physik, 133, 297, 1952.

19. Van Wijngaarden A. and Hastings L.
    Can. J. Phys. 45, 2239, 3803, 4039, 1967.

20. Takagi T., Yamada I. and Kimura H.
    ref. 7, p. 17.

21. Kushiro Y. and Kobayashi T.
    ref. 7, p. 47

22. Hemenger P.M. and Dobbs B.C.
    Appl. Phys. Lett. 23, 462, 1973.

23. Streetman B.G., Anderson R.E. and Wolford D.J.
    J. Appl. Phys., 45, 974, 1974.

24. Demars D., Quillec M., Ravetto M., Marine J. and Guernet G.
    ref. 7, p. 235.

25. Bryant F.J. and Nahum J.
    J. Rad. Effects, 31, 106, 1977.

26. Chandler P.J.
    D. Phil. thesis, Sussex, 1977.

27. Mattern P.L., Thomas G.J. and Bauer W.
    J. Vac. Sci. Technol., 13, 430, 1976.

28. Patrick L. and Choyke W.J.
    Phys. Rev. B8, 1660, 1973.

29. Townsend P.D. and Kelly J.C.
    "Colour Centres and Imperfections in Insulators and Semiconductors", Sussex Press, 1973.

30. Pooley D.
    Brit. J. Appl. Phys., 17, 855, 1966.

31. Saidoh M. and Itoh N.
    J. Phys. Chem. Solids, 34, 1167, 1973.

32. Hughes A.E. and Pooley D.
    J. Phys. C, 4, 1963, 1971.

33. Davenas J., Perez A. and Dupuy C.
    J. de Phys. C7, 37, 531, 1976.

34. Chassagne G., Hobbs L.W. and Serughetti J.
    J. de Phys. C2, 38, 229, 1977.

35. Kaufman J.V.R.K.
    1961, Private communication.

36. Destéfanis G.L.
    Doctoral thesis, Grenoble 1978.

37. Thevenard P.
    J. de Phys. C7, 37, 526, 1976.

38. Krefft G.B.
    J. Vac. Sci. Technol., 14, 533, 1977.

39. Arnold G.W., Krefft G.B. and Norris C.B.
    Appl. Phys. Lett., 25, 540, 1974.

40. EerNisse E.P. and Norris C.B.
    J. Appl. Phys., 45, 167, 3876, 5196, 1974.

41. Evans B.D. and Hendricks H.D.
    ref. 8 paper IV-2.

42. Peercy P.S.
    Proc. Int. Conf. in Insulating Crystals, Gatlinburg 1977,
    ORNL Conf. 771002, p. 335.

43. Saidoh M. and Townsend P.D.
    J. Phys. C, 10, 1541, 1977.

44. Chandler P.J. and Townsend P.D.
    To be published.

45. Tamir T.
    (Ed) "Integrated Optics", Springer-Verlag, 1975.

46. Special issue of IEEE Trans. Microwave Theory and Technol. 23,
    n° 1, 1975.

47. Tien P.K.
    Rev. Mod. Phys. 49, 361, 1977.

48. Schineller E.R., Flam R.P. and Wilmot D.W.
    J. Opt. Soc. Amer. 58, 1171, 1968.

49. Webb A.P. and Townsend P.D.
    J. Phys. D, 9, 1343, 1976.

50. Kaminow I.P. and Carruthers J.R.
    Appl. Phys. Lett., 22, 326, 1973.

51. Shah M.L.
    Appl. Phys. Lett., 26, 652, 1975.

52. Ranganath T.R. and Wang S.
    Appl. Phys. Lett., 30, 376, 1977.

53. Destéfanis G.L., Townsend P.D. and Gailliard J.P.
    Appl. Phys. Lett., 32, 293, 1978.

54. Garvin H.L., Garmire E., Somekh S., Stoll H. and Yariv A.
    Appl. Optics, 12, 445, 1973.

55. Valette S., Labrunie G., Deutsch J.C. and Lizet J.
    Appl. Optics, 16, 1289, 1977.

56. Evtuhov V. and Yariv A.
    IEEE Trans. Microwave Theory and Technol., 23, 44, 1975.

57. Nishimura T., Aritome H., Masuda K. and Namba S.
    Japanese J. Appl. Phys., 13, 1317, 1974.

58. Magee T.J. and Lehmann M.
    ref. 9, p. 112.

59. Schneider I., Lehmann M. and Bocker R.
    Appl. Phys. Lett., 25, 77, 1974.

A GENERAL SURVEY OF THE PANEL DISCUSSIONS

G. Marest

Institut de Physique Nucléaire (and IN2P3)
Université Claude Bernard Lyon-1
43, Bd du 11 Novembre - 69621 Villeurbanne, France

We have attempted to write a brief summary of the different round tables which took place in the mildness of the Corsican nights during this summer school. This has been made possible owing to registered magnetic tapes and the comments of :
I. BERKES, H. BERNAS, J. DAVIES, G. DEARNALEY,
J. DUPUY, L. M. HOWE, S. T. PICRAUX, K. ROSSLER,
W. TRIFTSHAUSER and G. VOGL. It is a pleasure to acknowledge here their invaluable contribution.

Every round table was regularly attended, discussions were of a high level and sometimes so lively that recordings reproduce only a boisterous atmosphere.

The different proposed panel discussions were :

I - Radiation Damage Simulation
II - Collective Phenomenon ("Thermal Spikes")
III - Implantation of Light Atoms
IV - Some more Physics of the extremely Disordered State
V - Comparison between Nuclear and Solid State Methods

I. RADIATION DAMAGE SIMULATION

This round table discussion, led by G. Dearnaley, was devoted to the use of ion bombardment as a means of simulating the irradiation damage effects brought about by fast neutrons in reactors and other systems. The problem of such damage has been recognized to be important since the first observation (at Dounreay) of the swelling of metallic components. This swelling is due to the progressive formation of voids, which are large vacancy clusters.

The presence of helium, produced by (n, α) reactions within the metal, has a considerable effect upon the nucleation of such voids. Both vacancies and interstitials are created by the recoiling ion which results from a fast neutron collision. However, there is a slightly greater tendency (by $\sim$ 1 to 2 %) for interstitials to be removed at sinks such as grain boundaries and dislocations, with the result that an excess of vacancies accumulates. Normally, vacancy clusters above a certain small size would collapse to form prismatic dislocation loops, but in the presence of small quantities of helium a three-dimensional form is stabilized. Beyond a critical size, the void can no longer collapse, and instead may grow to a radius of the order 100 Å, which is readily visible in the electron microscope.

Because the time scale for observation of void swelling in reactors is very long (5-10 years), some more rapid method of simulating the effect became very desirable. Ion bombardment, at energies comparable with that of the primary knock-on atom, enables the effects to be induced in a period of a few hours. In this way surveys of the behaviour of different structural materials for the construction of fast reactors could be carried out, and some understanding of the influence of different impurities and alloy compositions could be sought. The effects depend upon temperature and can be affected by mechanical stress, and so there is a great deal to be done in this field. The economic value of this knowledge, in the design of fast reactor systems or fusion plant, is so great that it dwarfs the cost of the research.

Bullough and co-workers have developed theories to explain the swelling of metals due to reactor irradiation. To a first approximation the percentage swelling is proportional to the product of three factors. These are : the displacement rate per atom K (dpa per sec) multiplied by the time t after a threshold period t_o ; a function S of the sink density which includes the dislocation density, and F(T) which is a function of the irradiation temperature. This

function shows a peak, and F also depends weakly upon the displacement rate. Thus, for a change from 10^{-6} to 10^{-3} dpa per sec the
peak temperature alters by 150° C in a typical case. This means
that in high bombardment rate simulations a correction must be
made for the form of F (T, K).

Using ion beam simulation methods it has been possible to
examine many different alloys and to establish that certain additi-
ves such as silicon and titanium will reduce the void swelling.
When the void density is high, it is possible to obtain a void lattice
or superlattice : this has been observed in Mo, Ni, Al and Nb,
and arises because of the long range interaction between voids. In
response to a question,   G. Dearnaley explained that the void
lattice is oriented consistently with the host metal lattice.

In reply to an enquiry about the percentage of helium in metals after neutron irradiation, Miss Bertram answered that it is
usually very small, perhaps 10 - 100 ppm.

A. Perez remarked that gas bubbles can exist in metals too.
G. Dearnaley mentioned that these can certainly be produced by
high intensity implantations of helium and other gases into metals,
but that voids must be regarded as essentially free from gas. A
discussion of the pressure which would exist within a gas bubble
inside a metal followed, and it was agreed that it is difficult to
assess this pressure.

J.P. Gailliard doubted that ion beam simulations are realistic
because neutron damage is homogeneous, but ion bombardment
will produce a layered structure in which the stress gradient can
itself affect the clustering of vacancies. G. Dearnaley recognized
that such simulations are not perfect but nevertheless have proved
a remarkably powerful research tool for resolving problems about
voids, and for selecting materials for reactor construction. As
time goes on, more direct comparisons can be made with neutron
irradiated material.

Many problems have yet to be resolved : for example, the
way in which certain elements can reduce void swelling in steels
is not well established, and the way in which an applied stress
will affect void growth and creep behaviour requires further study.
J.A. Borders remarked that ion implantation techniques are not
sufficiently utilised in this research, and that they could provide
a means of resolving some of the problems. G. Dearnaley agreed
that the study of the interaction between implanted atoms and
defects is important for determining the properties of material

surfaces.

J. Davies started a discussion on nuclear and electronic energy loss in a collision cascade. If we may assume, in metals and semiconductors, that radiation damage is not created by electronic energy transfers, there we have a relatively simple situation. We can calculate the deposited energy distribution taking into account the amount of energy which escapes from the surface by the process of sputtering. The calculated and experimentally measured damage profiles have been compared for the case of 300 keV Ar^+ bombardment of silicon, and the agreement between them is good. However, if the distribution of damage is rather well reproduced by the theory, the density of damage in silicon appears to be greater than the calculated are. It has also been shown that for a monotomic Bi^+ ion of 50 keV energy and for 25 keV diatomic ions of As_2^+ (which proved to have the same cascade volume) the deposited energy ν (E) in the cascade is twice more important for Bi^+ than for As_2^+ . The value of ν (E) depends upon the ion energy : it is about 34 keV for 50 keV As_2^+ with a total number of displaced silicon atoms $N_D \simeq$ 4500 whereas for monotomic 25 keV As^+, ν (E) = 17 keV and N_D = 1300.

For a same energy per atom there is significantly more damage created by diatomic ions compared to the one created by monotomic ions[1]. Similar enhancement in sputtering yields has been observed with molecular ion bombardment. J. Davies explained this situation by the fact that damage and sputtering are determined by an "energy spike" when sufficient energy density is deposited within the individual cascade volume. The thermal spike effect is a sort of collective phenomena very important in heavy ion implantations. For a 100 keV diatomic ion implanted in a range of about 100 Å the temperature raise associated with the energy deposition in the small cascade volume is about 100 K. The temperature increase is proportional to the range of the heavy ion ; if range increases from 100 to 200 Å , the temperature increase is about 800 K. The quenching time of the thermal spike is very fast, about 5 picoseconds for insulators, 0.1 picosecond in metals and 1 nanosecond in semiconductors. In Chalk River it has been found enormous evaporation or sputtering of cross-over gas film ; it is believed to be due to a sort of evaporation in the thermal spike process by heavy ions.

This problem of energy spike raised many discussions and various opinions ; it has been decided to speak again of this stimulating subject in the next panel discussion.

REFERENCES

1. D.A. THOMPSON and R.S. WALKER, Rad. Effects
    <u>36</u>, (1978), 91

II. PANEL DISCUSSION ON "THERMAL SPIKES"

The concept of "thermal spikes" was mentioned briefly in an Introductory lecture[1] on Basic Implantation Processes under the heading of High-Density Cascades (i.e. Section 3.2). This topic provoked such vigorous discussion among the participants that a panel discussion on thermal spikes was subsequently organized under the chairmanship of Dr. Reintsema. This panel consisted of several (short) scheduled contributions by Drs. Bernas, Davies, Gibson, Reintsema and Rossler, interspersed by a very lively and (at times) heated discussion which makes it extremely difficult to prepare a written account that is both cohesive and chronological. This written summary therefore attempts to report the consensus view developed during (and after) the panel discussion. For this reason, we make no claim to have included all viewpoints or topics covered during the session, nor do we attempt to credit specific statements or ideas to individual participants.

To provide a general framework for the panel, Dr. Gibson outlined for us the four basic time zones that exist when a heavy ion is implanted into a solid (based on an invited lecture given recently[2] by Dr. Kelly). These zones and their properties are summarized below :

(1) Prompt collisional regime : 10^{-15} - 10^{-14} seconds. This consists of the first few violent collisions between the incident ion and the substrate atoms. Most of the reflected ions (figures 1 and 2 of reference 1) and occasionally an energetic sputtered atom are emitted during this period.

(2) Slow collisional regime : 10^{-14} - 10^{-12} seconds. This is the linear collision cascade regime (section 3.1 of reference 1) in which energetic knock-on effects such as defect production and sputtering occur.

(3) Prompt thermal regime : 10^{-12} - 10^{-10} seconds. This tran-
     sitional regime between (2) and (4) is by far the most complex
     and so far a satisfactory theoretical treatment has not been
     found. The time scale is too short for Maxwell-Boltzman
     statistics to be applied or for the coupling between atomic
     motion and electronic excitation to reach equilibrium. Hence,
     a true "temperature" cannot be assigned. On the other hand,
     the time scale is long enough for significant atomic motion
     to become distributed to all atoms in the cascade.

(4) Slow thermal regime : > 10^{-10} seconds. Here, the time
     scale is long enough for energy sharing between atomic
     motion (phonons) and electronic excitation, and normal
     thermal conductivity considerations can be expected to apply.
     Unfortunately, in most metal or semiconductor implanta-
     tions, the quenching rate is rapid enough to have already
     brought the cascade volume almost back to the substrate
     temperature before this stage; in some insulators a signifi-
     cant temperature rise (several hundred K) can persist for
     10^{-10} seconds and, if so, then collective processes such
     as melting, or evaporation may occur[3].

     As noted above, the first two zones are adequately treated
by linear collision cascade theory and zone 4 (when it exists) can
probably be described by standard thermal considerations. Most
of the present interest therefore centers on zone 3.

     The anomalously large number N_D of displaced atoms
observed in low-temperature heavy-ion implantation of Si and Ge
provides strong experimental evidence for spike effects, as
discussed by Davies and Howe[1]. In figure 16 of reference 1, the
slope of the N_D vs ν (E) curves at low energy and high Z_1 (where
the deposited energy density within the cascade is highest) can be
an order of magnitude greater than the predicted N_{kp} value given
by linear cascade theory. In fact, for the Tl$^+$ case in figure 16,
this initial slope corresponds to an effective displacement energy
of only ~ 0.8 eV/atom, wich is quite comparable to the heat of
melting. At high incident energies, where the deposited energy
density around the ion track is much smaller, all curves bend
over towards the same slope as the N_{kp} line, i.e. towards a
displacement energy of ~ 14 ev/atom. Presumably, in the high
density case, a pseudoliquid zone persists long enough within
each cascade to randomize the local orientation - and the subsequent rapid quenching would then produce an amorphous region.
Electron microscopy studies on very low dose implants has

confirmed the creation of such amorphous regions with almost 100 % efficiency (i.e. one zone created/incident ion) and with a size comparable to that of the predicted cascade, i.e. 40-80 $\mathring{R}$ in diameter.

In metals, high-density cascades do not produce any large increase in N_D; on the contrary, there is usually a gradual decrease in N_D with increasing Z_I. However,~ this could still be quite consistent with the existence of a short-lived spike effect, because metals and semiconductors exhibit drastically different quenching behavior. Rapid quenching (splat cooling, vapour deposition, etc.) of pure metals invariably produces crystalline material, whereas the same rapid quenching of semi-conductors or of certain metal alloys produces an amorphous material.

One interesting aspect of the low-dose Tl implant in Si is the fact that > 80 % of the implanted ions end up on substitutional sites, even though most of the cascade volume has been converted into an amorphous 'blob'. This would indicate that the deepest region of each cascade, where the Tl^+ has come to rest, quenches back to crystalline silicon or (alternatively) is quenched rapidly enough that the original lattice structure is not destroyed. In metals too, a very high substitutional fraction of the implanted ions is often observed.

Transmission electron microscope investigations in Au indicate that the nature of the defects produced during ion bombardment is quite dependent upon the deposited energy density whithin the collision cascade[4]. At energy densities of $\lesssim 2$ eV/atom, mainly interstitial type defects are produced whereas at energy densities $\gtrsim 2.5$ eV/atom vacancy type defects are predominant. This difference in behaviour may possibly be associated with spike effects. Spike effects have also been invoked to explain the irradiation-induced disordering of ordered alloys[5]. However in the ordered alloy case, the propagation of replacement collision sequences can also produce disordering. Hence, in general, it is difficult to determine the relative important of these two different disordering mechanisms.

There was considerable discussion about the various dynamical processes involved : the volume, quenching time, temperature distribution, mass transport, diffusion, etc. The predicted quenching time for a spherical drop of molten metal, 100 $\mathring{R}$ in diameter and surrounded by the cold metal lattice, is less than 10^{-12} seconds. Hence, as noted earlier, one cannot apply concepts such as "temperature" in the normal thermodynamic sense. Furthermore, one may often require two completely different 'temperatures' to describe the atomic vibration level and the electronic excitation ("hot" atoms but "cold" electrons, or vice versa). It is also important to recognize that 10^{-12} seconds is an extremely short time scale for processes such as mass transport or diffusion to occur : for example, even if one uses a typical liquid value of $\sim 10^{-3}$ cm^2 sec^{-1} for the diffusion constant D, the maximum diffusion length in 10^{-12} seconds is less than 10 Å !

The panel discussion ended with a very lively debate on the subject of "temperature" and especially on whether or not any of the existing hyperfine techniques could provide information on the excitation level within a cascade during the $\sim 10^{-12}$ second quenching time. No definite conclusions were reached, but this was obviously one of the key areas in which (hopefully) some of our experimental techniques will eventually provide information.

REFERENCES

1. J.A. DAVIES and L.M. HOWE, Introductory lecture on Basic Implantation Processes,   p. 7 of this book.

2. R. KELLY, Proceedings of the International Conference on Ion Beam Modification of Materials, (Budapest, 1978) in press.

3. J. BØTTIGER, J.A. DAVIES, J. L'ECUYER, N. MATSU-NAMI and R. OLLERHEAD, (Proceedings of the International Conference on Ion Beam Modification of Materials, (Budapest, 1979) in press.

4. M.O. RUAULT, B. JOUFFREY, J. CHAUMONT and H. BERNAS, in Applications of Ion Beams to Metals, (Plenum Press, New York, 1974) p. 459.

5. G.R. PIERCY, J. Phys. Soc. Japan 18 , supplement III, 169 (1963).

6. M.A. KIRK, T.H. BLEWITT and T.L. SCOTT, Phys. Rev. B 15, 2914 (1977).

III.   ROUND TABLE: IMPLANTATION OF LIGHT ATOMS

     The leader of this round table was Dr. S. T. Picraux who
began with general considerations regarding the migration of
defects in metals. As the temperature of a metal is increased
the consecutive moving defects are the interstitial, vacancy and
substitutional species. The diffusion coefficient for interstitials is typically very large at room temperature
$(D \sim 10^{-5}$ cm^2/sec) so that the characteristic times to migrate
a characteristic length $\bar{\ell} \sim 1000$ Å in an implanted layer is
$\tau \approx (\bar{\ell}^2/D) \approx 10^{-5}$ sec.

     The discussion was initially restricted to H and He
implantation and the trapping of such light interstitials by
defects. Comparisons between physical and chemical trapping
are summarized in Table I.

     Dr. Picraux discussed results obtained for D and He
trapping in W, Mo, and Cr. At relatively low temperatures D or
He atoms can be trapped by the lattice defects and subsequently
released at higher temperatures. In Fig. 1, examples are shown
of the trapped lattice location for 30 keV D and 60 keV ^{3}He ions
implanted to fluences of $\sim 10^{15}$ at./cm^2 in W. The ^{3}He(d,p)^{4}He
nuclear reaction was used for the channeling effect lattice
location studies of both D and ^{3}He.[1]

     The implanted D atoms occupy the tetrahedral interstitial
site in W whereas the He is located in neither a simple interstitial nor a pure substitutional location. The ^{3}He atoms are
believed to be trapped at lattice vacancies in a configuration
consisting primarily of three He atoms trapped at a vacancy with
a possible smaller component consisting of two or other multiples
of He atoms trapped at vacancies. The three atoms occupying a
vacancy lie in a (100) plane forming a nearly equilateral triangle.

     E. V. Kornelsen[2] has attempted to correlate the several He
binding states observed in a thermal desorption experiment with

Table I.   Physical and Chemical Trapping of H and He

	Physical Trapping	Chemical Trapping
Occurs for	H, He	H
Origin	e.g., strain energy of lattice relaxations	e.g., band energy as percentage covalent bond formed
Energy of binding	$\sim 0.2 - 0.8$ eV (H) $\sim 1 - 4$ eV (He)	0 - 11 eV e.g., Si(H) ~ 2.4 eV
Trap centers	vacancy dislocation grain boundary impurity	chemical unit impurity
Example	Cr(D) (D + vacancy?)	Si(H) (Si-H covalent bond)

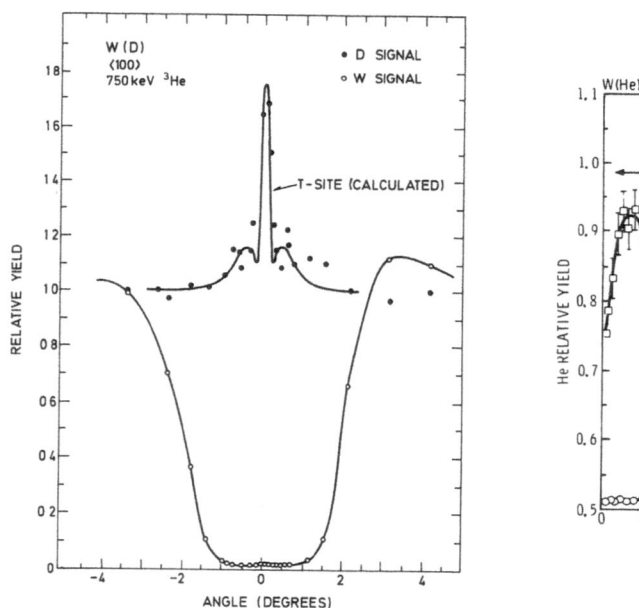 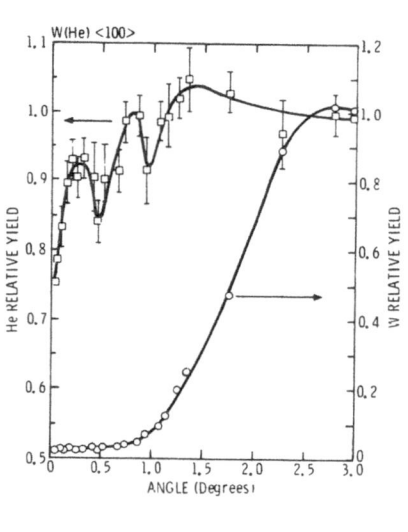

Fig. 1   Axial channeling <100> scans for D and ^{3}He implanted in W.

the particular defect trap configurations. A single He atom occupying a free vacancy escapes at 1560 K implying a binding energy of 4.0 eV. For two or more He atoms in a vacancy there will be significant compressive strain in the surrounding lattice.

Dr. Picraux showed that hydrogen trapping in W and Mo could be enhanced by pre-injection of He to create additional defect trap centers. The trapping increases as the number of damaging ions is increased. From the release of the hydrogen upon annealing it is possible to determine the binding of the hydrogen in these traps. For Mo it was shown that a broad temperature dependence of the detrapping process exists, implying a distribution of trap energies. The release centered at $\sim 200^{\circ}C$, indicating the migration plus trap activation energies are high (of the order of 1 to 1.5 eV) compared with typical hydrogen trapping energies in metals (~ 0.2 eV to 0.8 eV).

Dr. H. Bernas presented a brief survey of the "recoil implantation technique," since this can lead to light interstitial introduction during heavier ion implantation. The technique consists of the introduction of impurities into a substrate after elastic collisions of bombarding heavy ions with atoms of a thin layer deposited on the surface of the substrate (Figure 2). The recoiling atoms from a surface layer will be localized into a zone near the surface due to small penetration depths and the compositional change of this layer can lead to the formation of compounds such as oxides, hydrides, or carbides.

Ion implantation causes damage to the surface itself and a certain number of surface atoms can be ejected ($\sim$ 2-3 atoms per incoming nitrogen or oxygen ion). This sputtering results in the formation of vacancies which can be annihilated by mobile interstitials. The evolution of the implanted layer depends on the solubility of the interstitials, on their diffusivity and on the stability of compound phases which they may form with the substrate atoms. For light ions as boron or carbon very low self-sputtering coefficients permit well defined layers to be produced. Dr. Bernas restricted himself to the presentation of results concerning the production of carbides due to carbon implantation into different metals. By nuclear reaction techniques the carbon can be detected and by transmission electron microscopy the aggregation of carbon can be observed. A study of the image contrast indicates the defects consist of interstitial carbides. Also the structure of the damage can be studied by Mössbauer spectra. When carbon is replaced by boron the hyperfine field acting on iron sites must change and it has been shown that an increase in hyperfine field distribution is observed as the implanted boron concentration is increased.

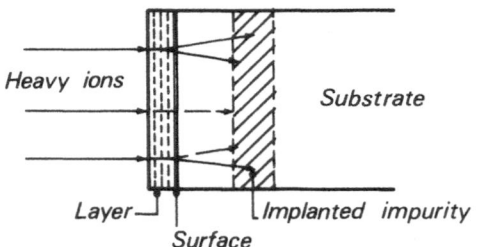

Fig. 2 Schematic of recoil implantation.

Finally Dr. Rossler summarized some of the relevant
parameters and mechanisms involving interstitials in hot atom
chemistry. This discussion is outlined in Table II.

Table II - Hot Atom Chemistry

Categories

1. Oxyanions, oxides, sulphides
2. Metal complexes, double complexes
3. Halocomplexes, organometallic complexes

Parameters influencing chemistry

1. Sterical and geometrical factors of the new compound
2. Bond energies
3. Electronegativity (with respect to atom host)
4. Charge state of the projectile

Mechanisms of chemical reactions

1. Substitution (replacement + reestablishment of a bond)
2. Attachment (to double bonds, free electron pairs: O-H)
3. Abstraction (compound formation with atoms from intact units)
4. Defect combinations
    a) Vacancy at covalent compound + interstitial
    b) Interstitials combine with oxygen

References

[1] S. T. Picraux and F. L. Vook Applications of Ion Beams to
    Metals, Ed. by S. T. Picraux, E. P. EerNisse and F. L. Vook
    (Plenum, NY, 1975), p. 407.

[2] E. V. Kornelsen, Rad. Eff. 13, 227 (1972).

IV.   ROUND TABLE : "SOME MORE PHYSICS OF THE
       EXTREMELY DISORDER STATE"

         D. Quitmann, as leader of this round table first
presented a number of different instances in which we can have
highly disordered systems (amorphous metals, liquid metals,
metal-chalcogenide systems, etc.) Various speakers were invited to present the current situation regarding the properties of
amorphous systems, characterization of the amorphous state,
and the behaviour of liquid metals.

         G. Dearnaley considered a number of areas where practical
applications of very highly disordered systems are possible. The
first was in the field of corrosion. During such attack there is a
movement of some atomic charged species through the film and
the problem is to resolve the mechanism by which the growth of
the layer takes place. Grain boundaries may play an important
role in corrosion because the transport of anions or cations is
often very rapid at such defects. Dislocations in the film may
also   accelerate migration. An amorphous film is free from
these extended defects and fast diffusion paths may therefore be
absent. During 1977 the Sussex group, under Venables, studied
the diffusion of copper (normally a fast diffusant) in amorphous
systems containing Ni, B, etc., prepared by splat quenching.
The method used was high resolution Auger spectroscopy. The
diffusion coefficient of copper in this matrix proved to be extremely low. In Japan, Masumoto and Hashimoto have confirmed the
remarkably high corrosion resistance of amorphous alloys of
Fe-Cr-P-C systems. Phosphorus, boron, silicon and carbon were
the non-metallic species investigated.

         Because dislocations do not move in these materials as in
crystalline metals, the mechanical properties are very different.
There are prospects of forming very hard and wear-resistant
surfaces in this way, but brittleness may prove to be a problem.

         Next G. Dearnaley considered the problem of catalysis. One
model for this supposes that two types of site, denoted A and B,
exist where the reactant species can become adsorbed. After
chemically combining, the energy release brings about their
ejection from the surface. It is desirable to produce many adjacent sites A, B and this can be done only by an intimate mixing
of the two species. The use of amorphous materials allows this
to be achieved, with a different radial distribution function from
that otherwise obtainable. This more randomly structured surface may have increased catalytic activity, because more pairs may

exist at the correct spacing. Irradiated catalysts have been found
to have a higher activity, and ion implantation may provide a means
of creating a mixture of insoluble species. As an example, the
system copper-ruthenium is a useful catalyst with high specificity
for certain hydrocarbon reactions. It is possible to implant Ru into
Cu to produce a highly disordered system. Thermal annealing
will bring about an ordering and segregation of the system and there
may be an optimum from the point of view of catalytic behaviour.

It will be interesting to study the magnetic properties of
highly disordered systems, for example based upon ferromagnetic
constituents. Nickel implanted with boron or dysprosium has been
shown to become amorphous (W. Grant and the Salford group) and
the same may be true of cobalt implanted with rare earth species.
A totally disordered material would be very soft magnetically,
and it is possible to consider tailoring the magnetic properties
of a thin film by ion implantation.

The techniques discussed at this summer school, such as
hyperfine interactions, channelled backscattering etc. should be
very useful for the study and characterisation of these disordered
systems.

W. Triftshäuser pointed out the attention on amorphous
and liquid metals studies. Compared to pure metals in which
atoms occupy regular lattice sites, liquid metals can be considered as a crystalline distorted state with different kind of
structures of short range order (20 to 40 atoms). A metallic
glass is obtained when we go with a very high quenching rate
(10^6 - 10^8 K/sec) from a molten liquid state to a frozen liquid
state. The amorphous configuration of these metallic glasses can
be stabilized by various compounds (P, Si, As, Zr) in rather high
concentrations (10 - 40 %). These metallic glasses are produced
by cooling down very quickly a drop of a liquid metal through a
moving piston to obtain a glasslike structure, but the only way
to produce pure amorphous metals is the evaporation onto a cooled
target at helium temperature, in order to stabilize the amorphous
state.

Different methods can be used to characterize an amorphous
state : density measurements (not very sensitive), X-ray or
neutron scattering measurements, electrical resistivity measurements, positron annihilation methods, hyperfine interaction
techniques as Mössbauer effect or perturbed angular correlations.

W. Triftshäuser discussed different characteristics of
the metallic glasses. Electrical resistivity measurements or

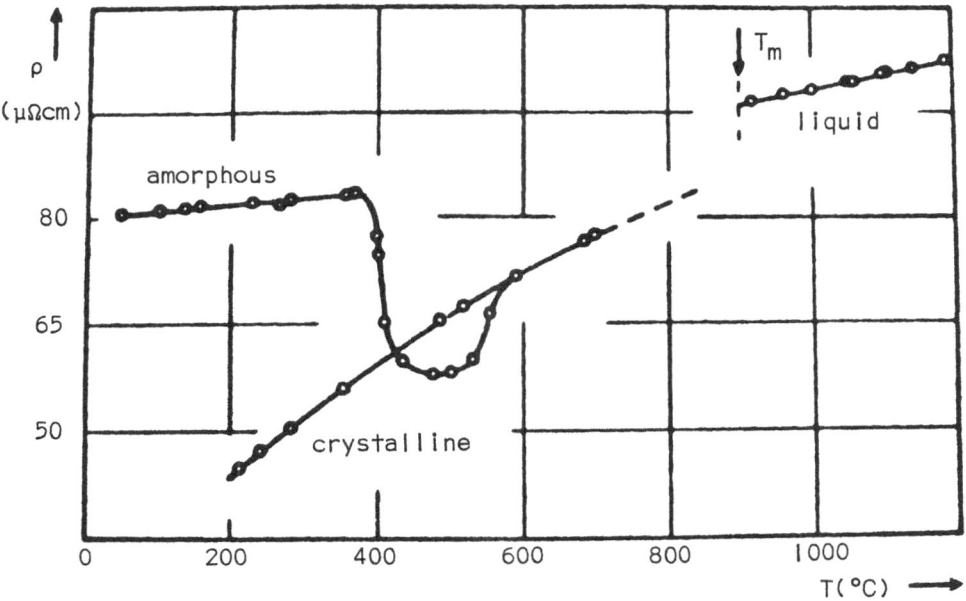

Figure 1 : Electrical resistivity of $Pd_{81}Si_{19}$ in the glassy, crystalline and liquid state[1]

measurements of the specific heat versus temperature permit to define a conversion temperature corresponding to the transition from the amorphous to the crystalline state (figure 1)[1].

Such experiments have been performed on $Pd_{81} Si_{19}$ which is one of the most simple materials because there are only two constituents. This conversion temperature is explained by the scattering of the conduction electrons by lattice distortions.

Some mechanical properties of metallic glasses are : high strength and hardness due to their structural properties without dislocations and a very low or null corrosion.

The angular distribution of annihilation photons which is related to the momentum distribution of electrons in the material is different for disordered systems and for crystalline states. In liquid metals trapping of positrons is observed. However, in metallic glasses in which we have no vacancies inside no trapping of positrons so far has been detected. Figure 2 shows the angular distribution of annihilation photons in aluminium at two temperatures[2]. The curves are normalized to equal areas, i. e. the same number of annihilation processes.

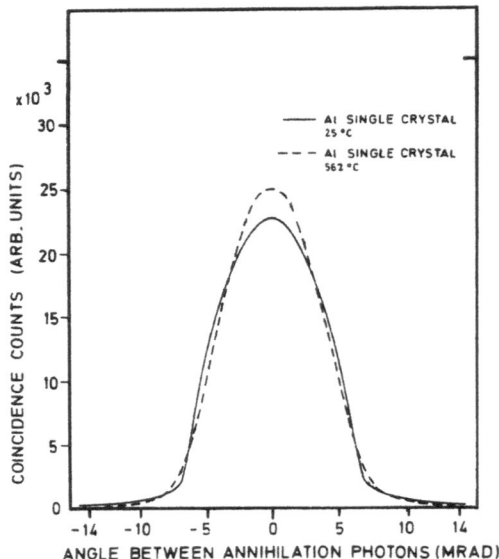

Figure 2.: Positron
angular correlation
curve of aluminum
single crystal at low
and high temperature.

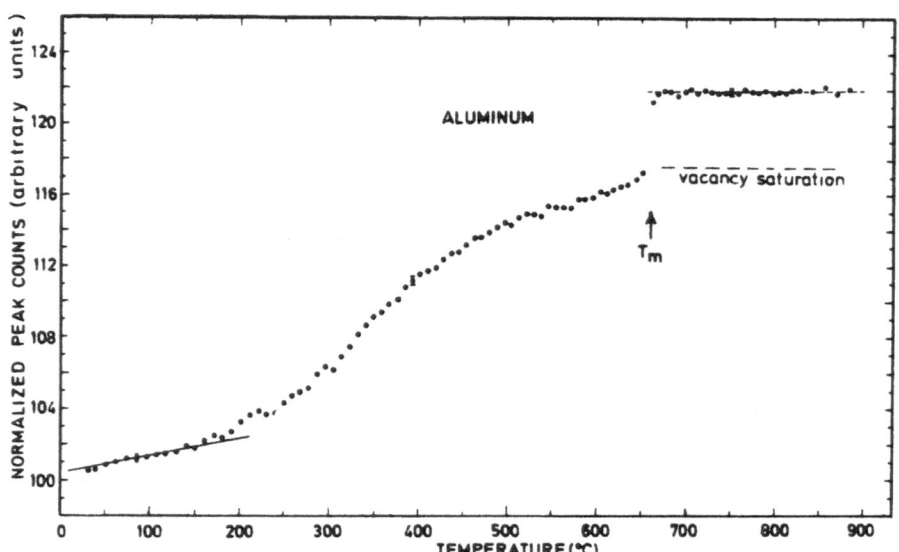

Figure 3 : Coincidence counting rate at
the peak of the angular-correlation curve
as a function of temperature.

Figure 3 shows the temperature dependence of coincidence counts at the peak of the angular correlation curve[3]. In amorphous metals trapping of positrons is very likely because the spongelike structure of the material ; in figure 3 we observe a change in the peak counts when we go from the solid state to the liquid state. The differences in the momentum distributions of the photons result from changes in the electron momentum distributions seen by the positron. From the analysis of the data in the solid phase information about vacancy formation energies can be obtained. In diluted alloys values for vacancy - impurity - binding energies can be deduced. Binding energies up to 0.4 eV have been observed.

A remark of G. Schatz concerned hyperfine measurements on various compounds. Isomer shift measurements in ^{119}Sn Cu have shown a quadrupole splitting twice more in the amorphous phase than in the polycristalline phase and an electric field gradient fifty per cent higher in the amorphous phase. By perturbed angular correlation measurements on ^{111}In Au samples a same difference in the quadrupole interaction has been observed for the amorphous and crystalline phases.

Mrs. J. Dupuy presented different considerations about liquid alloys and ionic liquids. Information can be obtained by means of neutron diffraction measurements on the local structure and the possible heterogeneity of disordered systems. By isotopic substitution neutrons permit to study multi component systems. In this way the local order in the following ionic and ionocovalent liquids : K Cl, Na Cl, Rb Cl, Cs Cl, Ag Cl, Cu Cl and mixed metal compounds such as Ag-Ge, Li-Pb, Li-Ag, Cu-Sn have been recently studied at Grenoble high flux reactor.

The coherent intensity elastically diffused by a N disordered atom or ion system made up of two different α and β particle types with C_α and C_β concentrations respectively is a linear function of b_α, b_β coherent diffusion amplitudes of these atoms and of the correlations between density fluctuations , $S_{\alpha\beta}(k)$:

$$I_{Coh}(k) = N \sum b_\alpha^2 c_\alpha + N \sum_\alpha (c_\alpha c_\beta)^{1/2} b_\alpha b_\beta [S_{\alpha\beta}(k) - \delta_{\alpha\beta}]$$

(1)

These $S_{\alpha\beta}(k)$ are therefore directly obtained through experimentation if we can vary b_α or b_β for a same compound (use of isotopes).

By Fourier transform we obtain the distribution probabili-
ties of α and β atoms in the real space. Thus for a binary system
unlike $g_{\alpha\alpha}(r)$ and like $g_{\beta\beta}(r)$ distributions will be obtained and
will allow a description of the local order. Some examples are
given on the specificities of this local order. Concerning the
ionic liquids (e. g. Na Cl,[4]) or ionocovalent liquids (e. g.
Cu Cl,[5]) :

- the correlations between distributions disappear after
the second coordinance sphere (diameter ~ 6 Å) (fig. 4 and fig. 5).

- the charge alternance is respected to ensure the local
electronegativity.

- the order on the nearest neighbours is not very different
from the one which existed in the solid state (interatomic
distance and number of neighbours), excepted for iono-covalent
liquids in which we observe a reduction of the coordinance number.

- there is an equivalence between the distribution of ions of
the same sign except for the iono-covalent liquids ; Cu Cl presents
an original distribution on Cu^+ (figure 5).

- the agreement with simulation experiments for ionic bound
liquids using a Coulomb interaction potential is rather good.

The coherent intensity diffused by a liquid can also be written
in terms of number fluctuations S_{NN} and concentration fluctuations
S_{CC} (or of charge fluctuations in the case of electrolytes)[6] :

$$I(k) = N [(\bar{b})^2 S_{NN}(k) + (\Delta b)^2 S_{CC} + 2 \Delta b . \bar{b} S_{NC}] \qquad (2)$$

with $\bar{b} = c_\alpha b_\alpha + c_\beta b_\beta$ and $\Delta b = b_\alpha - b_\beta$

This has been astutely used to study Li compounds containing
^{7}Li (e. g. Li - Ag and Li - Pb)[7]. This element having an isotope
with a negative diffusion cross-section, $\bar{b} = 0$ for a determined
concentration. A diffraction measurement gives the value of the
concentration fluctuations directly. It has been possible to display
and study the formation of compounds and the deviation with random atomic distribution.

The interest of the formula (2) consists in its limit at high
wavelength ($k \rightarrow 0$). Indeed the $S_{NN}(0)$ and $S_{CC}(0)$ structure
factors respectively represent the thermal fluctuations of the
number of particles $<\frac{\Delta N^2}{N}>$ and of the concentration $N<\Delta c^2>$.
They can be expressed in thermodynamic terms : partial volumes,

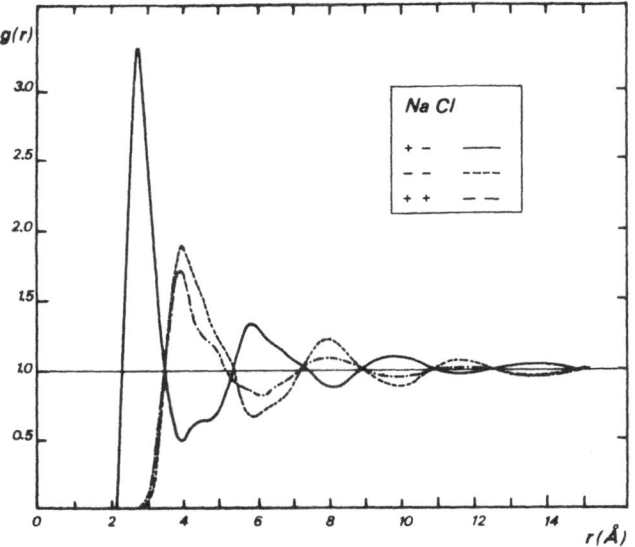

Figure 4 : Partial radial distribution function for molten Na Cl.

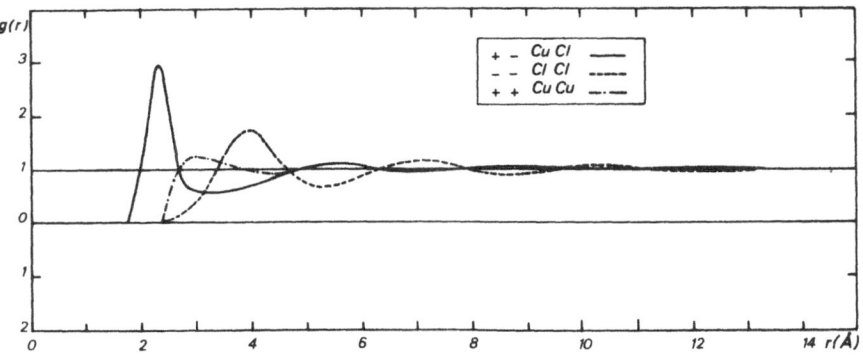

Figure 5 : Partial radial distribution function for molten Cu Cl.

isothermal compressibility and second derivative of the thermodynamic function G. Studies at small transfer moment could be
used to give information about the local atomic order.

It is also possible to study a particular order-disorder
type near a critical point because at the demixtion point of a
mixture $\frac{\partial^2 G}{\partial x^2} \to 0$ and consequently the concentration fluctuation $S_{CC}(0) \to \infty$. At small transfer moments, that is to say
for long distances, the interactions are expressed in the form
$\frac{1}{r} \exp(-\frac{r}{\xi})$ with ξ the concentration fluctuation range.
We can approximate $S_{CC}(k)$ by

$$S_{CC}(k) \ \alpha \ \frac{A}{k^2 + \frac{1}{\xi^2}}$$

Measurements at small transfer moment neutrons permit
the study of the demixtion critical phenomenon, the determination of critical laws, the temperature and concentration range
of the critical area[8, 9].

REFERENCES

1.   H. J. GÜNTHERODT, in Festkörperprobleme, Advances in
      Solid State Physics, Vol XVII, p. 25
      J. Trensch (ed), Vieweg, 1977

2.   W. TRIFTSHAUSER and J. D. Mc GERVEY, Appl. Phys.
      5 , (1974), 177

3.   W. TRIFTSHAUSER, Phys. Rev. B12 , 1975, 4634

4.   F. G. EDWARDS, J. F. ENDERLY, R. H. HOWE, D. I. PAGE
      J. Phys C 8 , (1975), 3483

5.   S. EISENBERG, Thèse de spécialité - Lyon 1977 (unpublished)

6.   A. B. BHATIA, D. E. THORNTON, Phys Rev. B 2 , (1970)
      3004

7.   H. RUPPERSBERG, H. EGGER, J. of Chem. Phys., 63
      (1975), 4015

8.   E. S. WU, H. BRUMBERGER, Phys. Letters 53 A (1975)
      475

9.   J. F. JAL, P. CHIEUX, J. DUPUY, J. Appl. Cryst., 11
      (1978) , 610

V.   "COMPARISON BETWEEN NUCLEAR AND SOLID STATE METHODS"

The purpose of this panel discussion was to state to what extent nuclear and solid state methods are complementary or overlapping. G. Vogl begun by introducing the different techniques in which hyperfine interaction (HFI) methods are already to become solid state methods.

I. Berkes summarized the more useful features of each HFI methods in order to specify their applicability (see tables 1 and 2). The given values are order of magnitude estimations for the practical usefulness of hyperfine methods, subject to improvements by experimentator's keenness and technical development. The angular distribution measurements on accelerator beam (time dependent perturbed angular distribution TDPAD, integral perturbed angular distribution IPAD and stroboscopic observation of perturbed angular distribution SOPAD) are not introduced ; the accelerator beam creates by itself lattice defects and therefore these methods give information rather on the dynamics of lattice defects than on the structure itself.

The hyperfine frequencies are defined as $\omega_i = \dfrac{\mu N g Bi}{\hbar}$ for magnetic dipole and $\omega_i = \dfrac{e Q V_{zzi}}{\hbar}$ for electric quadrupole interaction. In this latter case the asymmetry $\eta = \dfrac{V_{xx} - V_{yy}}{V_{zz}}$ can enter as an additional parameter (see Recknagel's lecture). A pair of ω_i and η_i accounts for two quantities in the estimation of the maximum possible components i. For further comparison of hyperfine methods cf. H. de Waard[1].

At this time J.A. Borders claimed we did not have to speak only of comparison between methods but, on a given material problem, of how must we choose the right method. Therefore G. Vogl chose   an example : mixed dumbbell interstitials in f.c.c. metals to explain the difficulty in deciding if methods are overlapping or complementary. Aluminium, copper and molybdenum alloys have been investigated and, in the case of aluminium alloys many methods have been used.

L.M. Howe presented investigations by means of the backscattering technique of the interaction of impurity atoms with point defects introduced by He^+ ion irradiation at 35-70 K into ($\lesssim$ 0.1 at % Mn, Zn or Ag Al alloys) aluminium alloys containing $\lesssim$ 0.1 at % Mn, Zn or Ag. For low dose irradiations each trapped Al interstitial atom forms a < 100 > dumbbell configuration with the impurity atom[2]. When a small impurity

Table I   More useful features of hyperfine interaction methods
(see comments in the text)

Method	Nuclear Magnetic Resonance (NMR)	Mössbauer Effect (M E)	Time Dependent Perturbed Angular Correlation (TDPAC)
Detection Method	Macroscopic	Direct counting	Coincidence counting
Measured quantity	Intensity near resonance : $I(\omega)$	Absorption spectrum $I(\omega)$	Time spectrum $f(t)$
Derived quantity	Resonant frequencies ω_i Their weights α_i Relaxation times T_1, T_2	Resonant frequencies ω_i Their weights α_i s-electron density (relaxation time)	Frequency-amplitude Spectrum : $F(\omega) = \int_0^\infty e^{i\omega t} f(t)\, dt$ $\rightarrow \omega_i$ and α_i
Maximum possible components i being separated	Any, if $\delta\omega_i > \Delta W$	5	3
Energy resolution $\Delta W(eV)$ but $\Delta W > 6, 6 \times 10^{-16}/\iota(s)$	10^{-11}	10^{-8}	10^{-9}
Minimum number of sample atoms	10^{20}	10^{12}	10^{11}
Approximative number of possible sample nuclides	Any	30	15
Best sample atoms (b. s. a)	H	^{57}Fe, ^{119}Sn, ^{129}I ^{181}Ta	^{57}Fe, ^{100}Rh, ^{111}Cd ^{169}Tm, ^{181}Ta, ^{187}Re
Lifetime of sample state ι	∞	> 10 ns	3 - 1000 ns
Temperature range (K)	Any, but better efficiency at low temperature	< 77 K, but any for b. s. a.	Any
Measuring period	1 h	0. 5 d	1 - 20 d
Price of device (k $)	< 20	< 20	20 - 50
Remarks		The host atom can be either the sample atom or its radioactive parent.	The host atom is the radioactive parent of the sample atom.

Table II More useful features of hyperfine interaction methods (see comments in the text)

Integral Perturbed Angular Correlation (IPAC)	Low Temperature Nuclear Orientation (N.O.)	Nuclear Magnetic Resonance on Oriented Nuclei (NMR/ON)
Coincidence counting	Direct counting	Direct counting
Rotation angle of angular correlation pattern, proportional to Mean value of $< \alpha_i \; \omega_i >$ ω_i : resonant frequencies α_i : their weights	Angular distribution of emitted radiation of oriented nuclei Resonant frequencies ω_i Their weights α_i	Intensity near resonance $I(\omega)$ Resonant frequencies ω_i Their weights α_i
1	2	As for NMR
10^{-9}	10^{-8}	10^{-9}
10^{10}	10^{9}	10^{9}
Many	Nearly any for $Z > 25$ Few below $Z = 25$	As for N.O. Hyperfine magnetic field needed
Many	Metals	Metals
< 10 ns	> 3 h	> 3 h
Any	< 0.1	< 0.1
$0.5 - 5$ d	< 1 d	< 1 d
< 20	> 50	> 50
The host atom is the radioactive parent of the sample atom.		

atom traps a <100> dumbbell self interstitial, the solute atom can become incorporated into the dumbbell, forming a mixed dumbbell. These mixed dumbbells are stable up to stage III recovery (200 - 250 K for Al and Cu) in which vacancy-interstitial annihilation occurs[3].

W. Petry showed Mössbauer spectra of irradiated at low temperature (4 - 100 K) with fast neutrons and fast electrons dilute aluminium alloys containing a few ppm ^{57}Co. After annealing treatments a new line in the Mössbauer spectra appeared, ascribed to the trapping of interstitials at the ^{57}Co atoms. The large isomer shift of this new line is explained with a mixed dumbbell configuration in aluminium. Following the temperature dependence of the Debye-Waller factor f it is possible to study the dynamics of this mixed dumbbell. The rapid decrease of f accounts for diffusion jumps of Mössbauer atoms between six positions in an octahedral cage[4, 5, 6].

G. Vogl concluded that each technique has its limitations : ion backscattering needs rather high impurity and therefore rather high defect concentrations; Mössbauer spectroscopy is sensitive to very low concentrations but is not so suitable for yielding geometric information. These methods are over-lapping to some extent, they are rather supplementary. G. Vogl described positive features of HFI methods with their usefulness :

- truely microscopic to study atomistic problems (single atoms, point defects, onset of agglomeration, ...)

- sensitive for different lattice sites

- sensitive to short time range (nanoseconds) ; so possibility of studying the fate of implanted atoms in displacement cascades.

But G. Vogl noted that HFI methods have limitations and problems :

i) It is difficult for solid state physicists who are not in the field of HFI to understand the full meaning of HFI results. It is suggested that scientists working in the field of HFI should try to make contributions where open questions are seen by other solid state physicists. Simple problems should be studied first.

ii) There is no theory available for the HFI but it is badly needed.

iii) Usually HFI measurements are experimentally time-consuming. Therefore preference should be given to using efficient nuclear states such as ^{57}Fe and ^{111}Cd, etc...

iv) HFI methods are "short-sighted" and therefore not useful for studying bulk effects. It is suggested that HFI methods be compared with other solid state methods, preferentially by using the same samples.

J.A. Borders remains convinced that everybody wants to promote his own technique but in fact there is no confrontation among data obtained from different techniques.

N. Stone answered that from a theoretical point of view solid state physicists brought nothing to HFI physicists. Dederich's theory on mixed dumbbell is an exception. We seem to try to understand the circumstances of damage in a lattice and we have not heard where things like dislocations have a background in theory : why a dislocation should lie in $< 100 >$ in a fcc lattice.

S.T. Picraux admitted the solid state theory is in a very primitive state, e.g. at the conference in Nice every person is trying to calculate the vacancy in Si. It is the simplest thing to calculate.

W. Triftshäuser basically agreed there should be a theory but at the moment everybody knows it is quite difficult. We have to try to get the dumbbells by means of different methods for the same systems and combine them. We ultimately ask from the theoreticians to put up a theory to include all the facts we know about the system.

H. de Waard made different comments during this round table and I think I summarize his conviction with these selected sentences :"Each of the presented methods has advanced physics at some point, every technique produces something new and will contribute more and further to the advance of our knowledge of the solid state and the damage in it and so on. It is very hard to predict where this will happen, it does not depend on the technique, it largely depends on the people who do it. "

REFERENCES

1.   H. de WAARD, in "Hyperfine Interactions Studied in Nuclear
      Reactions and Decay", edited by E. Karlsson and R. Wäp-
      pling (Almqvist and Wiksell, Stockholm, 1975), p. 157

2.   M. L. SWANSON and F. MAURY, Can. J. Phys., 53 ,
      (1975), 1117

3.   N. MATSUNAMI, M. L. SWANSON and L. M. HOWE,
      Can. J. Phys., 56 , (1978), 1057

4.   W. MANSEL, G. VOGL, and W. KOCH, Phys. Rev. Let.,
      31 , (1973), 359

5.   G. VOGL, W. MANSEL and P. H. DEDERICHS, Phys. Rev.
      Let., 36 , (1976), 1497

6.   G. VOGL, W. MANSEL, W. PETRY and V. GROGER,
      Hyp. Int., 4 , (1978), 681.

CONCLUDING REMARKS

D. Quitmann

Institut für Atom- und Festkörperphysik

Boltzmannstr. 20, D1000 Berlin 33, Fed.Rep.Germany

In the last two weeks, we all have been engaged in a very vivid and diversified undertaking. Simplifying matters somewhat, it had two main aspects – the study of hyperfine interactions of radioactive nuclei implanted into solid materials, and the consequences which implantation has for the properties of solid state materials. Finding in between these two fields the areas of mutual interest is not trivial, and in this sense the summer school was at times close to a research meeting. Here, the posters exposed and the panal discussions have certainly contributed a lot. – But after all, learning what one does <u>not</u> know, and how to study it, is no loss of time for anyone.

In order to delineate a little more clearly the areas of overlap I want to mention explicitly a few problems related to implanted atoms where I feel progress has been achieved through the combined effort of both sides, solid state studies and hyperfine interactions of radioactive atoms. Consider the following list as a personal and arbitrary collection of examples.

1. Nature of single defects in cubic metals, interstitial configuration, interstitial versus substitutional site – with resistivity, theory, channeling, backscattering, perturbed angular correlation (PAC), Mössbauer effect contributing.

2. Reactions between defects in metals, annealing of radiation damage – resistivity, detection of agglomerates, PAC, Mössbauer effect, Muon Spin resonance.

3. Clustering, short range order – X-ray and neutron scattering,

495

transmission electron microscopy, Mössbauer effect, perturbed angular distributions (PAD).

4. Diffusion in solids - tracer, β-PAD, NMR, Mössbauer effect.

5. Vibrational amplitudes and bond strength for an impurity atom in a lattice - X-ray or neutron scattering, Mössbauer effect, PAC.

6. Diamagnetic impurities in ferromagnetic metals, conduction electron spin polarization - NMR, nuclear orientation, theory.

These problems were taken up more or less often during this school especially in the lectures of Stals (1,2), Howe (1,2), Vogl's panel (1,2), de Waard (1,3,5), Bernas and Borders (3), Quitmann (4), Recknagel (5), Stone (6).

That one tries to set up such a list only underlines the fact that the gain solid state physics has had from hyperfine interactions of excited nuclear states is mostly in specific, detailed problems. (I think, though, that the vast field of applications of the Mössbauer effect has to be left out for such a strong statement.)

One of the reasons   for this "restricted applicability" may be, that the solid state properties of immediate interest are very often averages over at least 10^3 atomic distances (mechanical, optical, corrosion properties, even those of implanted layers; precipitates; strain). The sensitivity of the hyperfine interactions extends, on the other hand, usually over very few atomic distances (say less than 10: dipole-dipole interaction; spin polarization of conduction electrons; electric field gradient; electron density at the nucleus). Most of them are, in addition, sensitive to the symmetry of the surrounding.

If one uses excited states to measure hyperfine interactions, a lot of effort has to be invested for spectra with moderate energy resolution. This by the way also gives a hint to the concentration on metals which was rather obvious in the present meeting: in insulators EPR, NMR, optics etc. are applicable, and then they are usually much more effective and versatile.

At several occasions in the past, however, the work on hyperfine interactions of radioactive atoms seems to have been the decisive stimulus for a reconsideration of questions in solid state physics: interior magnetic fields, intraatomic electron density as related to chemical binding or neighbouring perturbations in a metal, low temperature solid state physics $<10^{-2}$K, Debye-Waller factor may serve as examples.

While there seems to be an easy and fashionable division between: applications, "useful" properties on the one hand and understanding, simple basics on the other, I would rather stress that communication between both is needed for progress in both areas. Just to illustrate this let us remember two problems mentioned but not even formulated in the present school: in the implantation process (with its many possible applications) one is talking about non equilibrium thermodynamics (basics) - lecture Davies. And: even the equilibrium phase diagrams (very important for materials production and properties) are only on the verge of being tractable in terms of interatomic interactions (basics) - lecture Picraux.

Where may one in the future expect a fruitful cooperation between classical solid state methods and hyperfine interaction measurements, on problems of implanted atoms, their location and agglomeration? I have tried to collect a few obvious questions from the talks:

1. Nucleation, the early stage(s) of precipitation, small clusters.

2. Short range order in equilibrium situations, and in amorphous alloys.

3. Dynamics of 1,2, or 3 defects, atomic migration.

4. Why has it so far been impossible to produce amorphous Fe or Cu?

5. Thermal vibration, its connection with the type and energy of binding.

6. Site, electronic state, and reactions of an implanted atom, chemical and metallurgical point of view; influence of very dilute impurities on mechanical and other properties; relaxation of the lattice around an impurity.

7. Electric multipole fields seen by an ionic shell (4f,...) as compared to the multipole fields seen by the nucleus; their dependence on strain.

8. Why has EPR of Fe in metals so far escaped observation?

9. Interfaces, like surfaces or extended defects; processes happening there (like diffusion).

10. Why is it that splat cooling and large dose implantation seem to produce essentially the same amorphous alloys although the cooling rates are of the order 10^6 K/sec vs. 10^{12} K/sec?

Of course, this is again a very personal list. Always expect the unexpected!

After a well attended school with vivid discussions and a

friendly atmosphere, held at a charming site, it is a pleasant task
to express the participants' gratitude.We all appreciate very highly
the effort of the organizers and of the camp staff and we acknowledge the financial support received. Let us terminate by thanking
all those who brought us here and who have helped us to learn and
understand.

PARTICIPANTS

Director and Co-Directors

A. PEREZ   Département de Physique des Matériaux,
                            Université Claude Bernard Lyon I, 43 Bd du
                            11 Novembre 1918, 69621 VILLEURBANNE-FRANCE

R. COUSSEMENT   Instituut voor Kern-en Stralingsfysika,
                            University of Leuven, Celestijnenlaan 200 D
                            3030 HEVERLEE-BELGIUM

A. CACHARD   Département de Physique des Matériaux,
                            Université Claude Bernard Lyon I, 43 Bd du
                            11 novembre 1918, 69621 VILLEURBANNE-FRANCE

G. MAREST   Institut de Physique Nucléaire, 43 Bd du
                            11 Novembre 1918, 69621 VILLEURBANNE-FRANCE

Scientific Committee

I. BERKES   Institut de Physique Nucléaire, 43 Bd du
                            11 Novembre 1918, 69621 VILLEURBANNE-FRANCE

J.A. BORDERS   Sandia Lab. Ion-Solid Interactions, Div.5111
                            ALBUQUERQUE, New Mexico 87115-U.S.A.

P. CAMAGNI   Centre Commun des Recherches EURATOM
                            Division de Physique-EURATOM, 21020 ISPRA
                            (Varese)-ITALY

J. DAVIES	Atomic Energy of Canada Limited, Nuclear Lab. CHALK RIVER, Ontario KOJ 1JO-CANADA
G. DEARNALEY	Nuclear Physics Division, UKAEA Research group, AERE Harwell, HARWELL, Oxfordshire, OX 11 ORA-ENGLAND
B. DEUTCH	Institute of Physics, University of Aarhus, DK 8000 AARHUS.C-DENMARK
H. DE WAARD	Lab. voor Algemene, Natuurkunde, University of Groningen, Rijksuniversiteit, Westersingle 34, 9718 CM-GRONINGEN-THE NETHERLANDS
A.N. GOLAND	Department of Physics, Brookhaven National Lab. , Upton, NEW YORK 11973-U.S.A.
J. GYULAI	Central Research Institute for Physics, Hungarian Academy of Sciences-1525 BUDAPEST P.O. Box 49-HUNGARY
Y. QUERE	C.E.A-S.E.S.I.-C.E.N. , B.P. 6-92260 FONTENAY AUX ROSES-FRANCE
G. SCHATZ	Fachbereich Physik, Universität Konstanz Postfach 7733-D 7750 KONSTANZ-WEST GERMANY
L.M. STALS	Limburg Universitair Centrum, Department W.N.F.-Universitaire Campus, B. 3610 DIEPENBEEK-BELGIUM
N. STONE	Clarendon Lab., University of Oxford-OXFORD, OX 1 3PU-ENGLAND
G. VOGL	Freie Universität Berlin and Hahn-Meitner Inst. für Kernforschung, Glienicker strasse 100, 1000 BERLIN 39. - WEST GERMANY

Organizing Committee

G. CHASSAGNE	Département de Physique des Matériaux, Université Claude Bernard Lyon I, 43 Bd du 11 Novembre 1918, 69621 VILLEURBANNE-FRANCE
J. DAVENAS	Département de Physique des Matériaux, Université Claude Bernard Lyon I, 43 Bd du 11 Novembre 1918, 69621 VILLEURBANNE-FRANCE

| C. DUPUY | Département de Physique des Matériaux Université Claude Bernard Lyon I, 43 Bd du 11 Novembre 1918, 69621 VILLEURBANNE-FRANCE |

| H. PATTYN | Instituut voor Kern-en Stralingsfysika University of Leuven, Celestijnenlaan 200 D 3030 HEVERLEE-BELGIUM |

| S. REINTSEMA | Instituut voor Kern-en Stralingsfysika University of Leuven, Celestijnenlaan 200 D 3030 HEVERLEE-BELGIUM |

Lecturers

| I. BERKES | cf. Scientific Committee |

| H. BERNAS | Institut de Physique Nucléaire, Université Paris Sud, B.P. 1, 91406 ORSAY-FRANCE |

| J.A. BORDERS | cf. Scientific Committee |

| A. CACHARD | cf. Co-Directors |

| R. COUSSEMENT | cf. Co-Directors |

| J. DAVIES | cf. Scientific Committee |

| G. DEARNALEY | cf. Scientific Committee |

| H. DE WAARD | cf. Scientific Committee |

| L.M. HOWE | Solid State Science Branch, Atomic Energy of Canada Limited, Chalk River Nuclear Lab. CHALK RIVER, Ontario K0J 1J0-CANADA |

| J. LETEURTRE | C.E.A.-S.E.S.I.-C.E.N. , B.P. 6, 92260 FONTENAY AUX ROSES-FRANCE |

| S.T. PICRAUX | Sandia Lab. P.O. Box 5800, ALBUQUERQUE, New Mexico, 87115-U.S.A. |

| D. QUITMANN | Fachbereich Physik Inst. für Atom und Festkörperphysik, Freie Universität Berlin, Boltzmannstrasse 20, WE 1A-1000 BERLIN-33 . WEST GERMANY |

| E. RECKNAGEL | Fachbereich Physik, Universität Konstanz D 7750 KONSTANZ-WEST GERMANY |

L.M. STALS   cf. Scientific Committee

N. STONE   cf. Scientific Committee

P.D. TOWNSEND   School of Mathematical and Physical Sciences
                             The University of Sussex, FALMER-BRIGHTON,
                             BN 19 QH-ENGLAND

Technical Assistance

J. DUPIN   Département de Physique des Matériaux
                             Université Claude Bernard Lyon I, 43 Bd du
                             11 Novembre 1918, 69621 VILLEURBANNE-FRANCE

Mrs M. EVERAERTS   Instituut voor Kern-en Stralingsfysika
                             University of Leuven, Celstijnenlaan 200 D
                             3030 HEVERLEE-BELGIUM

G. GUIRAUD   Département de Physique des Matériaux
                             Université Claude Bernard Lyon I,43 Bd du
                             11 Novembre 1918, 69621 VILLEURBANNE-FRANCE

J. PIVOT   Département de Physique des Matériaux
                             Université Claude Bernard Lyon I, 43 Bd du
                             11 Novembre 1918, 69621 VILLEURBANNE-FRANCE

Participants

G. ALESTIG   Department of Physics, Chalmer University
                             of Technology, Fack, S-40220 GOTHENBURG-
                             SWEDEN

O. ALSTRUP   Lab. of Applied Physics III, Technical
                             University of Denmark, Build. 307, 2800 DK
                             LINGBY-DENMARK

G.W. ARNOLD   Sandia Lab. Ion implantation Physics,
                             Div. 5112-ALBUQUERQUE, New Mexico, 87115-
                             U.S.A.

N. AYRES DE CAMPOS   Department de Fisica, University de Coimbra
                             COIMBRA-PORTUGAL

J. BELSON   Department of Electronic and Electrical
                             Engineering, University of Surrey,GUILDFORD
                             Surrey GU 2 5 XH-ENGLAND

M. BENMALEK   Centre des Sciences et de la Technologie
                       Nucléaire, Lab. Cristaux et Couches Minces
                       Bd F. Fanon, ALGER-ALGERIA

Mrs M. BERTI   Istituto di Fisica, Universita di Padova
                       Via Marzolo 8, 35100 PADOVA-ITALY

Mrs K. BERTRAM   Metallurgy Division A.E.R.E Harwell,
                       HARWELL, Oxfordshire OX 11 ORA-ENGLAND

A.G. BIBLIONI   Dept.de Fisica, Faculdade de ciencias
                       exactas, Universidad Nacional de la Plata
                       Calle 115 Y 49,C.C. n° 67, LA PLATA-
                       ARGENTINA

H.D. CARSTANJEN   Munich University, Sektion Physik, Amalien-
                       strasse 54, 8000 MUNCHEN 40-WEST GERMANY

G.F. CEMBALI   C.N.R. Lamel, Via Castagnoli 1, 40126
                       BOLOGNA- ITALY

Mrs E. CLAES   University of Leuven, Department Natuurkunde
                       Intituut voor Kern-en Stralingsfysika,
                       Celestijnenlaan 200 D- 3030 HEVERLEE-BELGIUM

H. DEICHER   Fachbereich Physik, Universität Konstanz,
                       Bücklestrasse 13, D. 7750 KONSTANZ-
                       WEST GERMANY

J. DELAFOND   Lab. de métallurgie Physique, 40 av. du
                       recteur Pineau, 86022 POITIERS-FRANCE

R. FASTENAU   Technical University Delft, Applied Physics
                       Department, Interuniversitary reactor Insti-
tut   tut, Mekelweg 15, DELTF-THE NETHERLANDS

K. FREITAG   Lehrstuhl für Kern-und Neutronenphysik der
                       Universität Bonn, Nussallee 14-16, D 5300,
                       BONN-WEST GERMANY

J.P. GAILLARD   C.E.N.G.-L.E.T.I.-M.E.P. , B.P. 85 X
                       38041 GRENOBLE CEDEX-FRANCE

G. GEHRINGER   C.N.R. Strasbourg Cronenbourg, Lab. de
                       spectroscopie nucléaire, 23 rue du Loess,
                       67000 STRASBOURG-FRANCE

G. GEVERS   Lab. de Physique du solide, Université
                             Bordeaux I, 351 cours de la Libération,
                             33405 TALENCE CEDEX-FRANCE

W.N. GIBSON   Department of Physics, State University of
                             New York at Albany, 1400 Washington Av.
                             Albany, NEW YORK 12222-U.S.A.

R. HAROUTUNIAN   Institut de Physique Nucléaire, 43 Bd du
                             11 Novembre 1918, 69621 VILLEURBANNE-FRANCE

O. HARTMANN   Institute of Physics, P.O. Box 530, 75121
                             UPPSALA-SWEDEN

P.G. HERZOG   Lehrstuhl für Kern-und Neutronenphysik der
                             Universität Bonn, Nussallee 14-16, D 5300,
                             BONN-WEST GERMANY

H. HURDEQUINT   Lab. de Physique des solides, Université
                             Paris Sud, 91405 ORSAY-FRANCE

R. JARJIS   Department of Physics Shuster Lab. -The
                             University- MANCHESTER M13 9 PL-ENGLAND

W. KEPPNER   Fachbereichphysik, Universität Konstanz,
                             Bücklestrasse 13, D. 7750 KONSTANZ-
                             WEST GERMANY

P.J.W. KIKKERT   Lab. voor Algemene Natuurkunde, Westersingle
                             34, 9718 GRONINGEN-THE NETHERLANDS

K. KROLAS   Jagiellonian University, Institute of
                             Physics, ul. Reymonta 4, CRACOW-POLAND

A. LOPES DE OLIVEIRA   C.E.N.G, Département de recherche fondamen-
(Brazilian)   dale, Lab. de chimie physique nucléaire,
                             B.P. 85 X, 38041 GRENOBLE CEDEX-FRANCE

A. MANARA   Centre commun des recherches, EURATOM,
                             Division de Physique, EURATOM, 21020 ISPRA
                             (Varese)-ITALY

A. MASOERO   Istituto di fisica, Universita di Torino,
                             Corso M. d'Azeglio n° 46, 10125 TORINO-
                             ITALY

A.G. MULLER	Institüt für Strahlen-und Kernphysik der Universität Bonn, Nussallee 14-16, D 5300, BONN-WEST GERMANY
L. NIESEN	Lab. voor Algemene Natuurkunde,Westersingle 34, 9718 CM GRONINGEN-THE NETHERLANDS
L.O. NORLIN	Institute of Physics, P. O. Box 530 75121 UPPSALA-SWEDEN
A. NYLANDSTED (Danish)	Bureau Central de mesures nucléaires, B - 2440 GEEL-BELGIUM
J. ODEURS	Instituut voor Kern-en Stralingsfysika, University of Leuven, Celestijnenlaan 200 D 3030 HEVERLEE-BELGIUM
W. PETRY	Hahn Meitner Institüt für Kernsforschung Glienickerstrasse 100, 1 BERLIN 39- WEST GERMANY
F. PLEITER	Lab. voor Algemene, Natuurkunde, Westersingle 34, 9718 CM GRONINGEN-THE NETHERLANDS
Mrs. M. RAFAILOVICH	Physics Department, SUNY-Stony Brook, STONY BROOK L.I. 11794-U.S.A.
L.P. REQUICHA FERREIRA (Portuguese)	Shuster Lab. Department of Physics, University of Manchester, MANCHESTER M13 9PL- ENGLAND
K. ROSSLER	Institüt für Chemie der Kernforshungsanlage Jülich GmbH, Inst. 1 Nuclear Chemie, Postfach 1913-D-5170 JULICH 1-WEST GERMANY
M. ROTS	Instituut voor Kern-en Stralingsfysika University of Leuven, Celestijnenlaan 200 D 3030 HEVERLEE-BELGIUM
H.J. SCHMIDT	Institute for Hot Chemsitry, Kernforschung Zentrum Karlsruhe, 757 Leopoldshafen, KARLSRUHE-WEST GERMANY
R. SCHULTZ	Max Planck Institüt für Plasma-Physik, 8046 GARSHING BEI MUNCHEN-WEST GERMANY

D. SEGERS   I.N.W. , Proeftuinstraat 86, B. 9000 GENT-
                               BELGIUM

U. SHRETER   Solid State Institute, Technion Israël
                               Institute for Technology, Technion City,
                               HAIFA-ISRAEL

J.C. SOARES   Faculdade de Ciencias, Lab. de Fisica, Rua
                               da Escola Politecnica, LISBOA 2-PORTUGAL

K. STEEPLES   University of Lancaster, Department of
                               Physics, BAILRIGG (Lancaster)-ENGLAND

S. STERN   New York University, Department of Physics-
                               Radiation and Solid State Lab.,4 Washington
                               Place, NEW YORK, N.Y. 10003-U.S.A.

A. STRABONI   Groupe de Physique des Solides de l'Ecole
                               Normale Supérieure, 2 Place Jussieu,
                               75005 PARIS-FRANCE

L. THOME   Institut de Physique Nucléaire, B.P. 1
                               91406 ORSAY-FRANCE

Mrs C. TOSELLO   Facolta di Scienze, Dipartimento di
                               matematica e fisica, Universita di Trento
                               TRENTO-ITALY

W. TRIFTSHAUSER   Hochschule der Bundeswehr München, SWE
                               Physik-Werner Heisenberg-weg 39, 8014
                               NEUBIBERG-WEST GERMANY

J. URBANEK   Université de Bordeaux I, Centre d'études
(Czechoslovakian)   nucléaires de Bordeaux - Gradignan, LE HAUT
                               VIGNEAU, 33170 GRADIGNAN-FRANCE

Mrs H. VAN SWYGENHOVEN Limburgs Universitair Centrum, Universitaire
                               Campus, Dpt W.N.F, 3610 DIEPENBEEK-BELGIUM

E. VERBIEST   Instituut voor Kern-en Stalingsfysika
                               University of Leuven, Celestijnenlaan 200 D
                               3030 HEVERLEE-BELGIUM

D. VISSER   Lab. voor Algemene Natuurkunde,Westersingle
                               34, 9718-CM-GRONINGEN-THE NETHERLANDS

R.E.J. WATKINS   University of Sussex, Department of Physics
                        BRIGHTON-BN 1 9QH-ENGLAND

F.C. ZAWISLAK   Istituto de fisica, Universidade Federal do
                        Rio Grande do Sul, Av. Luiz Englert s/n°
                        90 000, PORTO ALLEGRE -RS- BRAZIL